In Pursuit of Unicorns

A JOURNEY THROUGH 50 YEARS OF BIOTECHNOLOGY

UPDATED EDITION

In Pursuit of Unicorns

A JOURNEY THROUGH 50 YEARS OF BIOTECHNOLOGY

UPDATED EDITION

Tim Harris

SV Health Investors LLP

COLD SPRING HARBOR LABORATORY PRESS
Cold Spring Harbor, New York • www.cshlpress.org

In Pursuit of Unicorns: A Journey through 50 Years of Biotechnology, Updated Edition

Publisher	John Inglis
Project Manager	Barbara Acosta
Editorial Assistant	Danett Gil
Permissions Coordinator	Carol Brown
Production Editor	Kathleen Bubbeo
Production Manager/Cover Designer	Denise Weiss

Front and back cover images: The Unicorn Purifies Water (front cover) and *The Unicorn Rests in a Garden* aka *The Unicorn in Captivity* (back cover) tapestries were gifted by John D. Rockefeller, Jr. in 1937 to the Metropolitan Museum of Art Tapestry Collection. The tapestries are housed at the medieval branch of The Cloisters, New York (Accession nos. 37.80.2 and 37.80.6, respectively). Two in a series of seven, the works, believed to have been designed in Paris, were woven in the South Netherlands between 1495 and 1505. The images were donated to Wikimedia Commons as part of a project by the Metropolitan Museum of Art and made available under the Creative Commons CCO 1.0 Universal Public Domain Dedication.

ISBN 978-1-621825-61-6 (paper)
ISBN 978-1-621825-62-3 (epub)

All World Wide Web addresses are accurate to the best of our knowledge at the time of printing.

For a complete catalog of all our Cold Spring Harbor Laboratory Press publications, visit our website at www.cshlpress.org.

Contents

Photo section follows page 212

Preface to the Updated Edition

Given the surprising (to me) interest in the hardback edition of the book, I was asked by John Inglis whether I would be willing to update it for a paperback edition, including writing a new chapter to "catch up" on some of the changes in this fast-moving field. The answer was obviously "yes," and so we have added a Chapter 19 to the paperback version. Although not a comprehensive update on all the companies I mentioned before, this new material includes some more information about a few of the companies that I wrote about previously. I have endeavoured to follow the same style as in the hardcover version, including using boxes for some of the technical details and pithy footnotes as warranted.

In the first part of Chapter 19, I try to place the industry into today's context (autumn of 2025), where we find ourselves in the early years of a new U.S. administration that seems to be promoting certain activities in the biotech space (and elsewhere) that are often divorced from scientific thinking and facts.[a] Nowhere is this more obvious than in vaccine science. What is happening is completely contrary to what we have learned in the past and to what is recorded in Chapter 16. It is quite concerning for the future health of the U.S. population.

I have noted the interesting developments in Chinese biotechnology, which has affected our industry and mention several of the China-influenced new biotech companies. Some commentators think this move to the East is a fundamental shift and others think that it is temporary, as the United States has always been a centre of innovation. I am on the side of it being an historical shift.

There have also been several new companies formed. Most of these are in the AI space directly or they are using AI to help realize their drug discovery aspirations. As I mentioned previously, the role of recent Nobel Prize winners and the large VC funds are again part of this emerging story. The

---

[a] It seems that they are doing this simply to appeal to and appease a particular section of the voting population who believe things that they are told to believe on social media rather than finding out for themselves what the facts are—if they cared to do so.

concept of developing tissue-specific mAbs or antisense oligonucleotides (ASOs) by using "shuttling vectors" targeting transferrin and other receptors to get proteins into the brain is also an exciting area for new companies, some of which are AI-driven. I have also mentioned the more recent progress in the gene editing and gene therapy space, especially the trials and tribulations of the DMD company Sarepta.

Given the rather challenging financing environment that has continued since 2022, and the lack of exits for companies by IPO, I have focused on some of the acquisitions that have taken place between biotech companies and Big Pharma and some of the consequences of those mergers. Given where I live in Nahant, Massachusetts, I have also taken the liberty to reflect on major players on the biotech scene here in the Boston area (e.g., Vertex and Alnylam), and what they have been doing.

Once again, I thank John Inglis and his team at Cold Spring Harbor Laboratory Press for their support, Fintan Steele for his editorial skills, and James Sabry, now the CBO at Biomarin, and David Goldstein, CEO at Actio Bio, for providing their comments on the new chapter. They are both quite familiar with the original book and enjoyed reading it.[b]

Nahant, Massachusetts, October 2025

---

[b] Or so they told me.

Preface

> I will ask you to mark again that rather typical feature of the development of a subject; how so much progress depends on the interplay of techniques, discoveries and new ideas, probably in that order of decreasing importance.
>
> —*Sydney Brenner*[1]

We are witnesses to—and beneficiaries of—a remarkable revolution in health care, especially in emerging new therapies and even preventions for a host of human diseases. Previously unimaginable cell- and gene-based therapies—such as the mRNA-based vaccines for COVID-19—seem to emerge daily. None of these advances would happen without biotechnology.

This book is about that technology: in particular, the basic science that drives it, and the people and companies who transformed that science into the biotechnologies we benefit from today and into the future.

This book is not intended to be a comprehensive history of biotechnology (a term first coined by Hungarian engineer Károly Ereky in 1919).[2] It is instead more of a narrative of the stunning technological advances of molecular and cellular biology in the late twentieth and early twenty-first centuries, and their application to the development of the industry and its products—told through my lens on the companies that were set up to exploit them. I try to do justice to the contributions of many of the individual characters involved but cannot cover them all.

I also recognize that this topic is of great interest to many different people, from patients and their advocates to business entrepreneurs, scientists, and investors. I have chosen to provide descriptions of the technologies and some relevant anecdotes where appropriate, in separate boxes. I also try to define unusual and/or professional terminologies without interrupting the flow of the story I am trying to tell.

This book is a chronicle that reflects my very fortunate almost 50-year journey through the biotech business, including the companies and people I have met along the way. The story is most definitely *not about me*. However, the scope and content directly reflect my experiences, and I have kept the

narrative lively by incorporating various anecdotes and commentary in footnotes as we travel together.

EARLY DAYS

My journey started in the 1950s. When I was a child, I lived on Grimms Hill
near Great Missenden in a county called Buckinghamshire, 35 miles west
of London in England. When it rained, I used to go out to the road by the
house and make sure the water could run down past the driveway without
getting hung up by sticks and stones. On occasion I would redirect the water deliberately and see how long it would take for the water to remove the
block. Sometimes, if it were late afternoon, I would see an older man walking up the road wearing a visor. One day I asked my father, a cancer research
chemist at The Chester Beatty Cancer Research Institute at Pollards Wood
in Little Chalfont nearer to London, who the person was. He told me it was
Sir Robert Robinson,[3] a synthetic organic chemist who had been awarded
the Nobel Prize in Chemistry in 1947 for his investigations on plant products of biological importance, especially plant alkaloids. "One of the most
eminent people you will ever meet," my dad said.[a]

Sir Robert lived at the top of the road in a house called "Lebanon," I assumed owing to the cedar of Lebanon trees growing in the garden. I remember one day Sir Robert gave me a block of rock salt from Switzerland. I have
no idea why—but it had a lasting impression on me. Couple that with living
in a house with the journals *Nature* and *The Lancet* lying on the coffee table
and I guess it is no surprise that both my sister Suzanne and I developed an
interest in biology and chemistry from an early age.

Another important milestone was reading *The Chemistry of Life* by
Steven Rose published in 1966.[4] The diagrams of the molecules and the
introduction to biochemistry totally captivated me, and a dog-eared copy
became a close friend.

One week on a visit to see my father at the Pollards Wood labs, Roger
Kirby (who is presently president of the Royal Society of Medicine in
England) and I were allowed to extract nucleic acids from rat liver. Ken
Kirby, Roger's father who worked with my dad, had optimized a procedure

---

[a] That probably was true but not exclusively, as I got to meet and spend time with Sydney
Brenner, who was probably the smartest molecular biologist there has ever been and certainly the smartest I have ever met.

of organic extraction using phenol *m*-cresol mixtures to isolate nucleic acids. After ethanol precipitation I spooled the isolated DNA on a glass rod and precipitated the RNA separately. From then on, I was hooked on labs, centrifuges, and nucleic acids.[b]

I made the decision to read biochemistry instead of medicine at university. I was not that keen to treat ill people, which was what most doctors I met in the 1950s and 1960s did. The University of Birmingham accepted me: Samuel (Sam) Perry, an eminent muscle biochemist, was the Chairman of the biochemistry department. It was one of the best biochemistry departments in the country at the time, with a new dedicated six-story building.

After graduating with a degree acceptable enough to be qualified to do a PhD, I decided to learn some virology with Peter Wildy at the Medical School in Birmingham, where I completed the virology master's programme. I learned about most of the DNA and RNA virus families, including those that infect plants. My PhD was entitled "mRNA Synthesis in Herpes Virus Infected Cells." I did two years' experimental work for it after finishing the MSc in the summer of 1972, and I received my PhD degree in November 1974. Like the work described in most PhD theses, the experiments and their conclusions were not really very good nor very insightful, but I learned a lot about how to isolate and work with RNA.[c] I also learned about what "controls" were (i.e., controlling the vagaries of the experimental process by including different conditions).

I was never first in the class nor the best at anything that I did, but one learns to adapt to that. What I do have and always have had is energy, enthusiasm, and ambition. I have always made it a point to keep up to date with what is going on in the relevant research world around me by reading the literature assiduously. I have developed a deep appreciation and understanding of elegant experiments. Appreciating experimental elegance is an important motivator for a scientist, if only so one can reflect on how other scientists managed to do some of the amazing experiments that they did do. There are only certain skilled people who can either develop or can get complicated experimental techniques to work.

---

[b] What a privilege it is to know what one wants to do at the age of 16 and to have had a lifetime doing it.

[c] A fragile and interesting molecule that transfers information from DNA to protein as well as being intimately involved in the process of protein synthesis in a cell. Much more will be said about this molecule later in the book.

The PhD enabled me to get a job at the Animal Virus Research Institute in Pirbright Surrey (now the Pirbright Institute) to work on foot-and-mouth disease virus (FMDV) in the well-respected laboratory of Fred Brown. I learned about picornaviruses, published some reasonable papers, and got to spend a year in Eckard Wimmer's lab at the State University of New York in Stony Brook on Long Island in October 1977. After my return from the United States to Pirbright in the autumn of 1978, I was in a perfect position to appreciate the importance of recombinant DNA (and monoclonal antibody technology) and to use those techniques myself.

FORMAL BIOTECH JOURNEY

My "formal biotech journey" started in late 1978, when we cloned FMDV RNA and made plasmids containing DNA copies of virtually the whole virus RNA. We were amongst the very first groups in the world to do that: we may even have been the first. I was also fortunate to form a collaboration with a research group at G.D. Searle in High Wycombe where they were cloning chicken ovalbumin cDNA. The large group at G.D. Searle not only provided the first research scientists for the first biotech company in the United Kingdom—Celltech—but later Searle scientists also started British Biotechnology. I joined Celltech in Slough in March 1981 and ran their cDNA cloning lab for several years.[d]

In 1989 I moved to Glaxo Group Research, where we used recombinant organisms to screen for small molecules that affected a "drug target" and that perhaps could be developed into potential medicines after the application of the appropriate medicinal chemistry. Glaxo did some of the very first pharma–biotech deals. I subsequently moved to the United States in early 1993 to embrace genomics and "functional genomics" technologies at the biotech company Sequana Therapeutics in San Diego and run their R&D. Later, as the CEO, I led the company Structural GenomiX (founded in 1999) through its transition to SGX Pharmaceuticals in its exploitation of protein structure determination in drug discovery.

Human genome sequencing was a major focus for me at the National Cancer Institute (NCI)/SAIC-Frederick in Maryland, where I moved in 2006 after a brief stint as CEO at Novasite Pharmaceuticals, a G protein-coupled receptor company in San Diego. SAIC-Frederick was the company

---

[d] My second daughter was the first "Celltech baby."

that ran the FFRDC (Federally Funded Research and Development Center) on behalf of the NCI. I looked after the Advanced Technology Program for them, which among other things had a big interest in genomics. My career then took me to "translational sciences" and cell and gene therapy at Biogen in Cambridge, Massachusetts in 2011, followed by working on haemophilia as EVP of R&D at Bioverativ, a Biogen spinout, in 2017. Finally, after the acquisition by Sanofi, I moved on to T-cell biology, CAR T cells, and related technology at Repertoire Immune Medicines in 2020. I am currently a venture partner at SV Health Investors.

CHRONOLOGY AND CADENCE

This book generally follows my own career chronology. The first three chapters describe the early days and the formation of the recombinant DNA–based companies on both coasts of the United States and in the United Kingdom.[e]

Chapters 4 and 5 trace the development of monoclonal antibodies (mAbs), the companies that were founded on this technological innovation, and the companies and leaders who developed some of the most important "blockbuster" monoclonal antibody products.

In Chapter 6, I draw on my experiences at Glaxo Group Research in the early 1990s to describe what pharma companies were beginning to do with these new technologies and the strategic alliances that they formed. It is also at about the same time that small-molecule drug discovery for common diseases—once the core strength of pharma—began to be done by biotech companies using combinatorial chemistry and screening. This led to new drugs for both old and new targets—some of which I describe.

Chapters 7 and 8 are organized around the astounding advances in genetics and genomics technologies and their applications in the last part of the twentieth century and into the twenty-first. I first describe some of the early excitement around cloning the genes causing the major forms of Mendelian inherited diseases, such as cystic fibrosis, Duchenne muscular dystrophy, and Huntington's disease. Several companies were founded on these technologies,

---

[e] The United States and the United Kingdom were the centres for the development of biotechnology in the early 1980s, although there were some nascent company activities in the Netherlands, France, and Germany. This was as much cultural as it was access to the appropriate capital.

including Sequana Therapeutics, which would attempt to clone the genes involved in common diseases by this means. The Human Genome Project, and the public versus private race to sequence the human genome, is described, as it was responsible for many additional advances in biotech. However, rather than just revisiting a much-told story,[f] I focus on the businesses involved, both "public" (e.g., The Institute for Genome Research [TIGR]) and private (e.g., Human Genome Sciences, Celera, and Incyte).

The inevitable turn to functional genomics, or the science of understanding the primary function of the genes associated with diseases, is the focus of Chapter 9. Several companies were founded on the premise that less complex organisms like fruit flies and worms would reveal the function of unknown genes of relevance to human disease. At the same time other scientists (and their new companies) turned to uncovering function by understanding the three-dimensional structure of proteins that employed crystallography technologies, but which were informed by a genomics-driven approach in order to facilitate drug discovery. Structural GenomiX (SGX), which was started in 1999, were a great example of such a company that I was fortunate to be able to help to build, as related in Chapter 10.

Despite business concerns expressed by traditional pharma, several companies emerged that were dedicated to applying genomics deeply to get insights into rare diseases, and more than a couple of companies were focused on finding and developing medicines to treat such diseases (e.g., Genzyme). I consider several of these—and their business models—in Chapter 11.

In pursuit of even greater understanding of molecular mechanisms in biology and disease, several academic scientists developed technologies that could reliably alter the expression of genes by blocking the translation of the mRNA copy of the gene into its designated protein. Chapter 12 covers these "antisense" and small interfering RNA (siRNA) technologies, the signature companies who were founded to blaze that trail, and the products that were made from the application of these technologies.

The grail of directly replacing or repairing a defective gene, colloquially known as "gene therapy," contains the future-defining technologies of today's biotech industry. I trace the halting beginnings of gene therapy to the present day in Chapter 13, focusing on both the technologies used and their applications and the activities of some of the many companies started in the space. Gene therapy segues into cell therapy (Chapter 14), where the early

---

[f]From strongly different points of view.

days mirror to some extent what happened in gene therapy. Important new companies have been and are still being formed, leading to the development of new cell therapy products such as CAR T cells for the treatment of haematological malignancies.

The most direct form of gene therapy—the ability to modify DNA sequences almost at will both *in vivo* and *ex vivo*—has exploded with the discovery and utilization of CRISPR-Cas 9 gene editing. In Chapter 15, I describe several companies founded on both gene-editing technology and locked in patent battles, as well as the earlier-developed DNA-editing technologies and the companies that exploited them.

I try to capture the impact of vaccines developed using biotechnological techniques on human health with some historical perspective in Chapter 16. Here I primarily focus on the diseases, mostly caused by viruses, to illustrate how biotechnology has influenced vaccine designs. Specifically, I consider several of the biotech success stories such as the development of hepatitis B and human papilloma virus vaccines, and I acknowledge the companies and people most responsible for developing them. In addition, and I hope without being too repetitious to what has already been published, I cover the development of the more recent vaccines against SARS-CoV-2, including the mRNA-based vaccines.

Biotechnology is not simply science and its successful (or not) application by innovative companies. There are many stakeholders, personalities, and events that can impact that success. Founders, finances, leadership, oversight, regulatory bodies, patient groups, employees, geography, etc., all demand attention and all play a role in success or failure. In the last two chapters I endeavour to capture the soul, or "essence," of biotech, at least as I have assessed it over my journey through its peaks and valleys. In Chapter 17— Fortunes and Unicorns—I use many examples of different companies to illustrate the dark art of "valuation," and the challenges of determining, justifying, and maintaining a high valuation, particularly as a "unicorn."[8]

The valuation discussion in turn sets the scene for Chapter 18—the Essence of Biotech—in which I define what I see as the primary "traits" that determine success or failure of companies in this industry (as demonstrated by the companies and technologies mentioned throughout the book).

---

[8]A Unicorn is a company with a valuation over a billion dollars. Remember, it does not always work out for many once-promising companies and technologies: sometimes they end in failure.

Again, this is not a comprehensive review of technologies and companies, as it relies on my own experience for its structure and conclusions. As the lines between pharma and biotech have become increasingly blurred since the biotech industry started in the 1970s, this last chapter also includes thoughts on how things have changed or not changed. Ultimately, it is designed to leave you with more questions than answers about the future of the industry so you can think further about it.

PORTENTS

When I was considering writing this parallel of my journey with the history of biotechnology, I recalled a book entitled *Invisible Frontiers* by Stephen Hall about the race to clone the insulin gene. The book describes efforts to clone both a chemically synthesized insulin gene and one derived from mRNA as cDNA.[5] The book was out of print, but I found a copy online from a secondhand bookseller in the Bay Area. It arrived on my doorstep in a nondescript and bedraggled wet package. The next morning when it was dry, I opened it to find that this copy of the book had been signed by all the principal founders and first employees of Genentech (see Plate 1). Ex-Genentech colleagues told me that there were in fact several dozen copies of the book that had been signed this way for Genentech employees involved in the insulin project. Presumably, this copy had been given away by someone who did not completely appreciate the importance of Genentech for the industry generally or of the principals who signed the book. I took this experience as a clear portent for me to get on and write this book. I hope you enjoy reading it and learning from it as much as I have enjoyed writing it.

REFERENCES AND NOTES

1. From Brenner's own handwritten notes of a Speech (20 Mar 1980), "Biology in the 1980s," he gave at the Friedrich Miescher Institute in Basel, Switzerland. Reproduced in his article Brenner S. 2002. Life sentences: Detective Rummage investigates. *The Scientist* 16: 15.

2. Ereky K. 1919. *Biotechnologie der Fleisch-, Fett-, und Milcherzeugung im landwirtschaftlichen Grossbetriebe: für naturwissenschaftlich gebildete Landwirte verfasst.* P. Parey, Berlin.

3. Sir Robert Robinson: The Franklin Institute. https://www.fi.edu/en/awards/laureates/robert-robinson

4. Rose S. 1966. *The chemistry of life,* 1st ed. Penguin, London.

5. Hall SS. 1987. *Invisible frontiers: the race to synthesize a human gene.* AbeBooks, Victoria, BC.

CHAPTER 1

The Beginnings

It is not because things are difficult that we do not dare, it is because we do not dare that things are difficult.

—*Lucius Annaeus Seneca, ca. 10 BC in the reign of Roman Emperor Tiberius*

October 14, 1980, was a breezy and chilly fall day in New York. The NASDAQ opening bell rang as usual at 10 am. This also turned out to be an auspicious day for Genentech, a South San Francisco–based company formed in 1976 to pioneer the use of recombinant DNA technology to make therapeutic proteins in microorganisms. Genentech completed their initial public offering (IPO) on this day,[a] marking the public beginning of today's multibillion-dollar molecular biology–based biotechnology industry.[1,2,3] It was one of the most successful IPOs in history with the stock trading from the initial price of $35 to $88 per share within the first 20 minutes, settling back to $56 before the closing bell.[b]

Genentech was not the first modern biotechnology company founded. That honour belongs to Cetus, which was started five years before Genentech by Nobel Prize–winning physicist Donald Glaser, with his partners Ron Cape and Peter Farley.[4] Its initial funding came from Standard Oil to support its work on using microbial processes to produce chemical feedstocks, including propylene oxide and antibiotic intermediates. The company did not embrace recombinant DNA technology until after the founding of Genentech. Cetus raised $108 million in their 1981 IPO, the largest to

---

[a] The IPO was managed by Blyth Eastman Paine Webber and Hambrecht and Quist[4] (H&Q) a boutique Bay Area investment bank formed by Bill Hambrecht and George Quist in San Francisco. H&Q was later acquired by Chase Manhattan Bank, who then merged with J.P. Morgan to become JPMorgan Chase.

[b] It is remarkable that the stock never traded below the $35 initial offering price right up to when Genentech was acquired in 1999 by Hoffman-La Roche.[5]

that date and three times bigger than that of Genentech, giving Cetus a market capitalization of $500 million.[5]

From an investor perspective, the Genentech and Cetus IPOs marked the beginning of the modern biotech industry.[c]

WHEN DID IT REALLY BEGIN?

This question is difficult to answer. It is clear that the origins of the modern biotechnology industry can be traced further back well before Genentech and Cetus. Perhaps it began with the solving of the structure of DNA in 1953 by James Watson and Francis Crick,[6] for which they were awarded the 1962 Nobel Prize in Physiology or Medicine. Or maybe its origins lie in the discovery of messenger RNA (mRNA) by Crick and Sydney Brenner and others, the deciphering of the genetic code by Marshall Nirenberg, and the understanding of the concept of a gene as a unit of inheritance—work dominated primarily by phage genetics. These fundamental discoveries, although critical to biotechnology, are not really its beginning. They rather mark the beginning of molecular biology, as so elegantly described in Horace Judson's book *The Eighth Day of Creation.*[7]

Biotechnology purists will argue that it was the discovery of various enzymes to modify DNA including restriction enzymes (proteins that can cut DNA at specific locations) and understanding the problem of antibiotic resistance in bacteria that together led to powerful new tools that in turn defined the first biotech companies. The problem of antibiotic resistance in bacteria was well known in the 1960s and 1970s, but the mechanism of that resistance was not. It was clear that antibiotic resistance was associated with "plasmids"—large circles of DNA replicating inside the bacteria independently of the chromosomal DNA. These plasmids could be transferred between bacteria and often carried one or more of the genes that conferred antibiotic resistance on its host bacterium.

These observations, along with the ability to make specific cuts using the newly discovered restriction enzymes, provided the ability to make "recombinant" DNA plasmids quite easily (i.e., ones combining two or more different DNA sequences of interest). This novel genetic material could then be transferred (transformed) into *Escherichia coli*, and individual clones containing a single unique recombinant plasmid derived (see Box 1).

---

[c] It was also the start of the fairy-tale "magic porridge pot" industry, which keeps on giving generously to patients, investors, and the people who work in it.

BOX 1. GENE CLONING

Viruses like HIV use an enzyme called "reverse transcriptase" to turn their RNA-based genomes into DNA, which can then be inserted into the host cell genome and take over its usual functions in favour of the virus. Scientists take advantage of reverse transcriptase to make cDNA libraries. They extract mRNA from cells and use reverse transcriptase as the first step in making a cDNA. The cDNA-containing plasmid, which also includes the genes for antibiotic resistance, is transferred into *E. coli* and the bacterial colonies grown on antibiotic-containing plates. Clones that grow on the appropriate antibiotic containing plate contain plasmids with cDNA inserts.

Figure 1 shows the process of making a "copy" or a "complementary" DNA (cDNA) from a messenger RNA (mRNA) from a cell of some kind, inserting ("cloning") the cDNA into a circular piece of bacterial DNA (plasmid), and putting that cDNA-containing plasmid into *E. coli* where the cDNA can be "expressed" (i.e., a mRNA can be made by the bacteria from the cDNA sequence, and that mRNA can be translated into the protein specified by the sequence). For simplicity, this is shown as one step. Usually, the cDNAs were cloned and isolated and then the cDNA inserts engineered and transferred to an expression plasmid.

cDNA cloning was the most common approach used by early biotech companies to clone genes to make proteins in *E. coli*, although chemically synthesised genes and genes taken directly from the human genome were also used instead of cDNAs.

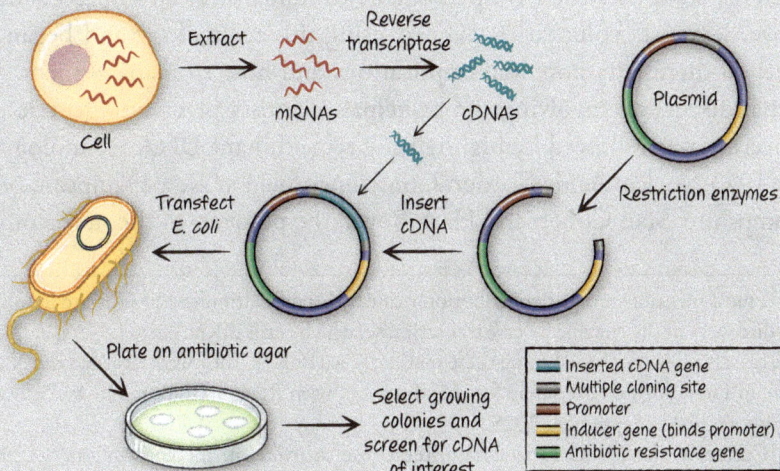

Figure 1. cDNA cloning into a basic expression plasmid.

This work began a new lexicon of terms like cloning, gene splicing, recombinant DNA, and genetic engineering.[d]

This new genetic engineering capability was species-agnostic: DNA from widely disparate organisms such as bacteria and humans could be cut and joined together in a single plasmid. These artificial constructs could then be transferred into bacteria that would make the human proteins coded for by those genes. This technology, more than any other, defined the emerging industry of modern "biotechnology." The first manuscript fully describing this technology was published by Stanley (Stan) Cohen and Herbert (Herb) Boyer in *The Proceedings of the National Academy of Sciences* in July of 1973.[8,e] The foundational patent on the technology was issued to Stanford University in 1980.[9,f]

CONCERNS OVER THE TECHNOLOGY

The artisans that were doing this work, in particular Professor Paul Berg at Stanford,[10] were very aware that this new technology had the potential to do harm through the inadvertent or deliberate cloning of genes causing cancer or antibiotic resistance into bacteria that could infect humans (e.g., the strains of *E. coli* that commonly reside in the human gut). Furthermore, these organisms can exchange genetic information with other types of bacteria, some of which are pathogenic to humans. In an unprecedented move, scientists voluntarily stopped using this technology and began to meet to discuss its potential implications and how to manage them. An initial discussion involving the principal scientists practicing genetic engineering was followed by highlighting recombinant DNA technology at the 1973 Nucleic Acids Gordon Conference, held at New Hampton, New Hampshire. Stan Cohen and Herb Boyer, the pioneers of the new cloning

---

[d] I must mention here the contribution made by David Baltimore and Howard Temin in the discovery of the enzyme reverse transcriptase, which turns mRNA into DNA ("cDNA") that can also then be cloned using these methods (see Box 1). They were awarded the Nobel Prize in Physiology or Medicine for this discovery with Renato Dulbecco in 1975. David Baltimore died in September 2025.

[e] Curiously, the Nobel Prize committee chose to overlook the contribution of Herb Boyer and Stan Cohen.

[f] It was filed one week before the publication deadline thanks largely to the persuasive powers of Neils Reimers, who went on to run the Stanford University Tech Transfer office. Stanford University chose to license this patent nonexclusively and it brought in hundreds of millions of dollars to the university.

techniques, presented some of their data at the meeting and suggested that the scientists should discuss the implications more completely. Maxine Singer and Dieter Soll were tasked with writing to the National Academy of Sciences (NAS) to ask them to set up a committee to investigate the issue and to lead communications, along with the National Institutes of Health (NIH), on the subject.[11]

A group of scientists sponsored by the NAS (called the Committee on Recombinant DNA) led by Berg, subsequently met at the Massachusetts Institute of Technology (MIT) in April 1974 to discuss concerns about the potential hazards of recombinant DNA (rDNA). A letter from the group that laid out their worries was published in *Science* in July 1974.[12] The letter called for an unprecedented temporary and voluntary pause on certain rDNA experiments, to buy further time to evaluate the risks of the technology and to avoid precipitating an "unanticipated hazardous event." Specifically, they asked scientists to defer experiments that involved inserting into bacteria genes that conferred either resistance to antibiotics or the ability to form bacterial toxins and the genes of animal viruses. They also recommended that caution be exercised before cloning genes from eukaryotic species, including humans, into bacteria.[g]

Many expressed concern that the guidance offered was way too vague, as well as to whether the temporary halt would stick. Several people questioned the risks associated with the cloning of eukaryotic DNA from *Drosophila* (the fruit fly), *Xenopus* (African clawed toad), or *Bombyx mori* (the silkworm)—experiments that had, in fact, already been done. Anticipating the uncertainties, the Committee on Recombinant DNA letter also called for NIH guidelines to be established and for a larger meeting to discuss potential hazards to be organised as soon as possible.

THE ASILOMAR CONFERENCE

The resulting meeting was the now-iconic Asilomar Conference, convened in February 1975. The conference was unusual in several ways. There were about 150 invited participants from 12 countries, including not only scientists but also reporters from the scientific and mainstream press, as well as selected representatives of church and state. It was made clear from the outset that the

---

[g]The genome DNA of eukaryotic organisms is contained in a nucleus, a structure that is absent in prokaryotic cells.

meeting would not consider the moral and ethical considerations of the technology, but rather the potential risks to humanity by employing it. It was clear that the scientific experts felt (probably rightly) that they understood the technology better than anyone else and were quite able to render the appropriate judgements in regulating themselves. Few believed that recombinant DNA technology was without some level of risk: the questions were what types of experiments were riskier than others, and what biological and physical containment conditions were appropriate for different types of experiments.

The Asilomar gathering resulted in a set of regulations that have stood the test of time. First, the participants agreed to change the blanket moratorium set up by the Berg letter and endorsed by the NIH to a system based on rational risk assessment. The questions of both the modification of the host organisms (mostly *E. coli*) and the cloning vectors were discussed at length. The attendees agreed that further research should focus on developing further disabled host/vector combinations that could not exist successfully outside the lab. The question of the scale at which the experiments were done was also addressed, given that physical containment was one of the key issues. Notably, the personal responsibility of the lab leaders was also stressed. In the end, the Asilomar participants established three categories of risk: low, intermediate, and high. DNAs from prokaryotes, bacteriophages, and other plasmids fell into the first group. Cloning animal virus genomes (whether DNA or RNA viruses) was considered an intermediate risk. Cloning eukaryotic DNA was classed as the highest risk, largely because very little was known about eukaryotic genes and gene expression at the time, making the actual risk very difficult to assess.

The Guidelines

The "high, medium, low" risk designations were viewed as "interim assignments," which could be revised upward or downward in the light of future experience. The highest level of containment meant the building of labs with special containment characteristics, such as negative pressure gradients, biological containment hoods, clothing changes, and showers. Other recommendations included a focus on training, education, and reassessment.[13,14]

It was very important for the group to set a realistic set of guidelines to avoid unnecessary legislation: guidelines could be altered in the light of new knowledge, whereas legislation, once it has been set up, is very difficult to change. The recommendations formed the basis of both the Recombinant DNA Advisory Committee (RAC) experimental assessment regulations in the United States and the GMAG (Genetic Manipulation Advisory Group) rules in the United Kingdom. Sydney Brenner, who was at the Asilomar meeting, took a very prominent role in drawing up the conclusions, which were important for the subsequent establishment of the recombinant DNA experimental guidelines in the United Kingdom.[15]

Asilomar undoubtedly helped to set the scene for the birth of the biotechnology industry on the west coast of the United States. In Cambridge, Massachusetts (MA), Harvard and MIT personnel worked hard to ensure recombinant DNA facilities adhered to NIH guidelines. This challenge was compounded by public opposition to the use of recombinant DNA technologies in Cambridge, including a moratorium on such experiments imposed in 1976. MIT faculty and their administration met with the citizens of Cambridge to help them understand recombinant DNA research, explaining how the NIH guidelines would ensure safety. By 1977, the scientific community won its case, when the City of Cambridge passed an ordinance adopting the NIH guidelines and lifting the rDNA moratorium. The competitive disadvantages of not participating in the biotech revolution—well under way on the west coast—had also become apparent.

Before introducing the biotechnology companies that were set up in those very early days and the products that they were trying to make, I need to help you get a bit more familiar with some of the cloning technology, especially as it pertains to making proteins in microorganisms like *E. coli*. An accessible (I hope) description of the cloning technologies is shown in Box 1. I will do this for the other technologies that we will explore together in upcoming chapters, as it is the best way to understand most of the companies and events. Those readers who either know about cloning technology or do not feel the need to know can skip on ahead to the description of the companies set up to exploit it.

THE FIRST COMPANIES AND THEIR PRODUCTS: BIOGEN, CETUS, GENENTECH, AND GENEX

Biogen

Biogen was set up in 1978 by Walter (Wally) Gilbert from Harvard and Phillip (Phil) Sharp from MIT. It started as a European company with headquarters in Geneva. It later would add a facility on Binney St. in Cambridge, MA. The venture capitalists (VCs) involved initially were Ray Schaefer, who knew Phil Sharp and Wally Gilbert personally, and who was one of the founding investors in the biotechnology industry, and Daniel (Dan) Adams, the head of venture capital at Inco, who had founded and had been the CEO at several companies (Adams had previously invested in Cetus and had made the first outside equity investment in Genentech).[h]

Because of the moratorium on genetic engineering in Cambridge at that time, Biogen did not initially have its own labs. Most of the early cloning work was done in the labs of the non-Cambridge founders and collaborators. In addition to Gilbert and Sharp, the Biogen scientific founders included Charles Weissman (Zürich), Bernard Mach (Geneva), Kenneth Murray (Edinburgh), and Heinz Schaller (Heidelberg).[i]

Cetus

Cetus was founded in 1971 by Ron Cape, Peter Farley, and Don Glaser. They initially focused on industrial microbiology and the development of antibiotics and chemical feedstocks but their focus altered after the development of recombinant DNA technology. The game really changed for Cetus when they raised $108 million in their IPO following that of Genentech. It was the largest IPO to that date, raising much more than Genentech, which had gone public in October 1980.

Genentech

In mid-1975, an out-of-work wannabe venture capitalist named Robert (Bob) Swanson, who was intrigued by the revenue-driving potential of the new recombinant DNA science, met a reluctant University of California San Francisco (UCSF) professor Herb Boyer for what Boyer thought would be a 10-minute discussion. The meeting turned into a several-hour marathon at Churchill's Bar in San Francisco, where the two men sketched out a plan for a genetic engineering technology (GenEnTech) company. Swanson and

---

[h] All of these investments would achieve a combined value of many millions of dollars.

[i] I remember visiting the original Biogen facility in Geneva for a job interview in 1980 before joining Celltech. At that time, it was an 8000-sq-ft lab in an old watch factory with about 16 scientists.

Boyer[j] both kicked in $500 of their own money to incorporate Genentech on April 7, 1976. In addition, Swanson, who had been "let go" earlier by the venture capital firm Kleiner Perkins (KP),[k] convinced his former employers, specifically KP founding partner Thomas Perkins, to invest $100,000 for 25% of the new company at that time.

Genentech's founders were truly undertaking the task of building something entirely new. They considered their initial late-1970s competition to be limited to the companies Cetus and Biogen and later Genetics Institute (founded 1981). Amgen, also founded in 1981, was chasing projects such as chicken growth hormone and cloning the plant genes that synthesised indigo (the dye used for blue jeans), and thus were not taken very seriously as scientific or business competition at that time (Chapter 2).[l]

Genex

Genex, which was formed in 1977 in Gaithersburg, Maryland, focused on the industrial application of biotechnology for enzymes, involving structural biology and nascent computational methods for structure prediction. Genex was founded by Princeton-based venture investor Robert (Bob) Johnston, with Leslie Glick as the CEO and Kevin Ulmer the most senior scientist. Genex somehow managed to broker a deal with the Bendix Corporation, which brought considerable dollars to the research table to fund their protein engineering aspirations for industrial enzymes. One of the projects was to make phenylalanine for making the artificial sweetener aspartame. Genex invested heavily in building a production plant for aspartame in Paducah, Kentucky, in an old Seagram's drinks cannery. Bendix was taken over by Allied Corp.[m] G.D. Searle, who owned aspartame already, had

---

[j]Initially there was a certain amount of political sensitivity and jealousy surrounding Herb Boyer and his associations with the company, especially from other UCSF faculty, who subsequently got into their own commercial ventures. Boyer was on the Genentech Board of Directors but deliberately kept a distance between the work being done in the Genentech labs and his lab at UCSF. Nonetheless, the animosity persisted.

[k]The company has been through several naming iterations. Biotech venture capitalist Brook Byers, who worked for Asset Management (another venture firm in the Bay Area) shared a workspace with Swanson but did not invest in Genentech. Byers later joined Kleiner Perkins and the firm became the law firm–sounding Kleiner Perkins Caufield & Byers. It is now simply KP.

[l]Early critics of Amgen turned out to be quite wrong.

[m]Who had absolutely no idea how to value or deal with the Genex project.

a relationship with Ajinomoto in Japan as commercial suppliers of the di-peptide, and Genex inevitably ran into financial difficulties.[n]

INSULIN: BIOGEN, GENENTECH, AND THE UNIVERSITY
OF CALIFORNIA SAN FRANCISCO (UCSF)

The first highly public research undertaking at Genentech was to try to produce insulin in bacteria through genetic engineering.

The insulin hormone protein was first isolated by Drs. Frederick Banting and John Macleod at the University of Toronto over 100 years ago as a substance from dog pancreas that could reverse diabetes symptoms in dogs after injection.[o] In 1922, the first human was treated successfully with insulin isolated from bovine pancreases.[16]

Insulin was also the first protein for which the complete amino acid sequence was determined by Frederick (Fred) Sanger in Cambridge (UK) in 1955. Sanger subsequently was awarded his first Chemistry Nobel Prize in 1958 for this technological tour de force. Dorothy Hodgkin in Oxford was also awarded a Nobel Prize in Chemistry in 1964 for determining the three-dimensional structure of insulin (and other macromolecules) by X-ray crystallography.

Despite its relatively small size as proteins go, insulin is somewhat complex. It is made up of two different proteins chains, an A chain of 21 amino acids and a B chain of 30 amino acids joined by two disulphide bonds.[17] It is made by the pancreatic β cells as a single 74-amino acid precursor called pre-proinsulin that contains a 31-amino acid C peptide that is chopped out to produce the final insulin heterodimer.

The pharmaceutical company Eli Lilly became the principal provider of insulin for human use, obtaining it from either bovine or porcine sources. Even though this was a massive breakthrough for many diabetes sufferers, allergic reactions could and often did occur in patients whose immune systems saw the animal-derived insulin (with some minor variations in amino acid sequence) as foreign. In addition, it was often challenging to obtain animal pancreases in sufficient quantity to meet the demand for insulin. It was not

---

[n] In 1985 they fired many of the staff in Maryland and closed the Paducah plant to focus on contract research.

[o] Banting and McLeod (but not Charles Best, who was a student) were awarded the Nobel Prize in Physiology or Medicine in 1923 for this discovery.

a big jump, even in the mid-1970s, to think about producing insulin using the new recombinant DNA techniques. However, the technical challenges were prodigious.

The Groups Racing to Clone the Insulin Gene[p]

Three main groups set out to make human insulin by recombinant DNA methods. The principal players were Genentech working with Keiichi Itakura, Art Riggs, and Roberto Crea at the City of Hope Hospital in Duarte, California and Herb Boyer at UCSF. A second group at Harvard led by Wally Gilbert was sponsored by Biogen, and a third group at UCSF (not involving Boyer) led by Bill Rutter and Howard Goodman in the same biochemistry department, jumped into the race. The City of Hope/Genentech team pursued a chemical synthesis method to make the insulin genes (i.e., they synthesised the insulin gene one DNA base at a time), whereas the other two groups attempted to make an insulin-encoding cDNA from human mRNA (see Box 1).

Genentech's City of Hope collaborators had already established that a gene coding for a small 14-amino acid hormone called somatostatin could be made by chemical synthesis and cloned into the bacteria *E. coli,* which would then make small amounts of the protein.[18,q] The stage was set for doing the same with the insulin A and B chains. This approach had the added advantage of not falling under the Recombinant DNA guidelines, because the DNA was synthetic rather than obtained from the human genome or via mRNA.

The Biogen/Harvard group led by Wally Gilbert, consisting of Argiris Efstratiadis, Lydia Villa-Komaroff, and Forrest Fuller, took the cDNA cloning route following the work of recombinant DNA pioneer Tom Maniatis, who had already cloned rabbit β globin this way. They focused first on cloning the rat insulin gene from mRNA isolated from a rat

---

[p] This race is recounted in detail in the book *Invisible Frontiers* by Stephen Hall (referred to in the Preface).

[q] The chemical synthesis of nucleic acids had been pioneered by Har Gobind Khorana. Solid phase synthesis had been recently developed by Bruce Merrifield, who was awarded the Nobel Prize in Chemistry in 1984. But it was not until 1982 that the first commercial DNA synthesizers became available from Applied Biosystems (the ABI 380A) and others, making the technology much more broadly accessible.

insulinoma, before attempting the human version. The Rutter–Goodman group, including postdocs Peter Seeburg and Axel Ullrich,[r] used the same general approach as the Harvard team.

Technical and other kinds of challenges were immense particularly for both cDNA-focused groups. The rat insulinoma as a source of mRNA turned out to be scarcely any better than that extracted directly from pancreas. The internecine politics in the labs at UCSF did not help either as it was clear to everyone that more than one team in the biochemistry groups at UCSF was trying to clone and produce insulin at the same time. Also, the Harvard group in 1976–1977 were forced to work remotely at Cold Spring Harbor, owing to the moratorium on recombinant DNA experiments that was in place at that time in Cambridge (MA). Recombinant DNA regulations notwithstanding, Ullrich (the UCSF Rutter–Goodman group) managed to clone most of the human pre-proinsulin cDNA into a bacterial plasmid called "pBR322," one of the preferred (from a regulatory perspective) cloning vectors.

A Trip to Porton Down

Efstratiadis and the Harvard team eventually succeeded in cloning rat insulin cDNA. But regulations required the use of a category 4 laboratory—the most secure laboratory possible—to clone the human version. They decided to approach the Ministry of Defence–operated Microbiological Research Institute (MRE) at Porton Down in Wiltshire in the United Kingdom.[19] MRE was the U.K. equivalent of Fort Detrick in Frederick, Maryland. It was directed at the time by my father Bob Harris, who was also a member of GMAG. He helped to organise the Biogen/Harvard visit, and Peter Greenaway served as their guide on the ground.[s]

As beautifully recorded in Hall's *Invisible Frontiers*, this was not a successful experience for the Biogen/Harvard team. The biological safety precautions were very onerous and commuting back and forth from Salisbury (the

---

[r] Seeburg and Ullrich both later joined Genentech to work on the cloning of human growth hormone.

[s] This was an interesting connection for me. I was not only quite familiar with Porton Down in the late seventies, but I also worked at SAIC-Frederick in 2006 and became familiar with Fort Detrick. The MRE midsummer garden parties were legendary. Although quite formal in a military way, the booze flowed liberally from each corner of the marquee that was set up. I remember them well and they were not to be missed.

nearest large city), was far from ideal. The team failed to meet their objective, succeeding only in recloning rat rather than human insulin cDNA.[†] In the end Biogen gave up on its efforts to clone the cDNA for human insulin. In the meantime, the Rutter–Goodman UCSF team, although they learned a lot about the molecular biology of insulin gene expression and protein formation, fell short of finishing first to clone it.

The chemical synthesis approach of Genentech won both the cloning and expression race.[20] A subsequent collaboration with Eli Lilly, who needed to remain relevant in the insulin market in this new rDNA age, led to the successful commercialisation of recombinant human insulin. Following rapid clinical trials, Humulin was approved in late 1982, only six years after Genentech was founded. Humulin is still sold by Eli Lilly alongside a faster acting version called Humalog, with combined 2020 sales of ~$4 billion.

GROWTH HORMONE: GENENTECH AND UCSF

Another obvious target for the early recombinant DNA companies was human growth hormone (hGH). The bovine version (like insulin) was made from cows but a human version was obtained by extraction and purification from cadaver-derived human pituitary glands. The Swedish company Kabi Vitrum was the only provider of growth hormone made this way.

Although the protein is relatively small (around 200 amino acids), the gene was still too big for the chemical synthesis approach used for cloning somatostatin or insulin. But its size was in the range for cDNA cloning. Peter Seeburg, who did his PhD research with phage geneticist Heinz Schaller in Heidelberg (a Biogen founder), came to Herb Boyer's lab at UCSF and ended up working with John Baxter on the cloning of rat growth hormone, from a rat tumour that overexpressed the hormone. The small team pursued the project in stealth mode, reporting the cloning and the comparative sequence of rat and human growth hormone in December 1977 in *Nature*.[21] The paper also described the pro-growth hormone precursor of 216 amino acids and showed how it was matured to the natural hormone of 190 amino acids after secretion. It should have been simple to express the cDNA in *E. coli*, but making a construct that expressed active protein in bacteria turned out to be less straightforward than expected. A paper describing the synthesis of

---

[†] Presumably from some contamination in their reagents.

the hormone in *E. coli* eventually was eventually published (also in *Nature*), but the amounts made were small.[22]

As Chairman of the UCSF Biochemistry department, Bill Rutter wasted no time protecting the intellectual property (IP) for the growth hormone cloning project. He also established a collaboration on growth hormone with Eli Lilly.

In September 1978, after the successful cloning and expression of human insulin, Genentech also decided to pursue human growth hormone and were very well aware of what was transpiring at UCSF. Genentech teamed up with Kabi Vitrum. After cloning most of the human growth hormone cDNA from mRNA, Genentech decided to use its chemical synthesis capabilities to make an hGH expression construct that would express mRNA in *E. coli* from a hybrid DNA molecule, consisting of chemically synthesised DNA coding for the first 23 amino acids of the protein, with the remaining 24–191 amino acids coming from cDNA. This was not only an elegant means to avoid having to look for a cDNA that expressed the whole protein, but it was also a smart move from an IP perspective, because the DNA was now semisynthetic and not derived solely from mRNA. This construct was also engineered for successful transcription and translation in *E. coli*.[23]

The ambitious and talented young scientists Peter Seeburg and Axel Ullrich (mentioned above) and John Shine (also at UCSF) decided to join Genentech, not only because they knew people there quite well, but also because they were all tired of the UCSF academic politics and the hierarchical infighting. They were also much encouraged by the attitude of the Genentech scientists and management in treating each other as colleagues and equals while providing the opportunity to publish their important results in the scientific literature.[u]

Their transition to Genentech, however, was fraught with intrigue. Contrary to commercial practice but consistent with academic practice, the investigators took their growth hormone clones with them to Genentech after a nighttime visit to the UCSF lab to retrieve them. The disputed origin of the clones resulted in a protracted lawsuit between UCSF and Genentech, settled by the latter paying the university $200 million in 1999.[24]

---

[u] Promotion of publication was quite unusual for both big and small companies at the time and remains so in some places still today.

From a commercial perspective the cloning of growth hormone was important for the upcoming planned Genentech IPO, as it led to a product called Protropin that underwent successful clinical trials demonstrating efficacy in pituitary dwarfism and people of short stature. Protropin, FDA-approved and launched in 1985, was a commercial success. The fact that, in early 1985, Creutzfeldt–Jakob disease—the human equivalent of mad cow disease (bovine spongiform encephalopathy)—had been seen in some patients treated with pituitary-derived human growth hormone certainly helped. Recombinant hGH (despite some allergic reactions in some patients) was safer than the natural material derived from bovine pituitaries, which was subsequently banned.

Following its separate collaboration with UCSF, Eli Lilly launched their own version of recombinant hGH (Humatrope) in 1987. Unlike the Genentech product, the Eli Lilly version was missing the amino acid methionine at the beginning of the protein, perhaps making the Eli Lilly product less antigenic. Despite that difference, Genentech did very well in the market competing with Eli Lilly, an established, traditional pharmaceutical company. A big and likely unwelcome surprise for the "big player," a result that has been repeated many times since then by the biotech industry.[v]

Going Public

The day of the Genentech IPO was as exciting for its scientists as it was for investors (Plate 3). All watched the ticker with bated breath. According to Mike Ross (the 10th employee at Genentech and now a managing partner at SV Health Investors), the day was filled with naïve conversation about liquidity and fortunes.[w]

There were many reactions to the IPO. The investors and VCs (e.g., Lubrizol, which owned 25% of Genentech at this point) were pleased with the liquidity, as it represented an opportunity for a real return and real cash. Others felt it was just a step along the way to building a completely new type of drug company. Neither view was wrong.

---

[v] I suspect it has something to do with the passion, commitment, and teamwork of small start-up companies.

[w] There was an earlier partial liquidity event for Genentech stockholders as some external people from Hollywood bought some Genentech stock from the staff before they went public.

The first building that Genentech leased was a 5000-sq-ft former dialysis center at 460 Point San Bruno Boulevard in South San Francisco. Over time other buildings on the street, now called "DNA Way," were taken over by Genentech. In Building 3, for example, there was a manufacturing suite where the first lots of recombinant t-PA (tissue plasminogen activator)[x] were made in 10,000-L fermenters in suspension Chinese hamster ovary cells using the dihydrofolate reductase (DHFR) amplification system. It was a replica of the plant that Boehringer Ingelheim, the Genentech partner in Europe for tPA, had built.

Like many companies after an IPO, Genentech did "follow on" rounds to raise more money. They were also one of the first companies to make use of Research and Development Limited Partnerships (RDLPs). This was a concept for off-balance sheet financing catalysed for use in biotech, having been used in the oil and gas industry before by Stephen Evans-Freke, as a tax shelter strategy. RDLPs were a way for the company to raise additional money from investors who invested specifically in the project that the partnership covered, in return for profits from any products that resulted from the partnership. Genentech raised money this way to finance the cloning of interferon-γ and tumour necrosis factor. RDLPs were also used by Centocor, Amgen, and Genzyme. They were subsequently disallowed via new accounting rules as a means of off-balance sheet financing for biotech companies.

Being the first "real" biotech company, Genentech worked on many cloning projects before anyone else, but they did not win all the cloning races.[y] Strange as it may seem, Genentech abandoned their erythropoietin (EPO) project (see Chapter 2) even though the scientists wanted to do it, based on what turned out to be a less than enthusiastic (and erroneous) marketing assessment from their commercial team.[z]

The Genentech Culture

The HoHos were a defining feature of the Genentech culture, regular parties held every week where all staff got together. I happened to be invited to one of these in 1980, where all the management team, including Bob Swanson,

---

[x]We will talk later about *t-PA*, a gene that my lab was involved in cloning at Celltech, in Chapter 2.

[y]They usually won the ones I happened to be working on at Celltech, so they were not only my bête noire but also the people I looked up to as being some of the best in the business.

[z]Not the first or last time has such a thing happened in either biotech or pharma.

wore pink tutus. It was an event where I felt unusually out of place as the "reserved" Englishman, at least for a while. Eventually, these company-sponsored events grew ever larger—persisting through the 1990 Roche acquisition—requiring considerable space and organising. Many early biotech companies had equivalent weekly functions for staff at which people tended to let their hair down.[aa]

Practical jokes were also popular. One true story concerned Mark Matteucci, one of the first chemists at Genentech originally from the Marv Carruthers lab in Boulder. Mark had bought a small pink car. He was very proud of it. One time when he was out of town the car was removed from its parking space, hidden, and replaced by a similar pink car that had been crushed at the local breaker's yard. You can imagine Mark's surprise when he returned to pick up his car after the trip to drive it home to find a pink cube with no wheels in its place. Such high jinks defined the "work hard–play hard" Genentech culture, which is missing from many start-up companies now.[bb]

Not everything at Genentech was magical. There were so-called "shit-storms," and many key mistakes made. The first volunteers treated with growth hormone, for example, got symptoms (chills, etc.) from endotoxin contamination of the product. It took a year to figure out why and to correct it.[cc]

INTERFERON: BIOGEN, GENENTECH, AND CETUS

Cloning Interferon

Interferon was discovered by Alick Isaacs and Jean Lindenmann in the United Kingdom in 1957 as a substance that reduced (interfered with) the replication of influenza virus in eggs and other viruses in plaque assays. By the late 1970s, Wellcome Research Laboratories were making lymphoblastoid interferon from large cell cultures. Today, several different types of interferons (and eight interferon genes) are known, called interferon α, β, γ,

---

[aa] Some of the behaviours would simply not be tolerated now (think the culture of the 1980s depicted in many movies like *Wall Street*[25]).

[bb] Sometimes for misguided political correctness.

[cc] Mike Ross, a protein chemist who looked after protein production from the very early days of Genentech helped to sort out the endotoxin contamination problem.

and λ, with several subtypes (especially for interferon-α), with different modes of action, and with different cellular receptors. These interferons are made by different types of cells but were originally defined as fibroblast interferon or leukocyte (white blood cell) interferon. The interferon genes were one of the first set of mammalian genes (or rather cDNAs) to be cloned and expressed in *E. coli* in anticipation that the recombinant protein could be a highly useful antiviral or anticancer drug.

Before joining Biogen, the Weissman lab had been trying to clone mouse interferon in a collaboration with Peter Lengyel, whom Weissman knew from Severo Ochoa's lab at New York University where they were postdocs together. Their approach was to clone cDNA from mRNA from mouse cells that were known to make interferon and use "hybrid selection" to screen for the *E. coli* clones containing the cDNA (see Box 2). When Weissman teamed up with Biogen, the members of the lab turned their attention to cloning human interferon by the same methods.

At Genentech, David Goeddel was also cloning human interferon in a collaboration with Roche and Sidney Pestka at the Roche Institute of Molecular Biology in Nutley, New Jersey. They had some protein sequence

---

BOX 2. FINDING INTERFERON CLONES

Hybrid Selection

In hybrid selection the plasmid DNA from the clones is immobilised in arrays on filters and mRNA from mouse cells hybridised to them. If a clone contains interferon cDNA, then the mRNA should hybridise to it. Recognition of the appropriate mRNA is done by washing off the mRNA from colonies on the arrays and translating it in *Xenopus* oocytes to look for any interferon protein activity (by antiviral assay) that had been translated from the mRNA. This would mark the appropriate "interferon cDNA-containing" clone. This is a very time-consuming and difficult assay.

Differential Hybridisation

Here the amount of radioactive cDNA from uninduced cells bound to clones is compared to that from similar radioactive cDNA made from "induced" cells. Induced cells were made by treatment with double-stranded oligonucleotides or virus nucleic acids, which were known to "induce" interferon synthesis (and increasing mRNA levels). Clones containing interferon cDNA would "light up" with the radioactive probe from induced cells but not the control. This was also a very time-consuming activity involving the picking of many thousands of recombinant *E. coli* clones to find the right ones.

but not enough to make appropriate short DNA "probes" to find their clones. Instead, they used differential hybridisation with radioactive cDNA made from cells that had been virus-infected to induce interferon synthesis (Box 2). Pestka was using the same sort of expression screen that the Weissman lab was using. When some protein sequence finally emerged, interferon-specific clones were found. In early 1979, when Biogen was running out of money, the pharmaceutical company Schering-Plough stepped in to rescue the work by buying 16% of the company (for $8 million), as well as to secure the rights to it (including the IP).

Science by Press Release

After filing the patent application around Christmas 1979, Biogen announced in a press release early the next January that the Weissman lab had succeeded in cloning the human α-interferon gene. According to reporter Nicholas Wade at *Science*, January 16, 1980, was "the date on which molecular biology became big business."[26] The interferon cloning news was also presented at a press conference at the Boston Park Plaza hotel by Weissman and Gilbert. *Time* magazine also ran a story about interferon as a potential anticancer agent.[27] It was called "science by press release" at the time, a practice generally derided by the industry but commonplace nevertheless.[dd] Schering-Plough stock went up on the news, although the intent was to promote Biogen and its cloning capabilities more than to enrich Schering-Plough shareholders.

Fibroblast (β-) interferon was cloned shortly afterwards by Tadasugu (Tada) Taniguchi in Japan, who had been a postdoc in the Weissman lab. Publications followed shortly thereafter in *Nature* and *Science* describing the cloning of both α1- and α2-interferon, β-interferon, and their amino acid sequence comparisons.[28] Both the Biogen and Genentech groups produced their interferons in *E. coli* using promoter-based plasmids (see Box 1).

Unusually, Genentech and David Goeddel came off second in the race to clone both α- and β-interferons, but they did succeed in filing their own patent applications. Patrick (Pat) Gray at Genentech cloned interferon-γ in 1981 ahead of anyone else. Interferon-γ, known as Actimmune, was approved by the FDA to treat chronic granulomatous disease (CGD) in children in 1990. InterMune subsequently acquired this drug from Genentech.

[dd] It still is common practice, and it is still derided by the hard-core science community.

Cetus was also working on cloning interferon(s) at the same time using by and large the same methods in a collaboration with Triton Biosciences. It subsequently cloned and developed β-interferon, which became (via Chiron) the Schering AG product Betaferon for treating multiple sclerosis.

The Commercial Potential

Genentech had their Roche connections and Biogen was working with Schering-Plough. The companies entered into cross-licensing deals for the respective patent applications, owing to the apparent size of the potential market for interferon as a drug. The cloning of α- and β-interferons was important for different reasons for both Biogen and Genentech. The milestone was a key contributor to Genentech's IPO in October 1980 (Plate 2), but the commercial success of α-interferon as a product was modest at best, despite the high hopes for it being an antiviral and anticancer drug for multiple indications. Both the Schering (Intron A, Biogen) and Roche (Roferon, Genentech) versions were approved in 1986 for hairy cell leukemia, a very uncommon blood cancer.

β-interferon (1a) had a very different outcome commercially. Schering AG–developed β-interferon, and Betaferon got approval before the Biogen product. It is still unclear precisely how β-interferon mediates its effect, but interferons generally affect the activity of the immune system, and β-interferon is known to affect regulatory T cells (T regs). Clinical studies in multiple sclerosis (MS) showed that β-interferon reduced the number of relapses in patients with RRMS (relapsing, remitting MS) and reduced inflammation. Avonex, the injectable Biogen product, was approved in 1996[29] with parallel orphan drug status given to the Schering AG product.

The Avonex Pen, a specialised prefilled syringe with a small and covered needle, now provides patients with the exact dose they need and is intended for easier self-administration. The Avonex Pen was approved in Europe and Canada in 2011 and in the United States in 2012. Avonex, which made Biogen commercially, is still used today as a very successful and important medicine for adults with MS, including clinically isolated syndrome (CIS), relapsing-remitting MS (RRMS), and active secondary progressive MS (SPMS). Tysabri was their second MS drug. Tysabri is a monoclonal antibody that blocks a cell surface protein on T cells to prevent them from migrating into the brain. It is extremely effective in combating the disease but has a serious side effect as it can reactivate John Cunningham (JC) virus in patients who have the latent virus in their brains and cause a neurological infection (progressive multifocal leukoencephalopathy [PML]) that is sometimes fatal. These two drugs were and still are a mainstay of the Biogen MS franchise.

In 1982, Biogen moved its headquarters from Geneva to Binney Street in Cambridge (MA). The company went public in 1983, selling 2.5 million shares at \$23/share in their IPO. In July 1987, Biogen sold its Geneva research facility to Glaxo to reduce expenses. It became the Glaxo Institute of Molecular Biology (GIMB).[ee]

CETUS PRODUCTS

Frank McCormick and I were at the University of Birmingham in the United Kingdom at the same time. I was the class of '68 while Frank was class of '69. We both read biochemistry, and crossed paths frequently. When I went to the State University of New York (SUNY) at Stony Brook in 1977 to work with Eckard Wimmer on polio virus, I ran into Frank again as he was doing postdoctoral work there with Seymour Cohen on polyamines in herpesviruses, after completing his PhD in Cambridge (UK) with Alison Newton. We even worked in the same building. We have been friends ever since I helped him to go back to the United Kingdom for a job at the Imperial Cancer Research Fund (ICRF) labs in London (see Box 3).

The ICRF lab at that time was heavily into understanding the function of a protein called p53, with many groups competing internally rather than collaborating. Peter Tegtmeyer at SUNY Stony Brook (coincidentally) had discovered p53 in SV40 transformed cell lines, but no one knew what the protein did. Many people made antibodies to p53 and to SV40 T antigen (a virus protein that transforms cells) and found out that p53 and SV40 Tag

---

[ee] As I was at Glaxo looking after Biotechnology, I remember the Institute very well and visited it frequently. Jonathan Knowles, who went on to run Roche R&D, was the GIMB Director. As part of the sale, Glaxo also obtained marketing rights to both the cytokines IL2 and GM-CSF (which you will learn more about later on in the book). Most of the more than 100 employees at Biogen Geneva, who had a great deal of molecular biology and genetics expertise, became part of Glaxo. The Institute moved to one research building in the late 1980s. After the Glaxo Wellcome merger in 1994, despite the talent that was there, Glaxo off-loaded the whole facility to Serono. This was a short-sighted decision to save money to increase focus on drug discovery, completely missing the point that diversity of research activities and thinking leads to much better innovation.

BOX 3. FRIENDS GOING TO JFK

Frank went back to the ICRF on a fellowship in February of 1978. As a good mate, I told him I would take him to JFK Airport in my rather beat-up old pale blue Plymouth Valiant (called Elizabeth after the Queen). Elizabeth had more than 175,000 miles on her clock (odometer) and cost me the princely sum of $250. The engine in this car had such a gap between the piston rings and the cylinders that if you really tightened the oil cap too much the dipstick would blow out. It was basically a wreck—but it ran for a while. It turned out on that very day I was taking Frank to JFK the petrol tank began to leak. I had no time to fix it then, but it was still drivable.

I picked him up at his apartment on that fateful day and found out that it was not just him going to the airport but his partner at the time (Judy) and their three dogs (Toulouse, Oggie, and Crackers). We got everyone in. There were tears and shouts and a growing smell of petrol. The dogs would be in quarantine in England for six months. As we got nearer to the airport, I asked Frank where the British Airways freight terminal was. "Over there somewhere" was the reply, as the deadline for checking in the dogs got ever closer. We circled the airport for 20 minutes and found it with about five minutes to spare.

---

formed a complex. At ICRF, Frank made an antibody to p53. With that antibody in hand, Frank returned to the States to work with Bob Tjian, who had pure T antigen at University of California Berkeley, to investigate the relationship between p53 and T antigen further. Rick Myers, who now runs the HudsonAlpha Institute for Biotechnology in Huntsville, Alabama, was a graduate student there at the time. Rick will appear again a bit later in the book.

Starting at Cetus

Frank went to Cetus in late 1980 after Genentech had gone public and Cetus was rebooted to work on recombinant interferons. Cetus had an impressive scientific advisory board (SAB) consisting of Stan Cohen, Francis Crick, Joshua Lederberg, Ham Smith, Tom Merigan (an interferon clinician), and Andrew Schally. Frank developed there a patented cell system for expressing recombinant proteins based on amplifying constructs containing the dihydrofolate reductase gene. Biogen infringed this patent and was sued by Cetus. Subsequently others also developed such amplifiable expression vectors, including Genentech and Genetics Institute (GI). After a chance meeting with the late Chris Marshall (who was at the Cancer Research Campaign

Labs in the United Kingdom) at a meeting in Spetsai in Greece, Frank got interested in the *ras* oncogene, and he developed a K-*ras* project at Cetus.

Cetus was an extraordinary place to do science and, of course, the company was not immune from the rather typical early eighties Bay Area culture of "partying." Cetus partied heavily and a visit there in early 1980 confirmed it for me. Apart from the great science and parties, one of the highlights I remember from that trip was Frank and I going to the movies to watch the film *The Long Good Friday* starring Helen Mirren, which had just been released. I think surreptitiously we both were hoping that Ms. Mirren (now of course Dame Helen Mirren) would disrobe as she had done before in previous movies.[ff]

Kary Mullis, who was awarded the 1993 Nobel Prize in Chemistry for discovering the polymerase chain reaction (PCR) and about whom many stories[gg] have been written, was at Cetus at the time Frank was there. He was using DNA polymerase and oligonucleotide primer extension to look for mutations in the β-globin gene, and for codon 12 and 61 mutations in K-*ras* (mutations that were oncogenic in tissue culture).

Kary Mullis realised that a newly synthesised DNA strand could be primed again with a complementary oligonucleotide and both strands could be amplified. He could never really get the concept to work very well, owing to contamination of the lab with β-globin DNA. Once a clean room was found and an able technician started to run the reactions, it worked. Heat was used to separate the complementary strands so another priming round could take place. Originally, DNA polymerase (which is heat-sensitive) would be added between each amplification step. It was Kary who realised that thermostable polymerases (like *Taq* polymerase from a thermostable bacterium) would enable continuous rounds of replication to be done. These efforts led to PCR. The protocols were developed at Cetus long before it was finally published in a landmark paper.[30] Kary was not even the first author, although he was on a later publication.[31,hh]

Neither IL2 (Proleukin) nor β-interferon was doing well enough at Cetus for it to continue as an independent company. It was taken over by Chiron for $360 million in 1991, and the PCR technology licensed to Roche

---

[ff] We were disappointed on that score, but it is still one of my favourite movies.

[gg] Some of them undoubtedly true.

[hh] A paper from the Khorana lab in 1971 in the *Journal of Molecular Biology* preempted the idea of PCR, but it was never taken up.[32,33]

at the same time. Wall Street perceived the deal very negatively despite the combined company having more than 20 potential products in clinical trials.

Biogen, Cetus, Genentech, and Genex were the major players in the new recombinant DNA technology in the late 1970s. Other companies existed or were formed, most notably those providing critical reagents and tools to the rapidly developing recombinant DNA technologists in industry and in academia. Collaborative Research (from whom I bought oligo-dT cellulose to purify mRNA via the poly(A) tail) formed a subsidiary called Collaborative Genetics. Bethesda Research Laboratories and New England Biolabs were also set up primarily to sell pure restriction enzymes and other important enzymes, such as DNA ligase and terminal transferase, needed to practice the art of recombinant DNA.[34,ii]

In early 1980 before any of the four major players had gone public, they had a *combined* valuation of ~$500 million, all based on perception and promise. Cetus had a valuation of $250 million and was trying to raise $55 million. Genentech had a valuation of more than $100 million, and Biogen also had a valuation of ~$100 million. Genex was valued at $75 million. These days single companies at that stage can easily have a valuation of $500 million, based entirely on promise and perception and nothing anywhere near the clinic. It was those valuations, however, that persuaded others to jump on the biotech bandwagon and build the industry in the United States, the United Kingdom, and other parts of the world.

REFERENCES AND NOTES

1. Jones S. 1992. *The biotechnologists: and the evolution of biotech enterprises in the U.S.A. and Europe*. Macmillan Press, London.

2. Robbins-Roth C. 2000. *From alchemy to IPO: the business of biotechnology*. Perseus, New York.

3. Hughes SS. 2011. *Genentech: the beginnings of biotech*. University of Chicago Press, Chicago.

4. Merrill RD. 1982. *First hand: starting up Cetus, the first biotechnology company—1973–1982*. Engineering and Technology History Wiki (ETHW). https://ethw.org/First-Hand:Starting_Up_Cetus,_the_First_Biotechnology_Company_-_1973_to_1982

5. Cetus IPO: SEC filing 1981.

[ii] Before pure restriction enzymes were available there was plenty of exchange of enzymes between labs where postdocs would be encouraged to make large batches of restriction enzymes that could be bartered and exchanged for others.

6. Watson JD, Crick FHC. 1953. Molecular structure of nucleic acids: a structure for deoxyribose nucleic acid. *Nature* **171**: 737–738.

7. Judson HF. 1996. *The eighth day of creation*. Cold Spring Harbor Laboratory Press, Cold Spring Harbor, NY.

8. Cohen SN, Chang ACY, Boyer HW, Helling RB. 1973. Construction of biologically functional plasmids *in vitro*. *Proc Natl Acad Sci* **70**: 3240–3244.

9. Bera RK. 2009. The story of the Cohen–Boyer patents. *Curr Sci* **96**: 760–764.

10. Baltimore D. 2023. Paul Berg (1926–2023). Father of genetic engineering. *Science* **379**: 1095.

11. Cohen SN. 2013. DNA cloning: a personal view after 40 years. *Proc Natl Acad Sci* **110**: 15521–15529. doi:10.1073/pnas.1313397110

12. Berg P, Baltimore D, Boyer HW, Cohen SN, Davis RW, Hogness DS, Nathans D, Roblin R, Watson JD, Weissman S, Zinder ND.1974. Letter: Potential biohazards of recombinant DNA molecules. *Science* **185**: 303.

13. Wade N. 1974. Genetic manipulation: temporary embargo proposed on research. *Science* **185**: 332–334. doi:10.1126/science.185.4148.332

14. Berg P, Baltimore D, Brenner S, Roblin III RO, Singer MF. 1975. Asilomar Conference on Recombinant DNA Molecules. *Science* **188**: 991–994. doi:10.1126/science.1056638; *Nature* **255**: 442–444; *Proc Natl Acad Sci* **71**: 1981–1984.

15. For a well-written account of Asilomar and what preceded it, see Cobb M. 2022. As gods: a moral history of the genetic age. Basic Books, New York.

16. American Diabetes Association. 2019. The history of a wonderful thing we call insulin. July 2019. https://www2.diabetes.org/blog/history-wonderful-thing-we-call-insulin

17. Weiss M, Steiner DF, Philipson LH. 2014. Insulin biosynthesis, secretion, structure, and structure-activity relationships. https://www.endotext.org

18. Itakura K, Hirose T, Crea R, Riggs AD, Heyneker HL, Bolivar F, Boyer HW. 1977. Expression in *Escherichia coli* of a chemically synthesized gene for the hormone somatostatin. *Science* **198**: 1056–1063. doi:10.1126/science.412251

19. Hammond P, Carter G. 2002. *From biological warfare to healthcare. Porton Down 1940–2000*. Palgrave Macmillan, London.

20. For a summary of the cloning story from chemically synthesised DNA: Riggs AD. 2021. Making, cloning, and the expression of human insulin genes in bacteria: the path to Humulin. *Endocr Rev* **42**: 373–380. doi:10.1210/endrev/bnaa029

21. Shine J, Seeburg PH, Martial JA, Baxter JD, Goodman HM. 1977. Construction and analysis of recombinant DNA for human chorionic somatomammotropin. *Nature* **270**: 494–499. doi:10.1038/270494a0

22. Seeburg PH, Shine J, Martial JA, Ivarie RD, Morris JA, Ullrich A, Baxter JD, Goodman HM. 1978. Synthesis of growth hormone by bacteria. *Nature* **276**: 795–798. doi:10.1038/276795a0

23. Goeddel DV, Heyneker HL, Hozumi T, Arentzen R, Itakura K, Yansura DG, Ross MJ, Miozzari G, Crea R, Seeburg PH. 1979. Direct expression in *Escherichia coli*

of a DNA sequence coding for human growth hormone. *Nature* **281**: 544–548. doi:10.1038/281544a0

24. Rasmussen N. 2014. *Gene jockeys*. Johns Hopkins University Press, Baltimore.

25. *Wall Street* is a 1987 American drama film, directed and cowritten by Oliver Stone and starring Michael Douglas, Charlie Sheen, and Daryl Hannah.

26. Wade N. 1980. Cloning gold rush turns basic biology into big business. *Science* **208**: 688–692.

27. *Time* magazine cover story. The big IF for cancer. Monday, March 31, 1980.

28. Taniguchi T, Mantei N, Schwartzstein M, Nagata S, Muramatsu M, Weissman C. 1980. Human leukocyte and fibroblast interferons are structurally related. *Nature* **285**: 547–549. doi:10.1038/285547a0

29. Biogen MS clinical trial: In one randomised Phase 3 trial, 158 people with relapsing MS were given Avonex (30 micrograms injected into the muscle once per week), while 143 received a placebo, for two years. Results showed that significantly fewer Avonex-treated patients experienced worsening disability after two years (21.9% vs. 34.9%), defined as an increase of at least one point on the expanded disability status scale (EDSS) that persisted for at least six months. Treated patients also had significantly fewer disease relapses per year—0.67 versus 0.82 with a placebo—and fewer and smaller brain lesions on MRI scans. In the other trial, 193 patients with CIS were treated with Avonex for three years, while another 190 received a placebo. The results showed that patients on Avonex were significantly less likely—by ~44%—to have experienced a second relapse and to progress to clinically definite MS at three years. Additional results suggested that the treatment decreased brain lesions.

30. Saiki RK, Scharf S, Faloona F, Nullis KB, Horn GT, Erlich HA, Arnheim N. 1985. Enzymatic amplification of β-globin genomic sequences and restriction site analysis for diagnosis of sickle cell anemia. *Science* **230**: 1350–1354. doi:10.1126/science.2999980

31. Mullis KB, Faloona FA. 1987. Specific synthesis of DNA *in vitro* via a polymerase-catalyzed chain reaction. *Meth Enzymol* **155**: 335–350.

32. Kagan W. 2021. Exponentially important: the scientific origins of PCR. *Nautilus*, August 2, 2021. https://nautil.us/exponentially-important-the-scientific-origins-of-pcr-238268; also see PCR entry in Wikipedia, https://en.wikipedia.org/wiki/Polymerase_chain_reaction

33. Dove A. 2018. PCR: thirty-five years and counting. Science, May 10. 2018. https://www.sciencemag.org/features/2018/05/pcr-thirty-five-years-and-counting

34. Wade N. 1980. Three new entrants in gene splicing derby. *Science* **208**: 690. doi:10.1126/science.208.4445.690

CHAPTER 2

The Second Wave

My journey through biotechnology began in 1978 at the Animal Virus Research Institute (AVRI) in Pirbright in the United Kingdom,[a] where I worked on cloning foot and mouth disease virus (FMDV) RNA with the goal of making a vaccine from VP1, one of the capsid proteins. We were not alone in that project: Genentech also had an FMDV program, working with the Plum Island Animal Disease Center in Orient, New York (the U.S. equivalent of AVRI).[b]

AVRI was a high-security laboratory, so it was straightforward to meet the genetic engineering guidelines that were in place to allow the cloning to be done. Previous data had suggested that isolated VP1 protein, one of the FMD virus capsid proteins, could elicit neutralising antibodies when inoculated into animals. Our first cDNA library of a few hundred clones was made from pure FMDV RNA.[1] The library covered virtually the whole virus RNA, including the piece coding for the capsid protein VP1. We did successfully make VP1 in *Escherichia coli*, but we never got any immunity to infection in animals: the VP1 "vaccine" did not elicit neutralising antibodies as was previously believed. This failure could have been predicted given that a naturally occurring nucleic acid–free virus-like particle (called 12S) did not elicit neutralising antibodies in cattle either. We all learned the lesson that the use of a single antigen may not be sufficient to induce an effective immune response.[c]

---

[a] Now called the "Pirbright Institute."

[b] Dennis Kleid, who was of the first employees at Genentech, visited us ostensibly to find out what we were doing. We showed them around the lab. Dennis seemed rather more interested in what my graduate student was doing than in our work.

[c] The Genentech FMDV program never went anywhere either.

CELLTECH: EUROPEAN BIOTECH EMERGES

Spurred on by the prospect of applying recombinant DNA to make therapeutic proteins like growth hormone and t-PA and tiring of government-based research, I joined Celltech in April 1981. Celltech was formed in November 1980 as another brainchild of Sydney Brenner, with close ties to the Medical Research Council.[2,3,d] The founding of Celltech was very important for biotechnology in the United Kingdom. It was a peculiarly British story. As Mark Dodgson remarks in his article on the formation of Celltech, some of the problems with the beginnings of the company were derived from the fact that it was conceived by a Labour government and born into a Conservative one.[4]

There were a lot of politics surrounding Celltech's formation. The negotiations to start the venture included the MRC and several government departments including the National Enterprise Board (NEB; formed by the Labour government to stimulate innovation in industry), the National Research Development Corporation (NRDC), the Cabinet Office, the Department of Trade and Industry (DTI), and the Treasury. Long-term strategic views were at odds from the beginning with the short-term thinking that demanded profitability in an unrealistic time frame.

Celltech was formed with a substantial (at that time) series A round of £12 million. This money was not from British or American venture capitalists (VCs), but from blue chip British investors who were persuaded to invest by the vision of a new industry that might change the world. They were sophisticated enough to appreciate the importance of the Genentech IPO in October 1980 in terms of return on investment. The NEB purchased 44% of the equity (£5 million). Four additional investors bought 14% each: Prudential Assurance, British & Commonwealth Shipping, the Midland Bank, and Technology Development Capital (TDC). The TDC investment was subsequently bought by Biotechnology Investments Limited (BIL), the Rothschild venture firm that Sydney Brenner advised.[4]

Both Recombinant DNA and Monoclonal Antibody Technology at Celltech

Celltech was unusual because it involved both monoclonal antibody work—primarily producing monoclonal antibodies at some scale—and

---

[d]Celltech's founding was, to some extent, a response to the failure to file intellectual property (IP) protection on the Milstein monoclonal antibody technology (see Chapter 4).

recombinant DNA technologies. Most of the start-ups at that time focused on either one or the other technology but not both. This dual focus later placed Celltech at the forefront of making recombinant antibodies.

Gerard Fairtlough was recruited from the National Enterprise Board to be the founding CEO. The founding Head of R&D was Norman Carey,[e] who brought several people with him from his research group at G.D. Searle in High Wycombe (later Monsanto), including Mike Doel, Spencer Emtage, and Mike Eaton. They formed the nucleus of the first Celltech R&D team in November 1980. This hiring was not without repercussions: G.D. Searle took exception to the move and attempted to sue Celltech, forcing the new Celltech team to file affidavits for the legal team to be among their first duties.

Celltech did not start in Oxford or Cambridge as there were at that time no "Oxbridge" science parks for the company to be placed on. Instead, Celltech was based in Slough on the A4 quite close to Heathrow Airport, across the railway line from the Mars factory and opposite a factory that made women's underwear.[f] We could smell the chocolate from the Mars bars being made from the lab.

Slough was described by the poet John Betjeman as "not fit for humans."[5] The town was also used as the fictional setting for the British sitcom *The Office*. Slough Trading Estate (our landlord then) is now one of the biggest commercial real estate companies in Europe. It is also the place chosen to build the world's second largest data hub. Slough has changed a good deal since Betjeman wrote his poem, and it continues to grow.

In early 1981, however, the Celltech building at 216 Bath Rd. was yet to open. We started our lab work at 250 Bath Rd. in an old Richardson-Merrell (of Vicks fame) research building with polished wooden benches.[g] The space was *absolutely nothing* like the tailor-made start-up spaces now available in the United Kingdom and the United States.

I was responsible for running the cDNA cloning lab at Celltech. If it moved, we cloned it. We cloned many cDNAs from both plasmid and bacteriophage libraries. Dog-eared copies of Tom Maniatis' *Molecular Cloning:*

---

[e] I still have a copy of the letter I wrote to Norman Carey in December 1980 asking him if he had any interesting opportunities in the new biotech company that he was now part of. I knew him from our collaborations with G.D. Searle when I was at AVRI. Fortunately for me, he did have an opening.

[f] You can imagine the comments made about that.

[g] Richardson-Merrell became part of Marion Merrell Dow and then Hoechst and Aventis and now is part of Sanofi.[6]

A Laboratory Manual (aka "The Cloning Bible") would be found on almost every lab bench[7] (Plate 3). Our first project was to clone the enzyme calf chymosin, the major enzyme in "essence of rennet" used to clot the milk in the first steps of cheese making.[h]

We cloned the cDNA coding for the enzyme successfully, and we were able to make the active enzyme. It was a great occasion when we all got together for a wine and cheese party, not only to eat the cheese that was made at the Dairy Research Institute in Shinfield near Reading using the recombinant enzyme,[i] but also to celebrate the paper recently published in the *Proceedings of the National Academy of Sciences (PNAS)* reporting what we had done.[8] In the end, the chymosin process was licensed to Pfizer's agricultural arm for a relatively small sum.

At Celltech, we cloned cDNAs coding for several important proteins including t-PA (as did everyone else), various metalloproteinases including collagenase and stromelysin, Factor I and Factor H, and glutamine synthetase (GS). The cloning of the metalloproteinases even got us a mention in *The Times* of London following a 1985 *Nature* publication.[9] The GS project was a good example of science that turned out to be important both scientifically and commercially.

Glutamine Synthetase (GS) Amplification System

The GS work started as an unsanctioned "skunkworks" project. Peter Rigby, who was at Imperial College, had just come back from Paul Berg's lab at Stanford. He was on the Celltech-Science Council, the equivalent of a science advisory board (SAB). The Council was chaired by Sir Michael Stoker. It included some important advisors including Sydney Brenner and César Milstein.

One day, Spencer Emtage and I were chatting with Peter over tea about amplifiable systems that might compete with the dihydrofolate reductase (DHFR) system currently used to amplify (i.e., produce large amounts of plasmid-derived sequences in bacteria) originally from the Bob Schimke

---

[h]The first mRNA extraction at Celltech was from mucosa of the fourth stomach of a 7-day-old suckling calf stomach (courtesy of a trip to another Agricultural and Food Research Council Institute at Compton in Berkshire, where these calves were slaughtered every week).

[i]I could not tell the difference between cheese made with natural chymosin compared to the recombinant enzyme, but the experts said they could. I had probably drunk too much wine to notice.

lab at Stanford but developed further by Randy Kaufman at GI and Frank McCormick at Cetus and used by Genentech. Peter suggested the gene coding for the enzyme glutamine synthetase could be amplified using a "transition state analogue" of the enzyme. We ended up collaborating with Richard Wilson's lab at Glasgow University who were already trying to clone the cDNA. Together we ended up cloning the GS cDNA from a large λ gt10 library.[10] Chris Bebbington and Spencer then made an amplifiable gene from the cDNA and a segment of genomic DNA and showed that this construct amplified genes as well if not better than the traditional system in use. It had the added benefit of being proprietary to Celltech. The technology was ultimately licensed to Lonza and is still available today as the "GS Xceed system."[11] This project made more money for Celltech in those early days than many of the other things they did.[j] It was also important for expressing recombinant antibodies and antibody derivatives (see Chapter 4).

Celltech's Financing

From a financial perspective Celltech was reasonably successful. They raised £21 million from collaborations with Sankyo (now Daiichi-Sankyo) on t-PA, calcitonin, and the related protein calcitonin gene-related peptide (CGRP). They obtained additional capital from a deal with Serono on human growth hormone and from various monoclonal antibody development contracts. The company worked as two parts: the research and development function and the cell products business based on the scaled-up production of monoclonal antibodies in 100-L and 1000-L airlift fermenters. This necessitated building a custom manufacturing suite (228 Bath Rd.) that was later sold to Lonza. Celltech was way ahead of the time in their ability to make antibodies at scale. Hybritech, a new monoclonal antibody company we will meet in Chapter 4, had kilograms of antibody made for it by Celltech.

It was not all high tech. Celltech also had a contract to make recombinant erythropoietin (EPO; more on this molecule later) for Johnson & Johnson (J&J) in adherent cells in large roller bottles, not unlike the roller bottles we grew virus in at AVRI. By the time Celltech had optimised this process they were producing three times as much EPO compared to the original roller bottle process.

---

[j]I have dined out on the skunkworks nature of that project many times.

I had the wonderful experience of going to Japan several times, mostly for the Sankyo collaboration. One trip was with David Gration, the new COO who came to Celltech from Wyeth. I went along as technical back-up to the finance team to present our end of year numbers to our Japanese collaborators. I have a very clear memory of watching the sun set behind Mount Fuji from one of the higher floors of the tower building where we were presenting to Barings.[k] I also remember a spectacular dinner hosted by Sankyo in the French restaurant Le Trianon at the Takanawa Prince Hotel.[l]

Celltech formed joint ventures (JVs) that were not very successful. One was with Boots to form Boots-Celltech Diagnostics (BCD) and another was in the microbial space with Air Products called Apcel. BCD did develop some important diagnostic tests using mAbs, including one for measuring α-fetoprotein for malignancy, but after a while both JVs were reacquired by Celltech. They subsequently off-loaded the diagnostics business to Novo in 1989.

The formation of those JVs did help to bring in more capital to the company. After the stock market crash in 1987, Celltech raised a further £42 million when BIL sold their stake to B&C. With a turnover of some £12 million, Celltech was marginally profitable, and had more than 400 employees by 1991. However, the MRC was not that happy with Celltech's progress. In a way they voted with their feet by helping to form another antibody company in 1989, Cambridge Antibody Technology, rather than rolling all their antibody engineering IP from the MRC labs into Celltech. Cambridge Antibody Technology was involved in the creation of both Humira and Benlysta. It was ultimately integrated under the MedImmune brand by AstraZeneca (see Chapters 4 and 5).

Dodgson makes an important observation in his Celltech history article[4]: in 1986, Celltech spent $5 million on R&D compared to Genentech's $80 million and Cetus' $32 million. But Celltech had also acquired or filed 44 patents and applications. I think these numbers represent value for money. They also contrast the cost of doing biotech in the United Kingdom compared to the United States. Celltech went public in December 1993, raising £50 million.[12] At the end of the day, it is difficult to determine how much money either the initial investors or the employees made from the company, but it was certainly nothing like the windfall that employees at

---

[k] This was well before the oldest merchant bank in Britain was brought to its knees by the rogue trader Nick Leeson in 1995.

[l] The crêpes Suzette was unforgettable, and the 50-year-old Armagnac was not bad either.

BOX 1. BRITISH BIOTECHNOLOGY

Another U.K. biotechnology company created with much fanfare after Celltech was British Biotechnology, formed in 1986 by Sir Brian Richards, another G.D. Searle alumnus. British Biotechnology was formed largely because Monsanto, who had by then acquired G.D. Searle, decided to close the High Wycombe research site. This was an opportunity to spin out the talent from there with Keith McCullagh as CEO. The company made a lot of noise as "British Biotechnology" and were backed by Rothschild's money in the form of Biotechnology Investments Limited (BIL). They also had a well-respected Board of Directors. Contrary to their predictions of becoming "another Genentech," however, British Biotechnology achieved considerably less than Celltech and had a rather dismal outcome, being quietly folded into Vernalis in 2003.

Genentech received. BIL did make a profit from their 1983 investment (see Box 1). I assume other investors made some money when the company went public, but personally I made virtually nothing from my Celltech stock.[m]

After several acquisitions (including Chiroscience for £700 million, Medeva for $915 million in 1999, and Oxford GlycoSciences for £103 million in 2003), Celltech became part of UCB in Belgium in a $2.25 billion transaction consummated in 2004.[13] To say this was a "cost-effective deal" is a bit of an understatement. It was in reality a very good and cheap deal for UCB. The antibody engineering technology that the company owned and had built over 10 years (see Chapters 4 and 5) was probably worth more than twice that on its own.

Genentech and other companies such as Genetics Institute and Biogen were the most prominent U.S. competition for Celltech: Genentech almost always won the cloning races that we were also in. Biogen was a key player but most of their early discoveries (e.g., cloning interferon-β and hepatitis B surface antigen) happened in the labs of their founders.

NEXT WAVE OF U.S. COMPANIES: GENETICS INSTITUTE

Several recombinant DNA–based start-ups also began their lives in 1981 on both coasts of the United States. One of the most notable was Genetics Institute, co-founded in Cambridge (MA) by Mark Ptashne

---

[m] I did not have many shares anyway, but that was not why I went to the company in the first place and, being a bit naïve, I did not think it mattered much anyway. To me then, the experience mattered much more.

and Thomas (Tom) Maniatis. Both Ptashne and Maniatis were Professors in the Department of Biochemistry and Molecular Biology at Harvard University. Ptashne, an international leader in bacteriophage molecular genetics, was one of the youngest professors to be tenured at Harvard. He had isolated the λ phage repressor gene and had cloned it and expressed it in *E. coli*, partly in competition with Walter (Wally) Gilbert's lab. Ptashne's lab was also working on the development of methods to produce recombinant proteins in bacteria.

Maniatis was well-known for his role in the development of two key recombinant DNA methods: cDNA cloning and making genomic DNA libraries. He was much in demand as an advisor to other developing recombinant DNA start-ups. While a professor at Caltech in 1978, Maniatis' lab was the first to clone human genes, specifically the four genes of the human β-globin gene cluster—truly iconic methods and papers.[14,15] In 1979, Maniatis founded and taught the first Cold Spring Harbor Course on "Molecular Cloning," which led to the molecular cloning laboratory manual mentioned earlier, in collaboration with the late Joseph (Joe) Sambrook (then the scientific director of the Cold Spring Harbor Laboratory) and Ed Fritsch, a former postdoctoral fellow in Maniatis' lab.

After the founding of Biogen, the Harvard Management Group (HMG; managers of the substantial Harvard endowment) approached Ptashne about starting a company in which the University would have an equity position.[n] Ptashne invited Maniatis, who had moved to the California Institute of Technology (Caltech) in Pasadena, California, to join a meeting with the HMG and a group of potential founding scientists, primarily from Ptashne's lab.[16] The interactions between the postdocs and the HMG did not go well, and the meeting ended without an agreement to move forward. The duo faced another setback when Derek Bok, the president of Harvard University at that time, introduced the Harvard Faculty Senate to the idea of the University holding equity in a faculty-founded biotech company.[o] The proposal was met with overwhelming opposition from the Harvard faculty. Although initially supportive, the negative faculty response persuaded Bok to give up Harvard's significant stake in a start-up, and to denounce

---

[n] HMG was not exactly overjoyed by Wally Gilbert's founding of Biogen with little financial benefit to the University.

[o] Unlike MIT, Harvard had not engaged in many entrepreneurial activities in the biological sciences.

all faculty biotech entrepreneurs.[p] Undeterred by the faculty response and Bok's rejection, Ptashne and Maniatis went forward with blue-chip venture capital support to start Genetics Institute.

Genetics Institute Gets Started

Maniatis returned to Harvard in 1980, and Genetics Institute officially started in April 1981.[q] The scientific/business strategy was simple: clone therapeutically important proteins by cloning and expressing the genes.[16] The name "Genetics Institute" was chosen to soften the "techy" image a bit owing to the controversy around recombinant DNA research in Cambridge (MA). The Founders had the foresight to hire a professional CEO (Gabriel [Gabe] Schmergel from Baxter Travenol) to raise money and guide the financial aspects of the company. Baxter Travenol at that time was full of young entrepreneurial talent looking for challenges outside a big organisation.[r]

Prior to the CEO's arrival, Ptashne's kitchen was the meeting place for strategic discussions, and business activities were carried out in the office of Thomas (Tom) Hexner (a friend and neighbour of Ptashne, and an initial Board member). The team managed to attract some top-tier VCs to invest in the company's series A round including J.H. Whitney (Benno Schmidt), Venrock (Anthony [Tony] Evnin), and Greylock (Dan Gregory and Henry McCance). Bill Paley, who was associated with the Whitney family and whose wife had recently died of cancer, also invested. The pre-money valuation was $6 million with a raise of $6 million for a post-money valuation of $12 million.

---

[p] Bok lamented the decision later when Genetics Institute went public, as Harvard would have had a share in those proceeds.

[q] A close friend of Tom Maniatis, Richard Axel at Columbia University, was invited to be a founding scientist, but he was already committed to advise the venture capitalist Fred Adler in his successful efforts to reorganise the Bethesda Research Labs (BRL), which became Gibco/Life Sciences. Axel, Michael (Mike) Wigler (Cold Spring Harbor), and Maniatis had recently collaborated to show that the cloned human β-globin genes could be stably introduced in human cells in culture using a transfection method developed at Columbia. This method later became a key step in producing therapeutic proteins and biologics in mammalian cells, the covering IP bringing nearly $1 billion to Columbia University in royalties over the years.[17]

[r] In addition to Schmergel, the "Baxter mafia" included Henri Termeer (who worked for Gabe as general manager of Baxter Germany, later becoming the CEO of Genzyme) and Robert (Bob) Carpenter, who founded Integrated Genetics (bought by Genzyme, who subsequently made Carpenter lead director on the Genzyme Board).

The founders had an almost equal share of the company, with 15% for future hires, 5% for the CEO, and 50% for the VCs—not so different from arrangements made today.

However, rather unlike today, the VC team took a long-term view and were not interested in micromanaging their ventures. They wanted to build an enduring company, rather than seeking a rapid return on their investment. None of this "biotechnology stuff" had been done before, so the scientists were given freedom to express themselves as well as the genes and proteins they worked on. Significant accomplishments came because the scientists were doing experiments and generating data to an overall plan of making therapeutically useful proteins without being overmanaged.[s]

Tom and Mark proceeded to identify and recruit an outstanding group of scientists trained in the best labs (see Box 2). Hiring turned out to be a

---

BOX 2. THE GENETICS INSTITUTE TEAM

GI built a small but very talented team that covered all the technologies that were needed to clone and express a range of eukaryotic genes. Their roster of talent was at least as good as that in Genentech in the early days.

Hires included Steven (Steve) Clark from Paul Berg's lab, who with Gordon Wong from Tom's lab invented expression cloning methods (i.e., looking for the protein from the gene rather than the gene in the clones); Rod Hewick from Leroy (Lee) Hood's lab (a co-developer of gas phase protein sequencing); Randy Kauffman from Phil Sharp's lab, who became a leader in expressing genes in mammalian cells; Ed Fritsch from Tom's lab, who would lead the effort to clone the erythropoietin gene; Chuck Shoemaker from David Baltimore's lab; Eugene Brown, from Gobind Khorana's lab, an expert in synthesizing DNA probes; and John (Jay) Toole and John Knopf, who would lead the effort to clone the Factor VIII gene. Tom also recruited Robert (Bob) Kamen, an accomplished cancer research scientist from ICRF in London to serve as the company CSO (Chief Scientific Officer). Glenn Larsen, who was a graduate student in Eckard Wimmer's lab in Stony Brook when I was there, was also hired by GI in March 1982 after finishing his PhD.[t]

---

[s]Like the porridge in "Goldilocks and the Three Bears": not too hot, not too cold, but just right.

[t]Glenn and I had a connection. I had spent October 1977 to October 1978, as a postdoc in Eckard's lab in SUNY Stony Brook on Long Island working on polio virus RNA structure, particularly the small protein attached to the 5′ end of the RNA called VPg, which Eckard and David Baltimore's lab had found simultaneously. Glenn and I overlapped and worked together.

difficult challenge, however, as a career in biotechnology was widely considered second-best to academia at the time.

The continuing moratorium on recombinant DNA in Cambridge (MA) meant that GI had to find lab space elsewhere in the Boston area. They settled on the Lying-In Maternity Hospital in Mission Hill in Boston, a run-down abandoned building in a not-so-good area of Boston. It became a race against time to get up and running, because Boston, too, was considering genetic engineering regulations. By setting up a lab bench and getting some experiments done quickly they could prove they were a going concern and could thus get "grandfathered" into any subsequent regulations controlling biotech in Boston.[16,u]

GI Space

The 14,000 sq ft of lab space at the Lying-In Hospital consisted of a kitchen, maternity wards, and partially renovated clinical labs—some being next to the old mortuary, which was apparently quite spooky when working there late at night on your own. Heading for home included a walk across the parking lot where robberies regularly took place.[v] Eventually, they occupied further space upstairs in the hospital, turning old operating rooms into labs and building a small manufacturing suite.

As with all biotech start-ups in the new biotech space, raising money continued to be a pressing necessity. Schmergel and his team did several nondilutive deals, including with Sandoz for lymphokines and cytokines. This was an umbrella contract covering 100% of the R&D costs for the GI program that covered many of the targets other companies such as Immunex and Amgen were cloning, including granulocyte colony-stimulating factor (G-CSF) and macrophage colony-stimulating (M-CSF) (see Chapter 3).[w]

---

[u] Apparently, one afternoon, Councilor Ed Flynn, who was on the Boston City Council and running for mayor, wanted to debate new biotechnology regulations on TV with Gabe Schmergel. He arrived at the hospital lab with his entourage of 20 people. Gabe knew he was not going to win anything in a conversation about the whys and wherefores of biotech with Councilor Flynn, so he left by the back entrance and the "interview" never happened.

[v] The company had a policy of chaperoning employees, which included carrying $100 in pocket in case of a mugging.

[w] Baxter Travenol invested $10 million in GI to own 9.9% of the company but did not get a board seat. The CEO insisted that there be no board seats for the corporations that invested in GI to avoid the appearance of being beholden to any one company.

GI had more IP than any other company in the nascent industry. They also built a great deal of credibility from their corporate deals, continuing to raise money at higher valuations. When they went public on May 19, 1986, with J.P. Morgan and Robertson Colman & Stephens as the bankers, they raised nearly $75 million at a good valuation.[x] Owing to their careful policy of raising cash from corporate collaborators, there was no "reverse split" of the stock.[18]

GI finally moved into tailor-made lab space in Alewife at the end of the Boston subway Red Line in 1983. This area is a bit distant from the Kendall Square/Technology Square hub near MIT in Cambridge (MA), but the rents were and still are cheaper.

CLONING OF TISSUE PLASMINOGEN ACTIVATOR (t-PA)

Glenn Larsen ran the t-PA cloning effort with Bob Kay, in collaboration with Cold Spring Harbor. As its name suggests, t-PA is an enzyme that converts plasminogen into plasmin, a blood protease that digests blood clots, and thus is potentially useful to treat blood clots in coronary arteries. t-PA was on the radar of almost every cloning company, including Celltech and Genentech. Glenn purified the protein himself in less than six weeks,[19] determined the amino acid sequence, and cloned the cDNA using short oligonucleotide probes inferred from the amino acid sequence.

The Genentech versus Wellcome t-PA Patent Case

The cloning of tissue plasminogen activator was the subject of a very large patent interference case in 1987 between Wellcome and Genentech/Boehringer Ingelheim. Wellcome had challenged the Genentech Europe issued patent primarily on the basis of obviousness. That is to say, Wellcome thought that what Genentech did lack an inventive step—meaning that anybody who wanted to clone t-PA would do precisely the same thing as Genentech did.

There were at least five groups trying to clone t-PA at the time, including Biogen, Celltech (my lab), Genetics Institute, Joe Sambrook's lab at Cold

---

[x]The S1 prospectus for the GI public offering was a document of 47 pages—nothing remotely like the (rather ridiculous) 400 pages you see for S1 prospectuses today. At the IPO they sold 2.5 million shares at $29.75, raising nearly $75 million and leaving the company with a valuation of $335 million.

Spring Harbor, and Genentech (led by Diane "golden hands" Pennica). All of us used the same basic approach of obtaining amino acid sequences from the purified protein and designing oligonucleotide probe mixtures that would cover the inferred nucleotide sequence of that part of the mRNA coding for those amino acids. All of us made cDNA libraries from mRNA extracted from the Bowes melanoma cell line, which was known to express t-PA. Wellcome pulled me into the trial as a "witness of fact" rather than as an expert witness, as my lab was one of the many that were cloning t-PA at the same time.

The approach my lab took appears in Judge Whitford's summary of the trial, taken from my affidavit filed on behalf of Wellcome in 1987: "It appears from a passage at the beginning of his examination-in-chief that Dr. Harris had previously used oligonucleotide probing. If he had proceeded to probe using pools, I have no reason to suppose that he would have been any less successful than Genentech were."[20] A *Nature* article by Dr. Pennica published after the date of the patent application to which Whitford later referred showed the DNA sequence, saving me the trouble of using a pool of probes so I could proceed directly to use one probe complementary to the known sequence. Whitford went on the say that in his cross-examination of me, Mr. Stephen Gratwick, the Genentech barrister, elicited the fact that "Dr. Harris was a man of inventive capacity, but my affidavit evidence was not in any material respect seriously challenged."[y]

This all seems a bit dry when read on the page. It was anything but dry when I was defending my affidavit in the high court in the dock, having sworn on the Bible to tell the truth, the whole truth, and nothing but the truth. Gratwick proceeded to grill me at some length about the time it took to use oligonucleotide probes and why I chose to use the probes that I did. I felt like he, a patent barrister, knew more about oligo probing of cDNA libraries than me, a molecular biologist.

The high stakes of the case were reflected in the involvement of England's best patent barristers. On the other side from Gratwick and Genentech was Robin Jacob QC (now Sir Robin Jacob, ex Lord Justice of Appeal in the patent court) representing Wellcome.[21] The tense atmosphere in the courtroom was heightened by the fact that several existing Nobel laureates were in the room including Paul Berg representing Genentech and James Watson

---

[y] This affidavit and all the others from the principals in the trial can be found in Sydney Brenner's archives at Cold Spring Harbor with the subsequent court of appeal proceedings.[22]

(Cold Spring Harbor). Sydney Brenner and Stanley (Stan) Cohen were also there. Sydney was an expert witness for Genentech both there and in their subsequent appeal of the negative verdict they received (they also lost on appeal).[22,23] Many other Fellows of the Royal Society were also present.[z]

Second-Generation t-PAs

To try to get around some of these IP issues and to improve the half-life of the recombinant protein in human blood, many groups, including Celltech, GI, and Genentech, made second-generation versions of t-PA.

Burroughs Wellcome licensed the GI t-PA program for the United States, and Suntory obtained rights for Japan. A phase 3 clinical trial of the longer half-life t-PA in 15,000 patients showed that there was a dose-dependent stroke risk that was deemed to be too high for this version of recombinant t-PA to be a viable product. Somewhat similar second-generation t-PA proteins were made by Sambrook and Mary-Jane Gething at Cold Spring Harbor and by scientists at Integrated Genetics. The most important second-generation t-PA was made by Genentech by substituting various amino acids in the protein to increase affinity for fibrin. This became the product TNKase, or Tenecteplase.[24]

BLOOD CLOTTING FACTORS COMPETITION: GENETICS INSTITUTE AND GENENTECH

In the Gospel of Luke from the Bible we hear about a woman who "issued blood" for 12 years before being healed by touching Jesus' robe. I was always curious about what that condition might have been. I ruled out haemophilia A, which is almost always found in males as it is a result of mutations (inversions and deletions) in the X-linked gene for blood clotting protein Factor VIII[25] (females can only have haemophilia A if their "good" copy of the Factor VIII gene is also inactivated in some way). But it is conceivable that this woman was suffering from von Willebrand disease, which is caused by mutations in a large gene on chromosome 12 coding for von Willebrand

---

[z] All together fairly intimidating for a guy of 37 whose only courtroom appearance before then was in Slough arguing for leniency (as I did not want to lose my driving license) for getting a speeding ticket doing 104 mph on the M4 in my black Saab Turbo on the way to Celltech.

factor (vWF). As it turns out, vWF forms a complex with Factor VIII in normal blood clotting processes.

Haemophilia B is also caused by mutations in an X-linked gene encoding another protein called Factor IX. Sometimes called Christmas disease (after the surname of the first person recognised with the disease in the United Kingdom), haemophilia B has a British royal family connection as Queen Victoria was a carrier and several of her male offspring had the disease.

Cloning Factor VIII

Recombinant DNA companies sought to clone both Factor VIII and Factor IX in the early eighties. Treatment for haemophilia A and B consisted of using clotting factors obtained from human blood. That desire reached a new urgency with the arrival of HIV in early 1981, when many haemophiliacs were infected from contaminated products derived from HIV-positive blood donors, before measures were put in place to screen plasma to remove HIV and other viruses.

Gabe Schmergel and his employer Baxter Travenol knew a lot about haemophilia, as Baxter was a purveyor of blood products across the world and made a range of plasma products under Schmergel's management. With his move to GI in April 1981, it came as no surprise that one of the projects GI took on was to clone Factor VIII. Genentech had initially tried to get Baxter to support their own Factor VIII cloning project, but the decision makers at the company had said no. But when Genentech heard that GI had elected to clone Factor VIII, they quickly followed suit, not wanting to be beaten by their fierce East coast rivals.

At that time, no one really knew what the protein consisted of except that it was large and complexed to vWF in blood. Indeed, Ptashne is said to have remarked that it would be impossible to clone Factor VIII because the protein was far too big. But Jay Toole volunteered to run the Factor VIII cloning project for GI. He hired a tall and quietly intense guy named John Knopf from Buffalo, who had not yet completed his PhD, to help. Knopf was a superstar. Knopf and Liz Wong (who subsequently cloned bone morphogenetic protein [BMP], a successful GI product used in complex bone fractures) turned the Ptashne impossibility into a reality. But it was far from easy.

Genentech and GI adopted essentially the same cloning strategies used for other proteins of interest. However, Factor VIII was much harder, because (as Ptashne remarked) the protein—and its gene—were very large, and much less was known about what tissue made the protein and at what abundance. Both groups had access to protein from various sources. GI researchers chose pig blood as a source on the correct assumption that the human and porcine Factor VIIIs would be highly related.

The company tasked Knopf with obtaining pig blood for running through their protein purification processes. GI was able to obtain enough porcine Factor VIII to obtain some amino acid sequence under Rod Hewick's guidance, using one of the first Applied Biosystems amino acid sequencers that GI had recently installed.

Meanwhile at Genentech, Gordon Vehar and his team were also purifying Factor VIII. Vehar had the advantage of having worked on bovine Factor VIII as a postdoc in Earl Davie's lab at the University of Washington in Seattle. Genentech also collaborated with Edward (Ted) Tuddenham at the Royal Free Hospital in London, whose lab had previously purified small amounts of human Factor VIII free from vWF. Somewhat ironically, both groups used their amino acid information to screen the λ human genomic DNA library made by Richard Lawn (who was now at Genentech) and Edward (Ed) Fritsch (who was now at GI) while they were both in Tom Maniatis' lab at Caltech (Fig. 1).[aa]

Using more than one set of probes for different DNA sequences from the same gene was a necessity, given that the full-length large Factor VIII gene, with its many exons and introns, would exist only in bits among many clones in a *genomic* library. But the partial genomic clones identified could be used subsequently to screen cDNA libraries to obtain a cDNA clone (i.e., no introns) coding for the protein that could then be expressed in mammalian cells.

Both groups found the project extremely difficult, with many false starts and many false positive clones to sort through. Each group had a pretty good idea of where the other lab was, because members of the respective teams knew each other from being postdocs in the same labs. It was very competitive. The game was between the "best west coast cloners" versus the "best east coast cloners."

[aa] The recombinant phages carrying the human genomic fragments were called "λ Charon 4" vectors after the boatman Charon, who in Greek mythology carried souls across the river Styx in Hades. I introduced Richard Lawn at a seminar he gave at Biogen in 2014 by showing the picture of the boatman Charon (Plate 4). No one in the audience appreciated or understood why I did that.

Genomic Cloning

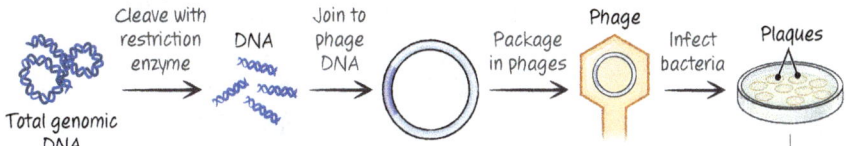

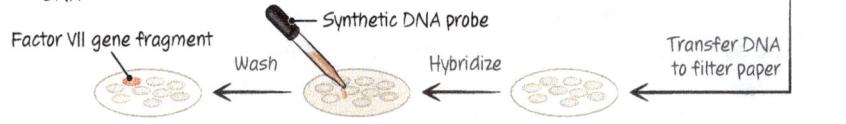

cDNA Cloning

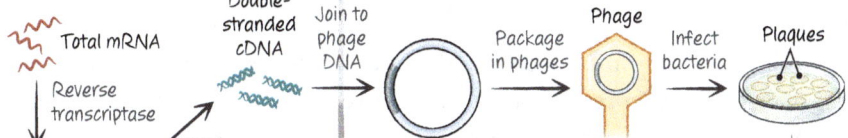

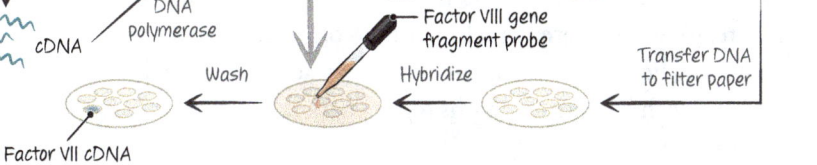

Figure 1. Cloning cDNA and genomic DNA for the Factor VIII gene.

Filing Patent Applications

By mid-1983 both groups had obtained short segments of the gene and subsequently cloned Factor VIII cDNA and expressed it as an active protein in mammalian cells almost simultaneously.[bb] On October 28, 1983, GI filed a patent application on "preparations of recombinant DNA which code for the cellular production of human and porcine Factor VIII and methods of obtaining such DNA and expression thereof in bacteria and eukaryotic cells." The application was based on a partial clone and outlined steps for deriving a full human sequence. Later the inventors added further sequence information in a "continuation-in-part" filing. Back on the west

---

[bb] The Factor VIII gene spans 186 kilobases (kb) of DNA, with 26 exons (coding regions) making an mRNA of some 9 kb coding for a protein consisting of 2351 amino acids. It was truly a "tour de cloning force" and has never been repeated. In some recognition of this effort, *Nature* published four papers in Autumn 1984 on Factor VIII—three by Genentech, and one by GI.[26]

coast Genentech was able to express its clones in early 1984: it also filed for patent protection covering the cloning, expression, and the full sequence of the human Factor VIII gene.

Commercially, things took a little longer to sort themselves out. GI remained committed to delivering recombinant Factor VIII to Baxter but the development phase of Factor VIII as a product was very protracted, largely because of stability issues with the protein. Genentech sold the rights to recombinant Factor VIII and turned its technology over to Bayer. Like the GI counterpart, the Genentech molecule also languished in preclinical development. The stability issues were solved by coexpression of Factor VIII with recombinant vWF, a technology GI had incorporated into the manufacturing process.

The IP situation for Factor VIII commercialisation was less complicated than many other recombinant proteins (e.g., t-PA). In their 1989-issued patent, GI had described a method for producing the Factor VIII gene. Genentech's issued patent included the complete gene sequence. A royalty-free cross-licensing agreement was negotiated that gave both GI/Baxter and Genentech/Bayer clear paths to the market.[cc] As we will see later in this chapter when we talk about the cloning of erythropoietin, this way of resolving a complex IP position was not always what happened when two groups had competing IP.

GI's Recombinate was approved for sale in December 1992. Bayer brought its product, Kogenate, to market in early 1993. Genetics Institute also made a "second-generation" Factor VIII product by removing the less-conserved B domain of the protein, which did not contribute to its clotting activity. This modification turned out to be very important, as the smaller size increased the efficiency of expression and significantly reduced the costs of production. Despite Genentech coming to the same conclusion and making similar B-deleted proteins, Bayer decided to forgo the development of a B-domain-deleted product. The FDA approved Genetics Institute's B-deleted product, under the trade name Refacto. It was marketed by Pharmacia and Wyeth and subsequently by Pfizer (after they acquired those companies).

Improvements to Factor VIII continue to this day. Factor VIII derivatives were subsequently made by adjustments to the manufacturing process for the existing products. In 2007, Biogen-Idec acquired a company in Waltham, Massachusetts, called Syntonix, who were making third-generation

---

[cc] Given the needs of the patients, this seemed like a reasonable way to proceed.

extended half-life versions of Factor VIII and Factor IX by combining the clotting factor proteins with the Fc region of an antibody. These became the highly successful products ELOCTATE (Factor VIII-Fc) and ALPROLIX (Factor IX-Fc). Over time, these products commanded a large share of the recombinant factor clotting market. Biogen spun these haemophilia assets out into Bioverativ, a company where I worked, that was acquired by Sanofi in 2018. An even longer-acting and very innovative form of recombinant Factor VIII, Efanesoctocog alfa (formerly known as BIVV001) was developed by Bioverativ and, later, Sanofi. Now approved, this medicine (ALTUVIIIO) features once-a-week dosing, a huge benefit for the haemophilia A population (see Chapter 11).

ERYTHROPOIETIN: GI VERSUS AMGEN

In 1981 the hormone EPO—a protein that stimulates red blood cell differentiation and proliferation—was on the "cloning wish list" of almost every start-up recombinant DNA biotech company, including GI, Amgen, Biogen, and Genentech. It was thought that the recombinant protein could be used to reverse the anaemia seen in kidney transplant and dialysis patients and perhaps many thousands of other anaemic patients. For GI, the EPO project was probably the most important one that the company undertook.

The principal academic player in the EPO saga was Eugene Goldwasser at the University of Chicago. He had been studying the relatively small (166-amino acid) protein for some time. He had developed biological assays for the protein, purifying it from urine of anaemic individuals. Takaji Miyake from Kumamoto University in Japan brought along some semipurified EPO when he visited Goldwasser's lab in the early seventies, starting a long-term collaboration to obtain purified EPO.

Goldwasser was originally contacted by Winston Salser at UCLA about collaborating with the new company he was starting called Applied Molecular Genetics or "Amgen" for short.[dd] George Rathmann from Abbott Labs became the CEO and Salser became chairman of the SAB.[ee] Amgen effectively "bought the services" of Goldwasser via stock options and a collaboration. The first protein sequence of the amino-terminal 26 amino

---

[dd] Applied Biosystems was originally part of the Applied Molecular Genetics business plan but was formed as a separate company.

[ee] I present much more about Amgen in Chapter 3.

acids from purified EPO appeared in early 1981. This was obtained by Rod Hewick, before he moved to GI and became involved in the cloning of EPO, tPA, and Factor VIII there.

Under the collaboration with Goldwasser, Amgen got "exclusive" ac-cess to the protein sequences. Wally Gilbert and Bernard Mach, two of the Biogen founders, had asked Goldwasser to provide some protein for sequencing, but their request was denied. Biogen tried other differential screening approaches to clone EPO cDNA but was unsuccessful and even-tually gave up on the project.

One of the problems with cloning EPO was the very low abundance of the mRNA from cell sources. Consequently, very large cDNA libraries needed to be made and screened. It was also not entirely clear what cells to use as a source of the mRNA. A strategy was developed that focused on cloning the gene from both genomic and cDNA libraries, as was used for Factor VIII (above).[ff]

Both Amgen and GI sourced as much protein as they could to get more amino acid sequences so that multiple different stretches of sequence could be derived and more than one pool of oligonucleotide probes covering the whole sequence could be made. GI bought EPO from Miyake and, with Rod Hewick and Ed Fritsch, cloned multiple EPO cDNAs via genomic fragments isolated from the now-famous genomic λ library made in the Maniatis lab at Caltech.

Patent Litigation

An important part of the litigation that happened later was a patent applica-tion that GI had filed on "purified EPO."[27] The Miyake protein was not as pure as supposed and further purification by high-performance liquid chromatog-raphy (HPLC) led to a single protein with a higher specific activity, indicating higher purity. GI filed claims based on this "essentially pure EPO protein" and its defined specific activity. GI also cloned EPO within weeks of Amgen, but Amgen were the first to file on the cDNA and the protein sequence.

With Upjohn as a partner in the United States, GI licensed the rights to EPO to Boehringer Mannheim in Europe and to Chugai (before they became part of Roche) in Japan. Amgen was working with J&J in Europe and with Kirin in Japan. The GI patent was upheld everywhere except in

---

[ff]In a genomic library the EPO gene would, in principle, be as abundant as any other gene.

the United States, so GI would be able to sell their product everywhere—except in the United States. Amgen sued and GI countersued. The Federal court upheld both patents, with the apparent intention of promoting a cross-licensing deal. But with all the commercial participants in the various countries involved the situation became very complicated.

In January 1989, the Federal District Court of Massachusetts ruled that Amgen had infringed the GI patent. At the same time, the court did not *disallow* the Amgen recombinant DNA patent. This led to a magistrate trial, the result of which was that *both* patents were upheld. GI then filed an injunction to stop Amgen selling EPO as it "clearly infringed" the pure protein patent.

Despite further requests for a cross-license arrangement, Amgen procrastinated and appealed the Massachusetts court decision. The case went to the Court of Appeals for the Federal Circuit (CAFC), a newly created Appellate Court for Patents. CAFC did not uphold the GI patent for technical reasons to do with the definition of the specific activity of the protein but did uphold the Amgen recombinant DNA patent in the United States.[88] The Amgen patent also had additional claims to all proteins based on the original cDNA sequence (i.e., all second-generation EPO proteins).[hh] In essence this ruling, which GI appealed, prevented GI from bringing their product to the U.S. market. GI hired an attorney to take the case to the Supreme Court, but they declined to hear the case.[27]

This patent battle set an unfortunate precedent, given that it was not decided on the prevailing law at the time but on technicalities. In Europe the situation was different with a modified Amgen patent with narrower claims (to a human genomic clone and monkey cDNA) being issued and full acceptance of the GI patent. This allowed both companies and their respective partners to move forward with their EPO products in Europe.[28]

A Difficult Future for GI

One litigation loss should not have been fatal for GI, but Wall Street took the decision very negatively. The stock market punished the company, with the stock losing half its value in an afternoon. Access to further financing

---

[88] This decision may have been based on a misunderstanding of the technical complexity of the oligonucleotide probing that was used to clone the gene.

[hh] This ruling in hindsight was clearly an overreach.

pretty much dried up completely. The normal standing-room-only presentations to investors at the usual investment banker meetings and full breakout sessions became a shadow of what they once were.[16]

GI, by then a company of more than 600 people, had invested heavily in the manufacturing capability necessary to make recombinant DNA products, including a new facility in Andover, Massachusetts and one in Rhode Island. With their research led by Patrick Gage, they were well on their way to being a small and well-integrated biotech company based on therapeutic proteins made by recombinant DNA methods, and they had a burn rate to match their lofty ambitions. They were, as demonstrated by the cloning of t-PA, EPO, and Factor VIII, one of the very best recombinant DNA companies out there.

But the stock market and investor reaction to the EPO rulings meant that the company either had to downsize radically or find another solution. Jack Stafford, the strong-minded and rather imperious CEO of American Home Products Pharmaceutical Division at the time, saw the need to revamp their research activities and believed that the acquisition of GI was the means to that end. In 1992, Wyeth (as the American Home Products company had now become) paid $700 million for 60% of GI with a five-year option (at increasingly higher prices each year) to buy the rest of it. The option deal left GI independent—not unlike the Genentech-Roche deal done in 1990, but rather less successful. In 1997, Wyeth bought GI out completely, largely owing to the financial ambitions and impatience of the small shareholders, who made up the remaining 40% ownership of the company.

In the now classic (but entirely wrong) way, Wyeth started to merge therapeutic areas, combine research activities with GI, and generally make the whole company more "integrated." And, at the same time, much more bureaucratic. The idea in principle was to bring the innovation and inspiration of GI to the entire Wyeth R&D operation. As you might imagine, this was not met with much enthusiasm by the pharma people. In fact, the complete opposite occurred. Many of the best people at GI saw the writing on the wall and had either left GI already or left after the full takeover.[ii]

Meanwhile, Fred Hassan—who was running the day-to-day business and was on the board of Wyeth—was hired away to fix the messy Pharmacia-Upjohn merger.[jj] Ironically, after Hassan sorted that challenge out, he became CEO of

---

[ii] Apparently, the GI scientists referred to Wyeth as the "lower GI" and to themselves as "upper GI"—a rather graphic representation of how they felt.

[jj] The merger was floundering owing to another huge cultural mismatch, this time between a rather dull midwestern company and a Swedish wannabe.

Schering-Plough, only for the company to be acquired by Merck shortly afterwards—an illustration of the rapidly shifting sands of pharma at that time.

There were big lessons to be learned from GI's demise: I am not sure that Big Pharma has learned them even now. First and foremost, the quality of R&D is entirely dependent on the quality of the R&D people and their ambitions, and these people need to be taken care of. Not just financially, but with an environment where they can do their work in as autonomous a fashion as possible. It is quite appropriate for some strategic guidance to be provided here and there, but one cannot do research and development by rote. Innovation can be killed with a single check mark.[kk] More consideration of the cultural issues will appear later in this book. Suffice it to say here that culturally at least, biotech R&D is full of people who want to achieve things *their* way—just what completing a PhD is designed to teach you to do.

Many GI alumni have gone on to do great things in biotech, from running R&D to running small and not so small companies as the CEO or COO. It is a testament to the quality of the leadership that was there from the start. Today, the desire to make quick money by many (but not all) VCs seems to outweigh the need to create sustainable organisations filled with people who want to make a difference and are prepared to go to extraordinary lengths to do so.

These early cloning stories also underscore the need for patience. It takes a good deal of time to apply new methods to make products for patients. Good people will get you there and great people will get you there faster.

In the end, patent litigation was GI's nemesis. The company should and could have become another Biogen, Genentech, or Amgen. Technically, they were probably better than any of them.

REFERENCES AND NOTES

1. FMDV is a picornavirus, a positive strand RNA virus and a mRNA with a poly(A) tail at the 3′ end.

2. 1980. British Biotechnology boat comes home. *Nature* **286**: 321.

3. 1980. Celltech set up. *Nature* **288**: 110–111.

4. Dodgson M. 1991. *The management of technological learning: lessons from a biotechnology company (De Gruyter studies in organization)*. De Gruyter, Berlin.

5. Betjeman poem: John Betjeman published his poem about Slough in 1937 in the collected works *Continual Dew*. Slough was becoming increasingly industrial, and some

---

[kk] Or, as in the unfortunate case of many large pharma companies, multiple check marks.

housing conditions were very cramped. In willing the destruction of Slough, Betjeman urges the bombs to pick out the vulgar profiteers but to spare the bald young clerks. He really was very fond of his fellow human beings. Slough is much improved nowadays and he might be pleasantly surprised by a stroll there.

6. A good history of how mergers contributed to the present big pharma companies can be found in the book: Kinch M. 2016. Autophagy. Chapter 10 in *A prescription for change: the looming crisis in drug development.* UNC Press, Chapel Hill, NC.

7. Maniatis T, Fritsch EF, Sambrook J. 1982. *Molecular cloning: a laboratory manual.* Cold Spring Harbor Laboratory Press, Cold Spring Harbor, NY.

8. Emtage JS, Angal S, Doel MT, Harris TJR, Jenkins B, Lilley G, Lowe PA. 1983. Synthesis of calf prochymosin (prorennin) in *Escherichia coli. Proc Natl Acad Sci* **80:** 3671–3675. Communicated by Sydney Brenner, March 23, 1983.

9. Docherty AJP, Lyons A, Smith BJ, Wright EM, Stephens PE, Harris TJR, Murphy G, Reynolds JJ. 1985. Sequence of human tissue inhibitor of metalloproteinases and its identity to erythroid-potentiating activity. *Nature* **318:** 66–69.

10. Hayward BE, Hussain A, Wilson RH, Lyons A, Woodcock V, McIntosh B, Harris TJ. 1986. The cloning and nucleotide sequence of cDNA for an amplified glutamine synthetase gene from the Chinese hamster. *Nucleic Acids Res* **14:** 999–1008. doi:10.1093/nar/14.2.999

11. Lonza. The GS Xceed® Expression System offers you a robust, fully integrated, and scalable mammalian system that can express a diverse range of biologic drugs. https://www.lonza.com/biologics/expression-technologies/gs-expression-system

12. Kenward M. 1993. Celltech completes IPO in Britain. *BioWorld* Dec 1, 1993. https://www.bioworld.com/articles/392483-celltech-completes-ipo-in-britain

13. Mitchell P. 2004. Celltech acquisition sends mixed messages. *Nat Biotechnol* **22:** 787.

14. Lawn RM, Fritsch EF, Parker RC, Blake G, Maniatis T. 1978. The isolation and characterization of the linked δ- and β-globin genes from a cloned library of human DNA. *Cell* **15:** 1157–1174. doi:10.1016/0092-8674(78)90043-0

15. Rita Allen Foundation. 2016. Tom Maniatis: mastering methods and exploring molecular mechanisms. August 9, 2016. https://ritaallen.org/stories/tom-maniatis-mastering-methods-and-exploring-molecular-mechanisms/

16. Conversations with Tom Maniatis, GI Founder and with Gabe Schmergal, CEO of GI at the time, both 2021.

17. Colaianni A, Cook Deegan R. 2009. Columbia Universities Axel patents: technology transfer and implications for the Bayh–Dole Act. *Milbank Q* **87:** 683–715.

18. Reverse split: A reverse split occurs when a small biotech company goes public if the number of shares outstanding (i.e., sold to private investors) is too many to justify selling more of them to the public at a price of $14–$16 each, the traditional price of a new public stock. A 4:1 reverse split would mean that if I had 1 million stock options in a company at an exercise price of $1, I would now have 250,000 options at an option price of $4. Given the need to sell stock as a private company the "cap table" often gets untidy and reverse splits at various levels become inevitable before going public. Investors generally like this information to be kept away from the employees for obvious reasons.

19. Conversation with Glenn Larsen, 2021.

20. Judge Whitford's summing up: [No. 24] 3 December 1987 [1987] R.P.C. IN THE PATENTS COURT. Before: MR JUSTICE WHITFORD. 7 July 1987, GENENTECH INC.'S PATENT.

21. E-mail communication with Sir Robin Jacob, patent barrister for Wellcome in the Genentech case, 2022.

22. Genentech appeal: 6 July 1989 [1989] R.P.C.: IN THE COURT OF APPEAL. Before: LORD JUSTICE PURCHAS, LORD JUSTICE DILLON, LORD JUSTICE MUSTILL. 27–30 June, 1, 4–7, 11–15, 18, and 20–22 July and 31 October 1988: GENENTECH INC'S PATENT.

23. Sun M. 1987. Companies vie over new heart drug: Genentech and Wellcome battle each other over patent rights to the clot-dissolving drug TPA, while other companies gear up to compete. *Science* **237**: 120–122.

24. Klausner A. 1987. Second-generation t-PA race heats up. *Nat Biotechnol* **5**: 869–870.

25. Centers for Disease Control and Prevention. 2023. How hemophilia is inherited. https://www.cdc.gov/ncbddd/hemophilia/inheritance-pattern.html

26. Factor VIII cloning papers. *Nature* **312**: issue 5992, November 22, 1984.

27. EPO litigation description: Rasmussen N. 2014. *Gene jockeys*. Johns Hopkins University Press, Baltimore.

28. Kalantar-Zadeh K. 2017. History of erythropoiesis-stimulating agents, the development of biosimilars and the future of anemia treatment in nephrology. *Am J Nephrol* **45**: 235–247. doi:10.1159/000455387

CHAPTER 3

The Wave in the West

The Genentech and Cetus IPOs in 1980 and 1981, respectively, coupled with The Bayh–Dole Act of 1980 that stimulated the patenting of federally funded university research discoveries and their out-licensing to commercial entities, contributed substantially to the formation of other biotech companies but particularly on the U.S. west coast.

APPLIED MOLECULAR GENETICS—AMGEN

The founders of Amgen were based at UCLA, but the company was started in the unlikely location of Thousand Oaks, about 40 miles north of the university over the hills from surf-central Malibu. The company began with a call from investment banker William Bowes to UCLA's Winston Salser in late 1980. Together, they managed to recruit both Lee Hood, a molecular technologist at the University of Washington, and Marvin (Marv) Caruthers, a biochemistry professor at the University of Colorado in Boulder, as founders.[a,1]

Initially, like Cetus, Amgen had a very wide range of potential applications in mind for the new technology, from the same recombinant proteins as everyone else to chicken growth hormone, oil-eating bacteria, and various vaccines. George Rathmann,[b] who was vice president (VP) of R&D at Abbott Diagnostics at the time, took a sabbatical year in California to "hang out" with the Amgen team and ended up becoming their CEO. He brought

---

[a] Caruthers had a big influence on the biotechnology industry. He and his research group developed methods for the phosphoramidite synthesis of DNA, which was useful not only for making small genes but also for making oligonucleotide probes. He also developed methods of RNA synthesis, the synthesis of DNA analogues, and applications of the resulting molecules.[1]

[b] Rathmann was a highly experienced executive who had spent time at Litton Medical Systems and at 3M before Abbott.

Philip (Phil) Whitcombe, a cardiovascular expert, to Amgen with him from Abbott. Nowell Stebbing was hired by Amgen from Genentech, where he was Head of Biology after leaving G.D. Searle (the same group that spawned Celltech and British Biotech). My sister Suzanne, an immunologist, worked for Nowell at Searle, and I recall that he had a penchant for driving vintage Bentleys, and drove one when he went back to the United Kingdom to run R&D at ICI before it became Zeneca and then AstraZeneca.

Within a few months of hiring Rathmann, the founding investors, including Bowes and his new firm U.S. Venture Partners with Franklin Pitcher (Pitch) Johnston at Asset Management (who were investors in several of the early biotech companies and where Brook Byers had worked), closed a series A financing of $19 million—the largest in biotech at that time. TOSCO, the shale oil company, put $3.5 million into this first round of financing. Another important investor was James (Jim) Blair at New Court, a Rothschild venture fund. Victor Rothschild was a friend of Sydney Brenner, who once again was instrumental in persuading Blair and Rothschild to invest $3 million in the round. Jim Blair later formed Domain Partners, which became an important investor in the San Diego scene. Owing to the connection with Rathmann, Abbott also invested $5 million in the initial round. Amgen went public in June 1983 in a deal led by Smith Barney, Dean Witter, and Montgomery Securities, just before the financing window essentially shut for three years. Seen as a high-quality biotech company with a diverse set of potential products, they raised $43 million, at a valuation of nearly $200 million. This was well before their successes with erythropoietin (EPO) and granulocyte colony-stimulating factor (G-CSF). Some of the founding investors still owned their Amgen shares 10 years later.[2,3]

The subsequent success of Amgen compared to the other biotech start-ups of the time is something of a paradox. Amgen was not considered by anyone to be an intellectual powerhouse or a cloning superstar company like Genentech or GI, but it could be argued that Amgen has been more successful than either of those companies. George Rathmann's single-minded leadership and the hiring of good people, including the quintessential biotech lawyer Alan Mendelson,[4] the EPO patent litigation decision (described previously), a bit of good fortune, and the business credibility to be able to raise money on Wall Street, all contributed to its surprising early success. Following their IPO, Amgen took advantage of the 1986 market window to raise $35 million in a follow-on stock offering, and in March 1987, they raised another $120 million.

Amgen's Products

Recombinant EPO (EPOGEN) was approved in June 1989 for the treatment of anaemia associated with chronic renal failure: three years after the Investigational New Drug (IND) application was filed! Even by today's standards, this was quick. The early clinical trials reported in the *New England Journal of Medicine* (*NEJM*)[5] demonstrated quite clearly that the recombinant protein worked as advertised and restored haematocrit in kidney transplant patients.

This was something of a mixed blessing for Amgen. They had retained rights for EPO for kidney transplantation in the United States, but they had licensed this right to Johnson & Johnson (J&J) outside the United States. J&J also had rights for all other indications worldwide. EPOGEN turned out to have a short half-life that compromised dosing, so Amgen later developed a second-generation product (called Aranesp), which could be dosed weekly and was strong competition for EPOGEN in kidney disease.

The results of Amgen's other cloning activities could also be described as "mixed."[c] One of the most successful products was G-CSF. The cloning of G-CSF and other colony-stimulating factors was on the agenda for many of the cloning companies, including Biogen.[d] The drug dosages chosen in the clinical trial by Amgen demonstrated that G-CSF stimulated neutrophils, restoring their numbers in patients after chemotherapy or transplantation without serious side effects. Rapid approval made NEUPOGEN (filgrastim, the recombinant G-CSF product) the second effective and—lucrative—Amgen product.

Amgen needed to acquire additional products to build the company. They bought Synergen (see below) for $240 million in December 1994 for its IL-1 expertise and IL-1-derivative product (Kineret). But Kineret could not compete with either REMICADE (Infliximab) from Centocor (later J&J) or Enbrel developed by Immunex in Seattle. Amgen's continued need to drive revenues, and the relative inability of Immunex to make Enbrel, led to Immunex being acquired by Amgen for $16 billion in 2002.

Amgen are one of the very few early biotech companies that remains independent.[e] Amgen are also unafraid of spending money to bolster their

---

[c] That means uninspiring.

[d] The G-CSF recombinant protein was licensed by Amgen from Michael Moore at the Memorial Sloan Kettering Cancer Center in New York.

[e] Amgen were very single-minded in remaining independent, despite acquisition interest from several large pharmaceutical companies.

pipeline. For example, in December 2022, it announced the acquisition of Horizon Therapeutics for ~$28 billion, acquiring a basketful of additional promising products (see Biotechnology Product Time Line [pp. 441–449]).

AN EXPANDED WEST: BOULDER AND BIOTECH

Synergen was another small company founded in 1981 after the Genentech public offering, although in the rather unlikely city of Boulder, Colorado. Larry Soll, Michael (Mike) Yaris, David Hirsh, and Larry Gold, all professors at the University of Colorado, started the company on the basis (according to Larry Gold in his inimitable way) that they were smart and could use recombinant DNA technology just as well as anyone else.[f] Larry Soll, who was the only one of the founders to leave academia, became Synergen's CEO.

The founders had connections to Warburg Pincus, a venture capital (VC) company that was highly aware of the Genentech IPO and wanted to get into the biotech business themselves. They ponied up $2 million to get Synergen off the ground. Money was part of the motivation for the founding: Gold wanted to create Synergen to make some money so he could send his kids to college.[g] The company went public in March 1986. One of their projects was to make a small proteolytic fragment of IL-1 as an IL-1 receptor antagonist for the treatment of sepsis. From a clinical trial point of view, this was not a distinguished choice, and the potential drug failed in the two clinical trials.

Despite the failures, the team at Synergen was good, and Amgen bought them largely for their IL-1 products and their pilot manufacturing facility. Amgen took over the Synergen lab in Boulder for a time, but getting good people to move to Boulder, unless they were avid skiers, was hard compared to the Bay Area or Boston. Boulder has become, nevertheless, a biotech nexus on a smaller scale. For example, Tularik, which was founded by Steven (Steve) McKnight and Dave Goeddel, was started there before it moved to South San Francisco. It, too, was bought by Amgen for $1.3 billion in 2004. Array Biopharma (now part of Pfizer) and Loxo Pharmaceuticals (now part of Eli Lilly), in addition to NeXagen and SomaLogic (both Larry Gold start-ups), were or are still based in Boulder.

---

[f] Marv Caruthers, mentioned above, was also a professor at the University of Colorado and was also associated with the launch of Synergen as well as Amgen.

[g] It took some time for them to get even to that point.

After Synergen, Larry returned to the Molecular and Cell Biology department at the University of Colorado as chairman. He subsequently started NeXagen with some of the Amgen proceeds to focus on oligonucleotides that could bind to proteins as potential therapeutics. To fund the company and to access a delivery system, NeXagen merged with Vestar, a company that had a proprietary lipid formulation for amphotericin B called AmBisome, to form NeXstar. The profits from this quite successful antifungal product funded further research resulting in the development of the SELEX technology (see Chapter 12). Gilead Sciences bought NeXstar in 1999 to get access to AmBisome. As part of the transaction, Larry secured the rights to all the SELEX research, and from this emerged SomaLogic (see Chapter 12). They developed very high-affinity "modified aptamers" to bind to proteins as selective biomarkers—the so-called SOMAmers. SomaLogic had a very successful IPO in 2021.

IMMUNEX

Chris Henney, an immunologist trained in the United Kingdom at the University of Birmingham (my alma mater), was at Johns Hopkins in the early 1970s. He was subsequently recruited to a faculty position at the Fred Hutchinson Cancer Center (The Hutch) in Seattle in 1978 to start an Immunology department, which became a challenge in and of itself. He recruited Steven (Steve) Gillis, who had just completed his PhD at Dartmouth and had developed cell lines overexpressing various lymphokines. At that time, these immunologically active "factors" were of much interest to the immunology community, although no one knew what these molecules consisted of from a molecular perspective.

Cloning these proteins with the new recombinant DNA technologies was an obvious choice to find out what they were and to make them. Except to most of the other faculty members at the Hutch, who thought that all this "cloning stuff" was rather trivial, and in addition in an Immunology department, you hired immunologists not gene cloners.[6,h] In other words, the old academic territorial rules were still in force. Henney and Gillis spent a huge amount of time trying to recruit the necessary molecular biologists to do the cloning experiments at the Hutch, but it was hard work.

---

[h] One exception was Bob Nowinski, who was head of Virology at the Hutch—later of Genetic Systems and Oncogene infamy (see Chapter 4) and co-founder with Chris Henney of Icos.

The first company to show an interest in recombinant cytokines was Ajinomoto, a Japanese company that sold monosodium glutamate. They wanted reagents to set up a cytokine lab and needed the Henney lab to coach their researchers in using sophisticated cytokine assays and other tools. Following a visit to Seattle in 1979, Ajinomoto hired Gillis and Henney as consultants, essentially doubling their salaries. This investment also enabled the pair of them to have access to other cell lines and to finally set up their lab to clone cytokines.

They also visited Nippon Roche with the hope of broadening their horizons.[i] But the push for the commercial exploitation of cytokines as therapeutics actually came from Roche in Nutley (the old Roche site in New Jersey).

Nippon Roche proposed to put $1 million a year into the Henney/Gillis Hutch lab, a significant amount of money. But first Chris and Steve had to get permission from the Hutch management, owing to the commercial rights issues that emerged in terms of who would own the cytokines once they had been cloned and expressed using Roche money in a Hutch lab. During the negotiations, various structures were suggested, including each of the eight Hutch principals (of which Henney was one) sharing the money equally. Another arrangement had the president of the Institute (who had done precisely nothing to help) getting a share.[6,j]

Let's Do a Company

Given that Genentech, Biogen, GI, and Amgen had all recently started or were starting, it did not take long before the suggestion of "let's do a company" was made.[k] Roche much preferred this idea because it simplified the business arrangements, and they were prepared to invest $2 million/year in such a venture. The Hutch leadership thought Henney and Gillis were both foolhardy and greedy at the same time and suggested that instead they do the company one day a week on—to coin a British phrase—"Wet Wednesdays," and to stay put at Hutch.[l]

---

[i] Henney had good Japanese academic contacts; his academic mentor Dr. Kimishige Ishizaka was Japanese and had won the Emperor's Prize in 1974.

[j] Not exactly what Henney and Gillis had in mind.

[k] Probably over a beer—that is how these things started then.

[l] Despite plenty of such wet days in Seattle, this was clearly unworkable.

Instead, Steve and Chris made the rounds of local potential venture investors. There was not much choice, as most investors were only prominent local businessmen—more like angel investors than venture capitalists. Stephen (Steve) Duzan, who would become the CEO of Immunex, was one of the businessmen they met. He advised that these local investors would not be there for the long term. Instead, he introduced them to some "real" west coast VCs, who by now were getting much more excited about the biotech space.

The new company was named "Immunology Experiments": Immunex for short. The first round of investment in 1981 was from four groups led by Thomas (Tom) Cable, Merrill Pickard, Hillman (a family fund), and Mayfield, who collectively put in $1 million, buying 1.5 million shares at 67 cents. Gillis and Henney had just over a million shares each at that point with about 5% of the company for Duzan as the CEO.[m]

One prominent visitor to the company in the early days was Sydney Brenner, who loved the whole concept. He came on behalf of U.S. biotech-investor Rothschild (Brenner consulted for Rothschild on various biotech projects).[n] Brenner's enthusiasm, which was second to none, led to Rothschild investing in the second Immunex financing round that occurred about a year after the first. The initial investors were joined by NEA (Richard [Dick] Kramlich), Rothschild (David Leathers from the United Kingdom), Jim Blair before he formed Domain Associates in San Diego, and Diamond Shamrock Ventures. They collectively bought 1.5 million shares at $1.67/share for a total of $2.5 million.

Moving Out

Chris and Steve left the Hutch (taking their postdocs with them) and set up shop on the Seattle waterfront at 51 University St. The building, owned by the Teachers Retirement Fund, charged the princely sum rent of $10 per square foot per month. Immunex occupied 15,000 sq ft in 1981 and 30,000 sq ft by 1982. Over time, Immunex ended up in nine separate buildings.

---

[m] There obviously was some equity reserved for management and for hiring staff.

[n] Brenner suggested that they hire John Simms, who was a postdoc in his lab in Cambridge (UK), to help address the recruiting challenge of bringing molecular biologists into an immunology lab.

As one of its employee perquisites, Immunex always paid for parking. But as the company grew, the cost also grew, to the point that the company was paying $1 million per year just for parking. So, the company bought the parking lots, adding the "parking lot business" to their cytokine cloning efforts.[6,o]

The main business model was simple: clone as many of the lymphokines and cytokines—whose activity could be detected by sophisticated immunological assays, haematopoiesis, and/or stem cell differentiation—as possible. They did this through a mixture of expression cloning (in which cDNA was cloned into a plasmid designed to express the cDNA insert into mRNA and be translated into protein) and more robust protein purification methods to inform the creation of oligonucleotide probes that would be specific for the mRNA (or cDNA) coding for the protein of interest.

In expression cloning, the clones that expressed the protein and the immunological activity of interest were detected directly in fractionated bacteria or by using immunological screening methods with antibodies to the cytokine of interest. The cloning was not that easy relatively speaking, although some cell lines expressed quite large amounts of these proteins. This was just as well given that most of the proteins had very high specific activities (i.e., lots of activity from not very much protein). Fortunately, Immunex was one of the very few cloning companies at the time started by people who were also real biologists rather than molecular biologists (gene jockeys) for hire, so it was easier for them than others.

There were several flirtations with mergers before Immunex went public. Amgen, who appreciated that cloning and expressing cytokines and lymphokines would lead to products and who had precious few of their own at that time, visited Seattle reasonably frequently, often with their CEO George Rathmann. Amgen proposed a deal that also involved Hugh McDevitt's lab at Stanford, but was turned down by the majority of the VCs on the Immunex Board. Another potential deal, this time merging with ZymoGenetics (a yeast systems–based company in Seattle set up by University of Washington professors Benjamin [Ben] Hall and Earl Davie [see Chapter 18]) also fell through.

Immunex had raised the grand sum of $3.5 million before they went public in 1983, raising about $15 million. Robertson Coleman and Stephens were the bookrunners for the offering, with Cable House on the right of

---

[o]The Seattle waterfront was becoming a popular tourist site, and the parking need was exceptional.

the IPO prospectus. They sold shares at $11 and did a 3-for-2 reverse split, leaving the founders with one million shares at the IPO, amounting to 3% of the company each. The VC shareholders owned more than 50% of the company, now valued at around $60 million. This was a pretty good return on investment (ROI) for the VCs on a year's investment.[p]

Immunex Projects

The company had 40 employees at the IPO, 16 of whom had doctoral degrees. They referred to themselves as "Lymphokines are us" after the Toys"R"Us franchise.[7] IL-2 for stimulating T cells was always intended to be the first product, a protein that Cetus and others were also cloning.

The CSF projects (competing mostly with GI and Amgen) were funded by a research and development limited partnership (RDLP) off balance sheet arrangement to the tune of $40 million over five years. Immunex focused on granulocyte–macrophage colony-stimulating factor (GM-CSF) rather than G-CSF chosen by Amgen. G-CSF was cloned in Australia in Donald (Don) Metcalf's lab at Walter and Eliza Hall Institute of Medical Research (WEHI) and licensed to Amgen via Memorial Sloan Kettering Cancer Center (MSKCC). With GM-CSF, the challenge was that Amgen had already established febrile neutropenia as the standard for G-CSF approval. GM-CSF (Leukine) had rather broader activity on macrophages, and it induced small amounts of IL-1, causing some fever. This biology unfortunately reduced Leukine, approved in 1991, to a marginal product compared to NEUPOGEN (G-CSF) approved at the same time.

It is likely that GM-CSF, cloned and expressed by Immunex first, would have been a much bigger product for them if the clinical trials had started at lower doses to avoid the IL-1 induction.[q]

Another financing vehicle, Receptech, was set up to clone various cytokine receptors (i.e., the proteins that bind to the cytokines and signal into the cell on which the cytokine has landed).[r] John Simms and Steven (Steve)

---

[p] It was not always that good in those early days.

[q] The G-CSF cloning done in the Metcalf lab was all from the mouse. The subsequent patent fight was about whether cloning the human protein was obvious from cloning the mouse protein. You would think that this was of course true, but the human G-CSF patents nevertheless prevailed. It was widely rumoured that the patent issue was one of the reasons why Metcalf did not get the Nobel Prize, which he richly deserved.

[r] Receptech also allowed the development of Immunex's TNF receptor product.

Dower cloned all members of the IL-1 receptor family by expression cloning, screening huge plates of clones blotting and probing with radioactive IL-1.[s] The overall strategy was to clone all members of the receptor family and file IP on every one of them (including any splice variants).

For a group of about 150 people, the research was amazingly productive and competitive, as reflected in the sheer number of cytokines and receptors cloned. Many of the molecules that are being tested today in the immuno-oncology space or antibodies to them (as checkpoint inhibitors) have their origins in Immunex science.[7] They cloned them, did minimal biology to get the *Nature, Science,* or *Cell* paper and the IP, and then they moved on, in a sort of "expression cloning gold rush." The technology refinement to expression cloning of cytokine receptors like the IL-1 receptor to other cell-bound molecules is a great example of the technical prowess of the group. The only real competitor for Immunex apart from GI and Genentech was DNAX. This company was founded in Palo Alto in 1980 by Arthur Kornberg and Paul Berg and Charles Yanofsky at Stanford with Alejandro Zaffaroni, who had started ALZA. They hired Ken-ichi Arai, whose team of postdocs, coupled with his energy and vision, industrialized expression cloning. Schering-Plough bought DNAX in 1982, focusing it on oncology and immunology and ultimately joining it up with Canji (a La Jolla start-up) in 1996—never allowing DNAX to be what it might have become.

Enbrel

The tumour necrosis factor (TNF) superfamily was especially fertile ground for Immunex from a product perspective. A piece of the TNF receptor would become part of the drug Enbrel.[t]

One of the most important decisions Immunex ever made,[8] which was first debated hotly at the Semiahmoo Resort where the Immunex team often met for off-site activities and beer, was whether Enbrel would work best as a monomer or as a dimer (fused with the Fc region of an antibody; see Chapter 4). The dimer was chosen (correctly), but it was not clear at first if it could be manufactured.

---

[s] This was how most expression cloning at Immunex was done until the X-ray film was replaced by phosphor imaging plates.

[t] Enbrel died as a product at least twice. In a sepsis trial, Enbrel had a worse outcome than placebo but strong preclinical data in rheumatoid arthritis allowed the R&D advocates kept it alive.

Later, it became "all about Enbrel" at Immunex. Approved in 1998, Enbrel became the fastest-selling drug of all time after launch. It is still used today and was a direct competitor to adalimumab (Humira), Abbott Labs (now AbbVie) anti-TNF monoclonal antibody that became available some three years after Enbrel had been approved (see Chapter 5).

The potential market for Enbrel was one of the leading drivers for the acquisition of Immunex by Amgen. It was not the only reason: manufacturing of Enbrel was an issue to ensure provision of adequate drug supply. It was only available initially by lottery—not good for patients impaired with severe arthritis of the joints. Abbott had Humira close to approval, so Enbrel needed the production boost that Amgen could bring it. The founders of the company had all long departed before Amgen paid about $16 billion for Immunex in 2002.

CHIRON

Chiron was founded in 1981 by Bill Rutter, chairman of Biochemistry at UCSF, when the competition to clone the insulin gene using cDNA or synthetic methods was under way between the Goodman lab, Herb Boyer and Wally Gilbert, and Genentech (Chapter 1). In Greek mythology, Chiron was held to be the "superlative centaur amongst his brethren" because he was called the "wisest and most just of all the centaurs." A little pompous perhaps, but in concert with the time in which Greek names and characters were being used to describe different aspects of biotechnology, from the technology to the companies.[a] Bill was chairman of the Board, Edward (Ed) Penhoet (from UC Berkeley) was president and CEO, and Pablo Valenzuela (UCSF) was VP Research.

In contrast to some other companies, Chiron's strategy was to focus initially on recombinant vaccines. They were aware of the Biogen hepatitis B surface antigen cloning by Kenneth (Ken) Murray in Edinburgh (see Chapter 16). Chiron formed a partnership with the Swiss pharmaceutical giant Ciba-Geigy to use genetic engineering to develop vaccines to treat, prevent, and diagnose diseases such as HIV, herpes, and malaria, and a project with Merck to improve on the Hep B vaccines being developed (see Chapter 16). They also had an interest in eye disease applications as a kind of niche market: big enough to be interesting but not too big to manage. In 1982,

---

[a] It was also consistent with some of the UCSF faculty views of themselves.

they moved into labs in Emeryville, renting space from Cetus. Chiron's major claim to fame scientifically was undoubtedly that of Michael Houghton, an Englishman, who had worked at G.D. Searle in the United Kingdom. His lab at Chiron identified, characterized, and isolated the virus causing non-A, non-B hepatitis (now called hepatitis C), which was causing a huge amount of pathological liver disease in many patients.[v] Chiron was the first to clone hepatitis C virus (HCV) in 1989 after six years of research, and the company filed many HCV-related patents in more than 20 countries.

After Cetus failed to get the initial approval of IL-2 and had spun off all the polymerase chain reaction (PCR) assets to Roche, it merged with Chiron. Hollings Renton from Cetus became president and CEO before assuming a leadership role at Onyx with Frank McCormick in 1992. After the acquisition of Cetus, Chiron split its operations into five divisions consisting of Cetus Oncology for cancer drugs, the Biocine Company for vaccines, Chiron Diagnostics for blood screening and other diagnostic tests (later sold to Bayer AG for $1.1 billion), Chiron IntraOptics for eye surgery, and Chiron Technologies for research and development.

It was not all plain sailing for Chiron. The company filed patent infringement lawsuits in Europe, Japan, and the United States against Roche over their HCV products.[w] A settlement was reached in which Roche agreed to buy the global semi-exclusive nucleic acid test patents for HCV and HIV from Chiron.

Chiron successfully developed several other products. Recombinant IL-2, a cytokine made by T cells that stimulates T-cell activity (cloned by Cetus originally) called Proleukin, was eventually approved in the United States for the treatment of renal cell carcinoma and metastatic melanoma, despite relatively high toxicity. By 1990, Proleukin was approved in several European countries, but delays to the approval of it in the United States precipitated the breakup of Cetus and its acquisition by Chiron. IL-2 was approved in the United States in 1992 and is still used infrequently today in both high- and low-dosage forms. It is most often used in combination with other T cell–based therapies—for example, tumour-infiltrating lymphocyte (TIL) therapy, as practiced by Stephen Rosenberg at the NCI and

---

[v] For this discovery, Houghton was one of the winners of the 2020 Nobel Prize in Physiology or Medicine.

[w] Robin Jacob, the patent barrister who featured in the t-PA case in the United Kingdom (Chapter 2), was also involved in this patent litigation.

by the company Iovance (see Chapter 14). Chiron also received approval in 1993 for β-interferon (1b), called Betaseron, now marketed by Bayer well ahead of the competing product Avonex, the Biogen β (1a) interferon (Chapter 1). Chiron acquired PathoGenesis in 2001 to get the first inhaled antibiotic (tobramycin) approved for treating lung infections in cystic fibrosis patients. They also launched a nucleic acid–based blood testing business with Gen-Probe in San Diego in 1998. This business was further developed to detect RNA and DNA viruses in donated blood plasma to ensure that it was free from associated pathogens such as Hep C and HIV.

In late 2005, Novartis[x] made a $5.1 billion offer to buy Chiron, completing the acquisition in April 2006. Subsequently, Novartis divested most of the former Chiron assets as part of a global reorganization, with Clinigen ultimately acquiring the Proleukin franchise.

REFERENCES AND NOTES

1. Marv Caruthers M. 2013. The chemical synthesis of DNA/RNA: our gift to science. *J Biol Chem* **288**: 1420–1427. doi:10.1074/jbc.X112.442855

2. Robbins-Roth C. 2000. *From alchemy to IPO: the business of biotechnology*. Perseus Publishing, New York.

3. Goozner M. 2004. *The 800-million-dollar pill. The truth behind the costs of new drugs*. University of California Press, Berkeley.

4. Cranmer J. 2021. Remembering Alan Mendelson, mentor to biotech lawyers, CEOs: revered lawyer made his mark at Amgen, Cooley, Latham & Watkins, and beyond. *BioCentury* October 2021. https://www.biocentury.com/article/640206/remembering-alan-mendelson-mentor-to-biotech-lawyers-ceos

5. Besarab A, Bolton WK, Browne JK, Egrie JC, Nissenson AR, Okamoto DM, Schwab SJ, Goodkin DA. 1998. The effects of normal as compared with low hematocrit values in patients with cardiac disease who are receiving hemodialysis and epoetin. *N Engl J Med* **339**: 584–590. doi:10.1056/NEJM199808273390903

6. Conversation with Chris Henney, 2021/2022.

7. Various conversations with Doug Williams, 2021, 2022, 2023.

8. Enbrel: For the afficionados, this is a recombinant fusion protein in which the TNF receptor is fused to part of an IgG1 antibody. The DNA sequence that codes for the soluble TNF receptor 2a that binds to tumour necrosis factor-α was linked to the DNA sequence that codes for the Fc end of IgG1.

---

[x]Novartis was created in 1996 by the merger of Sandoz and Ciba-Geigy. It had a minority ownership in Chiron from Ciba-Geigy's original involvement.

CHAPTER 4

Monoclonal Antibodies

M onoclonal antibodies (mAbs) have revolutionised the treatment of cancer and inflammatory and infectious diseases. They are also used to treat many common human diseases (with mostly bearable side effects), making this class of therapeutics incredibly important and valuable.

The first molecular structure of an antibody was independently determined by Gerald (Gerry) Edelman in the United States and Rodney Porter in the United Kingdom in 1959, work for which they shared the 1972 Nobel Prize in Physiology or Medicine. Their work is the foundation for many of the antibody companies of today. Box 1 provides an overview of the different kinds of antibodies referenced in this chapter and throughout the book.

BOX 1. THE IMPORTANT DIVERSITY OF ANTIBODIES AND ANTIBODY (IMMUNOGLOBULIN) DOMAINS

The structure of antibodies (the proteins are called immunoglobulins) is critical to their function. There are several classes of immunoglobulins but the most important therapeutically are members of the IgG family. As shown in Figure 1, the canonical IgG consists of two long heavy (H) chains and two shorter light (L) chains joined by disulphide (cystine) bridges. The Fab domain consisting of the variable (V) binding region and a subset of the constant (C) region of the antibody is at the top of the Y-shaped molecule. The Fc region makes up the carboxy-terminal portion of the molecule and is responsible for some of the other activities of IgG. There are other forms of antibodies like a dimeric IgA and a pentameric form called IgM.

Llamas, Camels, and Sharks

In some mammals, immunoglobulins with a structure different to the conventional dimers are found. Antibodies from the Camelid family of animals (including llamas, camels, and alpacas) produce another type of antibody, in addition to the dimeric versions consisting of only two heavy chains. These antibodies do not contain the CH1 domain, but they retain an antigen binding domain

---

67

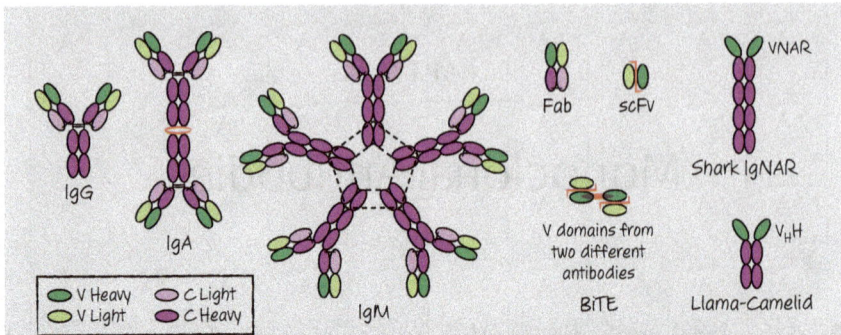

Figure 1. Major mammalian antibody types and engineered shorter forms used for many products.

called the V_HH region. These are known as single-domain antibodies or, when the binding domain is isolated, nanobodies. Sharks and other cartilaginous fish produce a heavy-chain-only antibody called an IgNAR, which has two VH domains. The variable domain of an IgNAR, the V_{NAR}, binds antigen as an independent soluble domain.

Derivatives of different immunoglobulins have been made by recombinant methods. The ones most relevant to biotech products both in this and other chapters are shown in Figure 1.

Different antigen binding domains consist of the same protein fold comprising the "immunoglobulin domain." Nature has put these together in different ways and it has been found that the domains can be put together in various orders using recombinant DNA methods. Another form of antibody that has been constructed (rather than being found naturally) are "bispecific mAbs." Milstein originally conceived of the idea of making antibodies with more than one binding domain, owing to the modular domain structure of the immunoglobulin molecule. You will read more about bispecific antibodies later in this chapter. They sometimes look a bit like two scFvs or two Fabs joined by a linker.

THE ORIGINS OF MONOCLONAL ANTIBODIES

César Milstein and Georges Köhler were working at the MRC labs in Cambridge, United Kingdom, in the mid-1970s, trying to find a reliable way to make a homogeneous source for any antibody that had a defined specificity.[a] In 1975, they made the first of these "mAbs" (literally from

---

[a] Polyclonal antibodies made by immunising sheep and rabbits were mixtures of antibodies. Antibodies from myeloma cells in culture were homogeneous, but their specificity was unknown.

one clone) by fusing plasma cells taken from the spleens of immunised mice with cells from a mouse cancer (myeloma) cell line (resulting in a "hybridoma" that produced a single type of antibody—see Box 2). They published the description of their work in *Nature* in the late summer of 1975.[1]

BOX 2. HOW TO MAKE MONOCLONAL ANTIBODIES

The MOPC 21 cell line was resistant to azaguanine allowing the selection of hybrid clones of spleen B cells fused with MOPC cells by a specific medium.

One of the keys to the initial fusion experiment was to be able to detect a specific antibody. In this case, sheep red blood cells were the antigen so a haemolysis test that Niels Jerne and Albert (Al) Nordin had developed could be used as an assay for the antibodies to that antigen. Hybridomas were initially propagated in mice in the peritoneal space as ascites, which could be harvested, and the mAb is obtained from the supernatants. Hybridomas could also be grown in tissue culture (Fig. 2).

Antigen is injected into a mouse. After a few weeks, the spleen cells are removed and fused with cells derived from a tumour cell line. The resulting hybridomas that make a defined antibody to the injected antigen are selected in a medium that only allows the hybridomas to grow. mAbs are derived from clones of cells from these hybridomas.

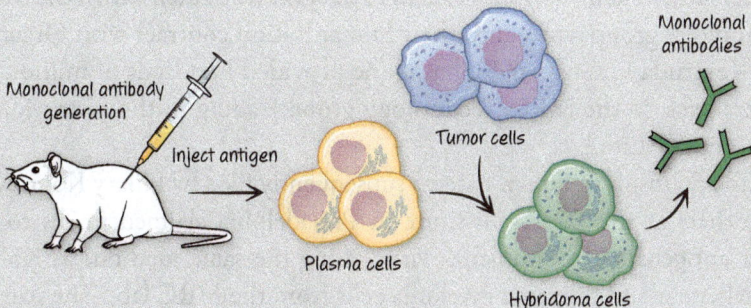

Figure 2. Making monoclonal antibodies.

(NOT) PATENTING THE DISCOVERY OF mAbs:
THE MRC AND THE NRDC

There was very little urgency from the investigators, the MRC or the NRDC (National Research Development Corporation), to file IP claims on the discovery of hybridomas and mAbs, despite knowledge of the discovery well before the publication. It is unclear why they did not. The submitted paper[b] clearly outlined the potential importance of the application of mAbs and mAb technology, and Milstein even communicated such to the MRC. The NRDC did not comment on the matter until well after the paper was published, and even then only in response to an enquiry from Milstein. The discovery was seen by the investigators more as a set of precise tools to test the Brenner–Milstein hypothesis that antibody specificity was defined by somatic mutation in the genes that make the antibody protein,[2] rather than for their commercial exploitation.[3] As a result, the mAb technology developed by Köhler and Milstein *was* commercially exploited by many companies over the succeeding years, making many millions of dollars for several of them.

My opinion (shared by the British government and Prime Minister Margaret Thatcher at the time) is that not filing IP on the discovery was a huge mistake. Monoclonal antibodies (now manufactured by recombinant DNA methods) are the single most successful set of products to have come out of the biotechnology revolution to date. Who knows what the financial return would have been to the MRC, if they had filed IP on the Köhler–Milstein invention? A small royalty on Keytruda, which the MRC Technology group wrote into their humanisation contract with Organon (see Keytruda in Chapter 5), has alone provided hundreds of millions of dollars back to the MRC Technology group leading to the formation of the company LifeArc.

To add insult to injury, patents subsequently filed by Hilary Koprowski and others at the Wistar Institute in Philadelphia claimed mAbs to tumour antigens and to influenza virus, using the same methods as Köhler and Milstein and the same myeloma cells from the MRC lab.[c] The patents were issued to the Wistar Institute in 1979 and 1980. Many articles have

---

[b] For this discovery Kohler and Milstein were awarded the 1984 Nobel Prize in Physiology or Medicine with Niels Jerne.

[c] It is unclear whether any stipulations on the commercial exploitation of the myeloma cells were made by the MRC or Milstein when the cells were sent to the Wistar.

been published about the U.K. monoclonal patent saga, including one by Nicholas Wade in *Science*.[4,5,d]

FIRST APPLICATIONS OF mAbs

The first practical application of mAbs was to replace the polyclonal antibodies that had been used for years in various diagnostic assays. Abbott Laboratories largely controlled the diagnostic assay business, through both their analysers and their reagent supply deals. The new approach was to develop mAb-based tests that would run on various analysers, regardless of who made them. In the United Kingdom, Sera Labs led the charge in making mAbs available, taking some of the heat off the MRC to provide reagents to everybody. By 1983, Sera Labs were offering more than 60 different mAbs for a variety of research applications, most of them made in mice.

Celltech's propitious move into recombinant antibody projects originated because the company was already working closely with the MRC/LMB to grow cells making CAMPATH-1 and other antibodies.[6] The antibody against the CAMPATH-1 antigen (CD52) (*Cam*bridge *Path*ology) was made by Hermann Waldmann in Cambridge, United Kingdom.[e]

One of the earliest sets of monoclonals to be used in more standardised immunoassays was for blood typing—that is, to determine blood type (A, B, or O) and Rhesus factor status. The production of these antibodies was licensed by the MRC to Celltech and was scaled up in cell culture when I was there.[f] Celltech had started scaling up mAb production in small-stirred vessels (Plate 5), in addition to their recombinant DNA activities. Ultimately, this effort turned into a cell culture business of its own, using airlift fermenters. The cell culture business was sold to Lonza.[7]

---

[d] It all seems a bit British and "rather unfortunate," with the outcome of "never mind, let us have a cup of tea." Making matters worse, investigators in the United Kingdom were not able to obtain or share in any royalties from patented discoveries. That has changed in the United Kingdom now.

[e] CAMPATH-1 is now a drug called alemtuzumab, approved for the treatment of chronic lymphocytic leukaemia in the United States in 2001. It was subsequently withdrawn from both the United States and the European markets in 2012 so that the drug could be relaunched at a different dose with additional IP protection by Sanofi (and at a higher price) as Lemtrada to treat multiple sclerosis as the antibody kills both B and T cells.

[f] I remember the cell culture lab at 250 Bath Rd., Slough, very well, as it was just across the hallway from my own lab. One of the other antibodies being grown was CAMPATH-1.

The first therapeutic mAb was approved by the FDA in 1986. Made in mice by Ortho Pharmaceutical, OKT3 (muronomab-CD3) targeted CD3, a differentiation antigen, on the cell surface of all T cells. It was used therapeutically in kidney transplantation patients to avoid the acute rejection by eliminating the T cells causing the rejection. Although effective, the antibody had a relatively short half-life and some side effects, including the formation of antibodies to the mouse protein itself. OKT3 was not commercially successful but is still used in research today, as a general anti-T-cell antibody.

RECOMBINANT ANTIBODIES AND THE IP

It quickly became clear to only a few of the recombinant DNA companies that B cells (and hybridomas) could be used as a source of mRNA to make recombinant cDNA libraries, from which cDNAs coding for the heavy and light chains of antibodies could be cloned, so that recombinant antibodies with defined specificity could be made. The two prime movers here were Celltech in the United Kingdom and Genentech in the United States. Their central roles are evident from the names of the now infamous sets of patents issued in 1983: the "Cabilly" patents, from City of Hope and Genentech (named after postdoc Shmuel Cabilly who worked with Art Riggs at City of Hope) and the "Boss" patent from Celltech (named after Mike Boss who worked in Spencer Emtage's lab at Celltech).[8,9]

Three City of Hope/Genentech patents[10] covered the expression of recombinant forms of antibodies in cell systems. Between them, they allowed claims to recombinant antibodies that, in effect, lasted for 29 years.[11] An interference settlement between Celltech and Genentech led to the issuance of another Genentech patent in 2001, extending the term of one of the original patents for an additional 17 years.

I became involved with the Cabilly and other antibody patents as an "Expert Witness" in the litigation between Genentech and MedImmune (who had developed Synagis, a recombinant antibody targeting respiratory syncytial virus infection in children). This case, which ended up before the U.S. Supreme Court, was exceptional for many reasons. For one, it raised the issue of the time it took for the U.S. Patent and Trademark Office (USPTO) to resolve the initial interference with the Boss patent, effectively buying Genentech additional time. Second, MedImmune was a licensee of the original patent. It was quite unusual for a licensee to bring a lawsuit against a licensor, as taking the license in the first place is considered a tacit acknowledgement that the claims of the patent being licensed are valid.

There are several things I remember about my involvement with this case. One was being paid very well for my services. Second was the arcane nature of the discussion around the claims—literally to the extent of whether it should be an "and" or a "but" in the claim language. Third, it soon became clear to me that I was a lot less useful when my views about what the state of the art was at the time differed from those of the lawyers on whose behalf I was working.

This case has been cited as one of the reasons for the move to a "first to file" rather than the "first to invent" system in the United States.[8] Another consequence of the extended patent life and the reason that Genentech/Roche fought so hard to maintain the patent coverage was the enormous amount of revenue Genentech was generating from the royalties they were getting from licensees, including from AbbVie for Humira. Not only that, but the City of Hope and its investigators, including Cabilly and Riggs, were making millions of dollars a year from the royalties. Not a bad deal and a lot more money than the involved individuals would have made if they had been Genentech employees.

MAKING MOUSE ANTIBODIES HUMAN: PROCESS OF HUMANISATION

One problem with using antibodies derived from mice for human therapy is the potential immune response of the human recipient to the mouse protein. Overcoming this issue required "humanisation"—that is, making the mouse mAbs more "human" in their primary structure. Several approaches were developed (see Box 3).

BOX 3. HUMANISED ANTIBODIES

A "chimeric" antibody is one where the antigen binding end (the top of the Y—Box 1) came from the mouse antibody and the back end (the Fc region) came from a human IgG. This had an advantage that either a human IgG1 or an IgG 4 constant portion could be used as the Fc domain, as these Fc regions have slightly different functions. More complete humanisation was done by moving the mouse complementarity-determining regions (CDRs) (responsible for antigen binding) from both the mouse heavy and light chains into an antibody consisting of human light and heavy chains (called CDR grafting—see Plate 6).

---

[8]The difference between the two is that in "first to invent," there has to be evidence, usually from laboratory notebooks of when the invention was made. It is simpler to use "first to file" as this represents the date of the filing of the patent application.

The United Kingdom was a centre for recombinant antibody humanisation. My group at Celltech was making recombinant antibodies and humanising them in the mid-1980s. Various known mAbs were humanised and tested to see if their affinity and activity were maintained. With a bit of variation in amino acid residues around the beginning and end of the variable regions (in particular, the CDRs where the specificity of an antibody is determined), the answer to the question was generally "yes."[12]

Another player in the humanisation effort was Protein Design Labs (PDL) in the United States, which was founded by Laurence Korn and Cary Queen in 1986. They were, as their company name suggested, experts in protein design, a necessary skill for this work at that time. They were involved in the humanisation of bevacizumab (Avastin), an anti-VEGF antibody (a Genentech product), and palivizumab (Synagis), the antibody directed against respiratory syncytial virus F protein made by MedImmune and approved in 1998. PDL went public in 1992, partnered most of their products, and lived off the royalties.[h]

A group at the Laboratory of Molecular Biology in Cambridge, United Kingdom, led by Gregory (Greg) Winter was also making humanised antibodies, by pioneering the use of phage display libraries. In this approach, DNA coding for the variable domains from recombinant human antibodies was expressed on the surface of bacteriophages, and those with the desired specificity were selected by binding to the required antigen on a solid phase.[i] The process results in both heavy- and light-chain binding regions derived from human DNA, which can then be turned into fully human mAbs. This technology was the core of the company Cambridge Antibody Technology, which was involved in making Humira (see Chapter 5). In 2000, Winter founded another company called Domantis to pioneer the use of "domain antibodies," which use only the active portion of a full-sized antibody. Domantis was acquired by GSK in December 2006 for £230 million.

"Humanising" Mice

Humanised mouse-derived antibodies (Box 3), although less immunogenic, still consisted of some mouse sequences. One of the most successful efforts

---

[h] PDL eventually became PDL BioPharma in 2008, a holding company that managed its patents and other IP until its liquidation by its shareholders in 2021. Not a bad outcome for the shareholders.

[i] Winter was awarded the Nobel Prize in Chemistry in 2018 for this work and was knighted in 2004.

to make completely human antibodies was by using transgenic mice, where the mouse immunoglobulin genes coding for heavy and light chains were fully replaced with their human counterparts. Several companies were set up to make human antibodies this way.

One of the first companies to develop a humanised antibody-producing mouse line was GenPharm, started in 1989 and financed by the venture capital firm Abingworth. They were acquired by Medarex in 1997, well before it in turn was bought by Bristol Myers Squibb (BMS). Abgenix, originally part of Cell Genesys (the first cell therapy company; see Chapter 13), had developed a similar mouse strain (Xeno mouse), where some of the mouse antibody genes had been replaced by human antibody genes. Abgenix worked with Immunex to make panitumumab (Vectibix), a fully human, high-affinity mAb against the epidermal growth factor (EGF) receptor for the treatment of certain forms of metastatic colon cancer. Amgen acquired Abgenix in 2005 for more than $2 billion with Vectibix being approved in 2006. Regeneron (see Chapter 6) had also developed "VelocImmune," a humanised mouse platform that they have used extensively to create their human mAb portfolio.

But by far the most sophisticated mouse platform for making human antibodies was developed by Kymab. This company was set up by Allan Bradley (one of the original Sequana SAB members; see Chapter 7) in 2009, based on his work at the Wellcome Trust Sanger Centre in the United Kingdom. They created several strains of transgenic mice using the stem cell technology that had been developed by Mario Capecchi, Martin Evans, and Oliver Smithies (2007 Nobel Prize winners). The Kymab mice had all their own antibody genes (not just a subset) replaced by the human DNA that contained all human heavy- and light-chain genes placed in the "correct" location in the mouse genome.[12] Kymab were bought by Sanofi in April 2021 for more than $1 billion as a platform for Sanofi's antibody pipeline that contains dupilumab, a mAb blocking the activity of IL-13 and IL-4 (DUPIXENT), approved in 2017 to treat atopic dermatitis. In 2020, 19 out of the 28 antibodies on the market were derived from humanised mice, the rest being made from phage display libraries.[13]

OTHER mAb TECHNOLOGY COMPANIES

There were originally three other companies set up to exploit mAb technology in addition to Celltech and Genentech: Hybritech in San Diego, Centocor in Philadelphia, and Genetic Systems in Seattle.

Hybritech

Ivor Royston, who was an investor through his venture capital (VC) firm Forward Ventures and later a Board Member at Sequana (Chapter 7), started Hybritech in 1978 with Howard Birndorf. Both men were academic scientists at UCSD, located in the VA Hospital near the university campus next to Interstate Highway 5. They were captivated by the opportunity that this new mAb technology presented. Hybritech was not only the first company dedicated to mAb technology but it was also the first San Diego biotech company, spawning many other companies that formed this active new biotech centre.[14,j] Birndorf was also a founder or helped to start many of the other local companies. One of the most important was Idec, founded with Ivor Royston. They developed the anti-CD20 mAb, rituximab—more on that later.

Like other mAb-based start-ups and encouraged by Brook Byers of Kleiner Perkins Caufield and Byers, who invested $300,000 in Hybritech, their initial focus moved over from providing research reagents to selling mAb-based commercial immunoassays for cancer screening. One of their most important diagnostic tests was developed to measure prostate-specific antigen (PSA) for prostate cancer screening, which became a successful product.

Tina Nova, one of the senior scientists at Hybritech, who helped to develop the PSA test, was quoted as saying that the culture of the company was a little like *Animal House* meets *The Waltons*.[k] Maybe the fact that Howard (Ted) Greene, the CEO who came to the company from Baxter Travenol, was—at 38 years old—the oldest person in the company had something to do with it. In addition to Greene, the dons of the Hybritech mafia included David Hale, who served as the president of the company. Tina subsequently became the second employee at Ligand Pharmaceuticals; the CEO of Genoptix, a local company that got sold to Novartis; the CEO of Molecular Stethoscope, a diagnostics company; and later the CEO of Decipher, a company that got acquired by Veracyte. She has been a standard bearer for San Diego biotech.

---

[j] As a person who became the CEO of a La Jolla–based biotechnology company (SGX Pharmaceuticals), I thought that the rather incestuous nature of the Hybritech alumni club was a little unhealthy. I was not backward in coming forward to say so either. It would probably have been better to have been a bit more politically astute.

[k] Having never seen either of these TV shows despite living in America for nearly 30 years, I am not quite sure what that means.

Hybritech was sold for ~$300 million to Eli Lilly in early 1986. The executives and the investors became much better off financially. Many companies, including Canji, Corvas, Gensia, Genta, Idec, Ligand Pharmaceuticals, Nanogen, Neurocrine, and Pyxis, can trace their origins back to Hybritech. Inbreeding in this case was not detrimental to success.[1]

Centocor

Centocor was founded in May 1979 by Dutch biochemist Hubert Schoemaker and a group of scientists and entrepreneurs led by Michael Wall, who had started Flow Laboratories in the 1960s. He had the initial idea of making mAbs for diagnostics. Hilary Koprowski at Wistar was associated with the company, and Centocor licensed The Wistar Institute patents. Centocor was more about product development than research and many of their products were in-licensed from academia: a different and more capital-efficient business model. Centocor went public in 1982, which brought in $21 million. By 1985, Centocor was generating about $20 million in revenue annually from product sales and contract research, with a focus on cancer diagnostics.

They were among the very first companies to think about using mAbs as "in vivo" products, initially as imaging reagents to "image cancer" and then as antibody–drug conjugates (ADCs) to carry toxic molecules to the cancer cells. With increasing sales, the company expanded to more than 400 people in 1990 and built a manufacturing facility in the Netherlands.

Centocor's first foray into therapeutic antibody development in 1986 was to in-license a mouse IgM monoclonal antibody called HA-1A (later called nebacumab [Centoxin]) from Stanford University, with the potential to treat sepsis caused by Gram-negative bacterial infection.[15] This was very like E5, a monoclonal IgM (edobacomab) that Xoma, a Bay Area monoclonal antibody start-up company, had licensed from UCLA. Sepsis was and still is an important unmet medical need, killing ~100,000 patients/year—a substantial potential market. The antibody being a mouse IgM (a pentamer vs. a monomer; see Box 1) was by no means straightforward to make or develop, compared to other IgG-based antibodies. It was rather insoluble and not very stable, but the company moved forward aggressively, spurred on in part

---

[1] This is likely the most important legacy of Hybritech, rather than the products they made.

by the success of products like recombinant growth hormone (Protropin) made by Genentech. To finance the endeavour, the management team raised several hundred million dollars in equity, debt, and off-balance sheet deals, with most of the money going to fund the extensive clinical trials needed to demonstrate that Centoxin was effective and safe.

Centocor got way out over their ski tips.[m] Things started well enough: the phase 1 study of the drug in 1988 (published in 1990) showed that the drug was well-tolerated, with no anti-drug antibodies (ADAs) seen.[15] A large and expensive 24-centre clinical trial followed, including a diagnostic test for the bacterial type causing infection. This showed that the drug was not only safe but also effective in reducing sepsis and mortality in Gram-negative bacterial infection. Centocor filed a product license application (PLA) with the FDA on the back of these results. Xoma had also filed for a PLA on a similar antibody they were developing with Pfizer.

Centocor was facing IP problems in the United States at the time, as the federal court in San Francisco had ruled that Centocor was infringing the Xoma patent. This had the unfortunate effect of focusing everyone's attention more closely on the Centocor clinical trials. At the same time, a few dog studies had indicated that Centoxin could be lethal under certain circumstances and did not, in addition, protect against sepsis. These results were both unexpected and alarming. The most important issue that Centocor faced though was that sepsis could be caused by both Gram-negative and Gram-positive bacteria. A diagnostic test would be needed for patients entering the trial, because those with Gram-positive sepsis would not respond. In small trials, this could be reasonably well-controlled with such a test, but in larger trials and in the marketplace that control would not be possible. Coupled with the average cost per patient of the therapy ($5650), it was considered unacceptable for patients who were never going to respond anyway to be treated.

From Centocor to "Centocorpse"

In early 1992, the FDA—with whom Centocor had a poor relationship, owing to misplaced expectations from the trial design—mandated additional clinical data. For Centocor, this was almost the straw that broke the camel's

---

[m] It was necessary, for example, to hire a sales force before the assumed FDA approval.

back, and the company became known as "Centocorpse," an aphorism I first heard from Sydney Brenner and entirely consistent with his sense of humour. It was also the name for Centocor adopted by Wall Street, as the company quickly lost $1.5 billion of its value after the FDA refused to approve the drug. Shareholder suits and allegations of federal securities law infringements quickly followed.

To their credit (and given that the company had other potentially valuable assets), the executives took stringent moves to save the company. They laid off hundreds of employees and, like the stock price, Centocor headcount went down markedly from 1600 employees to about 400. This event probably spelled the end of the road for Centocor as an independent company. Coupled with the subsequent failure of the Xoma antibody, it also spelled the end of mAbs for septic shock.

But it was not the end of the road for Centocor's pipeline of other products. With less than $150 million in the bank and a valuation in the $250 million range, Centocor reached a deal with Lilly, who bought a 5% stake in the company for $50 million, to help with the further development of Centoxin and for the co-development of a chimeric anti-clotting mAb, abciximab (ReoPro).[n] In June 1994, the FDA approved ReoPro, which was then marketed by Lilly. By 1997, the sales of ReoPro of $250 million per year were sufficient to keep Centocor afloat. In addition to ReoPro, Centocor also had a cancer mouse mAb (edrecolomab) called Panorex, an anticancer drug targeting the cell surface glycoprotein EpCAM (17-1A) expressed on epithelial cells found in various carcinomas.

But Centocor's most important product was an anti-inflammatory chimeric antibody raised against tumour necrosis factor (TNF), called infliximab (see more about this product in Chapter 5).

Centocor sold its diagnostic products to Fujirebio Inc. of Japan in 1998. It was acquired the next year for close to $5 billion, becoming a wholly owned subsidiary of Johnson & Johnson. Xoma, although burning through more than a billion dollars trying to develop their own antibody products, remained a relatively unknown member of the biotech scene for years. They eventually became a clearinghouse for in-licensed antibodies of various kinds, living off the royalties that they brought in.

---

[n] Abciximab bound to a glycoprotein complex called gpIIb/IIIa on platelets, preventing them from sticking together to cause a blood clot.

Genetic Systems

In his book *Gene Dreams*, Robert Teitelman chose Genetic Systems to write about as his example of a company exploiting biotechnology.[16,o] Genetic Systems was the third company set up to exploit mAb technology. It was founded by the two Blech brothers, Isaac and David, who, although they were aware of both Centocor and Hybritech, presciently believed that there was room for a third mAb-based company: especially if they could entice a hotshot to run it. The Blech brothers managed to seduce the virologist Robert (Bob) Nowinski, who, like Chris Henney (the founder of Immunex we met in Chapter 3), was at the Fred Hutchinson Cancer Center in Seattle, to become the CEO. That hire also decided the location for the new company (i.e., Seattle).

Nowinski was a very clever and motivated guy. He came (like the Blechs) from Brooklyn and had worked at the Memorial Sloan Kettering Cancer Center in New York with Lloyd Old. With his somewhat extravagant personality, his smarts, and his self-confidence, he became an important part of the Genetic Systems story. Like Centocor and Hybritech, Genetic Systems too was ultimately acquired by Big Pharma for close to $300 million, this time by Bristol Myers in late 1985 (shortly after the acquisition of Hybritech by Lilly).

Financing Genetic Systems

The financing of Genetic Systems was a good deal more complicated than either Centocor or Hybritech, including the use of warrants as part of the IPO financing.[18] As Teitelman discusses in *Gene Dreams*, warrants are generally positive because if the stock price goes up, it ensures more money comes into the company as investors exercise their "in the money" warrants. It can also be negative: too much warrant coverage when the warrants are at a price well above the prevailing stock price is just dilution waiting to happen, a major disincentive to new investors.

Things were made even more complicated by a relationship that Nowinski forged with Syva, the diagnostic arm of Syntex. In this deal, Syntex invested $40 million for 18% of Genetic Systems. With this corporate acknowledgement and the involvement of Allen & Co. (a well-respected Wall Street firm), Genetics Systems' "street cred" was established, and its future financing ensured.

°I am not sure that it would have been my choice, but Genentech, which would have been my choice, has been covered very well by others.[17]

The 1981 Genetic Systems IPO was very different from other bio-tech IPOs, as a large proportion of the stock was bought by retail rather than institutional investors. This profile tends to lead to stock volatility, as retail traders trade stock all the time on the margin (i.e., their decisions are not based on the fundamental science or other characteristics of the company).[P] Institutional investors used to analysing and dealing with the risks of biotech are much more reliable. The financing was also anything but straightforward. Genetic Systems sold one million "units" to the public at $1 per unit through D.H. Blair. A unit consisted of three shares of common stock, three class A warrants to buy additional stock at $3.25, plus an additional share of common stock and a class B warrant exercisable before June 1984, for another share of common stock at $5. In sum, this equated to a sale (if all warrants were exercised) of nine shares of common stock at about $3 per share.[q]

Genetic Systems focused initially on developing diagnostic tests for sexually transmitted diseases, to differentiate patients with gonorrhoea (for example) susceptible to penicillin from other bacteria or chlamydia that were not, which is important for babies of infected mothers. They were also working on herpesviruses; some strains of which (e.g., HSV-2) were sexually transmitted.

Edward (Ed) Clark was a prime mover at Genetic Systems in the early days. Ed had connections with the Milstein lab when working as a postdoc at UCL in London with Avrion Mitchison, one of the most respected immunologists in the world. Ed brought NS1 cells back to the United States with him, so Genetic Systems could make monoclonals with the Milstein cells. All the mAbs they made initially were propagated in mice as hybridomas. This called for the setting up of what was essentially an expensive "mouse factory," dedicated to producing the proteins.

Therapeutics: Oncogen

It did not take long before Genetic Systems thought about developing thera-peutic antibodies. Once again, their approach to that transition was anything

---

[P]Not really the kind of investor that a biotech start-up developing products using new technology can really afford to have.

[q]The reason I go into this much detail is because it was complicated and unusual in bio-tech at the time, and it led to some problems later.

but conventional and was different from both Centocor and Hybritech. Instead of making or in-licensing mAbs to potential therapeutic targets in cancer, they set up a subsidiary called Oncogen and persuaded George Todaro, a very accomplished and well-respected cancer research scientist from the National Cancer Institute (NCI), to come and run it. Todaro was a big name in oncology research. His peer group at NCI included Robert (Bob) Gallo of HIV fame and Edward (Ed) Scolnick, who went on to head up R&D at Merck. It was also during this time where work on RNA tumour viruses led to the discovery of oncogenes, for which Michael Bishop and Harold Varmus were awarded the Nobel Prize in Physiology or Medicine in 1989.

Oncogen, which occupied the second floor of the Genetic Systems waterfront building in Seattle, was, as succinctly put by Teitelman in *Gene Dreams*,[16] "a joint venture disguised as a limited partnership, where Syntex and Genetic Systems served as the two general partners." Syntex agreed to put $8 million over four years into Oncogen, with Genetic Systems adding another $1.5 million. For Todaro, this was a ticket to "science heaven" compared to the NCI, and a great opportunity to continue his work on growth factors and oncogenes in an environment where—at least initially—money was no object. He also negotiated a very good personal deal, with stock in both Syntex and Genetic Systems and with options to purchase more later. Formed in early 1983, it was not long before it had 60 people doing Todaro-directed research, including making mAbs to Oncostatin M and other growth factors. Oncogen hired the married couple Karl-Erik and Ingegerd Hellström, both very accomplished and experienced tumour immunologists, from the University of Washington.

The stock market absolutely loved the Oncogen deal. Genetic Systems' stock rose accordingly, creating the opportunity for a secondary offering without warrants, managed by D.H. Blair, Allen & Co., and Merrill Lynch. The company raised more than $20 million, leading to a market cap of $180 million. This represented a huge return (800×) for the original investors, who had bought in only two years before.

At this time, James (Jim) Glavin became the CEO of the company. This was a "name only" deal: it was clear to everyone that Nowinski still called all the shots. He was totally in tune with the "Seattle hip" culture of the company, including Friday breakfasts and the Genetic Systems equivalent of the Genentech HoHos. Nowinski—like Steve Jobs at Apple—liked to wear black clothes and he drove a Ferrari. His personality led to him going off the reservation consistently or, to mix metaphors, to "skiing off piste," forming potential relationships with all sorts of desirable and not so desirable business organisations. He apparently did not pay much attention to mundane things such as "accounts payable" or the rapidly escalating burn rate.

However, Nowinski did have a knack of recruiting top talent, bringing good hires to both Genetic Systems and Oncogen from his old lab at the University of Washington, the Fred Hutch, and other places. From an early stage, these people became directors of different departments in the company, although most of them were only just out of graduate school. They, in turn, hired people to work with them from their old academic labs who were all just as young and just as poor. Notable among the alumni were Jeffrey (Jeff) Ledbetter, part of a consortium that developed mAbs to many cell differentiation markers; Milton (Milt) Tam who helped to develop the mAb kits against sexually transmitted diseases (STDs); Kathy Shriver who led the team that made the first HIV diagnostic test; and Karen Nelson who made mAbs to human leukocyte antigens (HLAs). Pete Senter was a chemist at Genetic Systems who developed some of their mAb toxin conjugation methods, going on to work at Seagen to develop some of Seagen's mAb toxin conjugation methods. All these people were recruited quite early on.[19]

Selling the Company

One of the companies that Nowinski actively seduced was Bristol Myers. This pharma company had been contemplating for a while the importance of both recombinant DNA and mAb technology. Genetic Systems and Oncogen represented a way for Bristol Myers to capture both technologies at once, especially as it related to oncology, one of their focus areas. In early 1985, Bristol bought 33% of Oncogen for $13 million. With Genetic Systems and Syntex also increasing their stake, Oncogen would now receive more than $20 million over the next three years. In October 1985, shortly after the announcement that Lilly had bought Hybritech, Bristol Myers moved in to buy Genetic Systems and Oncogen. Bristol Myers agreed to acquire the company for $294 million,[r] consummating the deal on October 21 in New York in 1985 to close in 1986.

It was not an easy exit for the shareholders. Bristol Myers paid $15 million in cash to Syntex, for the one-third interest that they took in Oncogen when it was formed in 1983. The agreement also provided royalties to

---

[r] Not far different from the price Lilly paid for Hybritech a few months before.

Syntex on cancer diagnostics and treatments discovered by either Oncogen or Genetic Systems. The deal was not well-accepted by everyone at Bristol Myers. When questioned on the subject, the Bristol executives referred mostly to the necessity of having access to this new technology to be competitive. "Seems like a lot of money for a company not generating any cash but that is the price you have to pay" was allegedly the quote from one executive.[s]

Nowinski stayed with Genetic Systems for a few years after the merger, leaving in September 1989 to start another Blech investment, Icos, with Chris Henney and George Rathmann. Todaro ended up as President of Oncogen, which was now a Seattle Oncology Research Centre for Bristol Myers, with a senior R&D role at the pharma company, having made enough money to indulge his hobby of horse racing. Genetic Systems was subsequently divested by BMS to Sanofi in 1990, and the labs were closed.

In retrospect, it seems a pity to have focused so much on the financial aspects of Oncogen and Genetic Systems, rather than on their products. The reality is that there were few real products of note in the company, apart from their important STD mAb-based diagnostic tests. The investors made money, but BMS captured most of the value, in the form of the people who worked at BMS for a short time after the merger and what BMS subsequently did with the technology. Peter Linsley was an Oncogen scientist, who with Jeff Ledbetter developed Orencia, a CTLA-4 agonist, while at the Bristol Myers Seattle lab. Orencia was approved for the treatment of rheumatoid arthritis in 2011. By 2015, sales of Orencia were nearly $2 billion. Not too shabby for a drug created in an academic collaboration between Jeffrey (Jeff) Bluestone and Linsley. Bluestone later worked with James (Jim) Allison to develop mAbs to block CTLA 4 activity for checkpoint blockade for the treatment of melanoma (see Chapter 5).

Idec and Rituxan

Idec was set up in 1986 by former Hybritech founders Ivor Royston and Howard Birndorf in San Diego (see above), focusing on therapeutic mAbs rather than diagnostics. KP invested in the start-up, and Brook Byers became their first CEO and chairman of the company. Their most important product was unquestionably rituximab (Rituxan), a chimeric mAb for treating B-cell proliferative diseases made by recombinant methods in CHO cells.

---

[s] That comment could be applied equally well to several other acquisitions that have occurred in the industry since then.

The development of rituximab was another classic "skunkworks" project. It started off as "Pan B"—a project screening for mAbs that reacted with B cells. After the first Idec potential products stalled and with only about $10 million left in the bank, attention turned to the Pan B project, which had by then identified some mAbs that reacted with CD20, a protein found on the surface of most B cells. The antibodies seemed to act by depleting B cells, while sparing plasma cells and haematopoietic stem cells that did not express CD20. It was thought that B-cell proliferative diseases such as B-cell leukaemia and non-Hodgkin's lymphoma (NHL) might respond to a CD20-depleting antibody. The initial mouse antibody 2B8 was made into a mouse human chimeric antibody. By 1992/1993, the first clinical trials were initiated in NHL. An encouraging 6 out of 15 patients showed a clinical response.

For financial reasons, Idec needed to partner the programme. GSK turned down the opportunity but by deliberately sitting next to Kirk Raab (CEO of Genentech) at a fundraiser in San Francisco, Byers managed to persuade Genentech to take a meeting so that they could describe "Pan B" to them.[20] William (Bill) Rastetter, who later became the CEO of Idec, was a Genentech alumnus who recognised the potential importance of this Idec mAb. After the completion of a phase 2 trial in NHL, Genentech entered a commercial and clinical stage strategic alliance with Idec to help to develop and market the potential new drug. Genentech got more than 50% of the commercial rights to the antibody plus an interest in two other antibodies in the pipeline, for the princely sum of $9 million, more than half of which was used to buy Idec stock. Genentech bought additional stock for $17.5 million and agreed to pay more than $30 million in milestones and option payments.[t]

In November 1997, Rituxan was approved to treat NHL, the first new drug in 10 years for B-cell malignancies. Even back then this was a fast timeline for what turned out to be such an important product. With sales of $150 million in 1998, it was also the forerunner for the use of other mAbs for the treatment of cancer. It has now been used to treat more than six million people worldwide, not only for cancer but also for autoimmune diseases such as multiple sclerosis and inflammatory bowel disease.

Biogen and Idec merged in 2003 in the biggest biotech deal ($6.5 billion) since the Amgen acquisition of Immunex in 2001, creating a company with two one billion dollar drugs: Avonex for multiple sclerosis (MS) and

---

[t] This was one hell of a deal for Genentech and, as it turned out, for cancer patients.

Rituxan for B-cell malignancies. Wall Street did not like the deal initially and the stock of both companies slipped, but Idec was a very successful company with great returns for shareholders and patients alike, realised some 17 years after its founding.

ImClone Systems, Erbitux, and Insider Trading

One of the most obvious targets for antibody therapy was epidermal growth factor (EGF) to control the proliferation of cells that occurs when the growth factor binds to its receptor (EGFR). The hypothesis was that mAbs against the EGFR would block receptor activation by EGF, an approach pioneered by John Mendelsohn and Gordon Sato at UCSD.

It was a reasonable hypothesis as some tumours were known to depend on EGF for their growth. The EGF receptor had been identified by Michael (Mike) Waterfield's lab at ICRF in London as a tyrosine kinase, identical to an oncogene called erb-b found in chickens, following on from his work on platelet-derived growth factor (PDGF) and its relationship to the oncogene v-sis.[21] A chimeric Fc-exchanged version of a mouse mAb to human EGFR made by Mendelsohn (cetuximab) was shown to have activity in human cancer tissue culture assays and mouse human xenograft models.[u]

In the early 1990s, the NCDDG (National Cooperative Drug Discovery Group) at the NCI helped the work on cetuximab, and in 1999 ImClone Systems, a small biotech company in New York founded in 1984, commenced phase III trials of the mAb in collaboration with the German company E. Merck (now Merck KGaA). In 2001, BMS agreed to a $2 billion deal with ImClone to co-develop the mAb for the U.S. market. FDA approval was sought in November of the same year, but the application was not reviewed, owing to incomplete data in the clinical package, resulting in a huge drop in ImClone stock value and, ultimately, the indictment of Samuel (Sam) Waksal (the CEO) and Martha Stewart, a TV personality and friend of Waksal, who conspired to sell their stock before the FDA decision was made public. As a result, Waksal was sentenced to several years and Stewart to five months in federal prison for securities fraud.

After a two-year delay, ImClone finally submitted the results of additional studies with Merck KGaA, done to the liking of the FDA, and cetuximab (Erbitux) received approval for the treatment of metastatic

---

[u] An immunologically compromised mouse in which a human tumour is implanted.

BOX 4. CADUS: ANOTHER ImClone CLAIM TO FAME

ImClone spun out a yeast screening company called Cadus, with Jeremy Levin as CEO. The founders were Arnie Levine, Tom Schenk, and Sam Waksal, with technology derived from the Lefkowitz lab at Duke University. I met with Cadus while at Glaxo, as Cadus' yeast screens for GPCR activators were of some interest to us. Despite many corporate deals, Cadus was not successful, owing largely to a patent litigation dispute with SIBIA, who was also making and running yeast-based GPCR screens and had interfering IP. Although Levin could have settled the dispute for a reasonably modest sum, the Cadus board—which included Carl Icahn—decided to fight it out. At this point, Jeremy left the company. They lost the case in the appeals court. All the Cadus cash was put in escrow and the company was wound up. That cash found its way eventually into a Real Estate Investment Trust.[22] Many of the Cadus scientists found their way to Regeneron, including Andrew Murphy, who became EVP Research there.

---

colorectal cancer in 2004. This would not have happened at all without Waksal insisting that the ImClone Board take on the Erbitux project in the first place (see also Box 4).

Eli Lilly subsequently bought ImClone Systems in the fall of 2008 for $6.5 billion, after a bidding war with BMS. BMS continued to market Erbitux until 2015, when it was turned over to Eli Lilly, with Merck KgA continuing to sell the drug in Europe.[v]

Genentech and Herceptin

In 1998, former NBC science correspondent Robert Bazell published a book about the development of Herceptin, a mAb developed by Genentech.[23] His story is about the perseverance of several groups to make and develop a mAb therapy for breast cancer. First and foremost, it is a story about the patients—the women suffering from a deadly form of breast cancer—and what they managed to do to help to get the therapy on the market. It is also a story about an oncologist Dennis Slamon, who was

---

[v]The development of fully humanised antibodies to EGFR using transgenic mice by Abgenix has significantly reduced interest in Erbitux. One of the characteristics of these EGFR mAbs is that they work much better in EGFR-positive metastatic colorectal tumours that do not also contain KRAS mutations, as these cancers are thus more dependent for growth on EGF receptor signaling.

fixated on the role of HER2 (an oncogene) and believed with religious conviction—almost to the exclusion of everything else—that an antibody to this protein would work to treat breast cancer. It is also a story about Genentech and the rather disjointed and nonlinear way that drugs are often discovered and developed by larger companies, even one as proficient as Genentech.

Like Oncogen (above), the story starts with the Bishop and Varmus discovery of oncogenes. Oncogenes are human cancer-causing genes, derivatives of which were originally found in the genomes of avian and mouse retroviruses that caused cancer, such as murine leukaemia virus and Rous sarcoma virus of chickens. Robert (Bob) Weinberg's lab at MIT had discovered an oncogene called "neu" in a rat neurological tumour, showing that cells transfected by this gene gave rise to cancer in mice. The gene was cloned by Axel Ullrich from Genentech while working in London at ICRF, as part of a study to find genes related to the EGF receptor. It was called HER2 (human epidermal growth factor receptor 2) and as it turned out, it was identical to Weinberg's "neu" gene.

Slamon and Ullrich at Genentech subsequently discovered that HER2 was overexpressed in several breast and ovarian cancers. The Ullrich lab had transformed cells with HER2 DNA, and they asked the Genentech immunologists to make a mouse mAb against the protein, so they could correlate the HER2 protein levels on the surface of the cells with mRNA and DNA levels in the cells. They observed that when the mAb was added to breast cancer cells making HER2 in culture, the cells stopped growing, as did xenografts of the cells in mice treated with the antibody. This suggested, provocatively, that a mAb therapy for breast cancers overexpressing HER2 might be therapeutically useful. But it took quite a while to get Genentech management to take this potential mAb approach to cancer seriously.

The story as told by Bazell could be about many drugs that eventually got developed by the industry and were successful: serendipity certainly played a role. Genentech only took the project seriously when influential scientists at the executive level began to take an interest—but not before. Most notable in this respect were William (Bill) Young and Arthur (Art) Levinson. Steam was also rising from Dennis Slamon and some of the patients that he and others were treating who might benefit from such a drug, as biopsies showed they had tumours that expressed large amounts of HER2. Making sure that certain key opinion leaders (KOLs), including Larry Norton at MSKCC, embraced the concept was also an important part of the successful development of the drug."

Paul Carter, who came to Genentech from Greg Winter's lab at the LMB, was tasked with making the humanised version of the mouse antibody, and money was found to test it. In the meantime, Slamon had already begun testing the murine version in a few patients, with mixed results. He had been fortunate in making a contact through his oncology practice at UCLA with some wealthy people, including those at the cosmetics firm Revlon, who had an interest in promoting women's health. A Revlon endowment to his lab helped him with his HER2/neu research as well as with the successful lobbying of Genentech.

After several long and difficult phase III trials, Herceptin was approved by the FDA using the fast-track process in March 1998. The important new drug clearly worked in some breast cancer patients, especially those overexpressing the HER2/neu protein that drives the proliferation of the cells in those tumours.

Antibody–Drug Conjugates

A version of Herceptin attached to the cytotoxic agent mertansine (Kadcyla) is available from Genentech to treat metastatic breast cancer. This is a prime example of a newer class of drugs now called ADCs, an idea in fact conceived by Köhler and Milstein when they initially thought about applications of mAbs.

ADCs are basically therapeutic antibodies with a toxic compound or toxic natural product covalently linked to the antibody, making a kind of targeted cytotoxic drug or magic bullet (the antibody "delivers" the cytotoxic payload to the cancer). For all their promise, ADCs do have a somewhat chequered history, owing to the inevitable toxicity that accompanies attaching a powerful toxin or toxic chemical to an antibody and injecting it into people.

One of the first ADCs made was gemtuzumab ozogamicin (Mylotarg), an antibody attached to a calicheamicin derivative targeting CD33 on blast cells, created in a collaboration between Celltech and Wyeth in 1991 to

---

"Yet again, what appears scientifically obvious to some (Slamon) is not obvious to others (Norton), and big egos can get in the way of rapid decision-making, leading to time delays that can never be recovered.

treat acute myeloid leukaemia. This collaboration later produced a mAb conjugate called inotuzumab (Besponsa) for use in patients with acute lymphoblastic leukaemia (ALL) who are not suitable for a stem cell transplant or those not considered candidates for standard chemotherapy. These two mAbs were among the first ADCs to be approved. Both mAbs were conjugated with calicheamicin, a toxic antibiotic.

Coulter Pharmaceutical developed an ADC called tositumomab (Bexxar), directed against CD20 (like Rituxan). In this case, the antibody was linked to radioactive iodine 131 to irradiate and kill the B cells that the antibody is bound to. To cut a long story short, Coulter was acquired by Corixa, a company set up by Steve Gillis in 1994 after he left Immunex. Bexxar worked, but the radioactivity associated with it and the effects on the patients who took the drug caused issues which could not be overcome commercially, particularly in competition with Rituxan (which was also efficacious).[24]

The antibody parts of ADCs are now almost always humanised. Among the current generation of immunoconjugates gaining approval in 2011 was brentuximab vedotin (Adcetris), an anti-CD30 antibody coupled to bleomycin, developed by Seattle Genetics (Seagen) in a collaboration with Millennium, used to treat some forms of Hodgkins lymphoma. Seagen was co-founded by Clay Siegall in 1998 specifically to develop such antibody–toxin conjugates. The company has four commercial products, a deep pipeline, revenues of more than $2 billion/year, and employs more than 2500 people. Seagen agreed to be acquired in early 2023 for more than $40 billion by Pfizer, with the deal closing in December 2023.

Bispecific Antibodies

Another concept conceived by Milstein when he was thinking of monoclonal antibody applications was the possibility of safe and effective bispecific molecules (i.e., engineered "antibodies" that could bind two or more different molecules via two different binding sites simultaneously). Bispecific mAbs can be broadly classed into two groups (see Box 1), either very small proteins composed mostly of two binding domains or larger molecules that look more like classic IgG but with two or more binding domains. Bispecific antibodies are finding a niche in immunotherapy as an alternative to cell therapy (see Chapter 13).

BiTEs

A new class of anticancer treatments are the so-called bispecific T-cell engagers or BiTEs. The idea is to turn the attention of the immune system to the cancer by drawing in (engaging) the appropriate T cells. It is now quite easy to imagine how you could engineer an antibody to behave as a T-cell engager, by engineering a fragment recognising and binding to a tumour antigen (e.g., CD19 on B cells) at one end of the molecule and an antibody fragment that binds and activates a T cell at the other end by binding to, for example, CD3.

The Institute of Immunology in Munich had made such a BiTE around 2008. Patrick Baeuerle, who had been working at Tularik (the Dave Goeddel company in South San Francisco), had moved back to Germany to a job at Micromet. He went to visit the Institute to see what they were doing. He was completely taken in by the concept of BiTEs and impressed by the molecules that had already been made. Baeuerle persuaded Micromet to in-license them from the university, and with Lonza's help they created a BiTE called blinatumomab for the potential treatment of ALL.

Amgen had the foresight to buy Micromet in 2012 for about a billion dollars, when the antibody was in phase 2 clinical trials. The drug, called Blincyto, was approved in 2014 for BCR-ABL negative ALL, in both children and adults. Although effective, Blincyto has a short half-life, requiring the antibody to be infused every day for four weeks, the first nine days of which require hospitalisation: a significant burden for the patients. Nor is it cheap: the price without discounts is $175,000 per annum. Not surprisingly, the market is small.

Blincyto set precedent for the development of many other targeted T-cell bispecific antibodies. Genentech has developed a successor (i.e., potentially better) bispecific to Blincyto that uses the same binding sites but with much better properties and dosing regimens. In late 2022, the FDA granted accelerated approval to this bispecific antibody (mosunetuzumab-axgb, brand name Lunsumio) for treating follicular lymphoma. Lunsumio showed an impressive response rate in the clinical trial, with 60% of the (90) patients achieving a complete response with a median duration of nearly two years.

AstraZeneca recently paid $1.27 billion to acquire the BiTE TNB-486 from TeneoBio. TNB-486 targets the same proteins as Blincyto but is easier to use. It is currently in a dose-escalation study in relapsed and refractory non-Hodgkin lymphoma patients. AstraZeneca also picked up another bispecific mAb from a preclinical programme at Harbour BioMed, targeting solid tumours expressing the protein Claudin18.2, often found in gastric and pancreatic cancer.

Amgen paid $900 million in cash plus $1.6 billion in milestones, to bag TeneoBio itself, including its T-cell engager platform, which includes a BiTE in phase I development for castration-resistant prostate cancer. SV Health Investors recently participated in a $67 million series A round for Rondo, a cancer bispecific company founded by some of the TeneoBio team.

Recently, Regeneron reported two deaths in the clinical trial testing their bispecific mAb targeting PMSA on metastatic prostate cancer cells and CD28 on T cells, in combination with Libtayo, their anti-PD1 mAb. Remember that it was a CD28 superagonist mAb, TGN1412 from TeGenero that caused several deaths in a clinical trial run by Parexel at Northwick Park Hospital (located just up the road from Glaxo in Greenford, United Kingdom) in March 2006. This led not only to the demise of the company but also to the conclusion that CD28 is a T-cell activator that is difficult to manage therapeutically.

Janssen received approval in 2021 for a bispecific called amivantamab (Rybrevant), which targets two growth factor receptors for drug-resistant lung cancer. One end of the antibody binds to the mutated form of the EGFR found in non–small cell lung cancer patients, whereas the other end targets the c-met receptor. The purpose of this antibody is to block the activity of two growth factors at once on the same cell.

Immunocore in the United Kingdom has used its ImmTAC platform to develop a unique type of bispecific antibody, tebentafusp—Kimmtrak for immunotherapy purposes. It uses a T-cell receptor (TCR) protein to bind to a peptide presented by an HLA on a cancer cell at one end and with a T-cell engager at the other end. The importance of this approach for immunotherapy will become clearer in Chapter 14. For now, see Box 5.

It will be interesting to see how the competition between bispecifics and cell therapies plays out. The ease of manufacturing bispecifics and all the historical experience with mAbs would seem to me to give bispecifics a strong advantage over cell therapy, with the proviso of course, that they are safe. Many of the big players like Gilead and BMS are taking both approaches while others, like Regeneron and Amgen, are focusing more on bispecifics.

The winners will emerge from the clinical data. If there is clear and differentiated efficacy in terms of patient survival and lack of side effects, then that medicine is more likely to prevail and be successful.[25] Limiting patients

---

BOX 5. IMMUNOCORE

Immunocore was founded as a spinout of MediGene in 2008. They have raised considerable sums of money for a U.K. biotech, including $320 million in July 2015 in the largest ever private placement for a European biotech led by Fidelity, Woodford Investment Management, Malin Corporation, Eli Lilly, and RTW investors. In March 2020, they raised a further $130 million in a series B round and another $75 million in a series C financing. Immunocore went public in early 2021 when the markets were red hot, raising more than $250 million in a deal led by Goldman Sachs, J.P. Morgan, and Jefferies & Co. Immunocore's lead programme at the time was tebentafusp for uveal melanoma. They announced positive phase 3 data in 2020. In January 2022, the FDA approved the bispecific for HLA-A*02:01-positive adult patients with unresectable or metastatic uveal melanoma.

---

by either HLA type or tumour type or subtype of tumour will not be a problem, provided there is the clinical efficacy in that subset to warrant the extra work needed to find the appropriate patients.

More than 100 bispecific antibodies are now in development, but only a handful has been approved for use so far. The time does seem to be now for cancer-targeting bispecific mAbs.[26,27]

Noncancer Applications

Bispecifics are being developed for indications other than cancer. Roche's Hemlibra (emicizumab) was approved by the FDA for use in haemophilia A with or without Factor VIII inhibitors, in the fall of 2018 shortly after Bioverativ had been bought by Sanofi. This bispecific is an antibody painstakingly developed over many years by Chugai (now a subsidiary of Roche) in Japan that uniquely binds both the activated coagulation Factor IX and Factor X. The antibody, which is a modified IgG, brings these two factors into proximity so that Factor X is activated, a step normally mediated through Factor VIII (see Chapter 2). It is particularly suitable for use in patients who have developed inhibiting antibodies to Factor VIII. It is also sometimes used in place of long-acting Factor VIII in patients. Hemlibra has eaten into the sales of ELOCTATE, the longer acting Factor VIII-Fc molecule developed by Syntonics and Biogen and now marketed by Sanofi, and it is a successful and important drug for Roche/Genentech.[x]

---

[x]Many patients who develop Factor VIII inhibitors have to go through an expensive and time-consuming inhibitor tolerisation protocol. Using Hemlibra avoids this process.

A recent bispecific-mAb approval of note is Vabysmo (faricimab) for the treatment of wet or neovascular, age-related macular degeneration (AMD) and diabetic macular edema (DME). This Genentech bispecific targets both vascular endothelial growth factor receptor and angiopoietin 2, two proteins implicated in these sight-compromising eye diseases.

REFERENCES AND NOTES

1. Köhler G, Milstein C. 1975. Continuous cultures of fused cells secreting antibody. *Nature* **256:** 485–487. doi:10.1038/256495a0

2. Brenner S, Milstein C. 1966. Origin of antibody variation. *Nature* **211:** 242–243. doi:10.1038/211242a0

3. Alkan S. 2004. Monoclonal antibodies: the story of a discovery that revolutionized science and medicine. *Nat Rev Immunol Perspect* **4:** 153–156. doi:10.1038/nri1265

4. Wade N. 1980. Inventor of hybridoma technology failed to file for patent. *Science* **208:** 693. doi:10.1126/science.208.4445.693

5. Wellcome Trust Centre. 1997. Technology transfer in Britain: the case for monoclonal antibodies. In *Wellcome witnesses to twentieth century medicine*. Wellcome Trust, London.

6. Waldmann H, Hale G. 2005. CAMPATH: from concept to clinic. *Philos Trans R Soc B* **360:** 1707–1711. doi:10.1098/rstb.2005.1702. CD52 is found on the surface of mature T cells and binds to SIGLEC 10, a sialic acid–containing lectin on an immunoreceptor called ITIM.

7. Fairtlough G. 1989. Exploitation of biotechnology in a smaller company. *Philos Trans R Soc London* **324:** 589–597. doi:10.1098/rstb.1989.0070

8. Cabilly S, Riggs AD, Pande H, Shively JE, Holmes WE, Rey M, Perry LJ, Wetzel R, Heyneker HL. 1984. Generation of antibody activity from immunoglobulin polypeptide chains produced in *Escherichia coli*. *Proc Natl Acad Sci* **81:** 3273. doi:10.1073/pnas.81.11.3273

9. The Cabilly, Boss, and the '602 patents issued in 1983. Claim 1 of the '415 patent is directed to a process for producing antibody or an antibody fragment by two (what are now) conventional steps: transforming a host cell with a first and a second DNA sequence encoding variable domains of the immunoglobulin heavy and light chains, respectively, and independently expressing and producing the heavy and light chains as separate molecules. These two steps, generally, are widely used in production of many antibodies, which may explain why this patent is the focus of several challenges. Claim 1 of the '602 patent covers a method for large-scale production of "properly folded" polypeptides, while requiring that the host cells are cultured under conditions of high metabolic and growth rate, and when the polypeptide expression is induced, modulating the metabolic rate by reducing the feed rate of a carbon/energy source (e.g., glucose) and/or reducing the amount of available oxygen. This method (commonly known as fed-batch method) is currently widely used in many biologic drug productions.

10. Storz U. 2012. The Cabilly patents. *MAbs* **4**: 274–280. doi:10.4161/mabs.4.2.19253

11. Teskin RL. 2003. It lives for 29 years. *Legal Times.* November 3, 2003. 6b.

12. Lee E-C, Liang Q, Ali H, Bayliss L, Beasley A, Bloomfield-Gerdes T, Bonoli L, Brown R, Campbell J, Carpenter A, et al. 2014. Complete humanization of the mouse immunoglobulin loci enables efficient therapeutic antibody discovery. *Nat Biotechnol* **32**: 356–363. doi:10.1038/nbt.2825

13. Lu R-M, Hwang Y-C, Liu I-J, Lee C-C, Tsai H-Z, Li H-J, Wu H-C. 2020. Development of therapeutic antibodies for the treatment of diseases. *J Biomed Sci* **27**: 1. doi:10.1186/s12929-019-0592

14. Quigley K. 2018. Long live Hybritech. Four decades after Hybritech's founding, the company's spirit and entrepreneurism is felt throughout the life science industry in San Diego. *BioCom Lifelines* **28**: 4–10. https://s23.q4cdn.com/595160625/files/doc_downloads/publications/LifeLinesSummer2018.pdf

15. Marks L. 2012. The birth pangs of monoclonal antibody therapeutics. The failure and legacy of centoxin. *mAbs* **4**: 403–412. doi:10.4161/mabs.19909

16. Teitelman R. 1989. *Gene dreams: Wall Street, academia and the rise of biotechnology.* Basic Books, New York.

17. Hughes SS. 2011. *Genentech: the beginnings of biotech.* University of Chicago Press, Chicago.

18. Warrants are a way to give investors an option to buy shares in a company at a specific price later in time.

19. Information obtained on a visit to a Genetic Systems reunion in Seattle in June 2022.

20. Conversation with Brook Byers, 2022.

21. Downward J, Yarden Y, Mayes E, Scrace G, Totty N, Stockwell P, Ullrich A, Schlessinger J, Waterfield MD. 1984. Close similarity of epidermal growth factor receptor and v-*erb-B* oncogene protein sequence. *Nature* **307**: 521–527. doi:10.1038/307521a0

22. Conversation with Jeremy Levin, 2022.

23. Bazell R. 1998. *Her-2: the making of Herceptin, a revolutionary treatment for breast cancer.* Random House, New York.

24. See Kinch M. 2016. *A prescription for change: the looming crisis in drug development.* UNC Press, Chapel Hill, NC.

25. Sheridan C. 2021. Bispecific antibodies poised to deliver wave of cancer therapies. *Nat Biotechnol* **39**: 251–259. doi:10.1038/s41587-021-00850-6

26. Wang S, Chen K, Lei Q, Ma P, Yuan AQ, Zhao Y, Jiang Y, Fang H, Xing S, Fang Y, et al. 2021. The state of the art of bispecific antibodies for treating human malignancies. *EMBO Mol Med* **13**: e14291. doi:10.15252/emmm.202114291

27. Longo D. 2022. The expanding clinical role of bifunctional antibodies. *New Engl J Med* **387**: 2287–2290. doi:10.1056/NEJMe2208708

CHAPTER 5

Antibodies Become Blockbuster Drugs

One of the most successful antibody drugs ever made, which Cambridge Antibody Technology (CAT) in the United Kingdom helped to develop, was adalimumab (Humira), a fully humanised anti–tumour necrosis factor (anti-TNF) antibody derived from a phage display library.[a] The initial project to make a monoclonal antibody (mAb) to TNF at CAT was a collaboration with Knoll AG, part of BASF Pharma. They released positive results of the effect of this antibody on rheumatoid arthritis in 1998. Abbott acquired BASF Pharma in 2001 for $6.9 billion and, as a result, obtained adalimumab, the efficacy of which had not yet been determined compared with the other anti-TNF mAbs. It turned out to be a great deal: perhaps the best biotech deal ever. Humira was developed and marketed by Abbott Laboratories. In 2012, Abbott split into two companies, Abbott Laboratories and AbbVie, with AbbVie focused on maximising the Humira opportunity while discovering new biopharmaceutical products.

ANTIBODIES TO TNF

The interest in making antibodies to TNF did not start at CAT. TNF was discovered in 1975 by Lloyd Old at Memorial Sloan Kettering Cancer Center (MSKCC) as a small protein produced by macrophages that had an antitumour effect in mouse models and could kill sarcoma cells in vitro. Anthony (Tony) Cerami and Bruce Beutler[b] at Rockefeller University identified a substance that induced cachexia (wasting) in mice and was a mediator of inflammation and septic shock. They called it cachectin, but it was identified later as TNF.

---

[a] CAT was acquired by AstraZeneca in 2006 for £702 million and combined with MedImmune (which AZ acquired shortly afterwards).

[b] Beutler shared the 2011 Nobel Prize in Physiology or Medicine for innate immunity research.

Cerami was the first to propose that mAbs to TNF might have anti-inflammatory properties. He published a paper with Kevin Tracey in *Nature* in 1987, showing that anti-TNF antibodies prevented septic shock in baboons.[1] They had also filed patents on the idea before publication. These patents had a similar history to the Cabilly patents, and the IP ended up getting extended life. Mark Bodmer at Celltech, who heard about the work six months or so before it was published, was one of the first people employed by a biotech company to recognise the importance of this finding and immediately started immunising mice with human TNF to make mAbs from their spleens.[2] This led directly to the first ever study of an anti-TNF antibody in humans, published in the *Lancet* in May 1990.[3] It was not an elegant study, but it was extremely important because it broke, to quote Mark himself, "a taboo that systemic inhibition of TNF would never work."[c] Marc Feldmann, at the Charing Cross Sunley Research Centre near London, had shown that anti-TNF prevented IL-1 production in synovial cells from rheumatoid arthritis (RA) patients just a few months earlier.[4]

James (Jim) Woody at Centocor, who had been a postdoc in the Feldmann lab, and Steve Gillis at Immunex also recognised the importance of these papers. Centocor (see Chapter 4) obtained the chimeric antibody against TNF (infliximab, or Remicade) via a license from New York University (NYU), followed by the in-house development of golimumab (Symponi), a fully humanised anti-TNF antibody. The proof-of-concept (POC) clinical trial for the anti-TNF antibody was done initially in 10 patients in the summer of 1992. All patients responded to 20 mg/kg of infliximab, infused several times over two weeks. The responses were both clinical and biochemical (i.e., the mAb treatment reduced C-reactive protein). The positive results were disclosed in September 1992.[4] Infliximab was approved for use in Crohn's disease and RA in the United States in 1998 and in Europe in 1999. It is marketed by Janssen as "Remicade" and has returned billions of dollars in sales to its parent Johnson & Johnson over the succeeding years.

Pat Gray, who had left Genentech and was working in the Feldmann lab, somehow managed to clone one of the TNF receptor genes using amino acid sequences from which oligonucleotide probes were constructed.[5] The cloning of the TNF receptor led to the development of Enbrel at Immunex, a TNF receptor-Fc fusion protein (Chapter 3).

---

[c] For all the noise around TNF antibodies, it is strange that this paper is rarely cited.

Celltech's Anti-TNF Antibody

What about Celltech's anti-TNF antibodies? In partnership with Bayer, the mouse antibody was cloned and expressed in mammalian cells using the glutamine synthetase amplification system we had developed (Chapter 2), but it went nowhere. A humanised version (CDP 571), which was also part of the Bayer deal, did go into development for RA and inflammatory bowel disease (IBD), but the programme was ultimately returned to Celltech. CDP 571 was subsequently partnered with Biogen but did not survive there either.

Mark Bodmer at Celltech was asked one day what he wanted to do with some mice that had been immunised with human TNF some 18 months earlier. He suggested boosting them to see if higher-affinity anti-TNF mAbs could be derived from their spleens. It worked, with one antibody in particular (HTNF 40) having spectacular activity. Given the Bayer and Biogen history with this project, Celltech management was not in favour of any renewed programme. But Celltech had been making antibody fragments in *Escherichia coli* and they decided to use HTNF 40 as a source of heavy and light chains for these experiments.[2,d] After the acquisition of Celltech by UCB, this antibody became the basis for CDP 870—certolizumab pegol—a pegylated Fab that neutralises TNF. This mAb (called Cimzia) was approved for use in RA, IBD, and other inflammatory conditions. It has been very important for UCB, with billions of dollars of sales since its approval in 2008, making UCB a continuing legitimate player in the anti-TNF mAb space.

Anti-TNF antibodies are important tools for the treatment of almost all autoimmune diseases, including rheumatoid arthritis and inflammatory bowel disease. Humira was the world's first fully human antibody therapeutic, and AbbVie's aggressive marketing campaigns assured that Humira became the world's top-selling drug, with sales of more than $20 billion in 2021 (see Chapter 18). Humira is clearly a great example of a blockbuster mAb drug. But it is far from being the only one.

CHECKPOINT INHIBITORS, PD-1 AND KEYTRUDA

Pembrolizumab, or "pembro" as it is commonly known scientifically (brand name: Keytruda), is a humanised mAb blocking the protein PD-1 on T cells. PD-1 interacts with PD-L1, its receptor protein, to prevent T cells from

---

[d] They had been very effectively humanised by Dee Athwal, a former PhD student of mine.[2]

moving in to kill tumour cells. Pembro essentially blocks this interaction, freeing up T cells to attack.

PD-1 and PD-L1, members of a class of molecules called "checkpoint control proteins," are just one example of checkpoint control proteins to which new antibodies have been and are being directed.[e]

Keytruda works in a variety of cancers, especially those with a high tumour mutation burden (e.g., non–small cell lung cancer [NSCLC] in smokers and in melanoma), but it does not have universal efficacy. By February 2022, Keytruda was under evaluation in more than 5000 clinical trials and had been approved in multiple indications.[6] In 2023, the large number of approved indications (alone and in combinations) is one of the reasons why Keytruda has replaced Humira as the world's most important—and valuable—therapeutic mAb.

A Serendipitous Start

The story of the development of Keytruda is a good example of how serendipity plays a part in drug discovery. Some of the success of Keytruda can be attributed to the good judgement and leadership of Peter Kim and subsequently Roger Perlmutter, who were R&D chiefs at Merck from 2003 and 2013–2021, respectively. But it was really a comprehensive team effort over several years.

The original mouse mAb to PD-1 was made by Organon, a Dutch company based in Oss in the Netherlands, with research labs in Scotland in the United Kingdom and in Cambridge, MA. Organon had a programme to find antibodies that would block T cells in autoimmune disease by activating PD-1. Unfortunately, they did not find good agonist (activating) antibodies, which are generally hard to find. But they did find plenty of very active PD-1 blockers that they imagined might be useful in cancer by modulating the activity of T cells.[f]

The Organon team set to work around 2006 to make a humanised version of their highly potent and selective blocking antibody, in collaboration

---

[e]Many of which were cloned by Immunex—see Chapter 3.

[f]This was well before Jim Allison, the 2018 Nobel Prize winner for his studies on T cells, identified anti-melanoma, anti-CTLA4 antibodies, which were shown to activate T cells and to have antimelanoma cell effects in animals and subsequently in humans.

with the MRC Technology (MRCT) labs.[g] In 2006, the MRCT already had a proven track record in antibody humanisation from work at LMB and at Celltech, successfully humanising more than 30 antibodies. The MRCT used complementarity-determining regions (CDR) grafting (see Chapter 4, Box 3) to humanise the murine anti-PD-1 mAb from Organon. As part of the Organon contract, the MRCT had the foresight to include a royalty on future product sales of the humanised antibody. In 2016, the MRCT sold a slice of its rights to DRI Capital,[7] resulting in the realisation of $150 million that could be funnelled into new research projects.[h]

While the MRCT was still humanising the antibody, Organon was bought in 2007 by Schering-Plough for their women's health business for $14.4 billion. This acquisition included not only the PD-1 programme but also a BTK inhibitor that turned out to be more selective and more potent than ibrutinib (Imbruvica), synthesised originally by Axys chemists (see Chapter 7) for the treatment of chronic lymphoblastic leukaemia. In 2009, Merck bought Schering-Plough for $41.1 billion. After the PD-1 antibody programme wound up at Merck in 2009, it was considered such a low priority that the asset was placed on the "to be out-licensed list," a.k.a. "put on the shelf."

Medarex Shines the Light on Checkpoint Inhibition

BMS described very compelling antitumour effects in humans with both Yervoy (ipilimumab, an anti-CTLA-4 antibody) and nivolumab (nivo, brand name Opdivo, their own anti-PD-1 antibody). Both programmes came from Medarex, which BMS acquired in 2009 for ~$2 billion as part of their programme to develop antibodies that activated T cells using

---

[g] These labs were originally set up as the MRC Collaborative Centre (MRC CC) in the late 1980s and were housed in the old ICRF Laboratory of Environmental Carcinogenesis that my father ran in the late 1960s on Burtonhole Lane, next to the National Institute of Medical Research in Mill Hill near London. The first Director of the MRC CC was Chris Hentschel, who had worked with us all at Celltech and had been in Charlie Weissman's lab in Switzerland before that as a postdoc.

[h] To do this effectively, MRCT became LifeArc, now a venture firm dedicated to providing financing for new companies and some select academic projects. LifeArc has a new lab in Stevenage near the GSK campus. Their antibody engineering service has successfully humanised more than 65 antibodies, including four that are now medicines on the market and five in clinical trials within pharma pipelines.[8]

humanised mice. The deal involved $200 million upfront in cash, $400–500 million in lieu of milestones, and the rest for other antibody assets.

Medarex was set up in 1987 by a group of Dartmouth immunologists financed by Donald Drakeman at Essex Vencap, who became their initial CEO.[i] Alan Korman and Nils Lonberg made MDX-010, a humanised anti-CTLA-4 mAb. The Medarex team also used their humanised mouse platform to make nivolumab/Opdivo, which was being tested in three clinical trials in Japan in collaboration with Ono, a small Japanese pharma company.[9]

The Medarex acquisition very nearly did not happen. Jeremy Levin was the Head of Business Development at BMS at the time, having previously led business development at the Novartis Institutes for BioMedical Research (NIBR) under Mark Fishman. Levin thought that antibodies for the immune modulation of cancer were going to be a big deal, so he went out on a limb to persuade the executive team at BMS to buy Medarex "at all costs."[10] He was not wrong: the return on the Medarex acquisition amounted to well over $120 billion and set the scene for the development of all the checkpoint inhibitor antibodies and was also very important for cancer patients.[j] This windfall also put BMS in a position to buy Celgene in 2019 for ~$75 billion.

Ipilimumab/Yervoy was FDA-approved in 2011 for the treatment of refractory melanoma, following publication in the *New England Journal of Medicine* (*NEJM*) in 2010 of the successful phase III study, marking the beginning of the new therapeutic modality of checkpoint inhibitors.[11]

Merck heard through the grapevine that the nivolumab results were encouraging, which woke up the sleeping giant. They dusted off pembroluzimab (pembro/Keytruda) and rapidly moved it towards clinical trials, even though they were some years behind BMS and nivolumab. A particularly defining difference in the Merck programme was to select patients in their initial trials using a biomarker, in this case patients whose tumours expressed high levels of PD-L1 on their cells. They surmised that these would be the most likely to respond to the PD-1-blocking antibody. Merck reactivated Investigational New Drug (IND) development in mid-2010, targeting a filing by the end of the year. They duly filed the IND application in December, and the first clinical study began to enrol patients in early 2011.

---

[i] Genmab, run by Drakeman's wife, Lisa Drakeman, was spun out of Medarex as their European arm.

[j] More than half of the oncology clinical trials being done now involve checkpoint inhibitors in one form or another.

Exciting First-in-Human Studies

The early results from the melanoma cohort were particularly striking. Six out of the first seven patients enrolled had measurable responses, leading to an expansion cohort in melanoma. The study eventually included more than 600 patients with metastatic melanoma and a similar number of patients with lung cancer. It became one of the largest phase 1 trials ever done in oncology.[12] Merck also applied for and obtained "Breakthrough Designation" for Keytruda, enabling "fast track" communication of clinical trial data with the FDA.

The results for Keytruda and Opdivo for first-line NSCLC were reported in 2016. Merck's study was strongly positive, whereas the BMS study showed more equivocal results. Using the biomarker of PD-L1 levels had allowed Merck to prioritise which tumours to target, including head and neck squamous cell carcinomas (HNSCCs), bladder cancer, gastric cancer, and triple-negative breast cancer. Meanwhile, Bert Vogelstein and colleagues at Johns Hopkins noticed that Keytruda worked most effectively in patients with tumours with high mutation rates, such as NSCLC and melanoma. This suggested that it should also work in Lynch syndrome, a form of colon cancer with high mutation rates in key DNA repair genes called "MSI high tumours" more common in early-age-of-onset colon cancer patients. This observation further expanded the footprint of this mAb in clinical oncology.

Keytruda received its first approval in 2014 for advanced melanoma. It is now available in more than 95 countries for an expanding range of cancer indications. Merck is making very considerable returns from their Keytruda franchise.[k]

mAbs FOR ALZHEIMER'S DISEASE

Treatment of Alzheimer's disease (AD) with mAbs is a scintillating prospect from both a medical need and a commercial perspective. This possibility has been on the docket for both biotech and pharma for some time, but, until recently, with relatively little success.

β-Amyloid and the Amyloid Hypothesis

The amyloid hypothesis states that it is the accumulation of β-amyloid that causes AD. It is not wrong. Much of the evidence that still supports the

---

[k] These mAbs are clearly not checkbook inhibitors!

hypothesis comes from studying the genetics of familial forms of the disease. As I have stated many times (to anyone who will listen) and I reiterate here, genetics never lies—it just does not tell you the whole story. Nowhere is this truer than in the genetics of AD.[1]

Normally, Aβ is cleaved from the amyloid precursor protein (APP) by the enzymes β- and γ-secretase and does not accumulate in the cell. In older people or in those with genetic predisposition to or familial forms of AD, two forms of Aβ (Aβ 1–40 and Aβ 1–42) accumulate in brain cells and form fibrils. These forms are neurotoxic, increasing the levels of another fibril-forming and toxic protein called Tau. Mutations in the APP near the β- and γ-secretase cleavage sites affect processing of the precursor protein and cause early-onset AD in families with those mutations. Some older people in Iceland who have no plaque in their brains and no AD have protective mutations in APP that reduce the levels of Aβ 1–40 and Aβ 1–42.[13] A duplication of the APP gene on chromosome 21 in Down's syndrome and mutations in the presenilin gene, a component of γ-secretase, both occur in families with early-onset AD. Considered together, these data directly imply that β-amyloid deposition causes AD. That said, there are people walking around with heads full of amyloid plaque who never get AD, so the amyloid hypothesis is clearly not the whole story.

β-Amyloid is readily detectable by imaging techniques in affected brains, and its detection and removal have been a therapeutic objective in AD for some time. The most common approach has been to use mAbs with various specificities against β-amyloid to remove the amyloid plaque or at least the forms of Aβ that cause plaque.

The Neurodegeneration Companies

Athena Neurosciences, based in South San Francisco, was one of the first of a group of biotechnology companies founded in the 1980s to pursue treatments for neurodegeneration (including AD). It was founded by Lawrence (Larry) Bock and Larry Fritz, who were connected to Avalon Ventures in San Diego (Chapter 7) in 1986. John Groom, an expat from European pharma, became the CEO, with Fritz as VP Research.

---

[1]Despite attacks from some quarters, and with apologies to Mark Twain, "rumors of the amyloid hypothesis's death are greatly exaggerated."

As well as having an interest in β-secretase inhibitors to reduce the processing of APP, Athena pioneered immunotherapy for AD and Parkinson's disease (PD). The initial approach taken by Dale Schenk and his team was to make a vaccine against AD by using a β-amyloid peptide as an antigen. Even though the approach worked in transgenic mouse models of AD, it caused brain inflammation in a phase 2 trial in humans and the programme was terminated.

Athena had in-licensed several drugs in the neurodegeneration space and had built a sales force and a pipeline of proprietary products including Antegren (natalizumab), an α-4 integrin mAb. The company was acquired in 1996 by Elan, an Irish drug delivery company, for $625 million. Elan would bring Antegren (now called Tysabri), a very important mAb for treating multiple sclerosis, to Biogen. They were also responsible for making the first mAb directly targeting β-amyloid, called bapineuzumab (Bapi). Partnered with Wyeth, it failed in the clinic and the programme was terminated in 2012.

Aducanumab

A research collaboration between the Swiss company Neurimmune and the University of Zürich led to the identification of protective anti-amyloid antibodies both in healthy elderly people and in patients with slowly progressing dementia. The genes coding for these antibodies were cloned and the proteins were found to bind directly to amyloid in the patient brain tissue. One of these antibodies was named "aducanumab," a high-affinity, fully human IgG1 mAb that binds to a specific site on the Aβ protein. Following intravenous administration, the mAb crosses the blood–brain barrier, binds to brain amyloid, and, in concert with the immune system, removes the amyloid, as seen by molecular imaging. Neurimmune licensed the antibody to Biogen, who developed it further for clinical use.

What followed in the development of aducanumab by Biogen can only be described as unfortunate. The data derived from the two key clinical trials of the drug in patients with mild cognitive impairment (MCI) were complicated by the way they were handled and analysed and the effects seen at different doses. It was further confounded by the side effect of "ARIA" (amyloid-related imaging abnormality, a sign of vascular oedema[m]) that occurred at higher doses.

---

[m] ARIA was first reported in the earlier clinical trials of Bapi.

The equivocal nature and interpretation of the clinical data for aducanumab led to a polarisation of views. On the one hand, there were Biogen and Eisai (with whom Biogen was collaborating), the AD patients and advocates, and even some people at the FDA, supporting approval for this critical need. On the other hand, there were the FDA expert advisory group and key opinion leaders, who quite justifiably could not see that the drug worked from the initial data and analyses. It was clear that Aβ was removed, but this was not necessarily followed by any gain in cognitive function as measured by a multitude of tests in *most* treated patients, regardless of dose (although there was some slowing of cognitive decline seen in *some* patients).[14]

In early 2019, Biogen discontinued recruitment into the global phase 3 clinical trials, ENGAGE and EMERGE, which were designed to test aducanumab in patients with MCI. Biogen also discontinued the EVOLVE phase 2 trial and the long-term extension PRIME phase Ib trial. An independent data monitoring committee, in a predetermined "futility" analysis, had concluded that the trials were unlikely to hit their primary endpoint.

However, with 11 more weeks of blinded data available, Biogen reanalysed the data set and on the strength of that new analysis decided to approach the FDA for regulatory approval. The FDA found that the phase 3 EMERGE trial did meet its primary endpoint, as it showed a decrease in cognitive impairment, at least at one dose of the drug. Unfortunately, some of the interactions that Biogen had with the FDA over this effort were perceived as lobbying for approval, and not deemed helpful (although the neuro division at the FDA was, in fact, very supportive of Biogen).

Despite all the brouhaha, the FDA approved aducanumab in June 2021. The pricing of Aduhelm (the brand name it was given) at $56,000/year for the treatment as an i.v. infusion every four weeks, when the drug only worked in a subset of patients, was perhaps another misstep.[n]

Lecanemab

The story is even more interesting, because the Eisai mAb (lecanemab) against β-amyloid showed some reasonably good (and unexpected) responses in the phase 3 trial that Eisai was doing with Biogen.[o] Lecanemab is

---

[n] It turned out to be a big mistake, actually, as it resulted in very little use of the drug, owing to the lack of reimbursement by the insurance companies and Medicare.

[o] Lecanemab was part of the Eisai–Biogen mAb deal that included aducanumab.

a humanised mouse mAb that binds preferentially to amyloid protofibrils, preventing amyloid deposition, at least in mice. Not only did lecanemab antibody clear β-amyloid, but it also slowed cognitive decline by 27% in more than 1700 patients. On the strength of these data, lecanemab was FDA-approved in 2023 for AD patients with MCI. It is now being sold by both Eisai and Biogen as Leqembi, a biweekly infusion treatment. Time will tell whether it is cost-effective, but the commercial opportunity is clearly very sizeable given the large—and growing—numbers of AD patients with MCI.

Other β-Amyloid-Binding mAbs

Several other mAbs that are β-amyloid-specific have been tested clinically or are in preclinical development.[15] The most promising of these were gantenerumab from Chugai (Roche–Genentech), a fully human IgG1 antibody that binds with high affinity to a specific site on Aβ fibrils targeting plaque, and donanemab from Eli Lilly, which primarily targets the insoluble forms of β-amyloid by binding to an N-truncated pyroglutamate in the Aβ protein.

Unfortunately, data released for gantenerumab in November 2022 showed that the Roche–Genentech mAb failed to meet the predetermined cognitive end points. In contrast, in June 2021, the FDA granted Lilly's donanemab breakthrough therapy designation and Lilly submitted a licensing application in October 2021, under the same accelerated approval pathway that was used for aducanumab, with rolling submission of trial data.

Lilly released their phase 3 results for donanemab in July 2023 and the results have been published in the *JAMA*.[16,17] According to the Lilly press release, "Donanemab significantly slowed cognitive and functional decline for amyloid-positive early symptomatic Alzheimer's disease patients, lowering their risk of disease progression: nearly half of participants at earlier stage of disease on donanemab had no clinical progression at one year." These data would suggest that although the Lilly mAb is perhaps more potent than lecanemab, with a similar reported safety profile, it may cause more ARIA than lecanemab and does induce antibodies that neutralise the mAb—the so-called antidrug antibodies or ADAs. Nevertheless, it seems very likely that donanemab will be approved soon.

The observation that some of the anti-β-amyloid antibodies that do remove β-amyloid do not modify the disease does not, as some people insist, invalidate the amyloid hypothesis. As stated above, the amyloid hypothesis is about causation and not about cure. The fact that amyloid plaque is part of the cause of the disease does not mean that removing it will necessarily reverse the pathology. The damage may already have been done early, kicking off a cascade of neuroinflammation leading to irreversible neurodegeneration and dementia.

Other approaches to treatment will likely be necessary. SV Health Investors started a specific venture fund for investing in AD and other neurodegenerative diseases called the Dementia Discovery Fund (DDF)[p] (see Chapter 18). They have made investments in several new companies dedicated to the further understanding of the pathology of Alzheimer's and other neurodegenerative diseases and the development of new drugs to treat them.[q]

mAbs for Neuroinflammation

Alector, the first company that the DDF invested in, is developing mAbs to proteins known to be involved in the neuroinflammation that precedes AD. Alector was co-founded in 2013 by Adimab founder Tillman Gerngross, ex-Genentech scientist Arnon Rosenthal, Asa Abeliovich, and Erikk Anderson. OrbiMed and Polaris Ventures led the series A and B rounds of financing. They formed an alliance with Adimab to make the appropriate antibodies.[r]

Following additional financing rounds, Alector formed a partnership with AbbVie in 2017 around their mAb (AL002) that targets TREM2, a genetically validated protein involved in the early stages of AD (see Box 1). Alector's second mAb programme (AL001) targets the up-regulation of progranulin, reductions in which contribute to the pathology of frontotemporal lobe dementia (FTD), a neurodegenerative disease related to amyotrophic lateral sclerosis (ALS) that leads to behavioural changes and language difficulties in patients with the disease. AL001 (and a newer antibody AL101) elevate levels of progranulin by blocking the sortilin receptor responsible for progranulin degradation. AL001 is currently in a pivotal

---

[p] I am an advisor for them.

[q] The failure of mAbs directed to Aβ and Tau for AD (and α-synuclein for PD) to have disease-modifying activity so far may lead to a resurgence of interest in small molecules that enhance cognition as a therapeutic intervention. There are acetylcholinesterase inhibitors like Aricept, rivastigmine, and galantamine that increase acetylcholine levels at nerve synapses to promote memory, and N-methyl-D-aspartate (NMDA) receptor antagonists like memantine that enhance cognition.

[r] Adimab founded in 2007 (and another SV investment) is now one of the best companies in the world at making specific mAbs.

BOX 1. mAbs TO TREM2

AL002 is a humanised IgG1 mAb that binds TREM2 on microglia and activates signalling. It works by increasing phosphorylation of the kinase syk, which lies downstream of TREM2 and induces the proliferation of microglia. There is some competition in the TREM2 space. Vigil Neuroscience, an Atlas Ventures–sponsored Cambridge (MA)-based company, was founded in 2020 with a $30 million series A round and recently completed an IPO raising another $98 million. The company licensed intellectual property for TREM2-targeting molecules from Amgen in return for equity in the company. Vigil is also targeting TREM2 to obtain POC in a rare Mendelian disease called adult-onset leukoencephalopathy (ALSP) that affects about 10,000 people worldwide. ALSP is caused by loss of function mutations in colony-stimulating factor 1 receptor that shares the syk downstream signalling pathway with TREM2.

---

phase 3 clinical study in people at risk for or with frontotemporal dementia owing to a progranulin gene mutation (FTD-GRN). This mAb called latozinemab has received fast track designation and is central to Alector's recent deal with GSK.

PSORIASIS

Anyone who watches morning or evening television is probably aware that there are multiple mAbs now available to treat psoriasis and psoriatic arthritis.[s] Secukinumab (Cosentyx), which was approved in 2015 for Novartis, and Ixekizumab (Taltz) from Eli Lilly both target IL-17. Brodalumab (Siliq) from Amgen and AstraZeneca targeted the IL-17 receptor, but these companies decided not to move forward with this drug, owing to its potential side effects and the competition. There are several antibodies specific for anti-IL-23/IL-39, including Risankizumab (Skyrizi) from AbbVie, which is also being used to treat multiple inflammatory diseases.[16,t] Ustekinumab (Stelara) is an anti-interleukin IL-12/IL-23 antibody.[18]

This intense competition in just this *one slice* of the inflammation market clearly illustrates the therapeutic and commercial importance of mAb therapy. In 2021, more than $110 billion were generated in revenue for mAb-based products, a number expected to reach ~$180 billion by 2025.

---

[s] Although most viewers probably do not know that they are mAb-based drugs.

[t] Skyrizi was being particularly heavily (obnoxiously to me) advertised by AbbVie, largely because their best-selling drug Humira now has no prevailing IP patent protection.

Twelve of the top 20 drugs in worldwide sales in 2020 were the so-called biologicals, amounting to 62% of the aggregate sales of those 20 drugs (see Chapter 18). It is a wonderful demonstration of how one of the foundational technologies of biotechnology has completely changed medical practice and, more importantly, patients' lives.

ACADEMIC VERSUS COMMERCIAL RESEARCH

mAbs and their discovery and exploitation put the chiaroscuro of academic and commercial research into sharp relief. The argument that pure academic research should not be soiled by commercial considerations raised by the discovery of mAbs seems like it should be a bit passé now. But it is not, and it still ripples underneath much of biotech history. My argument has always been that academic research is a privilege bestowed on society by the ability for society to pay for it, along with many other things like good health care and social security. Any way that academic science can return investment to the "exchequer" should be actively encouraged. This does not mean that "blue sky" research is forever doomed, and everything must be applied. But it does mean that all academic investigators are under an obligation to consider how what they are doing might benefit society and whether it can be applied somehow in the future. I do not mean just ticking the "relevance" box at the end of a grant application either or paying lip service to the idea, as is often encouraged. Do not forget it is the taxpayers (i.e., you and me) who are paying for the activity of these scientists. Remember too that in the not-so-distant past, it was only independently wealthy people who could indulge their scientific curiosity. As we will see in a bit, the genomics revolution and the Human Genome Project throw these issues once again onto the table in sharp relief.

REFERENCES AND NOTES

1. Tracey K, Fong Y, Hesse DG, Manogue KR, Lee AT, Kuo GC, Lowry SF, Cerami A. 1987. Anti-cachectin/TNF monoclonal antibodies prevent septic shock during lethal bacteremia. *Nature* **330**: 662–664. doi:10.1038/330662a0

2. Conversation with Mark Bodmer, 2021.

3. Exley AR, Cohen J, Buurman W, Owen R, Hanson G, Lumley J, Aukakh JM, Bodmer M, Riddell A, Stephens S, et al. 1990. Monoclonal antibody to TNF in severe septic shock. *Lancet* **335**: 1275–1277. doi:10.1016/0140-6736(90)91337-a

4. Monaco C, Nanchahal J, Taylor P, Feldmann M. 2014. Anti-TNF therapy: past, present, and future. *Int Immunol* **27**: 55–62. doi:10.1093/intimm/dxu102

5. Gray PW, Barrett K, Chantry D, Turner M, Feldmann M. 1990. Cloning of human tumor necrosis factor (TNF) receptor cDNA and expression of recombinant soluble TNF-binding protein. *Proc Natl Acad Sci* **87**: 7380–7384. doi:10.1073/pnas.87.19.7380

6. Upadhaya S, Neftelinov S, Hodge J, Campbell J. 2022. Challenges and opportunities in the PD1/PDL1 inhibitor clinical trial landscape. *Nat Rev Drug Discov* **21**: 482–483. doi:10.1038/d41573-022-00030-4

7. DRI capital: DRI was founded in 1992, at which time its name—Drug Royalty Corporation—left little doubt about its intentions. Since then, the investment group has picked up rights to royalties on a long list of drugs, including Alexion Pharmaceuticals' ($ALXN) Soliris, Biogen's ($BIIB) Tysabri, and Regeneron Pharmaceuticals' ($REGN) Eylea. In 2013, DRI put together a $1.45 billion fund to continue snagging. The model has proven attractive to investors and the biotechs, large drugmakers, and nonprofits who own a stake in royalties. Royalty Pharma, which was founded four years after DRI, pushed the idea into the mainstream consciousness in 2014 when it paid the Cystic Fibrosis Foundation $3.3 billion for the nonprofit's royalty stream from Vertex Pharmaceuticals' ($VRTX) Kalydeco. In making the sale, CF Foundation, like MRCT, traded future earnings for a payout it could start reinvesting immediately.

8. Conversation with LifeArc and LifeArc website.

9. Drakeman DL, Drakeman LN, Oraiopoulos N. *From breakthrough to blockbuster: the business of biotechnology.* Oxford University Press, New York.

10. Conversation with Jeremy Levin, 2021.

11. Hodi FS. 2010. Improved survival with Ipilimumab in patients with metastatic melanoma. *N Engl J Med* **363**: 711–723. doi:10.1056/NEJMoa1003466

12. Kang SP, Gergich K, Lubiniecki GM, de Alwis DP, Chen C, Tice MAB, Robin EH. 2017. Pembrolizumab KEYNOTE-001: an adaptive study leading to accelerated approval for two indications and a companion diagnostic. *Ann Oncol* **28**: 1388–1398. doi:10.1093/annonc/mdx076

13. Jonsson T. 2012. A mutation in APP protects against Alzheimer's disease and age-related cognitive decline. *Nature* **488**: 96–99. doi:10.1038/nature11283

14. Sevigny J, Chaio P, Bussiere T, Weinreb PH, Williams L, Maier M, Dunstan R, Salloway S, Chen T, Ling Y, et al. 2016. The antibody aducanumab reduces Aβ plaques in Alzheimer's disease. *Nature* **537**: 50–56. doi:10.1038/nature19323

15. Van Dyck CH. 2018. Anti-amyloid beta monoclonal antibodies for Alzheimer's disease: pitfalls and promises. *Bio Psychiatry* **83**: 311–319. doi:10.1016/j.biopsych.2017.08.010

16. Sims JR, Zimmer JA, Evans CD, Lu M, Ardayfio P, Sparks JD, Wessels AM, Shcherbinin S, Wang H, Monkul Nery ES, et al. 2023. Donanemab in early symptomatic Alzheimer disease: the TRAILBLAZER-ALZ 2 randomized clinical trial. *J Am Med Assoc* **330**: 512–527. doi:10.1001/jama.2023.13239

17. Widera E, Brangman SA, Chin NA. 2023. Ushering in a new era of Alzheimer disease therapy. *J Am Med Assoc* **330**: 503–504. doi:10.1001/jama.2023.11701

18. Chaplin S. 2019. Monoclonal antibodies for the treatment of plaque psoriasis. *Prescriber* **30**: 35–40.

CHAPTER 6

Pharma, Biotech, and Common Diseases

GLAXO GROUP RESEARCH

I went to Glaxo Group Research (GGR) from Celltech in 1989 to exploit biotech methods as a part of traditional drug discovery. I believed that the new recombinant methods of expressing human proteins could be used for screens, three-dimensional structure determination of the proteins, or to make recombinant organisms for drug screening and testing. I also wanted to learn something about how Big Pharma "worked" compared to biotech, and I was not disappointed.

My job as Director of Biotechnology at Glaxo was, initially, to oversee three groups: genetics, screen development, and the high-throughput screening operation itself. Screening was done weekly and consisted of putting natural product extracts and chemical compounds from the Glaxo library through various assays on behalf of all the therapeutic areas that we supported (infectious diseases, GI, cardiovascular disease, respiratory diseases, and neuropharmacology).

Before my arrival, Glaxo had developed not only ranitidine (Zantac) for gastric ulcers but also inhaled β agonists for asthma: both important products for common diseases.[a] Glaxo launched another five new common-disease treatments while I was there, including sumatriptan for migraine (a 5-HT1A agonist), ondansetron for nausea induced by chemotherapy (a 5-HT3 antagonist), and salmeterol, a long-acting β_2 adrenergic receptor agonist (LABA) used in the maintenance and prevention of asthma attacks and for chronic obstructive pulmonary disease (COPD).

---

[a] One of the highlights of my stint at Glaxo was a visit to Jurong in Singapore, where most of the world's supply of ranitidine was made (in a five-step chemical synthesis). Seeing blue plastic barrels full of raw material, worth millions of pounds, was something of an eye-opener.

The neuropharmacology groups at Merck and Glaxo in the United Kingdom were at least as good as the top university pharmacology departments at the time, as demonstrated by their productivity and pharmacological understanding.

Receptor Complexity

One of the paradoxical observations in applying recombinant DNA technology to pharmacology was that the pharmacological definition of a receptor in a tissue sample (e.g., guinea pig intestine) was gradually replaced by an understanding of the individual receptors based on their cloning and sequencing. All the well-known G protein–coupled receptor (GPCR) family of proteins—against which many drugs were directed—became simplified by understanding the function of each one molecularly. At the same time, the pharmacology became more complicated because tissues generally did not have just one receptor type on the cells in them, but many related ones, including dopamine receptors, serotonin receptors, adrenergic, muscarinic acetylcholine receptors, and many of the other GPCRs.

Screening Organisms

Many of the screens that we used at Glaxo were developed using recombinant enzymes or receptors, or specially designed recombinant organisms (e.g., yeast). In general, the screen was based on a colour change that would be mediated by a molecular circuit that contained a protein—the function of which was to be turned on or turned off by the screened compounds (see Cadus—see Box 4 in Chapter 4). The natural products that were screened came from extracts isolated from soil and plants around the world, which many scientists collected on their travels.[b] These were curated and stored in the extensive compound library that Glaxo, like many pharma companies, had assembled over the years. The makeup of the compound library was obviously biased, based as it was on the chemistry programmes that Glaxo had pursued in the past, and thus contained many compounds that were never going to be useful. Nevertheless, we did find many interesting compounds through our screening assays, but only a very small number ever amounted to anything.[c]

---

[b] It was quite legal to do that then.

[c] The far bigger challenge was managing the expectations of the different therapeutic area heads and setting priorities for getting screens online or offline. The strongest voice did not always prevail in that conversation.

Strategic Alliances

Before 1989, Glaxo had not formed any large strategic alliances with any biotech company, although they had the facility in Geneva that they bought from Biogen. It was only in 1990 that the late (and much missed by me) Allan Baxter and Leslie Hudson, I, and our teams started to look at opportunities to form strategic alliances with some of the newer start-ups in the United States. Three were important: an alliance with Regeneron on neurological growth factors, an alliance with Icos on the enzyme family of phosphodiesterases (PDEs), and an alliance with Gilead Sciences on antisense oligonucleotides (ASOs).[d]

REGENERON

Leonard (Len) Schleifer, a neurologist and assistant professor at Cornell University in Manhattan, founded Regeneron (the name was derived from "regenerating neurons") in Tarrytown, New York, in 1988. He wanted to start a company to focus on the development of nerve growth factor (NGF), following the observation that if administered to the brain, the protein could reverse cholinergic deficits. Len had worked with Alfred (Al) Gilman, who was awarded the Nobel Prize in Physiology or Medicine in 1994, for the discovery of GPCRs at the University of Virginia in Charlottesville. Ira Black and Len wrote a grant to get money to work on the cloning of the human NGF receptor, which got reviewed by Eric Shooter at Stanford (an NGF expert).[1] The grant was not funded, but Shooter suggested to Len that they start a company to use NGF and other growth factors therapeutically.[e] They raised $1 million from George Singh at Merrill Lynch to finance the start-up.[f] Fred Middleton of Sanderling Ventures, who had been the CFO at Genentech in the very early days, led a small $6–$7 million series A round. This was followed by a $10 million investment from Sumitomo Chemicals. They were interested in growth factors as they had an interferon on the market made in cell culture. Sumitomo was originally going to split the money

---

[d] I will cover Gilead Sciences and its founding in Chapter 12, where ASO technology is discussed further.

[e] Regeneron was also interested in brain-derived neurotrophic factor (BDNF) and ciliary neurotrophic factor (CNTF).

[f] Len had previously met with the Blech Brothers, who were fresh from their success with Genetic Systems and Oncogene (see Chapter 4), but he did not feel that they were the right people to invest in the company he wanted to create.

they invested between three academic labs, but when they found out that the three scientists who led those labs (Eric Shooter, Hans Thoenen, and A.J. Aguayo) were all on the Regeneron scientific advisory board, they decided to invest the money directly into Regeneron instead.

Thirty-Four Years and Counting

One day in 1988, Len Schleifer cold-called Ronald (Ron) Lindsay, a Scottish neuropharmacologist in the United Kingdom who had cell lines that were making BDNF. A five-minute conversation turned into four hours, the upshot of which was that Ron left the United Kingdom to join the new company.[1] George Yancopoulos, a first-rate molecular biologist from Columbia University and an alumnus from the Bronx High School of Science,[8] joined Regeneron in 1989. Len and George are still working at Regeneron together now nearly 35 years later.

Ron Lindsay and Mark Furth, with whom I worked briefly at Glaxo Research in Research Triangle Park later, were two of the first scientists in the Regeneron labs in the old Union Carbide Plastics lab in Tarrytown—called the Spine because it went over the highway. With George, they created a Genentech-like science culture by publishing their groundbreaking science frequently in highly cited journals, a trend that continues to this day.[2]

Regeneron did their first corporate deal in September 1989 in the form of a joint venture with Amgen to develop drugs for Alzheimer's disease (AD) and Parkinson's disease (PD). The Amgen deal supported 25 people in the Lindsay lab, cloning and expressing numerous neurological growth factors. By 1991, Amgen had invested $18.5 million in Regeneron. Regeneron went public that year, raising $99 million in an oversubscribed deal selling 4.5 million shares at $22. Led by Merrill Lynch, it was the second-largest biotech IPO ever after Cetus in 1981, giving Regeneron a post-money valuation of $248 million.

Glaxo and Regeneron

Glaxo's strategic alliance on BDNF with Regeneron started in 1990, the third corporate deal for Regeneron and one of the first that Glaxo did with a then-small biotech company. I remember my first visit there very well.

---

[8]This prestigious high school has provided biotech with several of its leaders and has eight Nobel Prize–winning alumni, although mostly in physics. Admission to Bronx Science is very difficult.

The facility was nothing like the huge campus that it is now, and I recall that it was raining heavily. Jonathan Knowles from The Glaxo Institute of Molecular Biology in Geneva (GIMB) had set the meeting up with Ron, who had recently given a seminar in Geneva. The goal of the Glaxo alliance was to screen a proprietary Regeneron cell line to look for BDNF receptor–activating small molecules. This cell line was sent to John Houston in my Glaxo biotechnology group for them to screen. Nothing much came of that work for which Glaxo paid $10 million to Regeneron.

Resilient Company

The word "resilience" applies well to both Regeneron as an organisation and to the two principals, Schleifer and Yancopoulos. They did not start out wealthy, but they are both super-wealthy now. It is too easy to forget that Regeneron's first years, rather like those of Ionis and other now-successful biotech companies, were filled with setbacks, including the failure of their first neurotrophic factor trial in 1997. For some start-ups that would have been the end of the story, but Regeneron had work going on in other therapeutic areas and, in some of these, they eventually prevailed.

Regeneron's leaders were not short on good advice. Their scientific advisory board (SAB) was stellar, including four Nobel Prize winners: Al Gilman, who shared the 1994 Nobel Prize in Physiology or Medicine for studies on GPCRs; Michael Brown and Joseph (Joe) Goldstein, who shared the Nobel Prize in Physiology or Medicine in 1985 for studies on cholesterol metabolism; and Arthur Kornberg, who shared the Nobel Prize in Physiology or Medicine in 1959 for his studies on DNA polymerase. Roy Vagelos, who had just left Merck as CEO, became Chairman of the Board of Directors and stayed in that role for many years.

George and his team had been "homology" cloning as many kinases and kinase receptors as they could find and had developed an interest in EGF and vascular endothelial growth factor (VEGF) in immunological diseases and cancer. One such effort was to build a protein called Rilonacept (Box 1) that "trapped" the pro-inflammatory cytokine IL-1 in a protein complex.

Innovation

Regeneron has a history of technological innovation. Their VelocImmune mouse platform enabled the rapid discovery of humanised mAbs (see

---

BOX 1. RILONACEPT

Rilonacept (brand name Arcalyst) is a dimeric fusion protein consisting of the ligand-binding domains of the human interleukin-1 receptor component (IL-1R1) and IL-1 receptor accessory protein (IL-1RAcP) linked to the Fc region of an IgG (Chapter 4). This "cytokine trap" concept (i.e., isolating pro-inflammatory molecules) was further elaborated with Aflibercept, a recombinant fusion protein consisting of the VEGF-binding domains of the human VEGF receptors 1 and 2, fused to an Fc region. This product was developed in the early 2000s in collaboration with Bayer, and the drug Eylea for treating eye disease was approved in 2011.

Chapter 4). Many of Regeneron's mAbs were made by this process, including an anti-IL6 mAb (sarilumab, brand name Kevzara) developed in conjunction with Sanofi-Genzyme for the treatment of rheumatoid arthritis and FDA-approved in 2017.[h]

Regeneron also fully embraced the use of human genetics as a driver for the development of new medicines. The very large-scale human genome sequencing initiative that they set up with Geisinger, to look for the rare mutations predisposing to common diseases, was well ahead of its time. They have now sequenced more than 2 million genomes.[3] The discovery and clinical development of the IL-4/IL-13 inhibitory mAb dupilumab, which binds to the IL-4 receptor α subunit and became the blockbuster drug Dupixent as part of the Sanofi collaboration, was informed by human genetics. In this case, it was the realisation that there were genetic variants in the IL-4/IL-13 pathways (mediated by the shared IL-4 receptor α chain) involved in atopic dermatitis and asthma. Sequencing data is also driving Regeneron's initiatives in weight loss—specifically the role of a GPCR called GPR 75 in controlling metabolism.[4]

But nowhere is this genetic thinking more obvious than in Regeneron's development of alirocumab (Praluent), a mAb binding to the protein PCSK9 (proprotein convertase subtilisin/kexin type 9). It was first discovered in 2003 that gain-of-function mutations in this enzyme were responsible for familial hypercholesterolaemia in a French family.[5] Conversely—and most importantly—loss-of-function mutations in PCSK9 were shown to lead to very low cholesterol levels and a low incidence of cardiovascular

---

[h] This was the beginning of a long-term symbiotic relationship between the two companies.

disease.[6] PCSK9 was subsequently found to control the number of low-density lipoprotein receptors on the surface of cells.[i]

Amgen had also developed an antibody to the protein (evolocumab, brand name Repatha), and both Amgen and Regeneron filed for patent protection on their respective mAbs, ending up in U.S. patent litigation. In March 2016, a district court ruled that Regeneron's drug infringed Amgen's patents. Amgen obtained an injunction barring Regeneron and Sanofi from marketing alirocumab, but the litigation continued all the way to the Supreme Court. In May 2023, in a unanimous decision, the Court ruled in favour of Regeneron (i.e., they found that Amgen's patents were invalid). According to Regeneron, this ruling validates their long-standing view on this matter. It also "represented an unequivocal win for America's innovation economy, its scientists, and researchers."

Both Repatha and Praluent are available today after FDA approval in 2015, but in different markets. But the prevailing issue for both is not the IP: it is that they are probably most appropriate for use in the smaller market of patients with familial hypercholesterolaemia rather than as substitutes for the HMG-CoA reductase inhibitors like Lipitor and Crestor used widely for high-cholesterol levels in the general population.

Sanofi Deal

Regeneron announced an extended collaboration with Sanofi in 2015 to discover, develop, and commercialise new immuno-oncology drugs. The deal included $640 million upfront for Regeneron and $750 million for proof-of-concept data, plus $650 million for the development of the anti-PD1 checkpoint inhibitor cemiplimab (brand name Libtayo, approved in 2018).[j] Sanofi brought money and commercial clout to the table, whereas Regeneron provided the pipeline reinforcement that the French company desperately needed.[k] Elias Zerhouni, who had been Director of NIH, was hired as the

---

[i] These receptors bind to low-density lipoproteins (LDLs), which bind to cholesterol in the blood. The number of LDLs on the surface of liver cells determines how quickly cholesterol is removed from the bloodstream. PCSK9 cleaves LDL receptors before they reach the cell surface, so more cholesterol remains in the blood.

[j] In addition to Libtayo, this collaboration also led to the development of the important mAb medicine dupilumab (Dupixent) for atopic dermatitis (approved in 2017) and sarilumab (Kevzara) for rheumatoid arthritis.

[k] Owing to its history of trying to grow by acquisition rather than robustly investing in R&D.

President of R&D for Sanofi. He was well suited to the task as he was familiar with R&D and what it meant. He is also Algerian and speaks French, giving him the advantage of understanding the way that business is done "en France."

In 2019, Regeneron Pharmaceuticals was announced as one of the best stocks of the 2010s, with a total return of greater than 1800%.[7] Regeneron is perhaps best known publicly for their development of an antibody cocktail for the treatment of COVID-19 infection, which saved many lives at the beginning of the pandemic (see Chapter 16).

From their humble beginning in 1988, Regeneron is now an independent biotech company of more than 10,000 people with more than $10 billion in annual revenue. Regeneron does not have many strategic alliances with small companies but has formed relationships recently with Intellia and Alnylam, and they acquired Decibel for their gene therapy assets in July 2023. Regeneron is a biotech success story writ large driven by resilient, determined, and rather talented people.

ICOS

Icos was an important alliance for Glaxo and was a deal that was done in 1991 shortly after I arrived there. I knew Icos from interactions with Chris Henney,[8] and I had interviewed at Icos early in their history, before going to Sequana, but the head of R&D position, which was what I was after, was not on offer. I thought the idea of working with the Icos team and living on Mercer Island to be quite an attractive proposition. The science became of interest to Glaxo through contacts with Leslie Hudson, who was running Research at the Glaxo RTP site and was an immunologist. He knew Chris from interactions in the United Kingdom.

Icos started in the offices of Stelios Papadopoulos at PaineWebber. In 1989, David and Isaac Blech came to Stelios' office in New York to discuss a new company they were planning to start with Bob Nowinski, with whom they had worked at Genetic Systems. Nowinski wanted to start a company to study the effect of rare retroviruses that infect brain tissue and are the potential cause of central nervous system (CNS) disorders. The name "Icos" was chosen to represent icosahedron, the shape of the capsid of many viruses.[1]

---

[1] Stelios remembers when Nowinski said jokingly he wanted to do business with him because he was the only person on Wall Street who knew what an icosahedron was, as Stelios had been a structural biologist and had worked on bacteriophages.

The Blech Brothers came back to Stelios more than once over the next few months. By this time, they had enlisted the participation of Chris Henney from Immunex to run R&D.

The financing plan was to raise money for the company utilising a subset of PaineWebber brokers who had a client base of high-net-worth individuals, who had made prior investments in biotech companies. In the course of conversation with the Blech Brothers, George Rathmann's name came up as an example of someone who could generate the necessary excitement with the PaineWebber client base. The Blech Brothers subsequently came back with the trio of Rathmann as Chairman, Nowinski as CEO, and Chris as R&D chief for Icos. Following the Stelios strategy in the spring of 1990, Icos did a very exclusive private placement reg D filing, with individual units selling at $75,000 each. The Icos placement memorandum was written almost entirely by Papadopoulos.[9]

Icos raised $33 million at an $18 million pre-money valuation, including obtaining a $5 million investment from Bill Gates.[m] A year later, PaineWebber took Icos public at double the private placement price.[9]

Enzyme Families

The focus of Icos' research turned from brain pathogens to the study of enzyme families. Unlike most companies that worked on protein kinases (enzymes that add phosphate groups to proteins), Icos elected to work on phosphodiesterases (PDEs). PDEs were known to control multiple physiological processes by affecting the levels of cyclic GMP (cGMP) or cyclic AMP (cAMP), the so-called second messengers that pass signals from the cell surface to the interior of the cell. It was clear to everyone that there were many more than the three canonical PDE enzymes described in the textbooks. The way to get after them was to clone all the PDE-related genes, express and isolate the whole protein family, and test them to see what enzyme activity and substrate specificity each member had.[n]

Nowinski and Henney had already earmarked several great scientists to join the team, and they brought them on as soon as the financing closed. Ken Ferguson joined from Mike Wigler's lab at Cold Spring Harbor, with

---

[m] This was Gates' first biotech investment. He also joined the Icos board.

[n] Broadly speaking, the enzymes have specificity for either cGMP or cAMP and are tissue-specific.

whom Icos formed a collaboration. They recruited Michael (Mike) Gallatin from Stanford, Tom St. John from the Fred Hutch, and James (Jim) Hicks an ex–Cold Spring Harbor Laboratory yeast geneticist. They were very accomplished scientists. Icos rapidly grew to 70 people. Glaxo and the team, including me, met them in their new Bothell laboratories in 1992 to set up the PDE alliance led by Ken Ferguson.

Ultimately, this deal—which provided Glaxo with insights into this enzyme family and some compounds—was terminated by Glaxo CEO Richard Sykes, and the rights to all the PDE drugs included in the deal reverted to Icos. As it happened, one of these was a PDE5 inhibitor that had been developed in France by Glaxo France called tadalafil, which became known as Cialis, used to treat erectile dysfunction. Sykes apparently did not feel that Glaxo should be in the business of providing drugs for this indication, despite it being part of the Icos deal. Cialis was licensed to Eli Lilly and was the main reason Lilly finally bought Icos in 2007 for $2.3 billion.

OTHER PEPTIDES, PROTEINS, AND PROTEIN FAMILIES

Glaxo flirted with several other San Diego–based biotech companies, including Amylin and Ligand Pharmaceuticals (both founded in 1987), but no strategic alliances were built with either, although both companies went on to some success.

Amylin

Amylin Pharmaceuticals was founded and led initially by Ted Greene after his Hybritech success.[o] It was based on the biological properties of a small peptide hormone (amylin),[p] which is secreted from pancreatic islet cells along with insulin but in much smaller quantities. Amylin formulated the peptide so that it would work as a therapeutic in diabetes patients who needed better control of their blood sugar levels. The company raised $56 million in an IPO in 1992, led by Morgan Stanley and Alex Brown. However, the drug (pramlintide, brand name Symlin) was difficult to develop and showed only modest effects in diabetic patients, leading to a cratering of the Amylin

---

[o] Amylin's R&D was led by Timothy (Tim) Rink (who had worked with George Poste at SmithKline Beecham), who subsequently became CEO of Aurora, the screening company set up by Kevin Kinsella and Avalon Ventures that was sold to Vertex.

[p] Amylin was discovered by scientists at the University of Oxford in the United Kingdom.

stock price. Joseph (Joe) Cooke, a former Lilly executive and Amylin Board member, took over as CEO in 1998 and restructured the company.

Amylin needed to reduce their space from 120,000 sq ft to 30,000 sq ft as they had reduced the workforce by 75% during the restructuring. Cooke asked Joel Marcus, the CEO of Alexandria Real Estate Equities (ARE) (see Chapter 18) and the landlord of Amylin's La Jolla space, to help. ARE duly restructured the leases and the company continued with a smaller footprint. In exchange, ARE bought stock in the subsequent refinancing at 50 cents/share.

It was a struggle, but Symlin was approved eventually for diabetics with poor glucose control (in 2005, nearly 20 years after the founding of the company). At that time, Amylin had also in-licensed exendin-4 (a peptide from the venom of the Gila monster), which had a mechanism of action like glucagon-like peptide-1 (GLP-1), a protein that regulates both insulin and glucagon levels. Amylin developed exenatide, a synthetic version of exendin-4, which led to an agreement with Eli Lilly for $325 million to help to develop it. In May 2005, Byetta (the brand name for exenatide) was approved in the United States.

The approvals of Symlin and Byetta led to the eventual BMS acquisition of Amylin in 2012. It provided a tidy return to its investors, including ARE. It did not, curiously, lead to BMS being anywhere in the current weight loss business now dominated by Novo Nordisk with their GLP-1 receptor agonist semaglutide (Wegovy, Ozempic) and Eli Lilly with tirzepatide (Mounjaro), a GIP-GLP receptor dual agonist and the yet-to-come triple agonist (retatrutide) targeting the glucagon receptor in addition to GIP and GLP-1.[10]

Ligand Pharmaceuticals

Ligand Pharmaceuticals was founded as Progenix in 1987 by Brook Byers from KP. The other founders included Howard Birndorf from Hybritech who was the initial CEO (succeeded by David Robinson in 1991). Ligand focused on developing drugs based on the nuclear hormone family of receptors in collaboration with Ron Evans from UCSD, an expert in that family of proteins. Nuclear hormone receptors, which generally control the expression of different genes, were known to be drug targets for common diseases. Robert (Bob) Stein, who had overlapped with Joshua (Josh) Boger (the founder and CEO of Vertex) at Merck, came to Ligand in 1990 to run R&D. Ligand went public in 1992. Ligand was not especially successful following this R&D

strategy. In 2007, with a new CEO, the company's strategy changed to one of acquiring candidate drugs and forming partnerships to develop them further. The company acquired Neurogen, Pharmacopeia, Metabasis, and CyDex. This resulted in a portfolio of some 60 new drug candidates. *Fortune* magazine ranked Ligand as one of the "top 100 fastest growing companies from 2014–2016." The journey was not, however, without some serious issues arising along the way. But it shows that you do not always have a drug discovery organisation of your own to create shareholder value.[11]

ONYX

Cancer collectively is a common disease and is the leading cause of death after cardiovascular disease. Several biotech companies were set up to exploit the molecular differences between cancer cells and normal cells by targeting, for example, known oncogenes as drug targets.

Onyx Pharmaceuticals was started in 1992 by Frank McCormick and Gideon Bollag following the merger of Chiron Diagnostics and Cetus. Kevin Kinsella, at Avalon Ventures in San Diego, was convinced that molecular oncology was ripe for exploitation in a small company. He approached Owen Witte, Harold Varmus, Peter Howley, Douglas (Doug) Fearon, and Douglas (Doug) Hanahan, all big names in the oncogene field, to form a company to go after the oncogene Ras and tumour suppressor pathways. Avalon put in sweat equity, and Kleiner Perkins (Brook Byers) and Institutional Venture Partners (Samuel [Sam] Colella and Nancy Kamei) invested some real money. Chiron also invested, gaining a 40% shareholding in what they then considered to be a Chiron subsidiary—which it clearly was not. Frank was CSO and Hollings Renton, who had previously been the President of Cetus and the President of Chiron, came on as CEO.[q] The company went public in 1996, and Frank left soon afterwards.

Onyx was important because it developed an effective anticancer agent that inhibited the Raf kinase associated with the Ras pathway, changes which are implicated in multiple types of cancer. By screening small-molecule libraries in a Raf kinase assay (and using other assays to check for selectivity), the Onyx team (led by Leonard [Len] Post) discovered compounds that were both potent and selective Raf kinase inhibitors. Frank and Nancy

---

[q]Most of the management equity was held by the CEO and the COO (William [Bill] Gerber), who was a friend of Renton. This was an issue for Frank.

Kamei, who had joined Onyx as VP of business development by then, secured a deal with Bayer by persuading George Scangos (before he went to Exelixis) that the Onyx compounds were superior kinase inhibitors to those discovered by Sugen[r] and Tularik.[s,12]

Bayer Deal

The Onyx–Bayer deal was a 50:50 co-ownership deal that funded 25 people in the Onyx lab for five years. The dark side of the Bayer deal for Onyx was that Bayer started their own R&D programme because they wanted to find their own compound, free from Onyx encumbrances. This ended badly, with Onyx taking Bayer to court in a highly risky jury trial in California.[t] Onyx prevailed.

A poison pill preventing Onyx from being acquired was put in place prior to the trial. That was vacated after the verdict, allowing Amgen to step in 2013 to buy the company for $10.4 billion primarily to acquire their Raf kinase inhibitor drug sorafenib (Nexavar). As part of the acquisition, Amgen inherited a cell-cycle inhibition project focused on CD4/6 kinase inhibitors sponsored by Parke-Davis in Ann Arbor (a company later acquired by Pfizer). This delivered a second kinase inhibitor drug called palbociclib (Ibrance).

Onyx also pioneered the use of oncolytic virus therapy for tumours where adenoviruses engineered to remove a gene called *E1A* could kill cancer cells in which the retinoblastoma (*Rb*) gene had been deleted (normally the E1A protein binds to Rb to prevent virus replication). This was a pioneering and important initiative, but it did not stand much chance of going forward in a big company like Amgen, and in fact it never went anywhere. Onyx is an example of a less well-known and understated yet highly successful biotech company. It produced several important new medicines.

---

[r] Sugen was founded in 1991 by Stephen Evans-Freke and oncogene kinase experts Joseph (Joe) Schlessinger and Axel Ullrich (from Genentech) from which, rather cornily, the name was derived. Sugen went public in October 1994. Their kinase inhibitor sunitinib (Sutent) was later approved for use in renal cell carcinoma, and they were acquired by Pharmacia in 1999 for $650 million.

[s] Tularik was a Genentech spinout led by David Goeddel, which was acquired by Amgen in 2004 for $1.4 billion with several anticancer drugs in clinical trials.

[t] This was "risky" because jury trials of this kind are notorious for giving the wrong verdict, based largely on ignorance.

OTHER 1980s AND 1990s SMALL-MOLECULE BIOTECHS

Neurocrine Biosciences

Neurocrine is today considered a successful company, with two drugs on the market in 2019 and a market cap of nearly $8 billion. But it was a roller-coaster ride to get there, a bit like the rise and fall and rebirth of Centocor, but more painful, as it happened over a more protracted period.

Kevin Kinsella and his team at Avalon Ventures in La Jolla (founders of Onyx, Sequana, Aurora Biosciences, and several other successful biotech companies) were founding investors in Neurocrine in 1992. The academic founders were Lawrence (Larry) Steinman, a neurologist at Stanford, and Wylie Vale, an endocrinologist at the Salk Institute. Kevin Gorman joined as head of business development in 1993 and became the CEO, a testament to his passion, commitment, and ability to manage through failure.

The company's original focus, when Genentech alumnus Gary Lyons was CEO, was on neuropsychiatric disorders based on corticotropin and corticotropin-releasing factor (CRF) antagonists, the expertise of their science group. They initially partnered this programme with Janssen (J&J), which enabled the company to go public in 1996, raising $34 million. At the same time, they formed a strategic alliance with Eli Lilly on CRF-binding proteins and later with GSK to identify CRF receptor antagonists.

But a deal in 2002 with Pfizer to develop the insomnia drug Indiplon, a novel nonbenzodiazepine hypnotic they had discovered, was the start of the real fun. Pfizer paid Neurocrine $100 million up front, with the potential to earn further milestones. By 2005, Neurocrine employed nearly 600 people and was worth more than $2 billion and had (as is often the case) built their own opulent R&D facility and headquarters on the Mesa near La Jolla. Unfortunately, after submitting their IND application to the FDA for Indiplon, the agency issued a "non-approvable letter" for the 15-mg dose proposed. True to pharma form, Pfizer nixed their deal.

The company resubmitted an IND application in December 2007 with data from 5-mg and 10-mg formulations in December 2007. But the FDA required yet more data, and Neurocrine decided to stop the programme in the United States.[u]

---

[u] They continued to develop and commercialise indiplon in Japan in partnership with Dainippon Sumitomo Pharma.

These failures resulted in several rounds of layoffs, selling and leasing back their building,[v] and an uncertain future. By late 2009, they had 10% of the employees they once had, Lyons had stepped down as CEO, and there was only about $50 million left in the bank.

Fortunately, Neurocrine found another asset in their hopper. Elagolix was an oral gonadotrophin-releasing hormone (GnRH) antagonist for pain relief that had succeeded in phase 2b studies. But the FDA again had issues with the data, particularly a new "pain scale" that Neurocrine had proposed. Although they had to go back to phase 2 testing with a modified end point, they were able to form a licensing deal with Abbott Laboratories that paid them $75 million up front with $500 million in potential milestone payments.

Further deals and better news followed. AbbVie ran the pivotal phase 3 programme for elagolix in 2015 and in July 2018 elagolix got FDA approval and became the drug Orilissa. In addition, valbenazine, a selective inhibitor of the presynaptic vesicular monoamine transporter type 2 (VMAT2), was approved under the brand name Ingrezza to treat tardive dyskinesia, a frequent side effect of some antipsychotic and anti-PD drugs.[w]

These recent successes, with the blockbuster Ingrezza leading the way despite competition from Teva, have given Neurocrine a new lease of life and a robust future.[x]

Cephalon

Cephalon was founded in 1987 by Michael Lewis, the late Frank Baldino, and James (Jim) Kauer (all of whom had worked at DuPont) to work on neurodegenerative diseases. The company was initially involved in research rather than development and worked with a lot of partners. They did not stick to their original research strategy, which led to their ultimate success.

Cephalon's major claim to fame was the development and commercialisation of modafinil (Provigil) and its chiral derivative Nuvigil, cornerstones of a cobbled-together franchise in sleep disorders and other indications. Cephalon was also known for Actiq, an opioid agonist, and the epilepsy drug tiagabine (Gabitril), a GABA re-uptake inhibitor. By buying ChemGenex

---

[v] They ended up vacating the front of their building that they had once owned in its entirety and were just using the back of it.

[w] You may have seen the rather awkward television advertisement for this drug.

[x] Neurocrine is also working with Voyager Therapeutics on gene therapy approaches to PD (see Chapter 13).

Pharmaceuticals in 2011, Cephalon acquired omapro, a natural product approved for use in chronic myelogenous leukemia (CML) patients who were resistant to the tyrosine kinase inhibitors targeting *BCR-ABL*, like imatininib (Gleevec). Teva Pharmaceutical Industries, an Israeli pharma company, acquired Cephalon for $6.8 billion in 2011 for their portfolio of products.

Cephalon is a good example of one of the rules of the biotech game: successful products make for successful companies and lead to good outcomes for both investors and patients. Another unwritten rule is that biotech companies that develop small molecules can be used to treat common diseases in large numbers of patients, whether it be in cancer or in another therapeutic area, almost always get acquired by Big Pharma. These kinds of drugs are the traditional "sweet spot" for pharma, fitting into their established development and marketing platforms. Whether there is a good return or not on the acquisition investment is difficult to determine and is obviously product- and company-dependent. The Teva–Cephalon acquisition was not one of the most successful.

BIOTECHS TRYING TO OUT-PHARMA BIG PHARMA

After the initial surge of biotech, it is becoming increasingly difficult to distinguish biotech companies from pharma companies. It is worth touching on some of the recently formed companies derived from pharma assets that have been "spun out" into smaller "biotech-like" entities or are biotech companies trying to do better what pharma traditionally has done.

Cerevel Therapeutics

Cerevel was formed at the end of 2018, as a partnership between Bain Capital and Pfizer, essentially to take over several of the CNS assets that Pfizer no longer wished to pursue. Most of these potential drugs had been tested in humans to a degree but had failed for reasons more to do with the way the clinical trials were set up than any inherent lack of quality in the drug or the science. So, the company started with a rich set of assets, significant financial support, and some naysayers.[y]

---

[y] I admit to being one of those who was somewhat sceptical of the idea of a biotech company based on Big Pharma failures. It seemed to me that the drugs that were the furthest advanced and had failed so far in the clinic were much less interesting than some of the earlier-stage assets that had not yet been into humans. But I was wrong in my assessment of the potential of Cerevel, recalling the Kevin Kinsella quote "it may be dog food but if the dogs love it... ."

Their most advanced programme, tovapadon, is a once-daily, orally bioavailable dopamine D1/D5 receptor partial agonist to treat PD motor symptoms. Although not a new concept, it has not been easy to get D1/D5 selective compounds and most drugs are directed to the dopamine D2 and D3 receptors. The data so far are encouraging, and the company is in phase 3 studies with this molecule.

Cerevel's second most advanced compound, darigabat, is a positive allosteric modulator (PAM) of the α 2/3/5 GABA-A receptor for the treatment of anxiety and epilepsy. Again, hardly a new concept, but the phase 1 trials are encouraging. Their third most promising candidate is emraclidine, a PAM for the muscarinic M4 receptor subtype, for the treatment of schizophrenic psychosis.[z]

Cerevel has recently run into some headwinds, having pushed their planned outcome reporting for several clinical trials into 2024. They also replaced their founding CEO, Tony Coles, with Ronald (Ron) Renaud from Bain.

Karuna Therapeutics

Not many of you will have heard of Karuna Therapeutics, but it is a small-molecule-focused company with a market cap of $8 billion and an interesting history. Andrew Miller founded the company in 2009 while he was at PureTech Ventures. Steven (Steve) Paul from Lilly was the original CEO but William (Bill) Meury (ex-Allergan) has recently assumed that role with Paul as CSO. The company's research in finding a better-tolerated schizophrenia treatment is based on the idea that you can ameliorate the peripheral side effects of centrally acting muscarinic *agonists* by co-administering selective, peripherally acting muscarinic *antagonists*. Their lead drug is a combination of xanomeline (a centrally acting M1/M4 muscarinic agonist) and trospium chloride (a well-known peripherally acting muscarinic antagonist used for overactive bladder) with the brand name KarXT. The company had a stellar IPO raising $250 million in the fall of 2019, and the results of the phase 3 EMERGENT 2 trial announced in August 2022 have pushed the stock firmly upwards. Further data released in March 2023 confirmed the efficacy of the combination, and Karuna plans to file for FDA approval sometime in

---

[z] It is unknown yet whether their compound will avoid the kinds of side effects seen in other muscarinic receptor inhibitors.

the second half of 2023. If approved, it will be the first time in years that a new treatment option for schizophrenia will be available. BMS announced in December 2023 that it would acquire Karuna for $12.7 billion.

Which deal would you have done—Cerevel or Karuna? (See Chapter 19.)

Amylyx

Amylyx is another, somewhat different, story of the development of a combination of two known compounds into a new formulation that can be tested and used therapeutically as a single drug: in this case, for amyotrophic lateral sclerosis (ALS), a disease where there are almost no effective options for patients.[aa] Since its shoe-string founding in 2013 by Justin Klee and Joshua (Josh) Cohen, undergraduates in the Biomedical Engineering department at Brown University, Amylyx has raised about $234 million.[bb] Morningside Venture Investments led a $30 million series B round in 2020 and Viking led the $135 million series C in July 2021. The company went public in an oversubscribed offering, just before the "window" closed in early 2022 raising close to $200 million in a Goldman–Sachs-led deal.[cc]

Their lead (and really only) drug candidate was AMX0035, an oral co-formulation of two active compounds, sodium phenylbutyrate (PB) and taurursodiol (TURSO). AMX0035 is proposed to reduce neuronal death by simultaneously mitigating endoplasmic reticulum stress and mitochondrial dysfunction. Despite some manufacturing and formulation issues, the company was able to show the safety of the combination in a phase 1 trial. In a placebo-controlled phase 2 study (CENTAUR) of ALS patients enrolled at the 25 top ALS treatment centres in the United States, the combo drug showed some apparent improvement on measures of physical function and disease progression in treated patients. Amylyx subsequently submitted a new drug application (NDA) to the FDA based on these rather early data.

[aa] There is a history to this strategy of making combinations of known drugs: most fail. Qsymia (a combination of phentermine and topiramate for weight loss) is a successful example.

[bb] Klee and Cohen are "co-CEOs," an arrangement that seems to work for now. However, I cannot see that lasting if Amylyx becomes a fully paid-up commercial company.

[cc] Maybe it helps to have Rudolph (Rudi) Tanzi, an MIT neurodegeneration scientist and TV personality, and Nobel Prize winner Wally Gilbert as advisors.

Although unusual but not unprecedented, this NDA submission based on phase 2 data may have been the right move for the company. Phase 3 studies with more patients and more heterogeneity tend to weaken efficacy signals seen in phase 2 trials. But the company is proceeding with further studies of the drug in ALS patients, despite the FDA granting priority review and the advisory committee supporting the approval of the drug for the treatment of ALS. Interestingly, the company initially said it would withdraw the drug if continued efficacy is not seen.[dd] History suggests that this is probably an empty promise. Relyvrio has been approved in Canada and is sold for CAN$160,000/year, and it was approved by the FDA in the United States in 2022 and launched at a price of more than $150,000/year.[ee] The European Medicines Agency has decided against approval, at least for now. Subsequently Relyvrio did fail in phase 3 and was withdrawn from the market.

It is interesting to contrast the development and financing of companies like Karuna and Amylyx that are based on simple concepts of old drugs being used in new combinations against all the new technology and new target companies that have been set up. The risk profiles are entirely different. First, it is all about clinical trial execution, and, second, it is clinical execution following the discovery of a decent target and the identification of a therapeutic modality: extra steps that all have additional inherent risks and take a good deal of time.

You will have surely noticed that there is a good deal of hyperbole across almost all "neuro" biotech. It is in that context I must touch briefly on both Denali Therapeutics and Verge Genomics, where the hyperbole reference applies in spades but in rather different ways.

Denali Therapeutics

Denali is a neurodegeneration-focused company in the Bay Area. It is a spinout from Genentech[13] in which Denali cofounders Ryan Watts,

---

[dd] It is not clear whether this will include long-term survival data. Regardless, whether the drug really works is an open question, and Amylyx must be held to their "promise" to withdraw the drug if it does not show benefit in the ongoing phase 3 trial. Once again, the FDA—with the best of intentions for patients—has put itself between a rock and a hard place.

[ee] For a drug for this disease with marginal positive effects in patients, this must be seen as an egregious price. ICER, a nonprofit that estimates the cost-effectiveness of drugs, has said that a fair price for Relyvrio (the name of the Amylyx drug) would fall somewhere between $10,000 and $30,000 per year.

Stanford University ex-president Marc Tessier-Lavigne, and Alexander (Alex) Schuth had worked together. Robert (Bob) Nelsen at ARCH Venture Partners was one of the prime movers for starting Denali. Nelsen talked with Douglas (Doug) Cole at Flagship and Stephen Knight at F-Prime for several months before investing in the company with the three cofounders.[ff] The company was formed with much hullabaloo in 2013, announcing its mission to find and use new drug targets based on the latest genetics of PD and AD.[gg]

In fact, Denali is going after all the old favourite genetically validated targets for PD, including glucocerebrosidase (GCase), implicated by mutations in the gene GBA 1 in the disease, and LRRK2, a programme for PD licensed in from Genentech. Despite their founding claims, these targets are not (and were not when Denali started) new news.[hh] In 2020, Biogen announced a $1 billion deal with Denali for their LRRK2 programme, which "looks promising" but has recently been reprioritised.[ii] Denali has also developed a clever system for delivering antibodies across the blood–brain barrier (Chapter 19).

Verge Genomics

Verge Genomics is a relatively new company formed out of Y Combinator, the start-up accelerator. It is one of the several companies that seems to believe that state-of-the-art artificial intelligence (AI) technologies, combined with large data management and analytical capabilities, will discover new drugs. Verge was set up to exploit this AI genomics approach in the neuro

---

[ff]Denali not only raised the most money in series A of any biotech company at the time (more than $200 million) but they also had the largest IPO of 2017, raising $250 million ahead of Ablynx, the other big IPO of that year.

[gg]When I first heard about Denali, I was expecting to see some exciting new targets. A conversation with them at the annual J.P. Morgan conference, when I was at Biogen, was frankly disappointing in that respect.

[hh]Denali (and many others) is also working on Huntington's disease, a dominantly inherited disease caused by a repeat expansion in the *huntingtin* gene (see Chapter 7). It is a difficult challenge because it is not appropriate to knock out both the expanded repeat copy and the normal copy of the gene: some Huntingtin protein is required for normal function. Both Roche/Ionis and Wave Therapeutics have reported failures of their ASO approaches to Huntington's disease for just that reason.

[ii]A scare at one point about liver toxicity of LRRK2 inhibitors was ameliorated by altering the dose.

space, focusing on AD, ALS, and PD in the hope that their computational algorithms will predict drugs that work.[jj]

Verge was founded in 2015 by Alice Zhang, who left the UCLA-Caltech MD/PhD programme to do so. Zhang did not see why the trial-and-error process of drug discovery, especially screening large libraries of compounds or mAbs to find specific molecules, should not be made a little less stochastic. I understand that frustration, but whether the consolidated analysis of genomics and proteomics and other of the numerous "omics" technologies will get you to a drug candidate faster or not is an open question.

Indeed, I find the hyperbole around the use of AI in discovering drugs stunning. At its most hyperbolic AI "will discover (and develop) the drug for you without you having to do that much of anything experimentally." In fact, AI is just another tool and technology to help control the uncertain process of drug discovery. The ability to predict the three-dimensional structure of many thousands of proteins, as has been done by AlphaFold and Rosetta (see Chapter 10), does not mean that new drugs will emerge like Excalibur from a lake of data. People sometimes forget that drug R&D, whether it be in biotech or pharma or even academia, is an experimental science. Although a computer tells you what might work, you still must test it in relevant experimental systems, both in cells and in animals. The results of those tests are often unpredictable, and without predictability you cannot design computer programs to predict, at least not yet. In time, when the data sets of cells, animals, physiology, pathophysiology, etc. are big enough and consistent enough, AI may be able to discover a good target and drug candidate (Chapter 19).

Hardly surprisingly though, Verge raised their initial $32 million series A in 2018 from Tech Investor DFJ that invested in Tesla, X formerly known as Twitter, and Skype. This is not your normal biotech investor, so the usual rules do not apply.[kk] Additional funding came from WuXi Apptec, Agent Capital, and the ALS Investment Fund.[13]

Using computational and other genomic approaches, an enzyme called PIKfyve (phosphoinositide kinase, FYVE-type zinc finger containing) has been implicated in the pathology of ALS and other neurodegenerative diseases. Verge's computational methods suggested that this might be a good

---

[jj] I strongly believe that they will not, but they will certainly help. Looking at human data to suss out why some drugs work for neurodegenerative and other diseases and some do not is a good place to start.

[kk] They are going to find out rapidly that the drug discovery time lines are distinctly different from the technology companies they are used to.

enzyme target for the treatment of ALS. They are deep into the development of a specific small-molecule inhibitor of the enzyme, with a compound different to an existing inhibitor of this enzyme. It is now in first in human studies.

Jane Rhodes, with whom I worked at Biogen, was the CBO of Verge Genomics. With her help, they successfully formed an alliance with Eli Lilly in July 2021 for various targets in the neuro space, bringing in $25 million up front, and in 2023 another deal, this time with AZ, with nearly $50 million up front. In December 2021, Verge closed an oversubscribed $98 million equity financing. This series B financing was led by BlackRock and included Eli Lilly, Merck Global Health Innovation (GHI) Fund, Section 32, and Vulcan Capital, along with Verge's existing investors. There is certainly nothing "artificial" or especially "intelligent" about the amount of money they have raised.

REFERENCES AND NOTES

1. Conversation with Ron Lindsay, an early Regeneron employee, 2021.

2. Maisonpierre PC, Belluscio L, Squinto S, Ip NY, Furth ME, Lindsay RM, Yancopoulos GD. 1990. Neurotrophin-3: a neurotrophic factor related to NGF and BDNF. *Science* **247**: 1446–1451. doi:10.1126/science.247.4949.1446

3. Dunn A. 2023. CSO George Yancopoulos talks Regeneron's next chapter of genetics, weight loss drugs, and Alzheimer's disease. *Endpoints,* August 13, 2023.

4. Akbari P, Gilani A, Sosina O, Kosmicki JA, Khrimian L, Fang Y-Y, Persaud T, Garcia V, Sun D, Li A. 2021. Sequencing of 640,000 exomes identifies *GPR75* variants associated with protection from obesity. *Science* **373**: eabf8683. doi:10.1126/science.abf8683

5. Abifadel M, Varret M, Rabès JP, Allard D, Ouguerram K, Devillers M, Cruaud C, Benjannet S, Wickham L, Erlich D, et al. 2003. Mutations in PCSK9 cause autosomal dominant hypercholesterolemia. *Nat Genet* **34**: 154–156. doi:10.1038/ng1161

6. Hall SS. 2013. Genetics: a gene of rare effect. *Nature* **496**: 152–155. doi:10.1038/496152a

7. https://www.cnbc.com/2019/12/13/the-best-performing-stocks-of-the-decade-it-all-starts-with-netflix.html

8. Many useful conversations with Chris Henney about Icos, 2020, 2021, 2022.

9. Icos Investment Memorandum: 4.5 million shares of common stock. Preliminary prospectus April 18, 1991. See also Icos form S1, filed with the SEC October 25, 1991.

10. Jastrebof AM, Kaplan LM, Frias JP, Wu Q, Du Y, Gurbuz S, Coskin T, Haupt A, Milicevic Z, Hartman ML. 2023. Triple-hormone-receptor agonist retatrutide for obesity—a phase 2 trial. *New Engl J Med* **389**: 514–526. doi:10.1056/NEJMoa2301972

11. See Wikipedia on Ligand Pharmaceuticals. https://en.wikipedia.org/wiki/Ligand_Pharmaceuticals

12. Conversations with Frank McCormick, 2021.

13. Carroll J. 2015. Ex Genentech team garners $217 million to launch neurodegenerative player Denali. *FIERCE Biotech*, May 14, 2015.

CHAPTER 7

Genetics and Genomics

In early 1993, I took a three-month internal sabbatical from my job as Director of Biotechnology at Glaxo to focus on the potential of molecular genetics for drug discovery and to write a strategic proposal for Glaxo on what I thought they should do. I was not doing this alone: I was encouraged by Peter Goodfellow, then Professor of Genetics at Cambridge; Sydney Brenner, who advised Glaxo Group Research; and Richard Sykes, the new Glaxo CEO.[a]

My interest came from watching the growing use of recombinant DNA technology and from the emerging discipline of "genomics," with its ability to manipulate and dissect large chunks of the genome.[b,1] The thinking went that if you can clone genes involved in rare Mendelian inherited forms of disease, then it may be possible to find the genes involved in more common diseases that were known to have a genetic component, such as asthma, diabetes, osteoporosis, and schizophrenia.

THE GLAXO GENETICS INITIATIVE

I decided to write what I thought was a reasonable proposal for what could become the "Glaxo Genetics Initiative" and presented it to senior R&D management. The proposal was to build a programme of 200 people over three years to identify the genes involved in susceptibility to common diseases, to characterise the genes functionally, and to develop drugs against the so-called druggable targets.

---

[a] Also spurring this interest were visits to the Robert (Bob) Williamson lab at St Mary's Hospital in London.

[b] This included the construction of large overlapping clones of genomic DNA, in large vectors based on artificial chromosomes (BACS and YACS),[1] and the use of positional cloning (Box 1) to identify genes causing Mendelian diseases such as Duchenne muscular dystrophy (DMD), fragile X, Huntington's disease (HD), and cystic fibrosis (CF). This work was initially done in academic labs but soon drew interest from entrepreneurs and investors.

BOX 1. POSITIONAL CLONING

The ability to identify and isolate genes that are implicated in Mendelian inherited diseases, such as cystic fibrosis (CF), Duchenne muscular dystrophy (DMD), fragile X, and Huntington's disease (HD), by positional cloning was achieved in the late 1980s. David Botstein, Ray White, Mark Skolnick, and Ronald (Ron) Davis demonstrated that restriction fragment length polymorphisms (RFLPs) could be used to make a linkage map of the human genome.[2] Theirs was a landmark publication as it described a way to start to identify and clone genes based on genetics.

Positional cloning is defined as "using the inheritance of polymorphic markers in the genome with reference to a phenotype to find a region of the genome associated with that phenotype." RFLPs provided those markers, but they were quite difficult and cumbersome to assay and were soon replaced by microsatellite markers that could be sized by gel electrophoresis. These days mapping is done more precisely using single-nucleotide polymorphisms (SNPs) or by DNA sequencing (see Chapter 8).

To undertake a positional cloning project to clone a Mendelian inherited disease gene before the whole genome sequence was available, you needed several things—the families with the disease, the polymorphic markers with known positions in the genome (like RFLPs), the ability to track the inheritance of those markers physically, and access to libraries of genomic clones, some of which might be chromosome-specific (see Fig. 1).

As mentioned in Chapter 1, one of the first genomics companies to emerge was Collaborative Research in Boston, who had industrialised the use of RFLPs as genomic markers. Collaborative Research was started in 1961 and went public in 1982. They helped to establish a map of RFLPs across the human genome. Collaborative Research changed their name to Genome Therapeutics in 1994 and again to Oscient in 2004. They went out of business five years later.

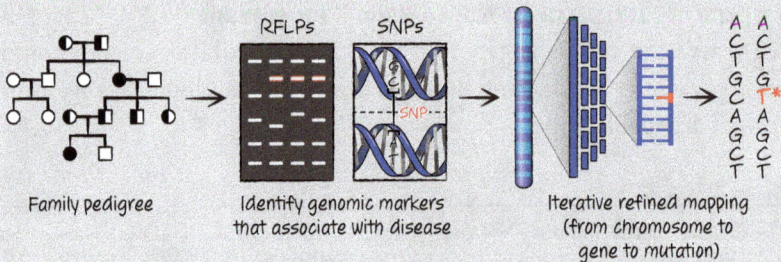

Figure 1. Positional cloning.

The response was interesting. Some of my colleagues were noncommittal, essentially wanting to see which way the official decision would go before committing either way.[c] Others were more supportive, whereas some were definitively against the idea, on the basis that finding targets is not finding drugs. The net result was that I was encouraged to set up a kind of pilot, on a shoestring of course, with five (not 200) people in the first year, and then maybe five more if warranted. It was not exactly the response I was looking for,[d] especially given that small companies dedicated to doing this work were beginning to emerge.

Ironically, after I left to go to Sequana Therapeutics, Glaxo did set up a Genetics Initiative at about the scale I had proposed. They brought in Allen Roses from Duke University to run it. Although Roses was involved in connecting *ApoE4* to the risk of the development of Alzheimer's disease,[3] he was not steeped in knowledge about either genomics technology or drug discovery.

CYSTIC FIBROSIS

The hunt for the genetic underpinnings of CF—mutations in the "cystic fibrosis transmembrane receptor" (CFTR) gene—is an elegant positional cloning story. It is also very pertinent to the history of biotech.[e,4]

Several groups were involved in the race to clone the CF gene including those of Francis Collins at the University of Michigan, Lap-Chee Tsui and John Riordan in Toronto at the Hospital for Sick Children, and Bob Williamson at St. Mary's Hospital in London. Williamson and colleagues had discovered, by co-inheritance studies of genome markers and the disease, that the gene for CF was on chromosome 7, somewhere in the many kilobases of DNA between a gene called *c-met* and a genetic marker called J3.11. Tsui and Williamson both narrowed the search area by discovering additional markers in between J3.11 and *c-met*, including a gene called *irp* (int-related protein), which was closest to the CF gene.

---

[c] Over the years, I have found that this is a politically expedient way to behave, but not one I either follow or condone.

[d] I called it "damning the initiative with faint praise."

[e] Vertex, a Cambridge (MA)-based successful independent biotech company, became successful owing to the work on the function of the CFTR and finding drugs to increase the function of the mutant protein that causes the disease. Two comprehensive books have been written about the history of Vertex.[4]

Tsui teamed up with Collins, who had some novel physical methods to align large genomic clones, and they "walked" (or rather "jumped") along the chromosome until they found DNA sequences that matched with something that was an appropriate candidate gene for the disease.[5,6] In the meantime, Riordan had made some cDNA libraries containing long cDNAs from epithelial cells from normal and CF patients. By a process of elimination and by putting other bits of data together, the investigators found a gene coding for a transmembrane protein with two ATP-binding domains.[f,7] In the first of these nucleotide-binding domains (NBD1) in CF patients, there was a prominent and frequent mutation at position 508 in the sequence, where the triplet of nucleotides coding for phenylalanine (called ΔF508) was missing in affected individuals. This was clearly a very plausible candidate gene, with a mutation in affected individuals that was not present in normal individuals.

This discovery was enormously important for the diagnosis of the disease. Many mutations that cause CF by a variety of means have now been found in the CFTR gene, usually involving either the function of the protein as an ion channel or the ability of the channel protein to move to the cell surface once it is made.

DUCHENNE MUSCULAR DYSTROPHY

DMD is a very severe X-linked recessive disorder (see Chapter 2). It is also one of the most common recessive disorders in humans. It is present in young males (age three to five years) as progressive muscle weakness, ending up in a wheelchair and ultimately death from respiratory complications, often before they are teenagers. I first encountered this disease early in my life, as my mother's best friend from school was the mother of a family where this disease was inherited, as she was a carrier of the mutant gene on one of her X chromosomes.

Earlier genetic linkage studies had shown that the gene lay on the short arm of the X chromosome, making positional cloning—the same approach as was taken for cloning the CFTR—a bit easier. Louis (Lou) Kunkel and Anthony (Tony) Monaco, a graduate student in Kunkel's lab at the Children's Hospital in Boston, identified a starting point on the X chromosome that appeared relatively close to where they expected the DMD gene to be. From there, they "walked" to the gene using cDNA libraries made from the muscles of affected individuals.

---

[f] As it happened, the Williamson lab had also found the gene but did not have a complete enough data set to declare victory.[7]

This was not an easy task: the DMD gene turned out to be very large, spanning more than 2.5 million nucleotides, including all the introns and exons.[g] The gene encodes a very long 14-kb mRNA transcript. Dystrophin (as the protein made from the DMD gene was named) was also very large with a molecular weight of 400,000.[h] The mutations causing the disease were shown to be deletions that led to splicing defects, producing aberrant mRNAs and inactive forms of dystrophin.[8,i]

HUNTINGTON'S DISEASE

Equally exciting studies were being done at the time by a collaborative consortium of scientists including James (Jim) Gusella, Hans Lehrach, David Housman, Nancy Wexler, and others to clone the gene responsible for the devastating neurodegenerative disease HD, a dominant genetic disease[j] that had been located to chromosome 4 by linkage studies in large Huntington's families from Venezuela. Unlike CF and DMD (but like many now-discovered disorders), HD is caused by the expansion of a CAG sequence repeat in the gene.[9] In the normal gene, there are 20 CAG repeats in a row. In the copy of the gene causing the disease, the CAG is expanded to more than 40 copies, resulting in a protein (huntingtin) with multiple extra glutamine amino acids (the codon CAG codes for the amino acid glutamine). Although there are still questions about how the mutant protein causes neurodegeneration, it is clearly a critical protein in regulating nerve cell development and function.

GENOMICS COMPANIES

The positional cloning of the genes implicated in CF, DMD, and HD not only led to the finding of many more disease genes from Mendelian inherited disorders but also led to the founding of several small companies dedicated

---

[g] The DMD gene locus is one of the largest in the human genome coding for a single protein.

[h] The sum of the atomic masses of all atoms in a molecule.

[i] The mutation data also explained why some mutations led to DMD, whereas others led to Becker muscular dystrophy, a milder form of disease. The Duchenne mutations caused out-of-frame exon "skipping" mutations resulting in inactive truncated proteins, whereas in Becker dystrophy, the exon skipping still occurred, but the reading frame was maintained, so a short form of dystrophin was still made. This will be relevant later when we talk about gene therapy for DMD (Chapter 13).

[j] You only need to inherit one copy of the mutant gene to get the devastating disease later in life.

to using the technology to find genes for more common diseases, in which genetic inheritance was known to play a role. Initially, four companies were set up: Mercator in San Francisco; Millennium in Cambridge (MA); Myriad Genetics in Salt Lake City; and Sequana in San Diego.[k]

One of the key differentiators for these companies was access to the families suffering from the disease of study. This was not just finding the families. It was sorting out all the legal and ethical issues, like consent to use their DNA for discovery, plus the need for good phenotyping (i.e., did the families really have the disease they were thought to have?). It was important to be sure that affected people did have the disease being studied and that the control samples did not. Although obvious, this was a lot more complicated than it might seem. Forming relationships with the academic physicians (and key opinion leaders or KOLs) who treated the patients was an important part of being successful.

MAKING THE MOVE TO SEQUANA

Despite Glaxo's dismissal of the Genetics Initiative proposal, my interest in cloning disease genes stayed strong. My ears perked up consequently when Peter Goodfellow and Sydney Brenner mentioned to me that these new companies were starting up. As someone who has always "put their money where their mouth is," I felt it necessary to reach out to the companies.[l] I had met with Mark Levin as he was starting Millennium in 1992, with a view to becoming R&D chief there, but that never really got any momentum. Instead, after speaking with Kevin Kinsella from Avalon Ventures in San Diego, who was the founding VC of Sequana (where Peter Goodfellow was a new scientific advisory board [SAB] member), I was persuaded to go there.[m]

---

[k] You might legitimately ask why venture capitalists (VCs) would want to invest in cloning disease genes, which are many steps away from effective drugs. The VCs were not just seduced by the technology: they also appreciated that the pharma industry had a paucity of drug targets on which to work and that genetics tools might bring them better targets that more directly affected the pathology of both rare and common disorders.

[l] It is all very well shouting loudly that genetics is important for drug discovery (as it clearly is), but if you really believe it, you have do something about it.

[m] I remember the first Sequana SAB meeting in Wengen, Austria, in the fall of 1993, when I was still at Glaxo. I had to be anonymous, so I pulled my sweater over my head as a joke, so no one could see who I was—hoping that would help.

My announcement of deciding to leave Glaxo to join Sequana had both predictable and unpredictable consequences. In trying to keep me at Glaxo (which I was quite gratified about), I was invited to meet Sir Mark Richmond, who was part of the senior R&D team. He proceeded to ask me "if I knew what I was doing by taking my wife and children to Southern California to a cowboy outfit like Sequana, when I could stay at Glaxo and be part of their plans." I did not appreciate the patronage. I also recall visiting Richard Sykes who had become the CEO of the company, at the Glaxo headquarters on Berkeley Square in London for tea. I knocked on the door to be greeted by a butler with white gloves who said, after a wait of a few minutes, that "Richard will see you now," and up the stairs we went.[n]

SEQUANA—GODDESS OF HEALTH AND WELL-BEING

Given the lack of support for the Genetics Initiative, I decided to move to the United States and Sequana. We arrived in San Diego in mid-December 1993 and spent the first three nights at the Lodge at Torrey Pines, before moving to Loma Larga Drive in Solana Beach near the tennis club, a house which we rented before we bought it.

Those who know La Jolla are probably thinking: "Wow ... the Lodge at Torrey Pines! How exclusive is that?!" Well, in 1993, it was not: it was a run-down motel with paper-thin walls by the golf club, but with stunning views across the Pacific Ocean. We were wondering what on earth we were doing there. At least the kids loved the "full English" breakfast despite the funny bacon.[o]

I got used to driving along Route 1 by the sea through Del Mar to get to Sequana, which was located on the bluff at 11099 North Torrey Pines Rd. I also got used to going for a run almost every day to Flat Rock and back through the Torrey Pines State Park.[p] Piattis's by the tennis club was a popular restaurant we frequented along with Il Fornaio in Del Mar and

---

[n] This was the time when a circulating joke was "What is the difference between Virgin Atlantic Airways and Glaxo?" The answer: Glaxo has more jets. I think there were eight at the time, underlining the profligacy of the 1980s. When he became CEO, Richard saw to it that there was only one: his. I travelled on it once, back to the United Kingdom from Research Triangle Park (RTP).

[o] After it was renovated and become what it is today, my second daughter decided to get married there! The wedding was a wonderful event if a trifle hot (temperatures that day being more than 100°F). It was also definitely not cheap.

[p] I revisit that experience whenever I go back to San Diego, but I tend to walk rather than run now.

Asian–American fusion Japengo's in La Jolla. I fondly remember one of our first dinners at Japengo's where my son, then age 6, fell asleep during dinner. I also remember introducing my eldest daughter to Sydney Brenner in the bar at George's at the Cove, telling her that this was the "most eminent person you will ever meet," reminiscent of what my dad said to me when I bumped into Sir Robert Robinson in 1960.[q]

Sequana became a positional cloning powerhouse, with too many very talented people to mention them all and an inspiring Scientific Advisory Board. The SAB members were chosen carefully by company founders Kevin Kinsella and Jay Lichter,[r] based on what they had achieved. It was probably one of the best SABs for a biotech company ever assembled, and the members have gone on to have stellar careers.[10] Members included Allan Bradley, Mark Boguski, Stephan Guttmann, David Housman, Peter Goodfellow, Hans Lehrach, Tom Marr, Tony Monaco, and John Todd. It is interesting to note that back then was no importance attached to ensuring any diversity in such a group. It would not have even been a consideration. Alan Buckler was a consultant who came to the meetings and subsequently became a Sequana VP, as did Nicholas (Nic) Dracopoli. A subset of the SAB and I used to do the Flat Rock run during SAB meetings. This exercise always started with us chatting about the day's proceedings, but as we neared home, the conversation stopped and the real running began: no one wanted to be second.

Sequana's business strategy was to provide useful and novel information about disease genes from different diseases of interest to pharma. Not so dissimilar to the strategy that Millennium took initially, although they managed to do somewhat bigger strategic deals than we did. Sequana formed alliances with Glaxo Wellcome on diabetes and obesity, with Boehringer Ingelheim on asthma, with Boehringer Mannheim on osteoporosis, and with Parke Davis–Warner Lambert to develop novel therapeutics for schizophrenia and depression by finding the genes involved in those diseases.

Samples and Collaborations

One major challenge we faced was obtaining the very large number of DNA samples we needed from well-characterised patient populations. This

---

[q] We do become like our parents after all.

[r] Kevin was the managing partner at Avalon Ventures in La Jolla and Jay was a geneticist from DuPont Merck (Jay had received the same resistance to human genetics research there as I had met at Glaxo).

required making collaborations with several academic groups that had access to sufficient curated patient samples. For example, we had collaboration with Philippe Froguel in Paris to obtain diabetes samples and with the Mt Sinai Hospital of the Samuel Lunenfeld Research Institute in Toronto to access DNA from patients with asthma, including those from all the islanders on Tristan da Cunha, where there was a very high incidence of the disease.

Whz 1 and Whz 2 and Tristan da Cunha

The island of Tristan da Cunha, the most remote inhabited island on earth (see Plates 7 and 8 for Tristan da Cunha island pictures), lies between South Africa and South America. It takes a week by boat from Cape Town to travel 1700 miles to get there, traversing the "Roaring Forties" on the way. Tristan da Cunha "belongs" to the United Kingdom as a British Overseas territory, and nearly all the 250 islanders[s] came from the British Isles originally. When the volcano erupted on the island in 1961, all the islanders were evacuated for two years to Southampton in the United Kingdom. It became apparent during their medical analysis that there was a very high incidence of asthma on the island.

One of the respiratory physicians, who was involved with the islanders at the time in the United Kingdom, was a Canadian doctor, Noe Zamel, who was now back in Toronto. Sequana contacted him to obtain access to the Tristan da Cunha population, their pedigree, and their DNA to hunt for the gene causing asthma in that genetically isolated population. The population was derived from William Glass, a British soldier in the garrison that was set up on the island to deter Napoleon escaping from St. Helena, another south Atlantic Island and British overseas territory some 1500 miles NNE of Tristan da Cunha.[t] Glass had 16 children, and according to the history, two sisters emigrated to Tristan da Cunha from St. Helena and introduced asthma to the island.

Sequana proceeded to genotype the DNA from all the islanders. Carrie Le Duc from the company went with Zamel to the island to collect some missing DNA samples, to confirm some of the asthma phenotypes and to check who was affected. To support the studies, Sequana formed a collaboration with Boehringer Ingelheim.

---

[s] With a total of seven surnames.
[t] I would not have thought that the distance was much of a deterrent to the French!

By a process of elimination, a couple of genes that we called *wheeze 1* and *wheeze 2* (but were subsequently renamed *ASTH1I* and *ASTH1J*) were found on chromosome 11, which segregated with the disease in the pedigree. Sequana filed a European patent application on the finding (see Plate 9 for structure of the locus), and a physical map of the region containing the genes was constructed. The patent was filed during the Arris Pharmaceuticals takeover in early 1998, and a publication on the genes was published later.[11,u]

Sequana also had a collaboration with the Foundation for Osteoporosis Research and Education in Oakland, giving the company access to more than 1000 patients with this disease. A research alliance with the Southwest Foundation for Biomedical Research in San Antonio further provided Sequana with access to a baboon colony, where there was comprehensive bone mineral density information and detailed baboon pedigrees. Unfortunately, many of the microsatellites that we used to construct our linkage maps worked for human DNA but did not work on baboon DNA.

Sequana Technology

The technology that Sequana developed was a rapid positional cloning platform consisting of several components that worked together impressively: collecting the DNA, genotyping using microsatellite sequence polymorphisms as markers detected by gel electrophoresis, physical mapping using genomic clones, DNA sequencing, mutation detection, and gene characterisation. A bioinformatics and statistical genetics group (run by Lon Cardon, now CEO of the Jackson Laboratory) provided the necessary computational glue to hold it all together. Visitors to the Sequana labs would see robots putting DNA into small wells in plastic plates and rows of gel electrophoresis machines used to separate the markers from individual samples.

Kevin Kinsella was a master of these lab tours as well as orchestrating all the very considerable Sequana public relations, a skill he no doubt inherited from his successful actor father, Walter Kinsella. We had all sorts of swag and gizmos to give visitors, including an empty CD box labelled "Human genotype— Know thyself," the idea being to convey the fact that your DNA genotype could fit on a CD.[v]

---

[u] Sequana also issued a press release of the finding before the paper was published, causing some consternation.

[v] It was a prescient precursor to the complete human genome sequence that emerged some 10 years later.

BOX 2. GENE CHIPS

Measuring SNPs across the whole genome has become the method of choice for studies associating genomic markers with implicated genes for many diseases (the so-called GWAS). It was much easier to use the "SNP chips" that had become available than RFLPs or the microsatellite variations that we used at Sequana. Many of these studies were published in *Nature Genetics*. They still go on today, with more and more genes being found to be associated with different diseases as more diseased people and more controls are used and the chips have a denser array of SNPs from across the genome that can be measured.

A SNP chip is essentially a DNA microarray where allele-specific oligonucleotides that will hybridise to one or other of the two alleles of a gene are printed onto a chip, with a detection system that records the binding of the fluorescently labelled fragmented DNA being tested. Chips were originated by Affymetrix, one of the most successful DNA array companies. Spun out of Affymax in 1993 in Santa Clara, by Stephen (Steve) Fodor, their "GeneChips" were generally referred to as "Affy" chips. The company grew by acquiring several companies, including the genetics company ParAllele BioScience for their novel probe technology, and went public in 1996. They are now a platform under Applied Biosystems, acquired by Thermo Fisher Scientific in 2016 for $1.3 billion.

Our technology platform predated by several years the development of SNPs and SNP chip technology (Box 2), which allowed for very large genome-wide association studies (GWAS) and accelerated the pace of gene discovery. These methods took over genotyping completely in discovering other genes involved in common diseases. Indeed, the work we did with the Tristan da Cunha families over two years could probably be done in an afternoon or two using GWAS and the molecular techniques of today.[w]

Sequana IPO

The Sequana IPO took place in August 1995 (see Plate 11 for Sequana S1 cover), led by Lehman Brothers and Hambrecht & Quist. The Sequana investors and Board of Directors were an impressive group of people, including The Carlyle Group (Richard [Dick] Darman, ex-Bush cabinet member), NEA (Thomas [Tom] McConnell), Sequoia Capital (Thomas [Tom] Stephenson), Morgenthaler Ventures (Robert Bellas), and the Sprout Group

---

[w] Technology relentlessly moves on.

(Kathy LaPorte). Irwin Lerner, ex-Chairman of the Board of Hoffman, La Roche Inc., was the Chairman.

I remember the IPO "road show" as a frenetic two weeks, visiting San Francisco, San Diego, Minneapolis, Los Angeles, Boston, Chicago, and New York in the United States, with a short visit to London and Zürich. Two things stick in my mind. One was making sure that we had multiple copies of the slides—no PowerPoint then—in case they got left on a plane (which of course they did). Second was not to be fooled when the investors who seemed most engaged and who paid attention—asking all sorts of interesting questions—were not those who ended up buying the stock. The opposite may even have been true: a sleepy investor (who may not actually have been asleep) might buy a good chunk of the offering.

The visit to London involved a "Lehman lunch." I asked Rodney Young, our Lehman VP, whether this was a formal lunch. "No" was his answer. Entering the room to waiters with white gloves and multiple sets of cutleries at each place setting suggested otherwise. I whispered to Rodney that this was what a formal English lunch looked like.[x]

The IPO got done, but it was not especially successful. We scaled back the offering of three million shares at \$12–\$14 to two million shares at \$9, with the stock trading up slightly in the aftermarket. The disappointing lack of enthusiasm might have been due to an announcement made by Millennium while we were on the road show that they had cloned the mouse "tubby" gene, one of the targets of our obesity programme.[12] The fact that the IPO got done at all was a blessing, as the markets were not generally receptive for new company stocks between 1993 and 1995.

In January 1998, Arris Pharmaceutical Corporation, which was a small molecule combinatorial chemistry drug discovery company in South San Francisco, bought Sequana.[y] I was generally unimpressed with the merged entity, which was named Axys Pharmaceuticals. The Axys management approach was focused primarily on cancer, not a real focus area for Sequana. That said, Sequana had formed a collaboration with Memorial Sloan Kettering Cancer Center (MSKCC) in 1997, in which both parties invested \$5 million into a joint venture (JV) called Genos. Jim Rothman was the

---

[x]At least it was not the curly edged white bread crustless cucumber sandwiches that, as you may have experienced, is an English lunch delicacy.

[y]Just over a year earlier (1996), we had bought NemaPharm for its functional genomics capabilities (see Chapter 9). Arris had also acquired Khepri in 1995.

MSKCC representative on the JV,[z] and I attended on Sequana's behalf.[aa] The idea was for Sequana to provide the JV with its gene discovery capabilities, including genome scanning, DNA sequencing, and bioinformatics, while MSKCC would supply a 30-year retrospective database of patient tumour-tissue samples.[bb]

The JV barely got off the ground before the acquisition, and Genos was not well-understood or supported by Sequana's new owners. Arris bought Sequana primarily for $50 million we had in the bank, not for our alliances or for our science. In the succeeding year, Axys took down—brick by brick—all the technology that my team had assembled in the previous six years and, in my opinion, all the value went with it. I did not think that the new management understood the technology or what it was we were trying to do. Axys eventually was bought by Celera in 2001. In the meantime, most of the best scientists had left the company.

As Kevin Kinsella liked to relate, Sequana was named after the Roman goddess who guarded the healing shrine established at the head of the river Seine in France, in the first or second century BC. Dedications were made to Sequana at her temple, in the form of stone and wood carvings of diseased body parts, in the hope of a cure. Clearly, we did not do enough to incur Goddess Sequana's long-lasting blessing or maybe we incurred her wrath instead. Or maybe we were just not in the right place at the right time.

MILLENNIUM PHARMACEUTICALS

Millennium—Sequana's nemesis—was started in early 1993 by Mark Levin, a venture capitalist at Mayfield who had spent time working on insulin at both Eli Lilly and Genentech in the early days. He invested in Cell Genesys, one of the first cell therapy companies (Chapter 13), and in the Dave Goeddel and Steve McKnight company Tularik. Like many others, Mark had become interested in using genetics and genomics in drug discovery. He assembled a first-rate team of Millennium founders, including Eric Lander from MIT, Jeffrey (Jeff) Friedman from the Rockefeller University, Raju Kucherlapati, at Albert Einstein, and Daniel Cohen, chief scientist of Genset in France.

---

[z] Rothman won the Nobel Prize in Physiology or Medicine in 2013 for his work on mechanisms of protein secretion from cells.

[aa] I still have the T-shirts for both Genos and Sequana (Plate 10).

[bb] The venture tended to do Rothman's bidding, much to the consternation of the late David Munroe (who was the Director of Genomics at Genos).

The genesis of the company was a 1990 meeting with Friedman, who introduced Mark to the concepts of cloning genes using familial inheritance patterns.[13] Mark then asked Raju to explain to him what LOD scores were[14] and suggested that a company be set up to exploit positional cloning activities. Raju and Mark met Kevin Kinsella several times in 1990/1991, including on the beach in La Jolla, but they decided to set up Millennium as a separate company rather than do a joint one with Kinsella and Avalon Ventures.[cc] An initial $8.5 million financing for the company was led by Mayfield.

Millennium pursued many of the same targets (e.g., diabetes and obesity) as Sequana, using the same technology. One key difference was that Millennium's business model was not to sell genetic information, but to use the gene information themselves to discover new drugs. They leased space close to MIT on Landsdowne St, Cambridge (MA), and within six months had hired 30 scientists to build their gene-finding platform.[13]

One of Millennium's many claims to fame was the number of lucrative disease-focused strategic alliances they formed with pharma.[dd] Their business development team was led by Steve Holtzman, previously at DNX, who put together the initial strategy. Millennium was careful to ensure that they maintained IP on as much of the drug discovery process as possible including the gene, the protein (the target), and any drugs and diagnostics that might follow.

They signed their first partner, Roche, in March 1994, to a five-year, $70 million deal to develop drugs to treat diabetes and obesity, using in part the Jeff Friedman connection. His lab had cloned the mouse and human genes coding for an obesity-controlling protein, called leptin. Millennium wanted to find the leptin receptor gene.[ee] In October 1995, a $50 million JV with Eli Lilly & Co. focusing on atherosclerosis was signed, and two months later, they signed a five-year, $60 million collaboration with Astra AB, targeting asthma and other inflammatory diseases.[ff] Steve Holtzman got the reputation as the "best deal doer" in biotech.[gg]

[cc] It is interesting to think about what such a joint company might have achieved.

[dd] The Sequana team would sometimes encounter the Millennium business development team, coming from or going to meetings with the same pharma companies.

[ee] In addition to other mouse obesity genes.

[ff] It is unclear to me whether this deal making helped or hindered Sequana's deal making. On balance, I think the competition helped.

[gg] Following Millennium, Holtzman went on to be CEO at Infinity Pharmaceuticals and then to Biogen, where we worked together.

Millennium IPO

Millennium's summer 1996 IPO was managed by Goldman Sachs and Robertson Stephens, raising $58 million in an oversubscribed deal. At the time of their IPO, Sequana (which had gone public at $9/share a year before) was trading at $22, and Myriad Genetics (IPO in October 1995 at $18) was trading at $30. At the time of the IPO, not only did Millennium have genetics and genomics programmes but they also had started drug discovery activities,[15] with interests in obesity, Type II diabetes, atherosclerosis, respiratory diseases, oncology, and the CNS.

Following an offsite executive meeting and some net present value (NPV) calculations, Millennium leaders realised that the only way to become a successful gene-driven drug developer was to acquire companies that already had potential drugs in their portfolio. Developing drugs from the "genetic" targets that they had uncovered in their collaborations, many of which were not being exploited by their partners, was not a viable business model.[16]

Acquisitions

The first such acquisition in early 1997 was Chemgenics, a company developing antibacterial compounds. At the same time, and partly to fund these acquisitions, Millennium formed a lucrative (>$200 million) relationship with Monsanto to apply their technology in the plant and agricultural crop space.

The subsidiary Millennium BioTherapeutics was also set up to develop therapeutic proteins, gene therapy, and antisense (siRNA) products. John Maraganore, who went on to be CEO at Alnylam, and Ron Lindsay from Regeneron were deeply involved in this new entity, and Eli Lilly signed a $70 million five-year deal to develop medicines from proteins, taking an 18% stake in the subsidiary.

Another spinout, Millennium Predictive Medicine, was set up to produce diagnostic tests. Bristol-Myers Squibb agreed to a $32 million, five-year alliance with this subsidiary to develop diagnostic tests in oncology. An additional $70 million strategic alliance with Becton Dickinson was also formed. These spinouts were eventually re-incorporated into Millennium, but they provided a good source of additional capital at the time.

Millennium's most extraordinary deal was with Bayer in 1998, designed primarily to accelerate Millennium's drug discovery aspirations. In this $465 million deal, Millennium was committed to delivering to Bayer more than

200 proteins derived from their genetics and genomics platform, across several disease areas, with Bayer gaining a 14% stake in Millennium.[hh,17] In this "cherry picking deal," Bayer could pick the targets they wanted, but those that they did not pick reverted to Millennium. As viewed from Sequana, the extent of this deal and how they managed to broker it seemed unbelievable. This deal also changed the Cambridge (MA) biotech scene and strategic deal expectations forever.

Millennium was probably the most prolific deal doer of any early biotech company, either before or since, and they rode their approach all the way to becoming a fully integrated drug discovery company. Very few biotech companies have been able to make this transition. But Millennium was particularly good at persuading others to work with them and give them money: a testament not only to the quality of the executive team but also to the research and development staff, and the shared "go-for-it" corporate culture.[ii]

Becoming a Drug Company

Millennium became a real player in the drug discovery and development business when they bought the oncology and inflammation company LeukoSite, Inc., in late 1999 for nearly a billion dollars, not only for its pipeline but also for its drug development capability. Out of this acquisition came a multiple myeloma drug called Velcade (bortezomib), which had been discovered by Julian Adams and his team at LeukoSite. Approved in 2003, Velcade is still highly important in treating multiple myeloma patients. However, Millennium's first drug on the market was CAMPATH 1 (see Chapter 2), used to treat chronic lymphocytic leukaemia. It also came with the LeukoSite acquisition.

In yet another spectacular deal, Aventis SA, a French–German company formed in 1999 from the merger of Hoechst Marion Roussel and Rhone Poulenc-Rorer (now part of Sanofi), signed a $450 million strategic

---

[hh] Stefan Lohmer, who is now the CEO of Axxam, a successful CRO, was the Bayer scientist who put the deal together.[17]

[ii] This was also reflected in some of the stylish things they did. Mark and Becky Levin held a lavish 25th wedding anniversary party in 2002, in their house (Gray Craig Mansion) near Newport, Rhode Island, a house they subsequently sold to the actor Nicolas Cage. I was invited and attended, and I was enamoured not only by the surroundings but also by the fact that Crosby, Stills, and Nash came to play and most of the company employees seemed to be there: it was a truly amazing event.

alliance with Millennium in 2000 to strengthen its access to genetics- and genomics-driven drug discovery. Under this agreement, the two companies would share profits in any drugs sold in the United States and Canada, and Millennium would receive a royalty in all other markets. To get access to top-class medicinal chemistry, Millennium also acquired Cambridge Discovery Chemistry, a British company.

Millennium continued to use partnerships and acquisitions to grow their business. After the Roche deal expired, the company was able to form a new partnership to develop drugs and diagnostic tests for diabetes and obesity with Abbott Laboratories for $250 million over five years. The acquisition of Cor Therapeutics for $2 billion, a cardiovascular drug discovery biotech company in the Bay Area in 2001, was a significant milestone in their journey to become a fully integrated product development company. Following the Cor purchase, Millennium had close to 10 drugs in human clinical testing, $400 million in estimated annual revenues, $2 billion in cash, and an experienced and talented workforce now focusing on drug discovery.

Mark Levin stepped down as CEO in 2005, more than 20 years after he started the company. His leadership and passion drove the company to excellence. In a December 2001 interview with *MIT Technology Review*, he had stated "Over the next five to 10 years, our goal is to become a company that's leading the world in personalized medicines, a company that is leading the world in productivity, a company with a value of over $100 billion, a company that has five to 10 products on the market that are making a big difference in people's lives, a company with the strongest pipeline in the entire industry."[18]

Despite the desire to maintain independence, any biotech company with the quality of Millennium inevitably becomes a takeover target. Takeda, a Japanese company formed in 1781 in Osaka, Japan, acquired Millennium in 2008 in a cash offer of $8.8 billion. By this means, Takeda established an R&D base in Cambridge (MA) and acquired several important oncology and inflammatory disease drugs, plus revenues of greater than $500 million/year (including that from Velcade).

It is interesting to contrast the success of Sequana and Millennium: both started at the same time and used the same technology, but with very different ultimate objectives. It clearly would have been better for Sequana shareholders if we had had a vision of becoming a drug discovery organisation and had avoided being bought by Arris. But the necessary combination of business development skills, R&D skills, and an executive team with the appropriate experience and vision was probably not there for us to have pulled that off effectively.

MYRIAD GENETICS

Another company set up about the same time as Sequana and Millennium was Myriad Genetics in Utah. They had a different view of the genomics world, thinking instead about gene information for diagnostic purposes from the start. Their CEO, Peter Meldrum, wanted to obtain a return on investment "before 10 years had gone by." The founders were Wally Gilbert and Mark Skolnick, a professor at the University of Utah who was well-known for the construction of the genetic linkage map in man using RFLPs. Skolnick had access to Mormon pedigrees and families for his research applications, which helped accelerate research. Myriad's major claim to fame was to identify and clone by linkage the *BRCA1* gene, mutations which can cause hereditary breast and ovarian cancer.[jj]

BRCA1 and Diagnostics

This discovery by a company set up to make money from such discoveries was not universally appreciated by academia. It also stole some of the glory away from Mary Claire King's lab at the University of Washington in Seattle. With access to families with early-onset highly penetrant cancer of the breast or ovary, the King lab had originally looked at the linkage of more than 170 markers and located the "cancer gene" to chromosome 17. Myriad published their findings in September 1994[19] and in December 1994, King and her collaborators published results based on a second cohort of families.[20,21] A second breast cancer gene called *BRCA2* was also found.

The so-called tumour suppressor genes like *BRCA1* and *p53* are known as the "guardians of the genome," as mutations in them allow uncontrolled cell division and cancer formation. At Sequana, I fondly remember hosting Mary Claire King for a seminar: Jeff Hall, who had worked on the *BRCA1* project in her lab in Seattle, invited her. We went for dinner afterwards in Del Mar and I insisted that she ride with me in my recently acquired 1989 Porsche 911 Carrera in red, with 22,000 miles on the odometer. Professor King was very graceful (and I think amused) when the car would not start after dinner.[kk]

Myriad Genetics went public on the back of the *BRCA1* gene discovery in 1994 and launched their first tests in 1996/1997 (well within the 10 years

---

[jj] This was the gene made famous by Angelina Jolie.

[kk] A little like computers, cars are only worth having when they work.

Meldrum had promised). Cloning of these two genes allowed the development of diagnostic genetic tests for families with early-onset cancer, which is precisely the business that Myriad then developed. Their patents, which also blocked others from entering this market, have recently expired. The blocking IP was not without considerable angst from a legal and commercial perspective.[22] Access to the database that Myriad has assembled on the consequences of having mutations in *BRCA1* (of which there are many) allowed them the ability to correlate almost any mutation with outcome and thus to drive more precise clinical decision-making. Myriad launched an oncology-focused pharmaceutical company (Myriad Pharmaceuticals) in 1999, which was run for several years by Adrian Hobden, who had worked with me at Glaxo in the early 1990s.

One of the most important classes of drugs for treating breast cancer is the poly (ADP-ribose) polymerase (PARP) inhibitors (e.g., Olaparib, developed by KuDOS Pharmaceuticals in the United Kingdom). These drugs work much better in patients with *BRCA1* and *BRCA2* mutations. The enzyme PARP repairs DNA after damage, and thus inhibition of the enzyme causes cells with DNA damage to die. The enzyme is "synthetic lethal" with respect to *BRCA1* mutations. In other words, cells that have *BRCA1* mutations must have active PARP to survive, so inhibition of the enzyme in breast cancer cells carrying *BRCA1* mutations makes them die preferentially. Myriad markets a companion diagnostic test to go with the use of this important medicine marketed and sold by AstraZeneca after they bought KuDOS in 2005. Recently, Myriad also launched the first polygenic breast cancer risk assessment score (RiskScore[23]), validated for women of all ancestries.

Within its genomic niche, Myriad can be considered to be a very successful biotech company.

MERCATOR GENETICS

Mercator Genetics was also founded in 1993 by David Cox, Rick Myers, and Dennis Drayna, before David and Rick moved on to the Genome Center at Stanford.[ll] I first met Rick when he was a graduate student in Bob Tjian's lab at UC Berkeley, when I went there to meet Frank McCormick in 1980.

The would-be founders talked with various VCs in late 1992 to find start-up money, including Mark Levin and Eric Lander (who formed Millennium

---

[ll] An important player in the public Human Genome Project (Chapter 8).

together), and with Kevin Kinsella (Sequana founder). In the end, they were financed by the unlikely group of Robertson Stephens, a small investment bank focusing on biotech that had previously taken various biotech companies public and did not usually become an investor in those small companies.[24] Mercator had about 30 employees and a good SAB, which included Michael (Mike) Bishop (awarded the Nobel Prize in Physiology or Medicine in 1989 for the discovery of oncogenes with Harold Varmus) and Arno Motulsky, in Seattle.

They were initially focused on using positional cloning to find the gene for hereditary hemochromatosis. They were successful, but it cost them close to $11 million to clone it. They went through several CEOs, including Kathy Behrens; Michael Forrest, ex-Syntex/Roche R&D; and Elliot Sigal, a scientist who went on to be a successful president of R&D at BMS.[mm] Mercator was acquired in 1997 by Progenitor, another genomics company, and quietly disbanded shortly afterwards.

deCODE GENETICS

As must be evident from Sequana's focus on isolated populations like Tristan da Cunha, a key to finding disease genes is access to relatively isolated populations where there is good historical genealogy and good health care and health records. Finland and Iceland both are notable in that respect. Given the population bottlenecks (e.g., from famine and infectious disease), there is usually a considerable "founder effect"[nn] seen in these populations and often an increase in recessive disorders.[25] This is true of Finland more than Iceland, but the structure of the relatively homogeneous Icelandic population, descended from very few Vikings originally, coupled with the pedigrees and the well-documented health records, represented a considerable resource for hunting disease genes.[26,27]

This potential was not lost on either Kevin or me. It was also not lost on Kari Stefansson, an Icelandic neurologist working at Harvard at the time. Initially, Kari wanted to study the genetics of multiple sclerosis in the 330,000 people who lived in Iceland. Kevin and I met Kari, who was purportedly interested in working with Sequana, several times in 1995 to access

---

[mm] Sigal's brother Irving, the senior director of molecular biology at Merck, was one of those killed in the Pan Am Flight 103 bombing over Lockerbie, Scotland, after attending a biochemistry meeting in the United Kingdom in December 1988.

[nn] Reduced genetic diversity that results when a population is descended from a small number of ancestors.

families in Iceland with multiple sclerosis (MS)[°°] and many other diseases. Unbeknownst to us at the time, what Kari was *really* interested in doing was to set up an Icelandic company to go disease gene hunting in competition with us and Millennium. He was just collecting intelligence on how we had set up Sequana and our research platform.

Visiting Iceland

Kevin and I were well-seduced when we first visited Iceland in early September 1996. It was barely above freezing, but the scenery was spectacular, and the University in Reykjavik faculty and the various government people we met were very engaging. The food and wine were delicious. Our visit is mentioned in a book (in Icelandic) about Kari and his formation of deCODE Genetics (see Plates 12–14), which subsequently took place in late 1996 with $12 million in venture investment from Atlas Ventures.

There was nothing really novel about their plan, given what Sequana and Millennium were doing. What was powerful was access to the Icelandic population. Kari had managed to finagle the system politically in Iceland, so that most of the Icelanders were comfortable with sharing their DNA and disease phenotypes, giving deCODE the necessary consent to work with the information. deCODE had access to the genealogical records of 800,000 living and deceased Icelanders, and their pedigree data, which were either confirmed or could be inferred from information in the Íslendingabók (Book of Iceland, a genealogical record stretching back some 1200 years). They also had access to the single-payer, high-quality health system records.

Owing to these resources, Hoffmann-La Roche decided to do a $40 million deal with deCODE to access the data and the database that they planned to derive. The deal was driven by Jonathan Knowles, the head of R&D at Roche who had previously worked in Finland, who knew some genetics, and who had previously run the Glaxo Institute for Molecular Biology in Geneva. Data protection, which might have been an issue was not, as secure confidentiality was the law in Iceland. With the investment, deCODE grew to more than 350 people and went public in July 2000 at a valuation of more than $1 billion, an Icelandic unicorn if you can picture such a beast (see Chapter 17).

[°°] I was personally interested in the MS project because my sister had recently been diagnosed with primary progressive MS, from which she subsequently died in 2020.

deCODE was finding genes implicated in many diseases, including Alzheimer's disease, heart disease, stroke, and schizophrenia.[pp] But as has been shown repeatedly by others, if it was hard to find the genes in the first place, then it was even harder and considerably more time-consuming to turn the targets into drugs. The company tried to bring their own drugs into the clinic in the mid-2000s, but without really having the resources necessary to pursue them. Even though other partnerships were formed with the likes of Merck and AstraZeneca, the company rapidly ran out of money. After a reboot financing in 2009, deCODE was bought three years later by Amgen for $415 million.[qq]

Even though the company was not able to deliver on their promise to use population genetics to find drugs, no one can argue that the science that deCODE did then and still does now (as a wholly owned subsidiary of Amgen) is anything other than top-tier. In 2002, they published a refined recombination map of the human genome using 5000 microsatellite markers: no small feat. Since the development of SNP mapping, GWAS (Box 3), and now cost-effective whole-genome sequencing, deCODE has contributed to the finding of many additional disease genes.[rr] These include an important study that confirmed that a mutation (A673T) in the proteolytic cleavage site of the gene coding for the amyloid precursor protein (APP) protects against AD, another point to show that amyloid hypothesis for the induction of AD is *not* wrong[29] (see Chapter 5).[ss]

23andMe

SNP mapping and disease risk became quite trendy when 23andMe, named after the 23 pairs of human chromosomes, was set up in 2006 as a "consumer genomics" company. That trendiness continues, as seen in the many advertisements for consumer genomics on the television today.

[pp] I visited deCODE's new facilities in 2015, which had everything that an erstwhile gene hunter might want.

[qq] I had just moved to Biogen at that time, but I would have recommended strongly to my management to buy deCODE as well, given the chance just for their genetic insights into multiple diseases.

[rr] Owing to the population structure, whole-genome sequencing of only tens of thousands of Icelanders allows you to impute the sequence of the remainder.[28]

[ss] One of their most recent achievements was obtaining the sequence of more than 150,000 genomes in the U.K. Biobank.[30]

BOX 3. USING 23andMe

I am not only a 23andMe subscriber but I also subscribe to Ancestry.com. Not surprisingly, and perhaps gratifyingly, a relationship was detected between me and my son William, who is also an Ancestry.com subscriber! It is quite fun. From the 23andMe data, I learned that I was an *ApoE4/E3* heterozygote, which means I have some amount of increased risk of getting Alzheimer's disease, but I had to ask specifically for that information. Otherwise, I learned nothing unexpected: all my SNPs really tell me is that I am of Northern European origin, which I knew already. But the data from these companies can also provide surprises, particularly in ancestry (old and recent!).[tt]

The general idea was to gather genetic information from individuals and build a massive database of the association between diseases that the individual consumers stated they had and known SNPs in the genome. Google founder Sergey Brin, who was married to 23andMe co-founder Anne Wojcicki at the time, had a personal interest in Parkinson's disease (PD) and provided early backing. The company did some quite important work on PD using their database, in collaboration with the Michael J. Fox Foundation. Besides Google, NEA, Genentech, and Mohr Davidow Ventures were initial investors.[uu]

23andMe's business model has vacillated between wanting to be a useful resource for pharma and giving genome-based health and ancestry advice to subscribers (Box 3). Some of this uncertainty derives from a run-in with the FDA, which was not happy with the health claims being made by their direct-to-consumer (DTC) health information business. Their business now looks much more like Ancestry.com, another SNP-based genetics data company whose genotyping is designed to find relationships between individuals in the database by looking for shared DNA segments.

Drug Discovery Aspirations

23andMe has had drug discovery aspirations for their growing database. They formed a relationship with GSK in 2018 for them to use their collected information for drug discovery. GSK invested several hundred million

---

[tt] Levels of unexpected nonpaternity are usually around 15% in various populations.

[uu] I visited them several times at their less than inspiring location in Sunnyvale while at Biogen, but we never managed to get a collaboration together. They now live in more opulent quarters on Oyster Point in South San Francisco.

dollars, gaining access to genotype and phenotype information for more than a million people.[vv]

Financing the continued life of 23andMe has not been straightforward. They did many rounds of private financing at increasing valuations, culminating in an $85 million series F round in December of 2020, capping a total private raise of $850 million. In 2021, they merged with VG Acquisition Corp. (a Richard Branson Virgin Group special purpose acquisition company [SPAC]) and began trading on the NASDAQ shortly afterwards. This is not necessarily the most expedient way to finance the building of a company, but they are still here in business 16 years after their founding. Their longevity has yet to be measured, but they have certainly done better than many other genomics companies (see Chapter 19).

PHARMACOGENETICS

Measuring changes (mutations) in different genes is important not only for drug discovery but also for drug development and drug usage. Many people forget that the drugs they take are metabolised by certain enzymes. Sometimes the drugs are metabolised by the same enzyme, causing what are called drug–drug interactions. Problems can also occur if there are mutations in the metabolic enzymes that speed up or slow down the metabolism of the drugs, because of raising or lowering the concentration of the drug in the bloodstream. There are many clinically important examples of this issue. For example, the breakdown of the antidepressant drug amitriptyline is influenced by two cytochrome P450 enzymes CYP2D6 and CYP2C19. If amitriptyline is metabolised too fast, you may need to adjust the dose or use a different drug. Conversely, if you break down amitriptyline slowly, you will need to take a lower dose.[ww]

Not all the genetic risks of common diseases are mediated by variation in common alleles.[32] Variation in rare alleles is also important, but the only way to find these rare variants is by DNA sequencing, the subject of the next chapter.

[vv] This was apparently larger than the data set available in the public domain that can be found in the U.K. Biobank or FinnGen.

[ww] Another good example is abacavir, a widely prescribed powerful reverse transcriptase inhibitor used to treat HIV. Unfortunately, in some individuals, unacceptable hypersensitivity reactions occurred. This was tracked down to polymorphisms in the Class 1 HLA allele B*57:01. Genetic testing to avoid abacavir hypersensitivity is cost-effective and is routinely used in countries where abacavir is still prescribed.[31]

REFERENCES AND NOTES

1. Monaco AP, Larin Z. 1984. Inside the race to BACS and YACS. *Trends Biotechnol* **12:** 280–286. doi:10.1016/0167-7799(94)90140-6

2. Botstein D, White RL, Skolnick M, Davis RW. 1980. Construction of a genetic linkage map in man using restriction fragment length polymorphisms. *Am J Hum Genet* **32:** 314–331.

3. Strittmatter WJ, Saunders AM, Schmechel D, Pericak-Vance M, Enghild J, Salvesen GS, Roses AD. 1993. Apolipoprotein E: high-avidity binding to β-amyloid and increased frequency of type 4 allele in late-onset familial Alzheimer disease. *Proc Natl Acad Sci* **90:** 1977–1981. doi:10.1073/pnas.90.5.1977

4. Two books describe the history of Vertex: (1) Werth B. 1995. *The billion dollar molecule: one company's quest for the perfect drug.* Simon & Schuster, New York: and (2) Werth B. 2014. *The antidote: inside the world of new pharma.* Simon & Schuster, New York.

5. Marx J. 1989. The cystic fibrosis gene is found. *Science* **245:** 923–925. doi:10.1126/science.2772644

6. Porteous DJ, Dorin JR. 1990. Cloning the cystic fibrosis gene: implications for diagnosis and treatment. *Thorax* **45:** 46–55.

7. Davies K. 2001. *Cracking the genome: inside the race to unlock human DNA.* The Free Press, New York.

8. Kunkel LM. 2005. 2004 WILLIAM ALLEN AWARD ADDRESS: Cloning of the DMD gene. *Am J Hum Genet* **76:** 205–214. doi:10.1086/428143

9. Lehrach H. 2001. Huntington's disease: from gene to potential therapy. *Dialogues Clin Neurosci* **3:** 17–23. doi:10.31887/dcns.2001.3.1/hlehrach

10. Sequana SAB: The SAB consisted of David Housman, an MIT Professor who co-led the group responsible for cloning the gene causing HD and the discovery of the expanded repeat mutation. He is a co-founder of Genzyme Genetics and Somatix Therapy Corp; co-founded Kenna Technologies in 2000 and serves as its Adviser; and founded Integrated Genetics in 1980, which was acquired by Genzyme in 1989. He has served as Chairman, Scientific Founder, and Principal Scientific Advisor of Variagenics Inc. since 1993. Peter Goodfellow was a Professor of Genetics at the University of Cambridge from 1992 to 1996 specialising in finding the genes involved in sex determination. He was a founder of Hexagen, which got acquired by Incyte, and subsequently he became Senior Vice President of Discovery Research, at GlaxoSmithKline, and subsequently worked for the VC firm Abingworth. Tony Monaco was in Lou Kunkel's lab and was responsible as a graduate student for cloning the gene for DMD. He was a postdoc at ICRF and became a faculty member at the Institute of Molecular Medicine at Oxford. He was also the Head of the Neurodevelopmental and Neurological Disorders Group at the Wellcome Trust Centre for Human Genetics. After being a pro vice chancellor at Oxford, Tony has become a very successful university administrator and was the President of Tufts University in Boston. John Todd is now Professor of Precision Medicine at the University of Oxford and Director of the Wellcome Centre for Human Genetics and of the JDRF/Wellcome Diabetes and Inflammation Laboratory (DIL). Until 2016, he was a Professor of Medical Genetics at the University of Cambridge and before that

Professor of Human Genetics and a Wellcome Trust Principal Research Fellow working on type 1 diabetes genetics and has provided many insights into the genetics of that disease. Hans Lehrach was head of the Genome Analysis lab at ICRF in London and also part of the Huntington's Gene Cloning Consortium. At ICRF, he designed and constructed one of the first picking and spotting array robots for fingerprinting. In 1994, he became the Director of the Institute for Molecular Genetics in Berlin and was a major player in the groups sequencing chromosomes 8 and 21. He also co-founded many German biotech companies in Germany. Alan Bradley was an assistant professor at Baylor College of Medicine in Houston. Bradley was appointed an HHMI Investigator in 1993 and Director of the Wellcome Trust Sanger Institute from October 2000 to April 2010, working on stem cells and using mouse genetics to make models of human disease. Mark Boguski, a bioinformatics expert, died in 2021. He was a pioneer in the fields of Bioinformatics, Genomics, and Precision Medicine. He had broad experience in government, academia, and industry. He was one of the original Medical Staff Fellows at the National Library of Medicine's National Center for Biotechnology Information and worked with the Human Genome Project from 1990 to 2003. He served on the faculties of Harvard Medical School and Beth Israel Deaconess Medical Center, where he founded the Genomic Medicine Initiative in 2009. He also held faculty positions at the Johns Hopkins University School of Medicine, the Fred Hutchinson Cancer Research Center, and the U.S. National Institutes of Health. Dr. Boguski served as the founding Director of the Paul Allen Institute for Brain Science and as Vice President of the Novartis Institutes for Biomedical Research.

11. Tugores A, Le J, Sorokina I, Snijders A, Duyao M, Reddy PS, Carlee L, Ronshaugen M, Mushegian A, et al. 2001. The epithelium-specific ETS protein EHF/ESE-3 is a context-dependent transcriptional repressor downstream of MAPK signaling cascades. *J Biol Chem* 274: 20397–20406. doi:10.1074/jbc.M010930200

12. Kleyn PW, Fan W, Kovats SG, Lee JJ, Pulido JC, Wu Y, Berkemeier LR, Misumi DJ, Holmgren L, Charlat O, et al. 1996. Identification and characterization of the mouse obesity gene *tubby*: a member of a novel gene family. *Cell* 85: 281–290. doi:10.1016/S0092-8674(00)81104-6

13. Conversation with Mark Levin, 2021.

14. LOD score: A LOD (short for "logarithm of the odds") score is a statistical estimate of the relative probability that two loci (e.g., a disease-associated gene and another sequence of interest, such as a variant or another gene) are located near each other on a chromosome and are therefore likely to be inherited together.

15. See page 28 of Millennium IPO S1 document. https://www.sec.gov/Archives/edgar/data/1903995/000121390022067279/ea166561-f1_millennium.htm

16. Watkins MD, Matthews SG. 1999. Strategic deal-making at Millennium Pharmaceuticals. *Harvard Business Review*, November 5, 1999. https://store.hbr.org/product/strategic-deal-making-at-millennium-pharmaceuticals/800032

17. Conversation with Stefan Lohmer, 2022.

18. Technology Review. 2001. Medicine's new Millennium: new information about genes and proteins promises precise diagnostics. Millennium Pharmaceuticals CEO is at the forefront of this medical transformation. *MIT Technology Review*, December 2001. https://www.technologyreview.com/2001/12/01/235343/medicines-new-millennium/

19. Miki Y, Swensen J, Shattuck-Eidens D, Futreal PA, Harshman K, Tavtigian S, Liu Q, Cochran C, Bennett LM, Ding W, et al. 1994. A strong candidate for the breast and ovarian cancer susceptibility gene *BRCA1*. *Science* **266**: 66–71. doi:10.1126/science.7545954

20. King MC. 2014. "The race" to clone *BRCA1*. *Science* **343**: 1462–1465. doi:10.1126/science.1251900

21. Hurst J. 2014. Pioneering geneticist Mary-Claire King receives the 2014 Lasker–Koshland special achievement award in medical science. *J Clin Invest* **124**: 4148–4151. doi:10.1172/JCI78507

22. Gold RE, Carbone J. 2010. Myriad genetics: in the eye of the storm. *Genet Med* **12**: S39–S70. doi:10.1097/GIM.0b013e3181d72661

23. Myriad Genetics website: The first and only genetic test to blend comprehensive assessment for hereditary cancer risk with the precision of genomic breast cancer risk assessment for patients of all ancestries. These scores are derived by looking at the composite number of SNPs that individuals possess, which are identical to those that across a population are known to be the risk allele. A high polygenic score indicated that at most of the disease relevant sites, the individual has inherited the disease-associated SNP and not the normal sequence at that locus.

24. Conversation with Rick Myers, 2021.

25. Founder effect: In population genetics, the founder effect is the loss of genetic variation that occurs when a new population is established by a very small number of individuals from a larger population. It was first fully outlined by Ernst Mayr in 1942: Mayr E. 1942. *Systematics and the origin of species*. Columbia University Press, New York.

26. Vogel G. 1998. Icelandic isolation pays off. *Science* **279**: 991.

27. Gulger J, Helgason A, Stefansson K. 2002. Genetic homogeneity of Icelanders. *Nat Genet* **26**: 395.

28. Imputation in genetics is the statistical inference of unobserved genotypes. It is achieved by using known haplotypes in a population. This allows the testing of the association between a trait of interest (e.g., a disease) and experimentally untyped genetic variants, but whose genotypes or sequence have been statistically inferred or "imputed." Genotype imputation can be used with dense SNPs and whole genome sequence information.

29. Jonsson T. 2012. A mutation in *APP* protects against Alzheimer's disease and age-related cognitive decline. *Nature* **488**: 96–99. doi:10.1038/nature11283

30. Halldorsson B, Eggertsson HP, Moore KHS, Hauswedell H, Eiriksson O, Ulfarsson MO, Palsson G, Hardarson MT, Oddsson A, Jensson BO, et al. 2022. The sequences of 150,119 genomes in the UK Biobank. *Nature* **607**: 732–740. doi:10.1038/s41586-022-04965-X

31. Davis DM. 2013. *The compatibility gene: how our bodies fight disease, attract others, and define our selves*. Oxford University Press, London.

32. Allele definition: An allele is a variation of the sequence of nucleotides at the same place in the sequence of a gene. If there is one nucleotide change in one place in the gene (e.g., from an A to a G), there will be two alleles of the gene—the A allele and the G allele. An individual may be a homozygote with two copies of either the A gene or the G gene. A heterozygote has one chromosome containing the A allele and one containing the G allele (AG).

CHAPTER 8

DNA Sequencing and the Human Genome

Recombinant DNA and monoclonal antibody technologies drove the initial wave of biotech companies and defined the start of the industry. The development of DNA sequencing technologies from 1972 on helped to define its continuation.

June 2000 is recognised as the date that the human genome sequence was "completed" in a tie between the National Institutes of Health (NIH)–Wellcome Trust publicly funded Human Genome Project (HGP) consortium and the "commercial" human genome sequence provided by Celera, a division of PerkinElmer. It was really only a first draft of the human genome sequence, being about 85% complete. Many mistakes and incomplete regions remained in both the public and private data sets. Indeed, it is only very recently that it can be said that the complete sequence is now known, including all the difficult-to-sequence repetitive regions near the centromeres of the chromosomes.[1,2]

Politics and hype aside, this fantastic achievement would not have been possible without enormous gains in the efficiency of DNA sequencing based on the original methods, as well as the development of completely new ways of sequencing nucleic acids ("next-generation sequencing, or "NGS") that are being employed today.[a]

This chapter will cover the various nucleic acid sequencing technologies that have been developed over the years and some of the companies that exploited them.[b] It is important to appreciate firstly how quickly the scale of

---

[a] It has been said that using the NGS methods of today the whole human genome sequence could be completed in two weeks. I think that rather depends on how you do the arithmetic, but I doubt it. Three billion base pairs are a lot of nucleotides to sequence, and there are many very difficult repetitive regions to work through, even with today's methods. This is also why it took 22 years to go from what was a draft sequence to the complete sequence.

[b] As always, from my perspective and experience.

sequencing went up (faster than Moore's Law[3]), how the costs came down, and who were the companies that developed the sequencing machines and capabilities of today. We will then go on to cover sequencing applications and the companies who applied the technology, and their relevance to drug discovery and to the biotech industry. For historical completeness, we review the role of private enterprise in the form of Celera Genomics in the Human Genome Project. It is a story that has been told before, but I have some anecdotes that have not been shared previously that should make it interesting enough to revisit.

ORIGINS

The first nucleic acid sequenced was RNA, not DNA. Robert Holley, then at Cornell, deciphered the sequence and structure of alanyl t-RNA in 1965.[c] Walter Fiers, who cloned interferon and was a founder of Biogen, was the first to sequence a complete gene—that coding for the coat protein of the RNA bacteriophage MS2, in 1972.[4]

DNA Sequencing Methods

After elucidating the amino acid sequence of insulin, Fred Sanger turned his own hand and that of his lab at the LMB in Cambridge (UK) to sequencing DNA. Fred did not publish very much, but when he did people took notice. The method that the Sanger group developed was based on using DNA polymerase to copy a DNA template.[5] Simultaneously, Allan Maxam and Walter Gilbert at Harvard introduced a method for DNA sequencing that was based on chemical modification of DNA[6] (see Box 1 for brief descriptions of both approaches).

I spent many happy hours running Maxam–Gilbert sequencing gels at Celltech: pouring the gels to avoid bubbles was something of an art! However, once the radioactive labels had been replaced by fluorescent ones, almost everyone used Sanger dideoxy sequencing, including the machines that were made to automate the process.

---

[c] He shared the Nobel Prize in Physiology or Medicine in 1968 with H. Gobind Khorana for his work on DNA polymerases and Marshall Nirenberg for deciphering the genetic code.

BOX 1. EARLY DNA SEQUENCING METHODS

See Plate 15.

Sanger Sequencing

In this method, DNA polymerase and the four normal deoxynucleotides (A, C, T, G) are added to a template DNA of interest. Also in the mix are low levels of one of the four radioactively (or, later, fluorescently) labelled "dideoxynucleotide" forms. If the dideoxynucleotide is randomly inserted by the polymerase in place of the "normal" deoxynucleotide as it copies the template, it terminates the growing chain at that spot. Sequencing was done in four parallel polymerase-driven syntheses each with a different dideoxynucleotide, and the resulting fragments were then separated by size in parallel lanes on polyacrylamide gels. The sequence could then be read off from the ladder of terminated products via an X-ray film of the gel. Depending on the quality of the gel itself, up to 800 bases at a time could be read by eye from the X-ray film.

Maxam–Gilbert Sequencing

The DNA of interest was radioactively labelled at one end and treated with chemicals that specifically modified both the purine and pyrimidine bases. For example, dimethyl sulphate primarily modified guanines, and hydrazine (with and without salt) affected the pyrimidine bases (T and C). The modified DNAs were then cleaved at the position of the modified base using piperidine. The concentration of the modifying chemical was controlled to introduce, on average, one modification per DNA molecule, so that on cleavage a series of radioactively labelled fragments of different sizes were generated. The fragments in the four reactions were then subjected to electrophoresis side by side on denaturing acrylamide gels to separate the fragments, which were visualised by the X-ray film. As with Sanger sequencing, the sequence could be read off the X-ray film from the smallest fragments to the largest ones as a ladder (see Plate 15).

FIRST-GENERATION SEQUENCING

Leroy Hood, who was a professor at the University of Washington, was intimately involved in a company called Applied Biosystems (ABI) set up in the early 1980s to automate protein sequencing, owing largely to the growing demand for protein sequencing (i.e., determining their amino acid makeup) to reverse engineer oligonucleotide probes for screening recombinant DNA libraries (Chapter 2). The first protein sequencing machine, called the ABI 470A, was made available in 1982. With Michael Hunkapiller and Lloyd Smith, Hood and ABI were also the first to fully automate the Sanger sequencing method (Box 2).

BOX 2. FIRST AUTOMATED DNA SEQUENCING MACHINES

The first automated DNA sequencer was the AB370A, introduced in 1986. The AB370A was able to sequence 96 DNA samples simultaneously and read about 600 bases per lane using slab gels, or about 500 kb per day. The successor instrument, the AB3700 PRISM, replaced the slab gel with a linear polymer in a capillary tube to separate the DNA, with a much higher throughput. More than a thousand of these PRISM machines were used in the initial sequencing of the human genome by both the HGP and Celera. Indeed, the only competition to the ABI machines was the capillary-based machines that had been developed by Molecular Dynamics, using the same sequencing principles. Many of these "MegaBACE" machines were bought by Incyte and several of the HGP Genome Sequencing Centers. Molecular Dynamics were later bought by Amersham International (who were subsequently acquired by General Electric). Despite some short-term sales successes, Amersham International were never able to get the market penetration afforded to ABI, anchored as they were by the intensely competitive HGP and Celera sequencing activities.

NEXT-GENERATION SEQUENCING

The development of the so-called NGS systems has some very interesting and diverse origins. Mathias Uhlén, a Swedish scientist and now Professor of Microbiology at the Royal Institute of Technology, Stockholm, and others first recognised that the pyrophosphate released during DNA polymerase-driven DNA synthesis could be measured by luminescence. Jonathan Rothberg and colleagues at 454 Life Sciences, a year 2000 spinout from CuraGen in Branford, Connecticut (near Yale University) in New Haven, first automated this approach.

454 Life Sciences

CuraGen was originally a genomic services business that Rothberg set up in 1996. They had several quite valuable government grants allowing them to develop differential display chips that could be used to analyse the abundance of RNA transcripts at some scale in etched glass capillaries. They had a large alliance with Pioneer Hybrid, an Ag biotech company that got acquired later by DuPont. CuraGen had also developed some sophisticated bioinformatics for analysing the data they generated. "Gene-calling" was developed for the transcriptional data and "path-calling" for protein–protein interaction data from both flies and yeast.[7]

One of Jonathan's children was born with the genetic disorder tuberous sclerosis complex, a rare disease in which benign tumours grow in many different organs of the body. This led to Rothberg thinking that a new kind of rapid and cost-effective genome-sequencing system could be very useful as an aid to diagnosing kids with genetic disorders and that became the focus of 454.

454 "pyrosequencing" (Box 3) started off as a skunkworks project in Branford, away from the main CuraGen labs. Their name was apparently derived from the number of a CISCO chip, important in the early days of the game-changing CISCO routers.[d] The 454 Bioinformatics team led by Martin Leach (who subsequently went to Biogen where I met him) was quite radical. The base-calling software and the computer hardware needed to analyse it was anything but trivial. The prime mover in this domain was Scott Helgerson who worked with Bradley (Brad) Carvey.[e]

Roche Diagnostics acquired 454 Life Sciences for $154.9 million in March 2007, and Brad moved on to PacBio (see below). Jonathan, despite coming from the wealthy family that started LATICRETE International (flooring materials) in Connecticut in 1956, became independently wealthy himself. CuraGen tried to become a small product company by developing a fibroblast growth factor product with drug developers they hired from Bayer. It failed to make it and the company got absorbed into Celldex in 2009.

BOX 3. PYROSEQUENCING

454 pyrosequencing consists of the cyclic flow of nucleotide reagents over a PicoTiterPlate. Each plate has approximately one million wells, each containing a bead carrying a unique single-stranded DNA fragment to which the nucleotides are added in a stepwise manner (and the fluorescence detected). 454 Life Sciences released their first next-generation solid phase sequencer (the GS20) in 2005. Three years later, with the GS FLX Titanium series of reagents, they created the ability to sequence 400–600 million base pairs per run with 400–500 base pair read lengths.

---

[d] Others claim that the name was derived from the temperature at which coins melt: 454 was certainly going to burn money.

[e] Brad was the brother of Dana Carvey (of *Saturday Night Live* and *Wayne's World* fame) and had built the first wire-wrapped "Video Toaster" for editing movies.[7] Dana apparently based his Garth Algar character on his "nerdy" elder brother.

SEQUENCING BY SYNTHESIS: SOLEXA AND ILLUMINA

One of the conceptual breakthroughs for very large-scale sequencing happened by accident, based on work done in Cambridge (UK) in 1998 by Shankar Balasubramanian and David Klenerman. They were interested in the kinetics of single-chain synthesis by DNA polymerase and devised methods to observe the synthesis of a DNA chain—one chain at a time, nucleotide by nucleotide. By being next door to the rich history of nucleic acid sequencing in Cambridge, they conceived (probably in the pub) a short-read solid phase sequencing method using reversible chain terminators, which they called "sequencing by synthesis" (Box 4). Their technology development ultimately led to Illumina and their Genome Analyzer instrument, acquired with their purchase of Solexa.

Illumina: The Genome Analyzer

The first Illumina (née Solexa) sequencer, the Genome Analyzer, was launched in 2006, giving scientists the power to generate as much as 1 gigabase (Gb) of sequence data in a single run.[f] The technology development

---

BOX 4. SEQUENCING BY SYNTHESIS

In sequencing by synthesis, DNA chains are fixed randomly to a solid support, and DNA synthesis is observed on each chain by the release of fluorescence whenever one of the four nucleotides is added to the new DNA strand. Solexa was formed with venture capital support from Abingworth to commercialize this approach in 1992, attracting a series A round of £12 million. Solexa subsequently acquired a company called Manteia that had developed a molecular clustering technology that reduced the cost of the optics needed to "see" the fluorescence at each step, and which enhanced the fidelity of the base calling. In 2005, Solexa acquired Lynx Therapeutics and became a public company. Lynx, which started in 1992, was yet another brainchild of Sydney Brenner and had a system called Megaclone, which transformed a sample containing millions of DNA molecules into one made up of millions of microbeads, each of which carried approximately 100,000 copies of one of the DNA molecules in the sample. This breakthrough was important for the subsequent development of the Solexa sequencing machine: Lynx's production and engineering teams essentially turned the Solexa prototype into a hugely successful commercial sequencing instrument. Solexa was acquired by Illumina in early 2007.

---

[f] For reference, the human genome is ~3.4 Gb.

set the scene for the often-shown sequencing PowerPoint slide depicting that "NGS data output had increased at a rate that outpacing Moore's law—more than doubling each year." Illumina were the clear leader in selling sequencing machines and sequencing reagents. They launched increasingly high-throughput and accurate machines and increasingly expensive reagents (see Plate 16 for a picture of the HudsonAlpha sequencing farm). They also developed smaller-throughput machines for "benchtop sequencing." By 2017, Illumina's NovaSeq platforms reached new heights in sequencing power for commercial platforms. Per run, the NovaSeq 6000 (S4 flow cell) generated an output of up to 3000 Gb of DNA sequence.[5]

Marco Island: Emerging Sequencing Companies and Third-Generation Sequencing

Representing SAIC Frederick, David Munroe and I went to the Marco Island Advances in Genome Biology and Technology (ABGT) meeting in 2010. We were particularly struck by the innovation that was happening in the sequencing field at that time, inspiring us to write up a meeting report that got published in *Nature Biotechnology* in May 2010.[8] It was not just the fireworks that inspired us, nor the scale-up developments of Illumina, Life Technologies, and Complete Genomics, but it was the prospect of long-read sequencing on emerging new large and small machines, including offerings from Pacific Biosciences, Ion Torrent, and Nanopore that were most exciting.

Complete Genomics

Complete Genomics was founded in March 2006 by Clifford (Cliff) Cliff Reid, Radoje (Rade) Drmanac, and John Curson. Inevitably, it was referred to by some, including the typically acerbic Sydney Brenner, as "Incomplete Genomics," but contrary to my expectations at least, their technology has

---

[8]I remember the new Illumina building in San Diego. It was up the hill from SGX Pharmaceuticals on Sorrento Valley Rd near UCSD, and my lunchtime runs would take me up a trail that led past their "campus." I mused often on whether to buy their stock, but because it never traded very well at the beginning, I never did. It was only after the Solexa acquisition in 2007 that the stock began to move. An investor who bought in at the Illumina IPO in 2000 and hung in there until 2015 would have made a greater than 1000× return. A rather painless way to turn a thousand dollars into a million dollars over 15 years.

BOX 5. MGI TECHNOLOGY

BGI (Beijing Genomics) has sequenced thousands of human genomes to date using their MGI technology, which is a legitimate competitor to Illumina. The technology is based on sticking DNA to self-adhering nanoballs, which then form arrays on which the sequencing takes place in specially designed flow cells. MGI recently launched a new commercial sequencing chemistry called HotMPS for its DNBSEQ-G400 sequencer. By adding all four labelled bases at once, HotMPS can achieve low error rates and low duplication rates, while being compatible with commonly used library preparation methods. It is unique in using very sensitive fluorescent antibodies rather than labelled nucleotides as the detection reagents.

---

endured and been developed. The company was acquired in 2013 by BGI Genomics, a Chinese genomics entity that has become one of the world's largest genomics service companies (Box 5).

THIRD-GENERATION SEQUENCING

Ion Torrent

Jonathan Rothberg's team at Ion Torrent (founded in 2007) produced the Ion Torrent Personal Genome Machine (PGM) that read DNA sequences by measuring the protons released at each (nucleotide binding) sequencing step. It was set up as an array of reaction wells supporting a proton-detecting semiconductor for sequencing. It was much more affordable than the Illumina offerings as it did not need the optics, lasers, and other paraphernalia that these machines needed. Ion Torrent claimed that they could do 100–200 base reads in one to two hours on an instrument not much larger than a microwave.

To market their wares, the company held a competition for the best sequencing proposals, with the winner receiving a PGM. As an invited judge, I picked a proposal for mutation detection in an oncology clinical setting. Life Technologies, now part of Thermo Fisher, acquired Ion Torrent for $375 million in 2010. Like Craig Venter, Jonathan Rothberg is into ocean-going yachts. He can afford it. His current 180-ft superyacht is called *Gene Machine*, which, like Craig Venter's yacht, has a lab on board: the not-much-smaller support boat is called *Gene Chaser*.[h]

---

[h] I most certainly do not begrudge him those acquisitions, but if it was me—it would be a racing car or cars.

Pacific Biosciences

"PacBio," as they are known (originally Nanofluidics Corp), is another pioneer in third-generation sequencing that really got going in 2004. The company raised nearly $400 million in six rounds of venture financing, making it one of the most capitalized start-ups in 2010, leading up to an initial public offering that October. Key investors included KP, Alloy Ventures, and the Wellcome Trust. In their initial public offering, PacBio sold 12.5 million shares at $16/share, raising an additional $200 million.

PacBio introduced a Single-Molecule Real-Time (SMRT) sequencing system using a zero-mode waveguide (ZMW) approach, which did not require amplification of the DNA before sequencing. A huge advantage of this method is its ability to do very long reads—initially of up to 1500 bases. The system works because the activity of DNA polymerase incorporating a single nucleotide can easily be detected within the volume of an illuminated ZMW (20×10^{-21} litres). When DNA and fluorescently labelled nucleotides are added to the guide, new strand synthesis occurs. The incorporation of a particular single fluorescent nucleotide can be detected directly by a laser below the ZMW.

The company's first commercial product, the PacBio RS, was sold to a limited set of customers in 2010 and fully released in early 2011.[i] There were several teething problems with the machine initially, largely because of the instability of the position of the laser, but the company addressed these issues. The PacBio RS II, with many more ZMWs, newly engineered polymerases, and updated chemistry, was released in April 2013 and in September 2013 Roche Diagnostics formed a $75 million partnership with PacBio to develop sequence-based diagnostic products. The Sequel System, with an approximately sevenfold greater capacity than the PacBio RS II, was introduced in 2015. It was followed a few years later (2019) by the Sequel II system that brought a further eightfold increase in throughput with even greater accuracy.

As a testament to the importance of long-read sequencing, Illumina tried but failed to acquire PacBio in 2018.[j]

---

[i] While at SAIC-Frederick, David Munroe and I persuaded the National Cancer Institute (NCI) to buy one of these first machines, thus becoming one of the 10 test sites for this technology in the United States.

[j] The Federal Trade Commission (FTC) opposed the merger, alleging that Illumina was expanding its monopoly in NGS systems and that future competition would be eliminated if the acquisition was allowed to go forward.

Oxford Nanopore Technologies

Another important third-generation sequencing company is Oxford Nanopore Technologies (UK). They have managed to build a small sequencing machine about twice the size of a computer flash drive (the MinION), in which DNA is squeezed through a biological nanopore. The machine uses the changes in electrical conductivity that occur during this process to read the nucleotide sequence as the single strand of DNA passes through the pore. In addition to the MinION, Oxford Nanopore offers the compact benchtop GridION (multiple simultaneous runs) and Flongle, an adapter for a single flow cell. Their nanopore technology can do both RNA and DNA sequencing, with reads up to more than 2 megabases!

Oxford Nanopore was formed in 2005 by Gordon Sanghera, James (Spike) Willcocks, and Hagan Bayley. In 2020, the company closed three separate financings raising more than $300 million, before going public in 2021 on the London stock exchange—unusual for a company of this type. They raised more than $250 million in an offering led by Bank of America, J.P. Morgan, Cazenove, and CitiCorp, attracting multiple tech investors that drove the company value to more than $6 billion shortly afterwards.

The ability to do very long and accurate reads played a critical role in the completion of the human genome sequence in 2022.[1,2] But long-read sequencing complements short-read sequencing and does not in principle have to be so accurate on a base-by-base level. The Oxford Nanopore devices came of age during the COVID pandemic, as they were used to sequence the coronavirus ribonucleic acid repeatedly to detect new variants.

EMERGING SEQUENCING TECHNOLOGIES

Several new trends are happening in sequencing technology in 2022 and 2023.[9] Illumina is facing significant competition as their IP position begins to run out of time. BGI, MGI Tech, and Illumina recently settled several of their various ongoing patent disputes.[k] Illumina paid Complete Genomics $325 million for a license to the MGI two-colour sequencing methods. As part of a cross-licensing deal, BGI will be able to use the Illumina "image mix" patents. PacBio announced plans in late 2022 for a short-read sequencer known as "Onso," a result of PacBio's $800 million acquisition of Omniome in 2021. PacBio is also launching a new long-read sequencer called Revio (Box 6).

---

[k]MGI was a very legitimate competitor to Illumina, which has traditionally had 80% of the sequencing machine market.

BOX 6. NEW ENTRANTS TO NUCLEIC ACID SEQUENCING

Ultima Genomics is one young company hoping to poach some of Illumina's market for short-read sequencing and at a cheaper price, even floating a potential future price point of $100/genome. The company has found new ways of distributing the DNA beads for the sequencing process so that a camera measuring synthesis can move in a circle like a compact disc player, along with new variations on the nucleotide synthesis chemistry. Repetitive regions will still be a challenge, but the sequencing of these regions will be helped by the long-read methods.

Other companies, such as Element Biosciences formed by Illumina alumnus Molly He, have developed a benchtop sequencer (AVITI) that can sequence three human genomes at a time, at a cost of $560 each. Singular Genomics, another San Diego–based company, have launched a four flow-cell machine with fast run times. And, in October 2022, Illumina revealed their new sequencing instruments (the NovaSeq X series), designed to maintain their leadership in NGS in the face of mounting competition. However, it may not be commercially available until 2024, giving some of these newer players a bit of a head start.[10]

SEQUENCING APPLICATIONS

The history of advances in sequencing technologies reveals a synergy with the development of very impactful applications of those technologies, primarily directed towards finding disease-related genes and drug targets.

Expressed Sequence Tags

Messenger RNA is the nucleic acid that is translated to make protein and, as its name suggests, is how information is transferred from the DNA genes to the proteins. Francis Crick defined the central dogma of "DNA to RNA to protein," but it is often forgotten that it was Sydney Brenner (with others—notably François Jacob and Jacques Monod) who not only suggested the existence of mRNA, but also did the experiments to prove it.[11,12] The work by Marshall Nirenberg, for which he was awarded the Nobel Prize in Physiology or Medicine in 1968, elegantly determined which triplets coded for particular amino acids, but the fact that the code was triplet in nature had also already been determined by Brenner and Crick.

As discussed in previous chapters, cDNA for cloning genes was made by reverse transcribing mRNA. Sydney Brenner proposed in 1986 that scientists should first sequence only the parts (some 2%–3%) of the genome that coded for proteins via mRNA. Craig Venter, who was at the NIH at the time, also appreciated that sequencing cDNAs would give information about the mRNAs that they were derived from. Sequencing of the protein-coding region of many genes via cDNA would thus generate amino acid sequences for a great many proteins, with multiple applications in science and medicine.

Sanger sequencing and the ABI 373A machines were used to sequence these cloned "expressed sequence tags" or ESTs, as they became known. It allowed comparison of the sequences of known genes in both humans and other organisms and contributed to the building of a database of coding sequences. There was also the opportunity to file IP on the new genes (or even bits of them) that were found this way. Venter and others were particularly keen to follow this approach.[13]

INCYTE GENOMICS

Incyte Genomics, Inc., incorporated in 1991 under the name Incyte Pharmaceuticals, were created by Schroder Venture Advisers to buy assets and technology from Invitron Corporation. Roy Whitfield, formerly the president of Invitron's subsidiary, Ideon, became CEO and Randy Scott, one of Invitron's founding scientists, became Incyte's CSO.

Incyte leaders believed (along with many others) that the new genomic technologies, especially finding genes and working out their function, would lead to new drugs. They also believed that EST sequencing was a cost-effective and fast way to find genes, and they used both the MegaBACE and the ABI machine to do their sequencing. With a goal of developing "gene-driven drugs" based on its own discoveries, Incyte planned to develop therapeutic proteins by partnering with Genentech to pay for human trials of their candidate drugs. However, the early data they obtained were disappointing, and Genentech backed out. Incyte then turned to large-scale EST sequencing, building a bioinformatics team that could construct an EST database to which drug companies could buy a subscription (Box 7). Incyte went public in November 1993 on the American Stock Exchange, selling 2.3 million shares at $7.50.[1] The proceeds were used to expand Incyte's high-throughput gene sequencing and analysis programs.

---

[1] The company later traded on the NASDAQ.

BOX 7. THE LifeSeq DATABASE

Incyte's first product was a database called LifeSeq, a tool that "promised to revolutionize the pharmaceutical industry's research." LifeSeq consisted of two parts: the database of DNA sequences and a database of where the sequences were found and their relative abundance (i.e., the tissues and cells in which they were expressed and at what levels). The primary starting materials were cDNA libraries made from the tissues and cell lines. Homology to known proteins with known functional information meant that a function could be tentatively ascribed to some of the new proteins. The data could also be used, for example, to define the difference in both quantitative and qualitative gene expression between diseased and normal tissue.

Incyte developed the database throughout the 1990s, and companies could purchase yearly subscriptions for access to the data. Pfizer was Incyte's first major subscriber (the agreement, valued at $24.8 million, was forged in June 1994). Incyte sold subscriptions on a nonexclusive basis, with the goal of making LifeSeq available to as many companies as possible. By January 1996, in addition to Pfizer, Incyte had signed up Johnson & Johnson, Abbott Laboratories, Hoechst AG, Novo Nordisk A/S, and Pharmacia & Upjohn. These subscribers brought in more than $100 million per annum in revenue. By the year 2000, when the human genome sequence was announced as "completed," Incyte had most of the top pharma companies as subscribers. Incyte then added rapidly accumulating public EST information to their database. They followed the then-standard practice of filing IP on those genes and sequences that they recognised as having value for drug discovery. This enabled them to make one version of LifeSeq available to all while keeping "LifeSeq Gold" for subscribers only. The database contained transcripts from thousands of genes, more than half of which were proprietary to Incyte (i.e., scientists could not access information about these genes from any other commercial source).

Given the inevitable democratization of gene sequence information, Incyte looked to diversify quite early.[m] In 1996, Incyte acquired a St. Louis company called Genome Systems for about $8 million in stock. Founded in 1992 out of Washington University in St. Louis, Genome Systems had built a "gene depot" of more than 100 industrial freezers, each containing half a million DNA clones stored in their own individual test tubes. The Incyte warehouse supplied the genetic material of all kinds of organisms, including humans, mice, flies, worms, fish, rats, dogs, plants, and bacteria to whoever requested them. Scientists at universities, government institutions, or private companies could purchase

---

[m] Their early diversification efforts included a 1997 collaboration with Smith Kline Beecham in a new company called DiaDexus, to develop gene-based diagnostic products. It was not especially successful.

individual DNA clones for as little as $22 for a common gene fragment or up to several thousand dollars for a critical, full-length gene.

Invitrogen in Carlsbad California (see Chapter 17) was really their only competitor in this "custom genomics" market.

Genome-wide association studies (GWAS) using gene arrays developed by companies like Affymetrix were becoming important in locating the variations in genes associated with susceptibility to various diseases (see Chapter 7). Incyte bought into this space in 1998 by acquiring Hexagen, an Abingworth Ventures financed British genetics start-up in Cambridge (UK) founded by Peter Goodfellow, for $41 million. Incyte acquired Hexagen primarily for the company's signature technology for the rapid identification of differences in DNA sequences from one individual to the next (Box 8).[n]

BOX 8. LATER INCYTE PRODUCTS

From 1997 to 1998, Incyte distinguished itself from most small biotech companies by being profitable. It also put considerable pressure on Celera (see below) to do the same.

Incyte Pharmaceuticals changed its name to Incyte Genomics, Inc. in 2000, to communicate the company's commitment to providing genomic information on a nonexclusive basis to biotechnology, pharmaceutical, and academic researchers worldwide. At the same time, Incyte launched its website, Incyte .com, and with it, the beginning of its e-commerce genomics program, known as LifeSeq GENE-BY-GENE.

With the new website, any scientist with Internet access could submit a question about a specific gene of interest. The "best view" of the gene sequence was available for free, and additional gene sequence information about that gene, as well as a physical copy of the gene, could be purchased via e-mail. Information about every gene in Incyte's database thus became equally accessible and affordable to all comers (a precursor to all the information you can find on particular genes and their relationships online today).

The business interest of most genomics companies after June 2000 (the publication of the human genome sequence) turned to trying to commercialise the proteins coded for by the genes. To that end, Incyte bought Proteome, a "proteomics knowledge company," to acquire their "BioKnowledge" protein function database. In 2001, Incyte formed a strategic alliance with Lexicon Genetics to access LexVision, a specialized gene function database. Lexicon had developed the Genome5000 program, in which the role and function of nearly 5000 genes were elaborated to pinpoint key targets for drug development. Incyte also held an extensive commercial portfolio of U.S.-issued patents for many full-length human genes and the proteins they encode.

---

[n] The Hexagen acquisition was set up by Incyte as a separate entity called Incyte Genetics.

Despite these "knowledge" acquisitions, it was clear by the end of 2001 that the genomics business was not going to maintain Incyte's revenue base. So, in what most of us thought to be a bit of a stretch for a genomics company with a database of potential targets, Incyte announced a complete restructuring to become a fully integrated pharmaceutical company. They signed key agreements with two pharmaceutical companies, Genentech and Medarex, to collaborate on drug development and they hired some very good and experienced industry professionals from DuPont Merck—Paul Friedman, Bob Stein, and Reid Huber—to run it. Over time, Incyte developed an impressive pipeline of innovative drugs for cancer and inflammation, by both acquisition and in-house research. Among the important drugs they have developed are orally bioavailable Janus kinase (JAK) inhibitors for multiple myeloma and other indications, and they acquired Ariad Pharmaceuticals to add ponatinib (Iclusig), a potent second-generation BCR-ABL inhibitor for both chronic myeloid leukemia (CML) and acute lymphoblastic leukemia (ALL) to their roster.

THE INSTITUTE FOR GENOMIC RESEARCH AND HUMAN GENOME SCIENCES

The commercial possibilities of the EST sequencing approach were clear to many others besides Incyte. Craig Venter, a Vietnam veteran and a bit of a maverick, clearly appreciated the potential value of the cDNA sequences that his lab at the NIH were generating. Despite considerable opposition from conservative NIH researchers, but in line with the Bayh–Dole Act of 1980, Venter tried hard to get the NIH to file patents on the sequences he was identifying. Although intellectual property (IP) was filed, the claims were rejected on the grounds of novelty, obviousness, and utility, and the Director of the NIH (Harold Varmus) decided not to continue to pursue the applications.° Venter's group published a major paper on ESTs in June 1991 in *Science*,[13] with other papers following. By applying this technology to brain and other cDNA libraries, Venter's team identified many human homologues of genes that had been found in other organisms.

Although Venter liked the press and the attention to his controversial activities, he was tired of NIH politics.

The late venture capitalist Wally Steinberg, who ran the venture fund HealthCare Ventures, encouraged Venter to leave NIH. The 1992 proposal

---

°Although these were mostly just thousands of partial sequences, I can see the utility argument but not the other grounds.

was for him to run the aptly named TIGR—The Institute for Genomic Research—which would continue to sequence ESTs and file patents, working alongside a new company set up by Steinberg called Human Genome Sciences (HGS), founded to exploit the findings to make drugs. William (Bill) Haseltine, a Harvard academic with a track record of developing HIV drugs, was chosen as CEO of HGS.

The personalities and backgrounds of Venter and Haseltine could not have been more distinct. Haseltine was east coast and Ivy League. I remember very well meeting him—or rather, not meeting him—at the palatial, no-expense-spared HGS headquarters in Rockville.[p] I had made an appointment to see him, but he kept me waiting for more than 45 minutes so I simply left. This seemed to endear me to him, because that was followed up by an invitation to meet him at his expansive apartment at the Pierre Hotel in New York.[q] I remember this meeting very well too, because the apartment was like something out of an old English country house with a stuffed leopard rug on the floor and other colonial décor.

I thought Bill would always be most happy wearing a suit. He was reasonably diplomatic but pleased with himself, in the way that a smart and aggressive politician might be. He was already super wealthy, having married Gale Hayman, who had built the Giorgio perfume brand. He was not shy about his academic prowess either. Four of his mentors were Nobel laureates, he had published more than 200 scientific papers, he was an inventor on some 50 patents, and his HIV research at Harvard had led to the first AIDS drugs that saved millions of lives.

Craig, on the other hand, was very west coast, having been brought up in Millbrae by San Francisco International Airport, and an avid sailor and rider of motorbikes. He was not especially politically correct, which tended to polarise situations and was not, at least at the beginning of the genomics revolution, particularly well off.[r] His experiences in the Navy in a field hospital in Da Nang in 1967 before and during the Tet Offensive in February of 1968 marked him deeply.[14]

---

[p] This facility won architectural prizes for the uniform design lines with glass walls: dozens of prints of famous paintings were to be found on the walls.

[q] Others tell me that my experience with Bill was by no means unique. Nevertheless, Bill was always very supportive of what I was trying to do.

[r] The story goes that he sold some of his HGS stock privately to upgrade his old GM car to a new Mercedes. His assistant Phyllis Filderman drove a bright yellow 380SL convertible, so he may have felt the need for an upgrade.[15]

Haseltine and Venter were alike in being very smart, extremely competitive, and very driven people. Venter wanted to build a first-class research institute: Haseltine hoped to build a new-generation pharmaceutical company based on genome sequences and genetics. *Business Week* ran an article on these "Gene Kings" (as they were referred to by the press) in May 1995.[16]

Unfortunately, these "Kings" did not like each other very much, even to begin with—a *sine qua non* for building a symbiotic relationship between TIGR and HGS—and it got rapidly worse. HealthCare Ventures invested $85 million in the two ventures and Venter was handsomely incentivized with 750,000 shares of HGS stock, so one could expect at least some alignment on how they worked together to derive DNA sequences and make money doing so.

SKB Investment

George Poste, the R&D chief at SmithKline Beecham (SKB), made a shrewd bet on the TIGR/HGS combo by investing $125 million in HGS to get access to the EST data.[17] It was shrewd from several points of view, including return on investment in the stock and access to the data. There was also the fact that HGS could form subsidiary deals with other companies to allow them access to the SKB data and genes that neither SKB nor HGS were pursuing. SKB could and did get more than their money back from their investment.

As part of this negotiation, Wally Steinberg demanded that Jan Leschly, the CEO of SKB and a former top-ten professional tennis player, play a game of tennis with him. Wally turned up with all the right gear and wearing white but hardly won a point.[s] According to George,[17] SKB recovered their investment in six weeks, so the game was worth it.

Poste also provided some sage advice to HGS and to other small biotech companies at the time that is still true today: the start-ups may be small and nimble like scampering mammals, but in contrast to Big Pharma (who the scamperers thought of as dinosaurs), the small companies do not have access to the resources, money, or that precious commodity—time—that are afforded to pharma. There was the risk that, instead of becoming a FIPCO (a fully integrated pharma company), which they had aspirations to be, they stood a much higher chance of becoming a NONCO. Poste also frequently remarked in his talks that the management of most of these small companies was homozygous at the "delusional gene" locus.[17]

---

[s] The difference between a good amateur and a professional in any sport is vast.

TIGR by the Tail

TIGR got going quickly. They installed more than 30 ABI 370As, attendant workstations, and a computer centre in Gaithersburg, Maryland in an old ceramics factory.[15] The core of the arrangement between HGS and TIGR was clear enough. TIGR would do the sequencing and pass the information to HGS so that they could develop products. HGS would fund activities at TIGR using the venture money and the SKB funding. There were concerns from TIGR about timely publications, given the time it took for HGS to review the sequences and, equally, concern from HGS that they did not have enough time to evaluate whether any of the gene sequence information had value for drug discovery, and thus whether to file IP.[t] What happened was predictable: HGS delayed decisions on many more genes than they should have and slowed publication, and the TIGR scientists moaned constantly about it.

HGS went public at the end of 1993, raising more than $30 million. Venter bought a new house with some of his proceeds (apparently one of the largest property transactions in Potomac at the time[15]).[u]

In 1993, HGS launched its own sequencing effort. The relationship between TIGR and HGS was clearly not working.

Although EST sequencing had been vindicated as an approach several times (e.g., to find the human *MSH2* gene—a DNA repair gene, mutated in Lynch syndrome), Venter saw no point in sequencing any more ESTs and turned his attention to whole-genome sequencing. With Hamilton (Ham) Smith, who won the Nobel Prize in Physiology or Medicine for the discovery of restriction enzymes in bacteria in 1978 (with Daniel [Dan] Nathans and Werner Arber), TIGR was successful in sequencing the genome of *Haemophilus influenzae.*[18] They did this by the then-novel technique of random shotgun sequencing, in which the genome was broken randomly into small overlapping fragments and, after sequencing the fragments, the DNA sequence compiled computationally from the overlaps. This was a totally different approach to the conventional mapping and sequencing of cloned DNA using markers as guideposts derived from positional cloning

---

[t] I can well imagine the scene: most of the gene families that are now targets for drug discovery were much less well-known then, so how could you know what was valuable or not?

[u] Later, he bought an 85-ft racing schooner, naming it *Sorcerer*, and skippered it to some impressive race wins. The yacht was slightly short of the length necessary to qualify for many of the races, but it was extended on the bowsprit so that it was eligible.[14]

(see Chapter 7). This became an important differential strategy in the race to complete the whole human sequence (see below).

Divorce Proceedings

The story goes that TIGR had to resort to legal action to prevent HGS from interfering with the publication of the bacterial sequence. Nevertheless, for what they were worth, patents were filed, and the sequence of the bacterial DNA published in 1995.[18,v] HGS and TIGR went their own ways, "divorcing" in the fifth year of the 10-year agreement. It was a scary time, as TIGR had to return the remaining endowment and there was no grant income of any substance.[15] Fortunately, the DOE came in with several projects and TIGR were reimbursed their corporate taxes retroactively, going back five years to the formation of the company.

At the end of the day, a combination of the publicly available EST data from TIGR and Incyte amounted to some 350,000 sequences. This included ESTs provided by Merck who worked in collaboration with the Waterston lab in St. Louis to make sure the ESTs they sequenced were available to the public. It could be interpreted that this was done largely to show up TIGR and Incyte's gene-patenting approach and to promote the myth that genome sequences belong to everyone as a matter of right (this will become important a little later in the chapter).

The number of genes (rather than transcripts) in the genome was always a debate. I remember estimates of anything between 20,000 and 65,000 varying from week to week. Finally, to find out accurately, we got the draft sequence of the human genome and the genome sequences of laboratory workhorses *Drosophila melanogaster* and *Caenorhabditis elegans*. The estimated number of human protein-coding genes stabilized at around 21,000, indicating that many genes code for several different versions of a protein via alternative splicing of the mRNA and other processes.

HUMAN GENOME SCIENCES

Among Haseltine's first moves as CEO of HGS was to hire the best and the brightest. These people, not too surprisingly, came from his lab at Harvard.

---

[v]With reference to irreverence related in the book *Biocode*,[19] Owen White, one of the TIGR scientists, inserted into gene number 1127 on the circular genome the words "Elvis lives." It can be seen in the publication.

They included Craig Rosen, who led the research team, and Haodong Li, a scientist from China, who did postdoctoral work in the Harvard lab. After their IPO in 1993, Haseltine raised about $2 billion in investments for HGS between 1999 and 2000, and their stock price soared. He retired as CEO in late 2004 with six new medicines on the way to human trials. Most of them did not make it. For example, the acquisition of Principia Pharmaceutical gave HGS access to the technology to make albumin-fusion proteins, which increases the half-life of a protein in humans. The best example of this class of drug is the Factor IX albumin fusion called Idelvion marketed for Haemophilia B by CSL Behring. To my knowledge, no HGS proteins of that kind were ever produced, much less marketed.

HGS can be remembered, though, for a monoclonal antibody called belimumab (Benlysta). Developed with GSK and part of a humanization deal with Cambridge Antibody Technology (Chapter 4), belimumab was approved in 2011 for the treatment of systemic lupus erythematosus (SLE) and lupus nephritis (kidney disease caused by SLE). This antibody binds to a protein called BLys, which controls B-cell function and works in some auto-immune diseases. In July 2012, HGS agreed to be purchased by GSK for $3.6 billion, giving GSK full access to belimumab and the rest of the HGS pipeline.

In addition to the drug development work, HGS applied for scores of patents on ESTs between 1992 and 1996 (as did Incyte and TIGR). At one point, at least 350 patent applications, covering more than 500,000 gene tags, were pending at the United States Patent and Trademark Office (USPTO). The single largest application contained 18,500 sequences.[w] Despite the assertions of George Poste and others that patenting ESTs was not different from the patenting of other biomarkers whose functions are unknown, the debate over IP and the filing of patents on ESTs, and partial DNA sequences with homologies to other genes with known or inferred function went on and on.

These debates were all in the context of the hyperbole around the actual value of ESTs and genes for drug discovery: big investments and big egos were at stake.[x] The value of the EST databases ultimately derived from the gene information contained in them, rather than any IP filed on the gene or

---

[w] The large number of filings on ESTs with no known function or utility was elegantly dealt with by the USPTO by limiting filings to 10 genes or less, which effectively priced the EST patenting process out of reach of the companies.

[x] *The Genome Directory* published by *Nature* as a supplement in September 1995 was the swan song for EST sequencing.[20]

the gene product. Most of the valuable IP were derived from using the gene or the protein to make a product (at that time either a small molecule or an antibody). Given that the gene could also be considered a "product of nature," IP for using the gene in gene therapy (Chapter 13) would have to hang on claims about the vectors containing the gene, rather than the gene itself.

This debate about the IP situation would continue to surface as the HGP began sequencing in earnest, and Celera was set up to "capture some of the value of the genes" following the same general business model as TIGR/HGS.

CELERA AND THE HUMAN GENOME PROJECT

Several very good books have been written about the history of the HGP, and the race between the public effort and Craig Venter's second company Celera to sequence the human genome. These include *Cracking the Genome* by Kevin Davies and *The Genome War* by James Shreeve.[21,22] Davies' book was published first, but Shreeve's book has additional information, owing to the author being invited into Celera to watch the show of the company's development in real time. There is also a rather miserable "after the fact" account of the public side of the story published in the *Journal of Molecular Biology* (*JMB*) by Maynard Olson, one of the major HGP protagonists.[23] I do not have a lot of additional history to relate, but I can add a bit of additional colour.

Although the idea to sequence the human genome had been around for a while, it was not until 1986 that a project began to take any real shape. Renato Dulbecco, the cancer virologist and 1975 Nobel laureate, had suggested the "Holy Grail" of the complete genome sequence, but it was Charles DeLisi at the Department of Energy (DOE) at Los Alamos near Santa Fe who initiated it operationally. The DOE was supposedly looking for another Manhattan Project, and they were certainly aware of the new automated sequencing technologies that were emerging.

The HGP, a multimillion-dollar project to obtain the sequence of the three billion nucleotides of DNA that make up the human genome, was formally launched in October 1990. Jim Watson from Cold Spring Harbor (the 1962 Nobel laureate) was hired to lead the effort.[y] Initially, there was very little interest from the pharmaceutical industry in the project, largely

---

[y] As it turned out this was not an inspired choice.

because it was thought by many that the 97% of the genome that did not code for proteins was not of much use for making drugs: so why sequence it?

Human DNA is contained in 23 pairs of chromosomes. The HGP initially consisted of a network of labs, each of which was assigned different chromosomes to sequence. The approach that was taken was a follow-on from positional cloning (Chapter 7)—that is, cloning the chromosomal DNA into a library of overlapping clones using yeast artificial chromosomes (YACs) or large bacterial artificial chromosomes (BACs), as described previously. This approach to sequencing the genome was entirely logical, based on the genomics experience from the positional cloning projects that had been started in the 1980s, and the existence of BAC and YAC libraries covering most parts of the human genome.

Jim Watson and Bernadine Healy, the NIH director at the time and responsible for funding the HGP with the Wellcome Trust in the United Kingdom, never saw eye to eye over how to deal with intellectual property that would be generated by the project.[z] In the end, Watson quit his role out of sheer frustration with the situation and Francis Collins from the University of Michigan took over. Collins was a positional cloning and genomics expert (Chapter 7) and way more ambitious and politically astute than his Christian country boy demeanour would suggest. These traits stood him in good stead, eventually leading him to preside over the NIH as the very successful director for 12 years.

The initial HGP sites included Washington University in St. Louis, University of Washington–Seattle, Stanford University, and the Whitehead Institute/MIT in Boston. The project started with $60 million, with an additional $20 million from the Wellcome Trust funding the project at the Sanger Centre in the United Kingdom. TIGR was unofficially part of the HGP, even though they did not receive any money.[aa]

CELERA

The idea of creating Celera in early 1998 as a private company to sequence the human genome came from PerkinElmer (PE). Tony White, their new CEO, was looking to revamp PE, which had acquired ABI in 1992 and,

---

[z] It was still a political hot potato driven by the EST IP debates.

[aa] By now TIGR was releasing all their data into the public domain, likely helping to secure TIGR's 501(c)(3) status, which was important for their further grant funding.

BOX 9. CELERA FINANCING

In early 1999, the old PerkinElmer, which was 90% owned by Institutional Investors, was reorganised into PE Biosystems and Celera. The latter went public as a tracking stock to PE in April 1999. For every two shares of PE that were purchased, investors got one Celera share. Celera went public at $17 and traded up to $18. Over a few weeks, the stock traded to as high as $30 before falling back down to $19. It was a relatively successful IPO, bringing hundreds of millions of dollars to PE. But investors were still confused about the business model.

---

more recently, PerSeptive Biosystems (an instrument company that made mass spectrometers).[bb] White saw increasing opportunities in the "genome tools business": thousands of first-generation ABI sequencers had been sold over the previous 10 years, and he surmised that the setting up a company to exploit genome sequencing (and a footrace with the public HGP) could only be a good thing for sales of their newer sequencing machines (see Box 9).[cc]

PE believed that dedicating several hundred of their machines to the task of sequencing the human genome would enable it to be done reasonably quickly and cost-effectively.

They needed someone to run the show, and no one was better qualified to do so than Craig Venter. Initially reluctant after the TIGR–HGS debacle, Venter accepted the offer and Peter Barrett from PE became CBO. Venter named the new company "Celera" ("Speed"). PE would own 80%, Venter 5%, and TIGR 5%, with the remaining 10% for the employees.

Celera started business in early 1998 in a 200,000-sq-ft facility in Rockville, Maryland (Box 9). As opposed to positional cloning and sequencing, they chose instead to use shotgun sequencing (Box 10), the approach used to sequence bacterial DNA.[dd] Three hundred ABI PRISM machines

---

[bb] PerSeptive Biosystems was founded in 1989 by Noubar Afeyan as a young postdoc fresh out of MIT. He is now (see later in Chapter 17) Chairman of Moderna and the CEO of Flagship Pioneering, a large venture capital company.

[cc] Including a capillary machine (the ABI 3700 PRISM) that was in development to compete with the Amersham MegaBACE 1000 machine. The PRISM would retail at some $300,000 per machine.

[dd] TIGR had used shotgun sequencing to determine the *H. influenzae* genome, but it would obviously need to be scaled up drastically for the human genome. It was also predicted to save the money, time, and effort involved in the cloning and mapping, which were going to be needed for the more logical HGP approach.

BOX 10. RIDING SHOTGUN

Shotgun sequencing meant making very large libraries of cloned human DNA fragments and, without ordering them, sequencing the fragments, and using advanced computational methods to assemble the complete sequence from overlapping sequences. There were proponents of this approach, most notably geneticist James (Jim) Webber and the aptly named Gene Myers, who had demonstrated computationally that this was possible.[24] Granger Sutton and Gene Myers wrote "Celera Assembler" (CA) for the 64-bit Compaq platform. It is unlikely that Celera would have succeeded at all, if not for their computational investment. This is one thing that Craig Venter always appreciated and respected, even mentioning it at the White House press conference celebrating the "completion" of the project.

---

were installed, along with the large computational hardware and software (and human expertise) needed to assemble the sequence. There was no multi-parallel Cloud computing then, so it necessitated building a data farm as well as a sequencing machine mega suite. Compaq (with their Alpha-64 CPU chip) rapidly jumped into bed with Celera.[ee]

The heat generated by the computer centre and the sequencing machines was predicted to be huge. It turned out to be rather less than the heat generated about the whole concept of a privately sponsored company set up to compete with the publicly funded HGP, using an approach (shotgun sequencing) that the public effort leaders continuously remarked "would not work."

Opponents like Phil Green at the University of Washington argued that some DNA would never be successfully cloned in the first place (and you would not know it), coupled with the fact that the genome was so full of repetitive sequences that it would be impossible to order these parts of the genome by overlapping sequence. There was certainly truth to both those assertions, but Celera reckoned that a 10-fold coverage of the genome in terms of clone numbers and sequences would be sufficient to offset those concerns.

Celera Gets Going

Celera grew to 50 people by September 1998 and at least a proportion of the ABI machines they bought were up and running. The first crew was almost

---

[ee] Rumour had it that Compaq's entire marketing budget was dedicated to Celera's business.[15]

entirely from TIGR. Claire Fraser, Venter's second wife, who was named TIGR's president, had to fight to keep Venter from pillaging her team.[15] The TIGR campus was in the middle of physical expansion when Celera was forming. Because Venter wanted to run Celera and TIGR at the same time and from the same physical location, the new building was getting redesigned as it was being built to accommodate both. Ultimately, Venter was convinced by advisors that mixing a profit and nonprofit in the same location was a bad idea.[ff]

In October 1998, Celera was given a boost of sorts from the determination of the genome sequence of *C. elegans* (a nematode worm with which we will spend more time in Chapter 9), by Bob Waterston at Washington University and John Sulston at the Sanger Centre at Hinxton, near Cambridge in England. This tiny worm has 18,500 genes: it was the largest complete DNA sequence available of any organism to date.[25] Debate continued around whether the *C. elegans* sequence could have been done more quickly and less expensively by the shotgun methods being employed at Celera.[gg]

By this time, Celera was beginning to function more effectively. They held their first SAB meeting in February 1999, chaired by Rich Roberts from New England Biolabs.[hh] The SAB included Arnie Levine, the discoverer of the p53 oncogene from the Rockefeller University; Melvin (Mel) Simon, who developed BAC technology at Caltech; Victor McKusick from Johns Hopkins University, the father of medical genetics and an expert in Mendelian inheritance of diseases; Arthur (Art) Caplan, bioethicist from the University of Pennsylvania; and Norton Zinder, who was an emeritus professor at the Rockefeller University and who had been part of the phage group in the 1950s.[26]

By March 1999, Celera consisted of 350 employees, the majority of whom had been hired in less than a year. Note that this is a rate of almost one new hire a day. As anyone who has built an organization from scratch can tell you, that is an absurd hiring rate. Not only do many wrong people get hired into the wrong jobs, but many right people for the organisation get hired into the wrong jobs. New hire orientation and briefings essentially go out of the window, and people are generally left to fend for themselves. It is

---

[ff]This ill-conceived flirtation with co-location led to the integrity of the new TIGR building being compromised by the numerous costly design changes.

[gg]It probably could have been.

[hh]Roberts shared the 1993 Nobel Prize in Physiology or Medicine with Phil Sharp, for their discovery of introns and splicing in eukaryotic genes.

very Darwinian: only the strong survive under such circumstances. Maybe that was the unwritten plan.

As the company grew, the décor began to look appropriately more opulent. A full-service subsidized dining area with a French chef cooking breakfast, lunch, and dinner was built. However, less than half of the ABI machines in the lab were up and churning out data. Celera calculated that they needed 230 machines to be fully functional 24/7 to get the sequence coverage of the genome that they needed in a reasonably fast time.

By mid-1999, the relationship between the HGP and Celera had deteriorated markedly. It was clear from their public announcements that Celera was using the public data to help to organise the random sequences they were generating. But the HGP could not benefit from any of the sequence assemblies that were being made by Celera, because they were not being released.

One positive event amongst these machinations, however, was the effort to sequence the genome of the laboratory workhorse *D. melanogaster*. Craig had formed a collaboration with Gerry Rubin, a very accomplished fly geneticist and biologist, to obtain the *Drosophila* genome sequence on an accelerated time line. Although completed a bit later than anticipated, the full sequence was finished in late 1999.

A "*Drosophila* sequence annotation jamboree" was subsequently organized over 11 days in that November, where 45 *Drosophila* and other biologists pored over the assembled sequences looking for genes.[ii] They found an estimated 13,600 genes, many of which had counterparts in other animals including *C. elegans*. The project also clearly demonstrated that shotgun sequencing worked as advertised, even though the fruit fly genome is ~25 times smaller than the human genome. It was still an important milestone, and the iconic paper describing the work was published in *Science* in 2000.[27]

Bermuda

The decidedly antagonistic view that the Wellcome Trust demonstrated towards any money being made in any way from genome sequencing by ensuring immediate data release of the data seemed to me to be existentially paradoxical: all the Wellcome money had come from selling drugs in the first place (Box 11). For some of the individuals involved in the HGP, it

---

[ii] It must have been literally a blast (excuse the pun: Basic Local Alignment Search Tool [BLAST] being a computational way of comparing sequences).

BOX 11. THE WELLCOME TRUST AND THE BERMUDA ACCORDS

As the HGP got underway, the Wellcome Trust worked hard to ensure that there would be unfettered access to any sequencing data generated. They proposed this at a meeting in Bermuda in 1996, the result of which was the "Bermuda Accords."[28] All participants in the project had to declare that they would make all the genome information they obtained available publicly, with immediate release. These Bermuda Accords, imposed on all HGP contributors, were upheld with what was religious fervour.

Personally, I admit to finding this a bit odd. While still in the United Kingdom, I had my own run-in with the Wellcome Trust. Before I left for Sequana, I had tried to set up a company to find drugs against cell-cycle proteins with Paul Nurse (who received the 2001 Nobel Prize in Physiology or Medicine with Tim Hunt and Leland Hartwell for cell-cycle discoveries), but we ran into what was referred to as "Henry Wellcome's Will." Wellcome had set up the company that became Burroughs Wellcome (which eventually became part of GSK). His Will stipulated that all profits from the company, which after the development of some of the first drugs against HIV became quite large, could only be spent as his Will stipulated, via a trust fund called the Wellcome Trust. This organisation provided not-for-profit funding to universities and there was no easy way then to fund "for profit" enterprises such as biotech start-ups. There did not seem to be the "will" to do so, either.

This all changed in the early nineties. Owing to the size of the Trust generated by income from HIV drugs, the Will had to be changed by court action. The Wellcome Trust is now one of the world's largest investors in the life sciences, including a venture arm investing in very much "for-profit" biotechnology. They were a founder of the VC firm Syncona.

---

seemed to be more to do with their notions of morality than anything else. The desire to do "academic and unfettered" research without being beholden to anyone—even to those paying for it—seemed to them to be a God-given right. The HGP's professed selfless motives, as supported by the Bermuda Accords, simply did not ring true: Collins, Sulston, and their teams were clearly in a race for glory with Craig Venter and Celera; they all knew it, and the HGP did not want to lose the race any more than Celera did.

Celera set the pace and the public effort responded by increasing the money available and streamlining the project to five groups (the "G5") to do the bulk of the sequencing work. This consisted of Waterston at Washington University, Gibbs in Houston, Lander at the Whitehead/MIT, The Sanger Center, and the Joint Genome Institute at Los Alamos.

It did not help relationships that Venter liked to upstage everybody. For example, he appeared in *Time* magazine as "the Gene Maverick"[29] and was depicted in *Science* as a "genome combative entrepreneur." Such coverage did not endear him to either the HGP or to Tony White (the PerkinElmer CEO), who was essentially pushed to the sidelines. The interactions were further compromised when Celera found out that the HGP had been planning behind the scenes to bring Incyte into the HGP, when Celera and Incyte had also been talking about combining forces.[jj] Neither of these proposed collaborations ever materialised, but it just added to the mistrust that each side had for each other.

The infamous Genome Sequencing and Analysis Conference in Miami Beach in September 1999 organised by Craig Venter provides a snapshot of the sequencing field at the time.[30] One of the speakers was from the FBI and mentioned the Monica Lewinsky dress incident and the use of DNA typing for forensics in the presentation. This led to Craig and his wife Claire Fraser being summoned to the White House the next day for an explanation.[15]

By late 1999 and early 2000, the politics between Celera and the HGP had grown very ugly, with both sides showing no sign of anything other than intransigence. Various overtures were made by both Eric Lander and the U.S. Department of Energy (DOE) for the two groups to collaborate, but the Wellcome Trust vetoed most of them, based primarily on their addiction to the Bermuda Accords. By this time, the HGP had completely sequenced small chromosome 22, which was found to have 545 genes, and the even smaller chromosome 21, which turned out to have 225 genes. As chromosome 21 is the chromosome that is triplicated in Down's syndrome, this sequence was of considerable medical interest.[31]

By the end of 1999, Celera had four database subscribers, bringing in more than $100 million/year. In addition to Howard Hughes who bought access for their investigators, Amgen,[kk] Novartis, and Pharmacia-Upjohn had all signed up. It was thought by the subscribers that, for example, having access to the sequences of all the G protein–coupled receptors (GPCRs) in the genome would in principle be very valuable as many drugs bind to GPCRs (see Chapter 6). But even today many of these suspected GPCRs are just that: sequences of potential proteins that look like GPCRs, but no one knows what they do. They are called orphan GPCRs for a reason.

[jj] Incyte were very careful to make much less noise generally than either the HGP or Celera. They had built a successful subscription business on the back of EST sequencing and other genomics activities and flew majestically under the radar.

[kk] Amgen was specifically interested in secreted proteins, which were identified by a signal sequence that would enable them to be transported out of the cell.

Celera had also filed IP. This IP filing was done selectively as Venter had initially promised, only covering genes and proteins that might be targets for drug discovery. Accent on the word "might," because what pertains to GPCRs is true of other gene families like ion channels as well. A January 7th, 2000, press release from Celera essentially upped the ante yet again, announcing that completion of an assembled human genome sequence was on the horizon. The Celera stock price reached $240, representing a 25× return for an IPO investor in less than eight months.[ll] Venter's net worth (on paper anyway) rose to more than $700 million, bringing him a bigger yacht called *Sorcerer II*.[14] ABI shipped more than 1000 machines by end of the millennium, so they also made good and legitimate money from selling machines to both Celera *and* the HGP.

Clinton–Blair Statement

The out-of-context and ill-advised joint statement that President Clinton and U.K. Prime Minister Anthony (Tony) Blair made in March 2000, parroting the Bermuda Accords that "all genome sequence information must be freely available" led to several billion dollars being wiped off the value of HGS, Celera, and other biotech companies almost overnight.[32] The origin of the "advice" that led to the statement is not clear but can probably be traced to the Wellcome Trust and not to the NIH. In any case, it was one of the most egregious public comments made throughout the whole ignominious project. Venter even wrote a letter to *The Wall Street Journal* on the subject.[33] It was simply ignorant and arrogant not to think, or even perhaps care, about the economic consequences of advice like that.

Unfortunately, the joint statement had much more to do with politics than confusion. It also put the animosity between the HGP and Celera into sharp relief, and each party was essentially told by their respective powers-that-be to "fix it." Both sides had to come together in some face-saving way, most likely by declaring a tie. The dead-heat deal was brokered by Aristides (Ari) Patrinos, who ran the Joint Genome Institute on behalf of the DOE, and who was liked and trusted by both Craig Venter and Francis Collins. Eventually, the Wellcome Trust, who had co-sponsored the whole HGP financially, demurred and blessed the idea of the joint declaration since both sets of data would end up being published.

[ll] This is what is called in the venture business a "good return on investment."

An (Un)finished Sequence

The "it's-a-tie" deal was consummated on June 26th, 2000, at the now infamous meeting at the White House. The race to sequence the human genome was over. But it was not complete. There were mistakes and gaps in both sets of data, which were gradually resolved over time. The two "draft" sequences (because that is what they really were) got published in February of 2001 in *Nature* (HGP) and *Science* (Celera), respectively.[34,35,mm]

Despite the Clinton–Blair scare, big science technology and biotechnology (including genomics, proteomics, and metabolomics) captured the imagination of investors. The 2000 IPO market was unprecedented in the numbers of companies going public, total capital raised, and valuations at IPO.[36,nn]

After the Tie

In June 2001, Celera bought Axys, the company that had previously acquired Sequana (Chapter 7) for $170 million, presumably to become a fully integrated pharma company. The vinyl sulphone cathepsin K inhibitor that Axys had in collaboration with Merck (odanacatib) got as far as Phase 3 but was terminated for toxicity reasons. In 2002, the Axys (now Celera) chemists began to focus on making kinase inhibitors, including those that inhibited Bruton's tyrosine kinase (BTK), for autoimmune disease.[oo] For the BTK programme, they generated compounds that bound irreversibly to the enzyme ("covalent" inhibitors). Pharmacyclics, a chemistry company founded in 1991, bought out Celera's drug portfolio in 2006 for $6.6 million. This was the origin of ibrutinib (Imbruvica), the BTK inhibitor that Pharmacyclics developed for B-cell lymphoma, which led to their eventual acquisition by AbbVie for $21 billion in 2015—a legacy of sorts for Axys, Celera, and Sequana. Celera later got acquired by Quest Diagnostics for $650 million, including the royalty stream for ibrutinib.

[mm] The use of the new long- and short-read sequencing machines has now led to what we *can* call a complete human genome sequence of 3,054,815,472 bp, encompassing some 20,000 protein-coding genes and many others coding for regulatory RNAs of various types.[2]

[nn] The year 2000 IPO record was only surpassed in the 2017–2021 period, especially 2020–2021—paradoxically, during the COVID-19 pandemic.[36]

[oo] An entertaining book by Nathan Vardi about the path to the development of BTK inhibitors was published recently.[37]

Craig Venter went on to form the J. Craig Venter Institute (JCVI), a non-profit genome-sequencing organization, designed to be a kind of oversight organization for a collection of Venter interests.[PP] JCVI initiatives included trips around the world on *Sorcerer II* (courtesy of the DOE), collecting and sequencing organisms from the sea as part of a "Biodiversity Project." You can only imagine the fun they had with customs when you have a yacht equipped with a lab on board. Venter also formed Synthetic Genomics, focused on biofuels from algae and funded by the Exxon Corporation, and a company called Human Longevity based in San Diego. They had a mission "to discover and harness the technological and biological unlocks that amplify span of life, health, & high performance."[38,qq] More on this subject in Chapter 17.

WHOLE-EXOME SEQUENCING

A logical extension of EST sequencing and an alternative to the still-pricey whole-genome sequencing was just to sequence the exomes (i.e., the parts of the genome that coded for proteins). This could be done directly from genomic DNA and not via cDNA. Various exome selection techniques became available. As exomes represent only about 2% of the genome, this was a cost-effective way to obtain coding sequence information on individuals or groups of individuals, especially for potential disease understanding. At Biogen, we sponsored an exome sequencing study to find rare alleles of genes that are altered in amyotrophic lateral sclerosis (ALS) by sequencing the exomes of 3000 affected individuals and about 8000 controls.

This study was led by David Goldstein, who was founding Director of the Institute for Genomic Medicine at the Columbia University Medical Center and Professor of Genetics and Development. He is now CEO of a small San Diego company called Actio Biosciences, finding drugs for rare genetically defined diseases that have mechanisms that are also reflected in common disease (see Chapter 11).

[PP] Oversight of TIGR never came to be, as it was somewhat complicated by the fact that Claire Fraser was no longer Craig's wife. The TIGR Board emphatically said "No!"

[qq] I guess it is not surprising how many successful and wealthy people obsess about ways to prolong a healthy life. I am all for it personally but far from obsessed by it. None of us get out of here alive anyway.

I met David when he was at Duke University, and I was still at SAIC-Frederick. I heard him give a talk on hepatitis C and the role of the interferon lambda gene, variations in which have a large effect on controlling the response to the drugs that were used to treat Hep C. It was a very elegant piece of work.[rr]

Rick Myers at the HudsonAlpha Institute in Huntsville Alabama, whom I knew from Bob Tjian's lab, also participated in the ALS exome project (they had a Genome Sequencing Center at the Institute). Among the discoveries we made were mutations in a gene called *TBK1*, which was known to be involved in the process of autophagy, in which mitochondria and other organelles are removed in both health and disease. Before our work, the *TBK1* gene had not been implicated in ALS, although other genes involved in autophagy had been implicated, such as *p62* and *optineurin*.

Despite the novelty of the finding and the obviousness of both the target specifically and the process of autophagy for ALS pathology, the Biogen neuro group could not get excited about it as a target for drug discovery.[ss] They were certainly not as excited as I was. The net result of that was that I persuaded SV Health Investors, where I had been doing a little work as a Venture Partner, to start a company (called Rheostat) to target both *TBK1* in ALS and the autophagy process itself, as it had also been implicated genetically in Parkinson's disease. The company (renamed Caraway Therapeutics) continues today, focusing on ion channels and lysosomal function, both being key to the autophagy process. This is a good example of how companies can get started and how research scientists can help drug discovery and the patients.[tt] As a bonus, we also published a now highly cited paper in *Science* in March 2015: it was one of the first whole-exome studies for rare allele hunting in any disease to be published.[39]

FOUNDATION MEDICINE

Instead of whole-genome sequencing or even exome sequencing, Foundation Medicine was conceived in 2007 by Levi Garraway and Matthew Meyerson

---

[rr] David was told by many fellow geneticists before he started the work that, with the number of patients he could access, he would not have enough genetic power to find anything. But, as you may be aware, that depends on the "effect size" of the mutations. The effect size of this genome alteration was much larger than expected, and thus the discovery was made.

[ss] Largely because they did not find it themselves, I suspect.

[tt] And not much different from what happened earlier in the biotech era. Caraway has recently been acquired by Merck.

at Harvard Medical School and the Broad Institute as a company that could simply measure mutations in a set of about 300 genes, to aid in the diagnosis and treatment of cancer. Third Rock Ventures led a $25 million series A investment. Michael Pellini, whom I met while on the Board at OpGen, was the initial CEO and Evan Jones, the OpGen CEO, who also invested in Foundation Medicine from his Venture Fund (jVen), was on the Board.

The company released FoundationOne, its first diagnostic test, in 2012. The company also began partnering with pharmaceutical companies to analyse patient samples. By 2018, the company had more than 30 partnerships. Their second test, a haematological cancer assay called FoundationOne Heme, was released in 2013. The quality of the test and the information derived from it was quite outstanding, allowing a physician to get a very clear view of what the genetic nature of a particular leukaemia (for example) was, and what the best clinical course of action would be given the mutation profile.[uuu] Roche acquired Foundation Medicine in June 2018, which now operates as an independent subsidiary of Roche Diagnostics. The FoundationCore database contains more than 300,000 genomic profiles coming from the results of the company's assays, as well as information on more than 150 different subtypes of cancer.

Foundation Medicine is a very good example of how sequencing methods can directly affect the patient outcome when diagnosed with a life-threatening disease like cancer. It also demonstrates that new diagnostics companies can still be created and be successful: not everything has to be therapeutics-focused.

GENOMIC HEALTH

We have now entered an era where population-scale genome sequencing is being done, which should give us a population-level view of the relationship between genotype, a whole-genome sequence, and disease or other condition phenotype. Several such initiatives are under way. One of the groundbreaking ones was set up between Regeneron (see Chapter 6) and Geisinger Health in Pennsylvania. The DiscovEHR partnership pairs genome sequence obtained by Regeneron Genetics (a subsidiary of Regeneron) with patient health records from the Geisinger system. DiscovEHR has

---

[uuu] I have recommended Foundation Medicine to many people for their cancer or for that of a relative. It is potentially lifesaving information.

uncovered both rare and common genetic variants associated with various diseases or biomarkers of health and disease (e.g., kidney function). Some of the work has been published in *Science* and the *New England Journal of Medicine*. Their database of two million whole-genome sequences from patients with different diseases taken from the Geisinger client population is probably unique. Regeneron is using it every day in their genomics-driven research programmes. In May 2023, Kaiser Permanente (a California-based HMO) acquired Geisinger Health.

Another important initiative in the United Kingdom is Genomics England (Box 12), a company set up to obtain the sequence of at least 100,000 people in the United Kingdom, preferably those who are already in the U.K. Biobank[40] (a large long-term study collecting information about people's genetic predisposition to disease and their environmental exposures). Genomics England has focused so far on cancer and rare diseases. This data set will be an important resource not only for drug discovery but also for monitoring sequence with respect to disease outcome. It will be of great utility, for example, to help to understand variable responses to COVID-19, where some people got very ill and died, whereas others of the same age had fairly benign disease. This sequence information can also be coupled to deep "immune phenotyping" to understand immune status.

Sequencing is an important part of the biotech story, and not just owing to the characters involved. Sequencing of nucleic acids rapidly and at scale is now a key driver of drug discovery as well as diagnostics and personal health. It has indeed been a revolution since the automated DNA sequencers became available, driving the biotechnology industry to the next level.

BOX 12. GENOMICS ENGLAND

The company was spun out of Oxford University in 2014 by four academic scientists, including Professor Sir Peter Donnelly, who was director of the University's Wellcome Trust Centre for Human Genetics. Donnelly became CEO in 2017. The company employs about 130 people in the United Kingdom and in Cambridge (MA). The company has raised about $100 million. The IP Group, a technology incubator, seeded the company and other investors include Lansdowne Partners and Hambro Perks and Woodford Investment Management's patient capital trust. In 2018, Vertex Pharmaceuticals became a strategic investor with other U.S. venture capitalists and Oxford Science Enterprises.

POSTSCRIPT

I remember fondly one conversation with Sydney Brenner, glass of wine in hand, in his apartment in La Jolla, watching the sun set over the Pacific Ocean after the human genome sequences had been published. He had an interest in the Fugu genome, the toxic Japanese fish that is a delicacy there. The Fugu genome is eight times smaller than the human genome but has essentially the same number of genes. He told me that he had read some of the Fugu sequence and was reading the human DNA sequence from the beginning of chromosome 1 all the way to the end of chromosome 23. I was not completely sure that this was true, but in retrospect, I can imagine that it was. It sums up the man: he was so inquisitive about how the whole genome fitted together. I am not sure that he needed the smaller scale of the Fugu fish to help him to understand the human genome: he got it anyway. Now he would not need to read it on paper at all: it is all accessible online. Sydney, as you will note from my many references to him, had a great influence over the development of the biotech industry—way more than he is given credit for.[41]

REFERENCES AND NOTES

1. Lovell JT, Grimwood J. 2022. The road to accurate and complete human genomes. *Nature* **606:** 468–469.

2. Nurk S, Koren S, Rhie A, Rautiainen M, Bzikadze AV, Mikheenko A, Vollger MR, Altemose N, Uralsky L, Gershman A, et al. 2022. The complete sequence of a human genome. *Science* **376:** 44–53. doi:10.1126/science.abj6987

3. In 1965, Gordon Moore considered that every two years or so, the number of transistors on microchips would double. Moore's Law says that computational progress will become significantly faster, smaller, and more efficient over time, exponentially.

4. Min Jou W, Haegeman G, Ysebaert M, Fiers W. 1972. Nucleotide sequence of the gene coding for the bacteriophage MS2 coat protein. *Nature* **237:** 82–88. doi:10.1038/237082a0

5. Sanger, F, Nicklen S., Coulson AR. 1977. DNA sequencing with chain-terminating inhibitors. *Proc Natl Acad Sci* **74:** 5463–5467. doi:10.1073/pnas.74.12.5463

6. Maxam AM, Gilbert W. 1977. A new method for sequencing DNA. *Proc Natl Acad Sci* **74:** 560–564. doi:10.1073/pnas.74.2.560

7. Conversation with Martin Leach, ex–Bioinformatics lead at 454 and colleague at Biogen later, 2021.

8. Munroe DJ, Harris TJR. 2010. Third-generation sequencing fireworks at Marco Island: advances in sequencing platforms promise to make this technology more accessible. *Nat Biotechnol* **28:** 426–428. doi:10.1038/nbt0510-426

9. *Nature* Milestones, February 10, 2021.

10. 2023. The NGS race is on: souped-up sequencers vie for frontrunner status. *Genetic Engineering and Biotechnology News*, January 2023.

11. Brenner S, Jacob F, Meselson M. 1961. An unstable intermediate carrying information from genes to ribosomes for protein synthesis. *Nature* 190: 576–581. doi:10.1038/190576a0

12. Gros F. 1961. Unstable ribonucleic acid revealed by pulse labelling of *Escherichia coli*. *Nature* 190: 581–585. doi:10.1038/190581a0

13. Adams MD, Kelley JM, Gocayne JD, Dubnick M, Polymeropoulos MH, Xiao H, Merril CR, Wu A, Olde B, Moreno RF, et al. 1991. Complementary DNA sequencing: expressed sequence tags and Human Genome Project. *Science* 252: 1651–1656. doi:10.1126/science.2047873

14. Venter JC. 2007. *A life decoded*. Penguin, New York.

15. Conversation with Vadim Sapiro, ex–TIGR employee, 2021, 2022.

16. Carey J, Hamilton J, Flynn J, Smith G. 1995. The gene kings. *Business Week*, May 8, 1995, pp. 75–78.

17. Detailed commentary from George Poste in a conversation with him in the summer of 2022.

18. Fleischmann RD, Adams MD, White O, Clayton RA, Kirkness EF, Kerlavage AR, Bult CJ, Tomb JF, Dougherty BA, Merrick JM, et al. 1995. Whole-genome random sequencing and assembly of *Haemophilus influenzae* Rd. *Science* 269: 496–512. doi:10.1126/science.7542800

19. Field D, Davies N. 2015. *Biocode*. Oxford University Press, New York.

20. Nature. 1995. *The Genome Directory, Supplement to Nature. International Weekly Journal of Science*, 28 September 1995, Vol. 377, Issue No. 6547S. Macmillan Magazines Ltd., New York.

21. Davies K. 2001. *Cracking the genome: inside the race to unlock human DNA*. The Free Press, New York.

22. Shreeve J. 2004. *The genome war*. Alfred A. Knopf, New York.

23. Olson MV. 2002. The Human Genome Project: a player's perspective. *J Mol Biol* 319: 932–942. https://doi.org/10.1016/S0022-2836(02)00333-9

24. Weber JL, Myers EW. 1997. Human whole-genome shotgun sequencing. *Genome Res* 7: 401–409. doi:10.1101/gr.7.5.401

25. Hillier LW, Coulson A, Murray JI, Bao Z, Sulston JE, Waterston RH. 2005. Genomics in *C. elegans*: so many genes, such a little worm. *Genome Res* 15: 1651–1660. doi:10.1101/gr.3729105

26. For more on Norton Zinder, see Judson HF. 1996. *The eighth day of creation: the makers of the revolution in biology (commemorative edition)*. Cold Spring Harbor Laboratory Press, Cold Spring Harbor, NY.

27. Adams MD, Celniker SE, Holt RA, Evans CA, Gocayne JD, Amanatides PG, Scherer SE, Li PW, Hoskins RA, Gale RF, et al. 2000. The genome sequence of *Drosophila melanogaster*. *Science* 287: 2185–2195. doi:10.1126/science.287.5461.2185

28. The Bermuda Accords set out rules for the rapid and public release of human DNA sequence data. At the 1996 and 1997 summits in Bermuda, leaders of the HGP and other senior academics agreed on a set of principles requiring that all DNA sequence data be released in publicly accessible databases within 24 hours after generation. These "Bermuda Principles" were distinct from the typical practice in the life sciences of making experimental data available only after publication. The three principles retained originally were (a) automatic release of sequence assemblies larger than 1 kb (preferably within 24 hours), (b) immediate publication of finished annotated sequences, and (c) aim to make the entire sequence freely available in the public domain for both research and development to maximise "benefits to society."

29. Thompson D. 1999. Craig Venter: Gene Maverick. Craig Venter is a man in a hurry, and now all the genome mappers are operating on Venter time. *Time*, January 11, 1999.

30. Letter to Jon Eisen from Craig Venter for the 11th International Genome Sequencing and Analysis Conference, in September 1999.

31. Hattori M, Fujiyama A, Taylor TD, Watanabe H, Yada T, Park H-S, Toyoda A, Ishii K, Totoki Y, Choi DK, et al. 2000. The DNA sequence of human chromosome 21. *Nature* **405**: 311–319. doi:10.1038/35012518

32. Langreth R, King Jr RT. 2000. Biotech, genomics stocks plunge on fear U.S. may curb data sales. By *Wall Street Journal*, March 15, 2000.

33. Venter JC. 2000. Letter to the WSJ. Clinton and Blair shouldn't destroy our research. *Wall Street Journal*, March 21, 2000.

34. International Human Genome Sequencing Consortium. 2001. Initial sequencing and analysis of the human genome. *Nature* **409**: 860–921.

35. Venter JC, Adams MD, Myers EW, Li PW, Mural RJ, Sutton GG, Smith HO, Yandell M, Evans CA, Holt RA, et al. 2001. The sequence of the human genome. *Science* **291**: 1304–1351. doi:10.1126/science.1058040

36. Conversation with Stelios Papadopoulos, 2021, 2022.

37. Vardi N. 2023. *For blood and money: billionaires, biotech and the quest for a blockbuster drug*. Norton, New York.

38. Human Longevity. *Who we are*. https://humanlongevity.com/who-we-are

39. Cirulli ET, Lasseigne BN, Petrovski S, Sapp PC, Dion PA, Leblond CS, Couthouis J, Lu Y-F, Wang Q, Krueger BJ, et al. 2015. Exome sequencing in amyotrophic lateral sclerosis identifies risk genes and pathways. *Science* **347**: 1436–1441. doi:10.1126/science.aaa3650

40. Kaiser J, Gibbon A. 2019. Biology in the bank. *Science* **363**: 18–20. doi:10.1126/science.363.6422.18

41. Freidberg EC. 2010. *Sydney Brenner—a biography*. Cold Spring Harbor Laboratory Press, Cold Spring Harbor, NY.

CHAPTER 9

Functional Genomics

Exelixis coined the term "functional genomics" and obtained a trademark for it in the mid-1990s. The concept was meant to reinforce the point that you can obtain the DNA sequence of many genes and catalogue them, but that effort does not tell you very much about what the genes do, or whether they are good targets for a drug discovery programme, or even whether the genes could be used directly as therapeutics. So, it is no surprise that the primary driving force behind the development of functional genomics was the biotech and pharmaceutical industries.

Big Pharma appreciated the fact that the various sequencing initiatives would bring many gene sequences into view, but they needed some way of turning those sequences into drugs. You cannot do that without knowledge of what the genes do (their function). Jürgen Drews, who was president of global research at Hoffmann-La Roche before Jonathan Knowles, was not alone in contending that the pharma pipelines were not going to be able to keep up with the bottom-line growth rates demanded by their shareholders or to develop medicines rapidly enough for the patients that need them. He believed that understanding the function of genes was fundamental to the development of new medicines.

Essential to the thesis of functional genomics was the idea that less-complex organisms such as the fruit fly (*Drosophila melanogaster*), the nematode worm (*Caenorhabditis elegans*), and the zebra fish (*Danio rerio*) could help define gene function, owing to their genetic tractability, many measurable phenotypes, and utility in the lab as model systems. Many strains of laboratory mice were also available and were already being heavily used by both academia and drug companies as surrogates for human biology. However, given their relative complexity and long lives, working with mice to establish genotype (mutation in gene) to phenotype (diseases, hair colour, behaviours, etc.) and other correlations was time-consuming experimentally and not always possible.

A put-down comment that Craig Venter made to Frances Collins at the beginning of the race to sequence the human genome, "You can do the mouse," profoundly irritated the public Human Genome Project (HGP) principals. Yet it turned out to be quite important to have the mouse sequence to compare to the human.

Several companies were started in the early 1990s to exploit functional genomics in the mouse, but for several reasons, they took a long time to get going or never really got going at all. Most were involved with making transgenic mice as disease models for human diseases based on certain biochemical pathways. This chapter will mainly consider several of the functional genomics companies that were set up to exploit flies and worms to understand gene function, and what happened to them.

EXELIXIS: HOW START-UPS GET STARTED

It took much longer than the normal human gestation period of nine months to birth Exelixis, a company set up to exploit fruit flies in functional genomics: a bit of background will help to explain why.

There are several reasons why fruit flies are good experimental animals.[a] One is a short gestation period: the female will lay several dozen eggs on fermenting fruit, and they become flies in about 24–30 hours in a process that is easily monitored. A second reason is that Thomas Hunt Morgan had pioneered the field of *Drosophila* genetics in the first decade of the 1900s, work for which he won the 1933 Nobel Prize. Morgan et al.'s studies were extended by David Hogness and others to develop molecular methods to characterise the mutations that Morgan and others had made or discovered, making it straightforward to link the genotype of the mutant flies to their phenotypes, of which flies have many.[a]

One expert fly geneticist is Spyros Artavanis-Tsakonas, a Greek–American who was an assistant professor at Yale University in New Haven in 1990 (now Professor Emeritus of Cell Biology at Harvard University). He had interest in developmental biology and cellular development in the fruit fly, including a

---

[a]I did my first experiments with flies in a week-long *Drosophila* Genetics course in the Genetics department at the University of Birmingham. I admit to never really understanding then the attraction of looking at wing structure, eye colour, or the behavioural traits of flies in small milk bottles. I get it now and, as in so many things, I wish I had paid more attention at the time.

special interest in a gene called *Notch*, which he had cloned in 1985 and knew to be important in fly development. We now know that the *NOTCH1* gene makes a protein called Notch1, a member of the Notch family of receptors. Ligands like a protein called Delta bind to the Notch1 receptor, forming a complex that sends signals that control the development of many tissues in the fly.

Upon cloning the gene, Artavanis-Tsakonas also noticed that it had human homologues. By using fly gene-interaction screens, he was able to genetically dissect the Notch pathway and show that the same general signalling plan happened in humans. Much the same homology pathways exist in the fly with other important genes and pathways such as the RAS oncogene pathway. Even in yeast cells, there are RAS gene homologues to those found in human cells, and a simplified but homologous signalling cascade. Artavanis-Tsakonas thought that flies could be very useful in a functional genomics context.[1]

One of the people Spyros talked to about his idea was the entrepreneurial Jonathan Rothberg, who had been a graduate student and postdoc in his lab. Jonathan later formed CuraGen and the 454 Corporation (involved in gene sequencing; see Chapter 8). It did not take long for Jonathan to catch on, and he became very enthusiastic about the project. Artavanis-Tsakonas connected with Aristide Fronistas, a Greek businessman he had met at Stanford in 1979, to form a company to exploit flies for functional genomics, with *Notch* as the exemplar. Gerry Rubin, another "fly guy" at University of California Berkeley and who had been in Cambridge (UK) with Spyros at the Laboratory of Molecular Biology (LMB) on Hills Rd. in the 1970s, and Corey Goodman, a UC Berkeley professor, were also persuaded to join.

After being told by some that "more intellects are better than less," Artavanis-Tsakonas also signed up several members of the "Yale mafia" to provide more intellectual heft to the deal. This included immunologist Richard Flavell, who had recently left Biogen; Jon Morrow, a pathologist; Mark Mooseker, a cell biologist; and Michael (Mike) Snyder, a biochemist with genomics interests. Through Jonathan, they were introduced to Alan Jakimo, a "science groupie" lawyer at Brown & Wood. Artavanis-Tsakonas met Stelios Papadopoulos, a Greek cell biologist who had worked in David Sabatini's lab at NYU and had become a successful biotech banker and investor in the 1980s, for the first time.[2] Given their fluency in several languages (including cell biology, Greek, and English), it did not take long for them both to want the company to happen.

The basic premise was that fly genetics could be used to dissect multiple pathways and understand the links between disease genes and pathways in humans, by analogy to what had been done with *Notch*. Stelios called flies "little people with wings," a moniker which stuck for a while. Leslie Misrock, a patent attorney at Pennie & Edmonds whom Stelios knew, listened to the story with interest. They became involved in filing some Notch intellectual property (IP) and came up with TopoRythmics as a name for the company. Adriane Antler, who was with the firm and fortuitously had worked at the Rockefeller on EGF and the EGF receptor in flies, helped to put the patent filings together. The company became TopoGEN and then TopoGenetics with Gerry Rubin, Jon Morrow, Mark Mooseker, Richard Flavell, and Artavanis-Tsakonas as founders.[b] Like a lot of start-ups, initial meetings were held at the founder's house: in this case, the Artavanis-Tsakonas house in New Haven.

Access to Capital

In early 1992, the fun really started in earnest as the team began to look for venture capital (VC) with which to get going. They talked to a lot of people and kissed a lot of frogs, but no princes or princesses emerged. For example, Corey, Gerry, Alan, and Spyros met with Wally Steinberg at HealthCare Ventures (HCV) corp. Despite the usual Kabuki performance, nothing came of the meeting, even though the TopoGenetics business plan reflected accurately what they wanted to do. Mark Levin and others suggested that there were too many people involved and that a restart would be necessary if a company was ever to get invested in and off the ground.

Stelios got more involved in 1993 and introduced Spyros to Charlie Cohen at Creative BioMolecules, a Roberto Crea–founded Genentech spinout. They were working on bone morphogenetic proteins (BMPs) and were using fly genes to find the human homologues of the fly BMPs.[c] Charlie was keen to get involved, and he, Creative BioMolecules, PaineWebber (where Stelios worked), and Stelios himself put together a seed round of $1 million for the company to move forward and to pay Pennie & Edmonds for the work they had done.

---

[b] At a meeting in Crete in October 1991, Spyros persuaded Corey Goodman, a professor of neurobiology and genetics at UC Berkeley and HHMI fellow, to become another co-founder of TopoGenetics. Corey is now managing partner at VenBio, a venture capital firm in San Francisco.

[c] Of course, invertebrate flies do not have bones, but many signalling pathways and their constituent proteins are conserved across animal phyla.[3]

Tony Evnin at Venrock, who had financed many successful Biotech companies including Gilead Sciences (Chapter 12), loved the pitch and with JH Whitney, another of the legendary VCs around (Chapter 18) said they would invest. The founders produced many bona fides for these investors, including letters of support from Nobel laureates! But these investments never materialised owing to a "change of heart."

Another six months went by. TopoGenetics was history but the idea was not. Corey, Gerry, and Spyros reconfigured things as a band of three founders, with Charlie as *de facto* CEO. In April 1994, Sherry Reynolds, a venture consultant based in Research Triangle Park (RTP) whom I knew from Glaxo days while doing the Gilead deal, became President and COO for Exelixis—the new name of the company taken from exelixis (Εξέλιξη), the Greek word for "evolution" (Plate 17).

Some Money at Last

Atlas Venture, in the form of Jean-François Formela and Jason Fisherman from Advent International, were persuaded to finance Exelixis with the first VC-led round of $4 million. The company was incorporated formally in November 1994, starting life in 600 Kendall Square, right in the heart of Boston biotech. Jean-Francois became the acting interim CEO. With the new company now starting, the Howard Hughes Foundation claimed that Spyros, who was a Howard Hughes fellow, had strayed off the Hughes reservation and was not keeping them in the loop. This was despite Spyros ensuring that all letters about TopoGenetics and Exelixis were copied to Max Cowen at the Foundation: in the end, their concerns were allayed.

Hiring People in San Francisco[d]

Exelixis began to hire people, including Lynne Zydowsky as a project leader (she had worked with Leslie Hudson at Glaxo Research at RTP in the early

---

[d]I remember having dinner with Corey Goodman and Gerry Rubin in San Francisco during the H&Q conference (Chapter 18) in January 1995, to talk about the fly company because it intrigued me. They were on the lookout for both a CSO and a CEO for the company. I also remember visiting 600 Kendall Square shortly afterwards and meeting the people there. These visits did not go down very well with Kevin Kinsella, the CEO at Sequana, who found out about the dinner. It was not in any case a secret, but he thought I was moonlighting and looking to leave "his" company, which was not necessarily the case at the time.

1990s). During a meeting at Millennium, Spyros and Stelios met with Geoffrey (Geoff) Duyk, whom they subsequently recruited as CSO after the CEO was hired. Shortly afterwards (October 1995), George Scangos, who Spyros had known at Yale and who was another one of the "Greek mafia,"[e] was recruited from Bayer on the west coast to be the CEO of the new company. There was a little wrinkle to his hiring: for George to lead Exelixis, the company had to be in the Bay Area. So, they moved. At the time the company was 35 people in Kendall Square, and almost all of them moved to South San Francisco. True to the business plan, they focused on the Notch and the semaphorin pathways.

Focusing on Drug Discovery

By 2000, the company was focused more on drug discovery and development than on fly genetics and had built up a large chemical compound library. Exelixis went public in 2000, raising $118 million. Finances were helped by the company forming a broad alliance with GSK, which included $30 million upfront in cash and committed R&D funding. Perhaps, the most important and differentiated program that Exelixis developed was its mitogen-activated kinase kinase (MAPKK) inhibitor project. Also called MEK, this enzyme is part of the RAS signalling pathway, which has counterparts in flies (and yeast), so they knew the signalling pathway well. MEK is overexpressed in some cancers, and overexpression of MEK is part of the resistance mechanisms that BRAF-mutated melanoma cells use, to overcome the BRAF inhibitors (e.g., vemurafenib) used to treat them. Exelixis partnered its MEK inhibitor program with Genentech, with cobimetinib (the MEK inhibitor) being part of the collaboration. A year later its lead cancer drug candidate XL-184 (cabozantinib), a multispecific tumour kinase inhibitor, was partnered with BMS—the rights to which were later returned to Exelixis.

Scangos left Exelixis to become CEO of Biogen in 2010. I remember talking to him about leaving an exciting company like Exelixis to go to Biogen. He said (and I quote) "at least every January 1st I will not have to think once again about where I am going to get the $300 million, I need to fund the company until we have income from real products." Biogen, in contrast, had real products, bringing in billions of dollars annually.

---

[e] One of several biotech mafias I refer to throughout this book.

Mike Morrisey, who was head of R&D, took over as Exelixis' CEO and marshalled the successful elaboration of the strategy to develop and co-market their kinase inhibitors. Exelixis' first drug approval came in 2012, when cabozantinib (Cometriq) was approved for medullary thyroid cancer. Its development was not without issues: the drug initially failed in a prostate cancer trial but was shown to work in melanoma and subsequently in renal cell carcinoma. Exelixis now has several products on the market and multiple partnerships. Not a bad outcome for a company based on fruit flies circulating around your fruit bowl.[3]

NEMAPHARM

NemaPharm was a company cofounded by MIT developmental geneticist and Nobel laureate H. Robert (Bob) Horvitz and Carl Johnson from Cambridge Neuroscience in 1992 to use the nematode worm *C. elegans* as a model for drug discovery. At Sequana, we had been thinking about expanding the genomics platform to incorporate the means to understand the genes and proteins we were finding and to use them to discover genetically relevant drugs. The worm fitted the bill.

C. elegans was another Sydney Brenner special. After the central dogma of DNA→RNA→protein had been derived, Sydney was looking for ways to understand neuronal development in whole organisms. After thinking about several alternatives, he proposed to use *C. elegans* as a model organism.[4,f] Brenner chose *C. elegans* because it is a simple hermaphroditic nematode worm—a multicellular eukaryote that grows on Petri dishes, lives on a cheap diet of *Escherichia coli*, and is simple to work with in the lab.[g]

The worm genome consists of about 100 million base pairs and was the first multicellular organism to have its genome sequenced (Chapter 8). Its genes are arranged much like other eukaryotic organisms, including mice and humans. Many of the known worm genes and their pathways are similar

---

[f] Almost all the subsequent work on *C. elegans* can be traced back to Sydney's lab in Cambridge, where many worm investigators were either his graduate students or his postdocs.

[g] The 2002 Nobel Prize in Physiology or Medicine was awarded to Sydney Brenner, Bob Horvitz, and John Sulston for their work on the genetics of organ development and programmed cell death in *C. elegans*.

to those in more complex organisms, so it was a good model for functional genomics.[h]

I first met Carl Johnson and his partner Minka vanBeuzekom in Cambridge (MA), in the winter of 1995/1996. There was snow on the ground. NemaPharm was in a lab that seemed to be in a house rather than a lab building on Inman St. It was, in fact, the original lab that housed the Fleischmann Yeast company and had huge ceiling heights to accommodate the culturing vats that the old company used.[i] The doors on the elevator at 100 Inman St. were decidedly old school, the kind you see in dramatic thriller movie scenes with two doors that pull across the front to start the elevator. When Kevin Kinsella from Sequana came to the 100 Inman St. lab in his stretch limo, it could not turn into the parking garage, so his chauffeur had to circle the residential neighbourhood in the snow until the meeting was over.[5] Eventually, after Sequana bought NemaPharm and much to their chagrin, we moved them to a lab on Landsdowne St. in Cambridge (MA), next to the old Necco candy factory, which itself is now the headquarters of the Novartis Institutes for BioMedical Research.

Worm Screens

Like flies, there are a variety of mutations that can be derived in *C. elegans*, and the relationship between mutations in particular worm genes and the resulting phenotypes (e.g., mobility patterns) were just beginning to be understood.

NemaPharm had built a machine for screening nematodes in a 96-well plate for phenotypic changes (e.g., in response to potential new drug candidates). The NemaPharm worm dispenser was modelled on a flow cytometer.[6] These worms included knockout mutations in the normal genes. NemaPharm had two full-time personnel, each one generating one gene deletion strain of worm per week.

---

[h] Please note that I use the words "more complex." I got into trouble with Carl Johnson for referring to worms or flies as "lower" organisms and to humans as "higher organisms." Obviously, higher and lower depends on what you are measuring.

[i] Charles Louis Fleischmann founded the yeast company to provide live yeast cultures for the baking and brewing industry in the nineteenth century, revolutionizing how bread is made and marketed. He is also Elizabeth Holmes' great-great-great-grandfather, although Theranos was far from revolutionary.

NemaPharm was primarily interested in gene pathways, but not all pathways were appropriate targets. Many of the genes known to be involved in Alzheimer's disease, for example, including the presenilin genes that make proteins involved in processing the amyloid precursor protein (Chapter 7), can be found in the worm. NemaPharm looked at approximately one million mutated genomes, but they found no genes that could be knocked out to suppress mutated presenilin genes. In contrast, they were able to target the Ras pathway, which is conserved in *C. elegans* as well as in yeast and flies.

Another pathway of interest to NemaPharm was programmed cell death or apoptosis. The *C. elegans* phenotypes for mutants in this pathway were particularly robust. Selectively blocking cell death was thought at the time to be an important way to decrease damage in degenerative diseases or after a stroke. The apoptosis screen used *ced9*, the *C. elegans* homologue of the human gene *BCL2*. In the absence of *ced9* function, all *C. elegans* cells died. Mutations in other genes, such as *ced3* and *ced4*, which encode other components of the cell death machinery, would keep a *ced9* mutant animal alive, so it was thought that a chemical screen might be able to identify compounds that keep *ced9* mutant worms alive by working through *ced3* or *ced4*.

The whole concept was certainly innovative, but not without its problems. The first was that the worm dispenser was not all that reliable. Sometimes, the worms got stuck, and often more than one worm ended up in a well. Second, there are not that many worm phenotypes to screen against, and many mutations in different genes caused the same general phenotypic changes. Third, the compound concentrations had to be high as the worms did not ingest much of them.[6]

NemaPharm ended up being used much more as a target-identification technology rather than a drug-identification technology and was acquired by Sequana in July 1996. We formed a deal with Glaxo Wellcome, before they were GSK, to use this technology to study some of their genes of interest. This complemented a deal that NemaPharm already had with the cell death–focused company Idun Pharmaceuticals (also set up in San Diego by Bob Horvitz). Idun was later acquired by Pfizer, maintained as a subsidiary in La Jolla, and later sold back to Conatus (which was formed by the original Idun Founders).[j]

---

[j] A rather typical circular biotech outcome.

Sequana complemented the NemaPharm acquisition by forming fly collaborations with Charles Zuker and William (Bill) McGinnis at UCSD to focus on certain obesity genes.[k] The company also hired a mouse geneticist, who had made many transgenic knockout mice at GenPharm (Chapter 4). Sequana could now claim that it had active functional genomics programs using flies, worms, and mice! The screening activities we started also helped us to form a deal with Aurora Biosciences; the high-throughput screening company Kevin Kinsella had set up shortly after Sequana, to screen some of the proteins made by the genes we were finding. Aurora later as mentioned got acquired by Vertex for their compounds, one of which affected the function of the mutant CFTR for the potential treatment of cystic fibrosis (CF) (see Chapters 7 and 11).

I saw Carl and Minka much more after Sequana bought NemaPharm, which continued as an entity even after the Arris merger to form Axys. Carl died in 2021. I will always remember Carl's passion for his worm craft and his ability to go with the flow, even though the corporate world was not really his comfort zone. *C. elegans* genetics was his world.

Perhaps, more important than the use of worms in functional genomics was their importance in the development of RNA interference or RNAi technology. Feeding *E. coli* that contained RNAi specific for different worm genes or soaking the worms in a solution containing the RNAi became a rapid way to generate mutant worms for screening or for further study.[l]

MAXYGEN

The name "Maxygen" may be familiar: the company was founded in 1997 to use the technique of molecular breeding or "DNA shuffling" invented by the late Willem P.C. (Pim) Stemmer to make proteins with modified function (Box 1). Stemmer started his work in this area while at Hybritech in San Diego in the 1980s and then joined Affymax, the Alejandro (Alex) Zaffaroni combinatorial chemistry (CombiChem) company that was started in 1988 and sold to Glaxo in 1995.[7]

---

[k] "Coincidentally," Zuker subsequently became the Kevin and Tamara Kinsella chair of neurobiology at UCSD, before moving to Columbia University in New York.

[l] Andrew (Andy) Fire and Craig Mello won the 2006 Nobel Prize in Physiology or Medicine for their discovery of RNAi in *C. elegans*. This led to another successful branch to the biotech industry tree (Chapter 13).

BOX 1. DNA SHUFFLING

The protein engineering technique called "DNA shuffling" consists of moving linear pieces of genes (blocks of DNA sequence) around in combinations to make proteins with novel functions, mimicking natural DNA recombination. This technology rapidly accelerated the engineering of therapeutic proteins, vaccines, and industrial enzymes with new functions, but was very dependent on having biochemical assays for the new proteins that were both fast and selective. This development and the IP filed to protect was the basis for the founding of Maxygen, an Affymax spinoff. Simba Gill (who I worked with at Celltech and was CEO of Evelo) was their CBO. The company went public in early 2000, raising nearly $90 million in an oversubscribed offering. Maxygen formed many successful relationships with pharma and with the agricultural and industrial enzyme sectors, and they launched several spinouts to promote different applications of the technology, including Verdia (sold in 2004 to DuPont) and Codexis.

Stemmer left Maxygen in 2003 to start Avidia, a company subsequently bought by Amgen. He also founded Amunix in 2006 and was the first CEO of the company.[7] Amunix's core technology is called "XTENylation." XTENs are long, unstructured, hydrophilic protein chains that can be fused to other proteins to improve their half-life in circulation. This technology was important for the construction and development of BIVV 001, the long-acting form of Factor VIII that Rob Peters and others designed and made while at Biogen (see Chapter 11). Maxygen ceased operations in 2013 but was re-established in 2017 as a contract research organization dedicated to making better proteins by design for their clients.[m]

Exelixis and NemaPharm are two different stories with very different outcomes. In some ways, functional genomics was just a moment in time in biotech and very much a means to an end—that end being the ability to discover new drugs based on the genes and the proteins they coded for, found by large-scale cloning and sequencing. But as has been shown repeatedly that the genes on their own are generally not worth that much. It is what you do with them and the biological pathways they were involved in that counted for drug discovery.[3]

---

[m] A rather disappointing outcome, given their talent.

REFERENCES AND NOTES

1. Conversation with Spyros Artavanis-Tsakonas on the history of Exelixis, 2020, 2021.

2. Papadopoulos S. 2022. For the 'godfather' of biotech, saving Biogen is the final act of a singular career. *STAT*, August 2022.

3. Fishman MC, Porter JA. 2005. A new grammar for drug discovery. *Nature* 437: 491–493. doi:10.1038/437491a

4. Goldstein B. 2016. Sydney Brenner on the genetics of *Caenorhabditis elegans*. *Genetics* 204: 1–2. doi:10.1534/genetics.116.194084

5. E-mail conversation with Minka vanBeuzekom, Carl's partner, 2021.

6. Wells WA. 1998. High throughput worms. *Curr Biol* 5: R147–R148. doi:10.1016/s1074-5521(98)90174-0

7. Larrick JW, Schellenberger V, Barbas III CF. 2013. Willem 'Pim' Stemmer 1957–2013. *Nat Biotechnol* 31: 584. doi:10.1038/nbt.2630

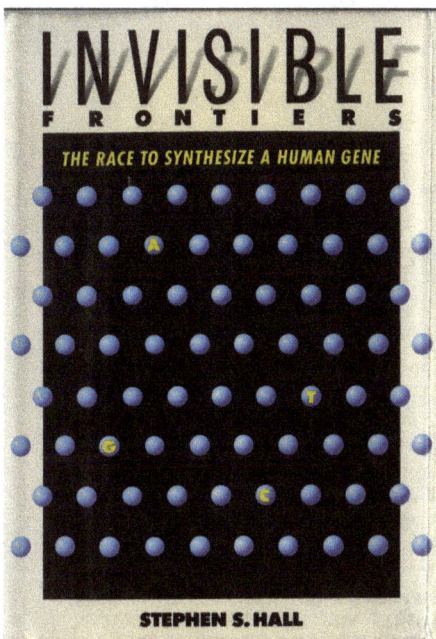

Craig

The last chapter will be written about you and Genentech's triumph in the pharmaceutical marketplace.

Regards
Herb Boyer
Dennis Kleid
Bill Young
Dave Goeddel
Axel Ullrich
Bob Swanson
Herb Heyneker

Plate 1. Front and inside covers of Stephen Hall book *Invisible Frontiers*. The signatures from *top* to *bottom* are Herb Boyer, Dennis Kleid, Bill Young, David Goeddel, Axel Ullrich, Bob Swanson, and Herb Heyneker. (*Cover reprinted from Hall [1987], Atlantic Monthly Press.*)

Genentech, Inc.
Genentech, Inc.
Genentech, Inc.
Genentech, Inc.
Genentech, Inc.

1,000,000 SHARES
COMMON STOCK

All the shares of Common Stock offered hereby are being sold by Genentech, Inc. (the "Company" or "Genentech").

Prior to this offering, there has been no public market for the Common Stock of the Company. See "Underwriting" for information relating to the method of determining the initial public offering price.

The Common Stock offered hereby involves a **HIGH DEGREE OF RISK.** See "Risk Factors" for information with respect to the Company's short operating history, uncertainty of financial results and capital needs and other risk factors.

THESE SECURITIES HAVE NOT BEEN APPROVED OR DISAPPROVED BY THE SECURITIES AND EXCHANGE COMMISSION NOR HAS THE COMMISSION PASSED UPON THE ACCURACY OR ADEQUACY OF THIS PROSPECTUS. ANY REPRESENTATION TO THE CONTRARY IS A CRIMINAL OFFENSE.

	Price to Public	Underwriting Discounts(1)	Proceeds to Company(2)
Per Share	$35.00	$2.25	$32.75
Total Minimum	$35,000,000	$2,250,000	$32,750,000
Total Maximum(3)	$38,500,000	$2,475,000	$36,025,000

(1) See "Underwriting" for information relating to indemnification of the Underwriters, shares reserved for sale to certain persons and other matters.

(2) Before deducting expenses payable by the Company estimated at $435,000.

(3) Assuming full exercise of the 30-day option granted by the Company to the Underwriters to purchase, on the same terms, up to an additional 100,000 shares to cover any over-allotments. See "Underwriting."

The shares are offered by the several Underwriters when, as and if issued by the Company and accepted by the Underwriters and subject to their right to reject orders in whole or in part. It is expected that delivery of the shares will be made on or about October 21, 1980.

BLYTH EASTMAN PAINE WEBBER
INCORPORATED

HAMBRECHT & QUIST

The date of this Prospectus is October 14, 1980

Plate 2. Genentech IPO S1 cover. *(Image courtesy of Stelios Papadopoulos, Chairman of the Biogen Board, 2014–2023.)*

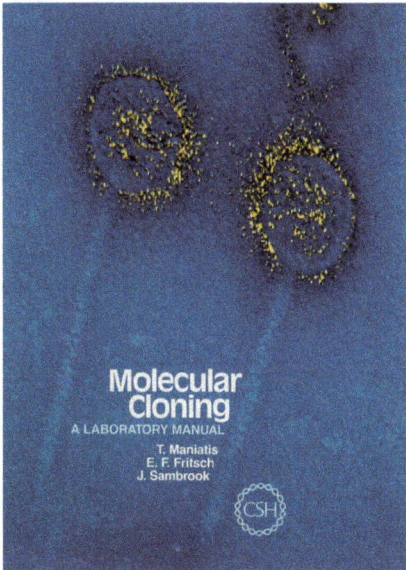

Plate 3. Cover of first edition of *Molecular Cloning: A Laboratory Manual.* (*Taken from Maniatis T, Fritsch EF, Sambrook J. 1982. Cold Spring Harbor Laboratory Press, Cold Spring Harbor, NY.*)

Plate 4. Charon the boatman ferrying souls across the river Styx. (*Figure created by Aceross Lord and reprinted with permission from Shutterstock.*)

Plate 5. Cesar Milstein and a stirred bottle for growing monoclonals. (*Reprinted from Rabbitts TH, Cell [2002], with permission from Elsevier © 2002.*)

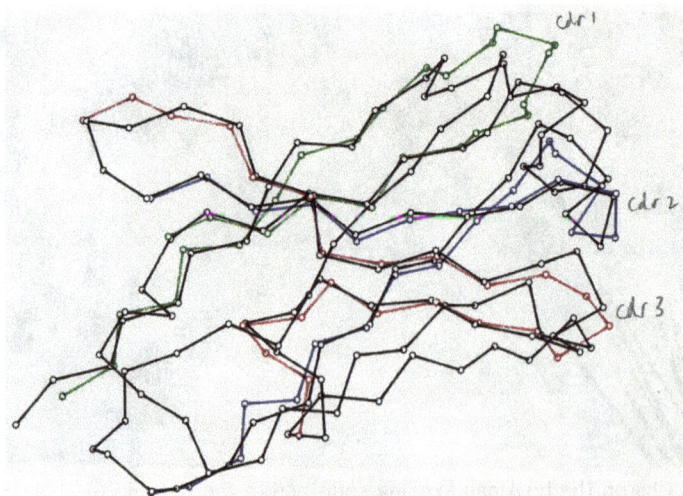

Plate 6. CDR grafting from the author's notes at the time. (*Photo from the author's private collection.*)

Plate 7. View of the island Tristan da Cunha. *(Photo credit: Tristan da Cunha Government. Image freely available in the public domain.)*

Plate 8. Welcome to the island Tristan da Cunha. *(Image from CNBCTV18 India; https://www.cnbctv18.com/travel/destinations/will-you-visit-this-remote-island-in-south-atlantic-all-you-need-to-know-about-tristan-da-cunha-17383441.htm)*

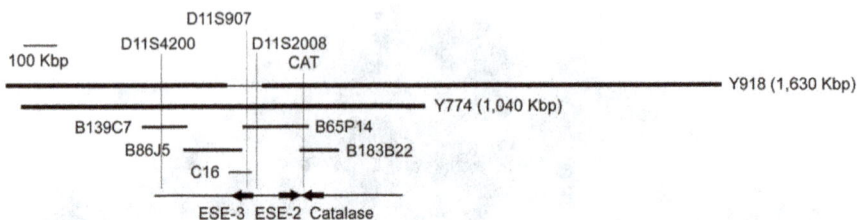

Plate 9. Physical map of the ESE-3 genomic region. Microsatellite markers are listed *across* the top. The *lines immediately below* the markers depict YAC clones. *Below* the YAC clones are BAC and cosmid clones. Genes are shown at the *bottom*, and the *arrows* indicate the direction of transcription. *(Figure reprinted from Tugores et al.* J Biol Chem **276:** *20397 [2001], doi:10.1074/jbc.M0109302000 under the terms of the Creative Commons CC-BY 4.0 license.)*

Plate 10. Sequana swag: Axys hat, Genos T-shirt, and Sequana polo shirt. *(Photo from the author's private collection.)*

Subject to Completion, dated June 14, 1995

PROSPECTUS

3,000,000 Shares

SEQUANA
THERAPEUTICS

SEQUANA THERAPEUTICS, INC.

Common Stock

All of the shares of Common Stock offered hereby are being sold by Sequana Therapeutics, Inc. (the "Company"). Prior to this offering, there has been no public market for the Common Stock of the Company. It is currently estimated that the initial public offering price will be between $12.00 and $14.00 per share. See "Underwriting" for a discussion of the factors to be considered in determining the initial public offering price. Application has been made to have the Common Stock approved for quotation on the Nasdaq National Market under the symbol "SQNA."

The shares of Common Stock offered hereby involve a high degree of risk. See "Risk Factors" beginning at page 5.

THESE SECURITIES HAVE NOT BEEN APPROVED OR DISAPPROVED BY THE SECURITIES AND EXCHANGE COMMISSION OR ANY STATE SECURITIES COMMISSION NOR HAS THE SECURITIES AND EXCHANGE COMMISSION OR ANY STATE SECURITIES COMMISSION PASSED UPON THE ACCURACY OR ADEQUACY OF THIS PROSPECTUS. ANY REPRESENTATION TO THE CONTRARY IS A CRIMINAL OFFENSE.

	Price to Public	Underwriting Discounts and Commissions(1)	Proceeds to Company(2)
Per Share	$	$	$
Total(3)	$	$	$

(1) The Company has agreed to indemnify the Underwriters against certain liabilities, including liabilities under the Securities Act of 1933, as amended. See "Underwriting."

(2) Before deducting estimated expenses of $575,000 payable by the Company.

(3) The Company has granted to the Underwriters a 30-day option to purchase up to 450,000 additional shares of Common Stock on the same terms and conditions set forth herein, solely to cover over-allotments, if any. If such option is exercised in full, the total Price to Public, Underwriting Discounts and Commissions and Proceeds to Company will be $, $ and $, respectively. See "Underwriting."

The shares of Common Stock offered by this Prospectus are offered by the Underwriters subject to prior sale, to withdrawal, cancellation or modification of the offer without notice, to delivery to and acceptance by the Underwriters and to certain further conditions. It is expected that delivery of certificates for the shares of Common Stock will be made at the offices of Lehman Brothers Inc., New York, New York, on or about   , 1995.

LEHMAN BROTHERS   HAMBRECHT & QUIST

, 1995

Plate 11. Sequana IPO prospectus cover. *(Photo from the author's private collection.)*

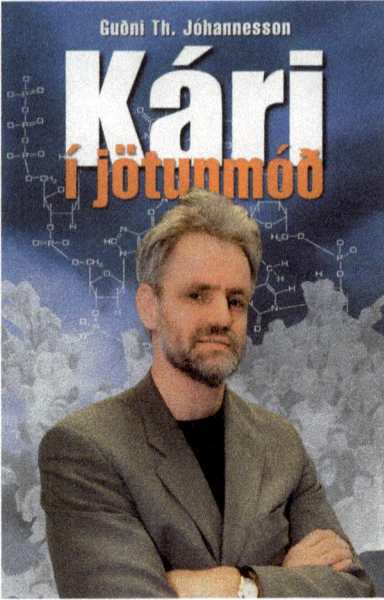

Plate 12. Cover of biography about Kári Stefánsson. *(Cover art © Guoni Th. Johannesson.)*

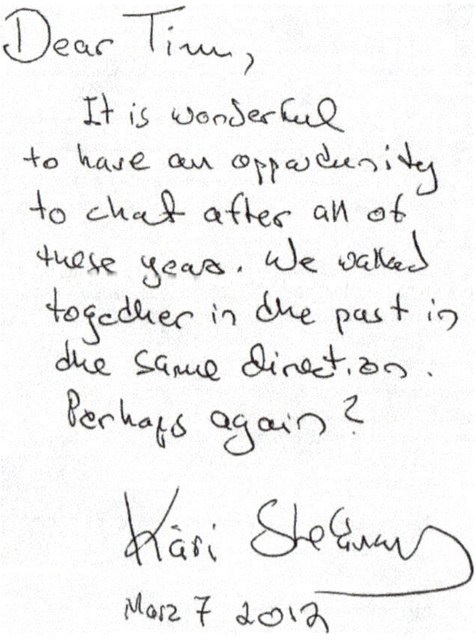

Dear Tim,

It is wonderful to have an opportunity to chat after all of these years. We walked together in the past in the same direction. Perhaps again?

Kári Stefánsson

Mar 7 2012

Plate 13. Note to me from Kári Stefánsson on the inside cover of the book. *(Photo from the author's private collection.)*

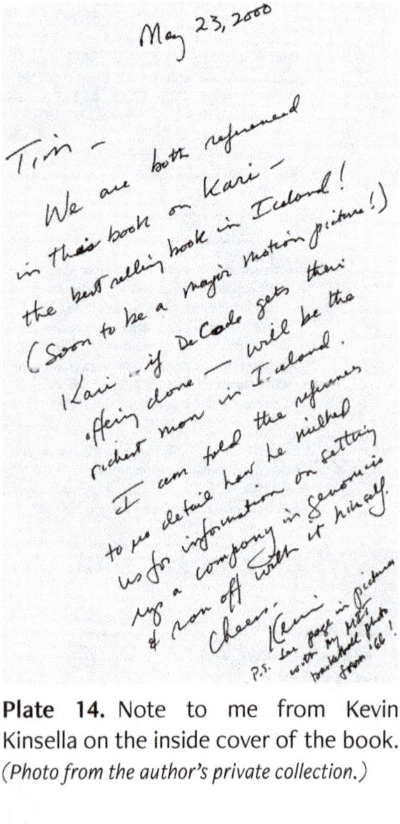

May 23, 2000

Tim —

We are both referenced in this book on Kari — the best selling book in Iceland! (Soon to be a major motion picture!) Kari ... if DeCode gets their offering done — will be the richest man in Iceland. I am told the references to us detail how he nickel us for information on setting up a company in genomics & ran off with it himself.

Cheers,
Kevin

P.S. see page in picture with my note translated photo from N.I.

Plate 14. Note to me from Kevin Kinsella on the inside cover of the book. *(Photo from the author's private collection.)*

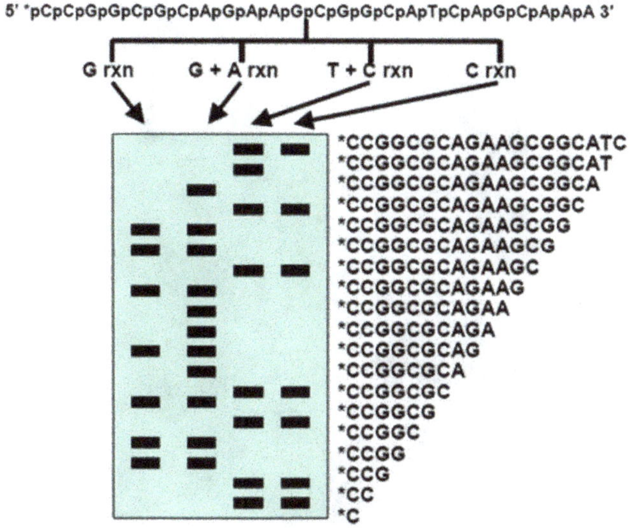

Plate 15. Depiction of a Maxam–Gilbert sequencing gel. *(Reprinted from Bisht and Panda, DNA sequencing: methods and applications [2014], Fig 2.1, with permission from Springer Nature.)*

Plate 16. Sequencing facility in the Genetics department at Stanford University at the time of the human genome sequencing project. *(Picture courtesy of Richard Myers.)*

Plate 17. Exelixis founders and key scientists. From the *top left to right and down*: Stelios Papadopoulos, Jean-François Formela, Gerry Rubin, Corey Goodman, Spyros Artavanis-Tsakonas, Charlie Cohen, and Sherry Reynolds. *(Art created by and reprinted with permission from Spyros Artavanis-Tskanonas, Harvard.)*

Plate 18. The original SGX crew. *(Left to right)* Janet Newman, Ken Schwinn, Tom Peat, Diane, Tim Harris, Mark Bergseid, Linda Grais, Sarah Dry, Chad Smith, Sean Buchanan, and our HR Consultant. *(Photo from the author's private collection.)*

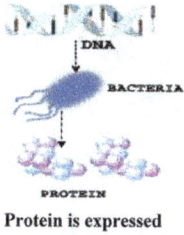

SGX Technology

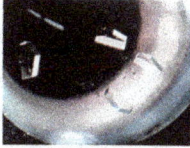

DNA

BACTERIA

PROTEIN

Protein is expressed

Proteins are crystallized

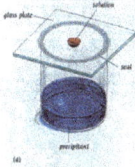

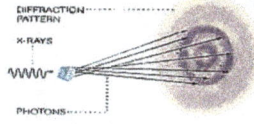

DIFFRACTION PATTERN

X-RAYS

PHOTONS

Crystals are analyzed with X-rays

Images courtesy of New York Times, July 4th, 2000 and have been edited

7

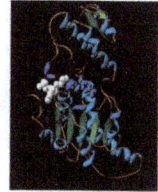

The structure is solved

Plate 19. The SGX process from the Structural Genomix company literature. *(Photo from the author's private collection.)*

Plate 20. The APS in Chicago. *(Image by Argonne National Laboratory.)*

Plate 21. Tim Harris, Kevin D'Amico, and Linda Grais during construction of our beamline. *(Photo from the author's private collection.)*

Plate 22. The majestic Allston Genzyme manufacturing facility. *(Courtesy of Nick Wheeler.)*

Plate 23. Signing the Glaxo–Gilead strategic alliance deal in RTP, July 1990. Seated are Michael Riordan (*left*) and Charles Sanders (*right*). (*Photo from the author's private collection.*)

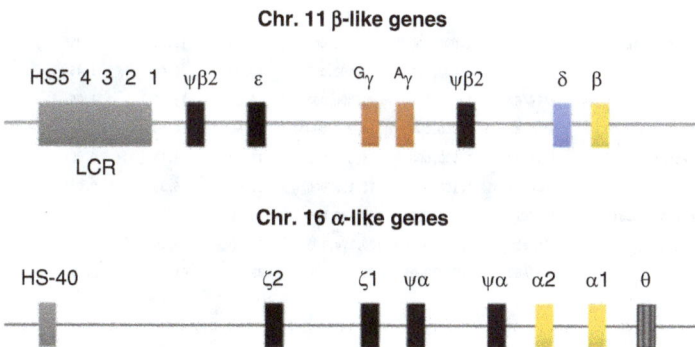

Plate 24. The α- and β-globin loci. (*Reprinted from* Clinical Molecular Medicine, *Fig. 18.2.1,* © 2020 with permission from Elsevier.)

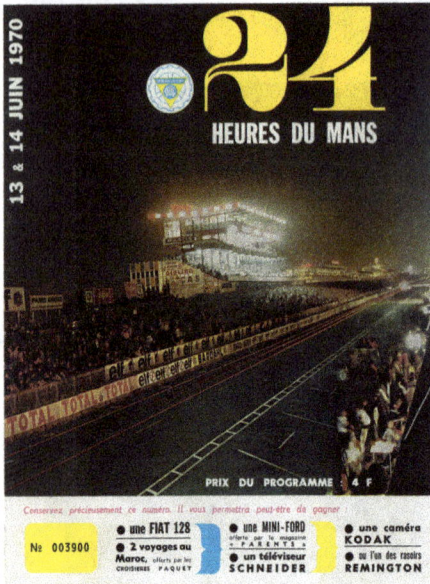

Plate 25. The cover of the 1970 Le Mans Race programme. *(Image from 1970 Le Mans 24h Race programme, Automobile Club de l'Ouest. Author's own programme.)*

	Company	Issue Date	Total Shares Offered	Total Dollar Amount of Offering	Offer Price Per Share	Current Market Price (6/28/83)	Gross Under-writing Spread	Gross Spread as % of Offer Price
	Selected Biotechnology Issues Initial Public Offerings							
1.	Amgen	6/17/83	2,350,000	$42,300,000	$18.00	$13.00	$1.17	6.50
2.	Damon Biotech	6/3/83	2,400,000	40,800,000	17.00	14.75	1.11	6.52
3.	Cambridge Bioscience	3/31/83	1,000,000	5,000,000	5.00	12.25	0.425	8.50
4.	Biogen	3/22/83	2,500,000	57,500,000	23.00	15.00	1.45	6.30
5.	Centocor	12/14/82	1,500,000	21,000,000	14.00	10.75	1.05	7.50
6.	Genex	9/29/82	2,000,000	19,000,000	9.50	16.50	0.75	7.89
7.	Collaborative Research	2/11/82	1,300,000	14,300,000	11.00	13.75	0.85	7.72
8.	Hybritech	10/28/81	1,100,000	12,100,000	11.00	26.50	0.825	7.50
9.	Monoclonal Antibodies	8/4/81	450,000	4,500,000	10.00	17.75	0.80	8.00
0.	Cetus	3/6/81	5,224,965	120,174,195	23.00	16.50	1.35	5.86
1.	Genentech	10/14/80	1,000,000	35,000,000	35.00	41.50	2.25	6.42

Robertson, Colman & Stephens
June 30, 1983

Plate 26. Early IPOs. *(Photo from the author's private collection.)*

Plate 27. The Sequana Lab at 11055 North Torrey Pines Rd. in the very early days. The Sequana logo is visible along with those of the other tenants. *(Photo courtesy of Alexandria Real Estate Equities, Inc.)*

Plate 28. The Seattle Light and Power building on Lake Union, Seattle, c. 1921. *(Courtesy of the Seattle Municipal Archives, ID #1810.)*

Plate 29. The iconic ZymoGenetics Headquarters Building, as viewed from Lake Union. (*Photo by Eben Calhoun, who grants the right to use this work for any purpose without any conditions.*)

Plate 30. The British and Irish soccer team at the Greek meeting 2003. Picture taken after a match at Stelios' house when the meeting was held at the Sani Resort in Halkidiki, Greece. I am reliably informed that Greece won the tournament. (*Back, left to right*) Mark Bodmer (GSK), Bruce Carter (ZymoGenetics), Tim Harris (SGX), Chris Henney (Dendreon), and Arthur Higgins (Bayer Pharma). (*Sitting*) Bill Matthews (Deltagen) and Declan Quirke (Cowen and Co). (*Photo from the author's private collection.*)

CHAPTER 10

Structural Genomics

I t all started for me in the back of a limo in La Jolla in early 1999. I met there with Linda Grais, an ex–Wilson Sonsini lawyer with a medical degree (whom I knew from work at Sequana), and Seth Harrison, a young MD who had just started a venture fund called Apple Tree Ventures. On our ride, they told me about a company called "Structural GenomiX" that Wayne Hendrickson, a respected structural biologist, and Barry Honig, a respected bioinformatics and computational biologist, had started out of Columbia University. Their company would focus on "structural genomics," and they were looking for someone to run it.

I did not know what structural genomics was at that time. But it did not take long for me to find out, helped largely by an article written by Lawrence (Larry) Shapiro and Christopher (Chris) Lima in the journal *Structure*, in which they reported on the proceedings of a 1998 meeting on structural genomics held at the Argonne National Laboratory.[1] Shapiro and Lima, who have both gone on to have stellar careers in structural biology,[2] laid out the elements of the "new structural genomics paradigm" in their short paper reviewing the meeting.

I was obviously aware of the importance of structural biology for drug discovery: we used it at Glaxo for our HIV program. I also knew of the San Diego company Agouron Pharmaceuticals, which was formed in 1984 to exploit the potential of rational drug design. Their drug Viracept became the leading HIV protease inhibitor.[a] Vertex Pharmaceuticals were also founded on the use of structure-aided drug design to develop novel HIV proteases and other enzyme inhibitors.[3]

However, structure-enabled drug discovery was quite slow, especially compared to the time frame in which the medicinal chemists could make compounds. You had to wait weeks for the structural experts to determine

---

[a] Agouron were sold to Warner-Lambert in 1999 for $2.1 billion.

the shape of the target protein with the potential drug compounds attached: information you needed to guide you to what compounds to make next. It turned out that structural biology was more helpful retrospectively than prospectively.

The new idea was to combine genomics and rapid structure determination methods to build a large database of the structures of representative proteins, from as many types and classes of protein as possible. That this goal was even possible was based on the convergence of several technical advances. These included access to the growing DNA sequence information being obtained from several organisms, coupled with the ability to clone the genes and express proteins and folding domains of proteins in surrogate systems, such as *Escherichia coli*, yeast, and insect cells. Better and quicker protein purification methods and the ability to freeze protein crystals increased the ease of the logistics and the likelihood of obtaining suitable crystals for successful structural analyses.

The development of new, more rapid X-ray crystallography methods to speed up the whole process of 3D protein structure determination was also important. This included ways to label the protein internally to aid analysis, brighter and more focussed sources of X-rays, and the use of charge-coupled device cameras instead of X-ray film to capture the diffraction patterns. As important as the technology itself was the rapid development of computationally derived structure prediction methods.[b]

STRUCTURAL GENOMIX

The company itself did not start in the back of the limo that day, although it was mostly owing to the grit and determination of Linda Grais that the company got going at all. It started by chance: one of Linda's friends was dating Wayne Hendrickson, who mentioned to Linda that Wayne and Barry Honig were trying to start a structural genomics company. They had met Frederic (Fred) Bourke, an angel investor who had been involved with HGS.[c] They had already formed a company, but after some negotiation, the founders and

---

[b] AlphaFold and other computational innovations have taken structure prediction to a whole new level. But these breakthroughs only happened thanks to access to large numbers of structures that the structural genomics initiatives helped to enable.

[c] Fred Bourke was a founder with Peter Dooney of the high-end leather goods company Dooney & Bourke in 1975.

Linda felt that the more traditional venture capitalist (VC) route would be better than working with a wealthy handbag guy.

They met with Seth Harrison (of limo meeting fame) at Oak Venture Partners, who introduced them to Jean-François Formela of Atlas Venture (Seth's brother-in-law at the time) at the Robertson Stephens conference in New York in 1997. They agreed to fund the company with $250,000, provided Linda left Wilson Sonsini and dedicated herself to getting it going. She started her new work in an incubator space amidst the pimps, prostitutes, and drug dealers prevalent on 168th street in New York in those days, but rapidly moved to a more desirable (and safer) space that Punk Ziegel, the investment bank, had available in midtown.[d] Linda spent the next seven to eight months writing the original business plan and pounding the pavement looking for money. They needed a CEO, ideally with pharma and genomics experience, so they came to talk to me: in a limo.

Meeting with the Founders

After the limo meeting with Linda and Seth—which was thoroughly engaging—I met with Barry and Wayne. I immediately liked their honest and straightforward style as well as, of course, their obvious knowledge of the space. I also met with Chris Lima and Larry Shapiro, the authors of the structural genomics meeting proceeding paper that I used to educate myself. I appreciated that they could bring a lot of enthusiasm and vigour to the enterprise, as consultants to the company. I thought that the confluence of the technologies was enabling, and I undertook to work with Linda to try to get Structural GenomiX off the ground. Though not very imaginative, the name was better than the original name "Protarc." SGX, as we became known, was incorporated in July 1998.

At the time I was renting a 100-sq-ft office in Encinitas, having left Axys a few weeks earlier. Linda and I used this office to refine what SGX might do, and how we would go about doing it. We were driven by the idea of creating the best high-throughput genomics-driven structural biology technology platform available anywhere. We did think about drug discovery at the start—but not enough—as we were too enamoured with and focused on getting the platform up and running, including licensing some intellectual property (IP)

---

[d] Jeremy Levin had been using that space to get Physiome off the ground, but he no longer needed it.

Hendrickson and Honig had at Columbia. We decided to work on bacterial proteins to get started, a good idea technically for ease of expression and purification of the proteins. But it was a strategic mistake because therapeutics in that space was never going to be a compelling commercial opportunity. We should have started with human drug targets: we got there later.

Lab Space and People

The first task in early 1999 was to find lab space and to hire some people to do the work. We rented 7500 sq ft of space at 10505 Roselle Street in Sorrento Valley, near the railway station. We increased that to 17,500 sq ft after more space in the building had been renovated. The first employee after Linda and me was Sean Buchanan, a Cambridge (UK)-educated biochemist and postdoc at UCSD, who was intrigued with the project. Thomas (Tom) Peat and Janet Newman, a husband-and-wife team who had spent time in Wayne's lab, joined as employees four and five from Los Alamos. They were both crystallographers with experience in expressing proteins, making crystals, and determining their structure. Mark Bergseid joined as an experienced protein chemist, and Sarah Dry, an arts graduate, joined to help Linda with corporate communications (see Plate 18 for the original SGX crew). To help with protein expression, I managed to persuade an expert (Spencer Emtage, with whom I had worked at both Celltech and Sequana) to come and join us as VP of Molecular Biology and to join the management team. Jay Knowles, who had worked at Vertex when it grew from 60 to more than 300 people, joined us soon after as VP Business Development. We also got George Poste, ex-R&D chief at SmithKline Beecham, to be chairman of the Board.

Raising Money

Obviously, we also had to raise money and get a Series A financing done. Linda and I went to Palo Alto to talk to Prospect Venture Partners (PVP), a relatively new venture firm, where the late Alexander (Alex) Barkas (formerly of KP) was a managing partner. They were not on Sand Hill Rd. but in downtown Palo Alto, on the right side of the railway tracks. I was quite surprised to learn the day after our presentation to PVP that they were "in the deal." Just that quickly we had our Series A investors—Apple Tree Partners, Atlas Venture (from Boston), and PVP from the Bay Area. The Del Mar races were on at the same time, and we felt we were off to our races.[e] In December 1999, when we closed Series A, SGX were just 24 people.[4]

The SGX Process

The process we put together went from target identification to expressed purified protein to structure determination in a defined sequence, supported by a Laboratory Information Management System (LIMS) with extensive computational assistance (see Plate 19 for the SGX process). As well as Wayne and Barry, we assembled a first-rate Scientific Advisory Board consisting of prominent scientists in the field.[5]

Our first scientific advisory board (SAB) meeting was held in New York in mid-1999 with a celebratory dinner at the esteemed Four Seasons to mark the company formation. The second SAB was in San Diego in early 2000, when we really got going in earnest, bringing others into the company as special consultants including Sydney Brenner and Nobel laureate Joshua (Josh) Lederberg, whom Wayne and Linda knew. Chris Lima, Larry Shapiro,[f] and Theresa (Terry) Gaasterland also joined as consultants.

Hot Stuff

Structural genomics was the next hot thing in the genomics space, after DNA sequencing and functional genomics.[6] There was competition in the form of companies focused on real structure determination, such as SGX were doing, and from those using more *in silico* predictive methods. Astex, a structure determination company, were an Oxford Bioscience and Abingworth–funded entity set up in the United Kingdom in 1999, to focus on rapid X-ray crystallography methods and screening using compound fragments as starting points for drug discovery. They were started by some of the people I had worked with at Glaxo, including Harren Jhoti, who was the CEO. Sir Thomas (Tom) Blundell, an expert British biochemist, was an Astex founder. In the structural prediction space, two companies had

---

[e] We routinely took the company to a day at the Del Mar races every year after that investment. We even sponsored a horse that won in 2002.

[f] One of Larry's achievements was to obtain the 3D structure of the tubby protein, the gene we raced Millennium to clone at Sequana. Tubby turned out to be a member of a family of bipartite transcription factors.

a more prominent profile: Inpharmatica Ltd. in the United Kingdom and GeneFormatics in San Diego.[7]

We were arrogant enough to think that none of these were really competition, from either the drug discovery or the technology perspective. We certainly underestimated Astex: they were successful in using their platform to turn low-affinity small protein–binding compound fragments into drug leads. Simon Campbell, an excellent ex–Pfizer medicinal chemist who had discovered many drugs, was on their Board. After completing several deals, Astex were acquired by Otsuka in 2013 for more than $800 million.

Syrrx were set up just up the road from SGX. They were a 1999 spinout from the Genome Institute of the Novartis Research Foundation. Started by Raymond (Ray) Stevens, a structural biologist, and Nathaniel David, Syrrx were thoroughly legitimate competition, boasting a structure determination platform (including a robot-based crystallisation tool) combined with expert molecular biologists and crystallographers. Stevens, one of the first structural biologists to determine the 3D structure of a membrane protein (a G protein–coupled receptor—GPCR), was an expert at working with small protein crystals. Wendell Wierenga, an ex–Parke-Davis R&D executive, was appointed CEO, and Sam Colella from Versant Ventures was an early investor. In addition, Syrrx were already focusing on human drug targets.[8]

There were also several nonprofit consortia being set up to do structural genomics, one of which was sponsored by the Wellcome Trust and led by Alan Williamson, who directed the Merck public EST effort and the SNP Consortium. Williamson had become by that time a kind of anti-industry spokesperson, to ensure the academics got access to new technologies. I did not think this was entirely necessary.[9]

There was clearly a competitive need to move fast. SGX closed a series B round of $32 million in April 2000, led by the Sprout group. It included Lombard Odier, Index Ventures, and Vulcan Northwest (Paul Allen money) as additional investors.[10] We were fixated at the time on the number of structures we could determine—5000 in five years was the mantra—rather than focusing on drug targets and structure-based drug design of small-molecule drugs.[8]

---

[8]At that time, the SGX Board consisted of George Poste, as Chairman, Jean-François Formela (Atlas), Alex Barkas (PVP), Philippe Chambon (Sprout Group), Chris Henney (Dendreon CEO) as an independent Board member, and me.

Constructing a Beamline

The general excitement of the space was captured in an article by Andrew Pollack in July 2000 in *The New York Times.*[11] It captured the excitement well and included a side bar entitled "Analyzing proteins with X-rays, crystals, and some luck"—a title that captures the fact that whether a target protein expressed or crystallized was not always predictable. SGX were labelled as one of the many companies in this now "postgenomic space" and were the subject of many articles about structure determination moving from a cottage industry to an "Industrial Revolution."[12]

With the money from the series B in the bank, SGX made the decision to build their own beamline at the Advanced Photon Source (APS) at Argonne National Laboratory (in Lemont, Illinois), which was scheduled to be completed in November 2001 at a cost of $8 million (see Plate 20 for the APS in Chicago and Plate 21 for Linda, Kevin D'Amico, and me at the beamline during construction). This allowed us unfettered access to beam time, a precious commodity if you wanted to collect X-ray diffraction data quickly.[h] Many of the Big Pharma companies had their own beamlines at the APS, or they shared one. SGX got quite a lot of press in Chicago and elsewhere for this initiative, and the late William (Bill) Richardson, the Secretary of Energy at the time, met us and spoke at the ceremony marking the event.[13] Kevin D'Amico, who was instrumental in helping to get the whole thing set up on time and on budget, was appointed as VP of Synchrotron Radiation research at SGX.

SGX now had about 60 employees. We were fortunate to raise $45 million in a series C private placement led by Stelios Papadopoulos and his team at SG Cowen.[14] BA Venture Partners led the deal, which included OrbiMed and Vector Fund management and all the original investors. Louis (Lou) Bock joined the Board from BA Venture Partners and Vijay Lathi replaced Philippe Chambon from Sprout, representing the investment of what was now New Leaf Ventures.

Cystic Fibrosis Foundation Deal

Jay Knowles did an important strategic deal for us in 2001, which I thought at the time undervalued our technology. I was wrong. The deal was with the

---

[h] I went to the APS several times. The synchrotron is a very large circular electron accelerator that bends electrons with magnets, forming X-rays. You can tap into the X-ray beam like tapping into water from a circle of fast-moving stream (a beamline). You needed a bicycle to go around the facility—it's that big.

Cystic Fibrosis Foundation to determine the structure of CFTR—the protein that is mutated in cystic fibrosis (CF) patients, usually by a deletion of a phenylalanine at position 508 in the protein sequence (ΔF508; see Chapter 7). We would receive $13 million over five years, which is why I thought it undervalued our technology. Eventually, we did determine the structure of one of the nucleotide-binding domains of the protein (NBD1), which contains the phenylalanine deletion. It was recognized as an important piece of work.

SGX were gaining traction. There was some generic coverage of structural genomics that mentioned SGX in *The Wall Street Journal* and the *Financial Times* at that time.[15,16] A thoughtful piece by Deborah Erikson in the March 2001 magazine *Start-Up*, called "Selling Structure," outlined the prevailing view[17] and covered some other players in the space, including Affinium.[i]

Competition

Syrrx raised nearly twice the money that SGX had raised, had a valuation north of $150 million, and in February 2001 employed more than 75 people. I remember one day looking out of my office window, seeing a truck drive down Roselle Street advertising that jobs were available at Syrrx. I thought this was rather poor form, and I told Wendell Wierenga what I thought.[j]

Syrrx had formed a relationship with Molsoft LLC, a company who had developed methods for docking compounds computationally into solved structures. We felt we needed this capability, too, so SGX bought a software company called Prospect Genomics in April 2001.[k] This was a mistake. First, they were not in San Diego but based in the Bay Area. Second, they had their own agenda and did not want to move from it. Third, we paid way too much for them. I learned a lot about liquidation preferences from that deal, including the fact that liquidation also means acquisition.

---

[i] Integrative Proteomics Inc. (later called Affinium) had been set up in Canada in August 2000 with John Mendlein as CEO. He was another talented MD, JD who went on to start several companies including Fate Therapeutics (see Chapter 14) and is now a partner at Flagship Pioneering. They were using mass spectrometry and NMR, in addition to X-ray crystallography to look predominantly at the structure of bacterial and viral proteins.

[j] Internally, we came up with lots of novel ways to reciprocate but fortunately, we never carried any of them out.

[k] Bill Rutter (ex-Chiron and UCSF) was a founder of Prospect Genomics. He joined the Board of SGX. Stelios Papadopoulos joined at the same time and George Poste stepped down as Chairman and left the Board. Bill Rutter died in July 2025 aged 97.

Although we formed a relationship with Compaq and had a state-of-the-art LIMS, SGX had now moved away from the database model. We were starting to work in earnest on human drug targets such as kinases. We hired Stephen Burley from Rockefeller University to head up our R & D and Douglas (Doug) Livingston from Axys to run chemistry.

One memorable day, we held a "leukaemia day" to refine our thinking about targets for those diseases. We invited Brian Druker from Oregon to chair the meeting, which included Gary Gilliland from the Dana Farber in Boston and some other eminent haematology and oncology specialists. It was a stimulating day. As a result, we focused our attention on *Bcr-Abl* (see Box 1).

BOX 1. *Bcr-Abl*

Bcr-Abl is an oncogene formed when the gene for the amino-terminal part of the BCR protein from chromosome 22 is fused to the gene for the ABL kinase from chromosome 9, encoding an enzyme that is as a result permanently "on." This chromosomal translocation, called the Philadelphia chromosome, is found in all cases of chronic myeloid leukaemia (CML) and some cases of acute lymphoblastic leukaemia (ALL) (Fig. 1).

Ciba-Geigy and latterly Novartis had developed STI–571, as a relatively selective inhibitor of *Bcr-Abl*. It became the drug Imatinib (tradename Gleevec), approved in 2001 for CML.[18] One of the problems with the drug was resistance arising from mutations in the enzyme, the most prevalent of which was a threonine to isoleucine mutation at position 315 in the ABL protein (the T315I mutation), which SGX sought to address in our drug discovery work.

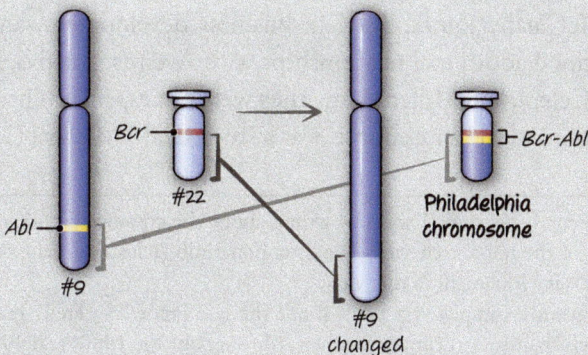

Figure 1. The *Bcr-Abl* translocation between chromosomes 9 and 22 (The Philadelphia Chromosome)

We decided to start a program to work on making a compound that inhibited both the wild-type and the Gleevec-resistant form of the BCR-ABL enzyme. The SGX Board did not consider this a commercially viable proposition. But SGX were one of the first companies to do this work, and by the time SGX had changed their name to SGX Pharmaceuticals in 2005, we had made many inhibitors with this profile, and even formed a relationship with Novartis in 2006 to develop them.[19] Unfortunately, the compounds were not good enough to go all the way. Fortunately for CML patients, other companies had developed second-generation compounds, such as dasatinib (BMS) and nilotinib (Novartis), which inhibited the T315I mutant enzyme as well as other resistant mutant enzymes that were arising. In June 2017, Takeda bought Ariad Pharmaceuticals for $5.1 billion to get access to their BCR-ABL inhibitor ponatinib, which specifically targeted the T315I mutation.[l]

Flying Too Close to the Sun

SGX were flying high but with perhaps a touch of Icarus. I had obtained a private pilot's licence at Palomar Airport in 2001 and decided it would be a good idea to take David Ewing Duncan, who was writing an article for *Wired* magazine about structural genomics companies, for a flight over SGX in a Cessna 172 while we talked about the company. The resulting article is quite fun, comparing SGX to Syrxx and containing some flying analogies.[20] However, I made a mistake during the flight by straying into Miramar (the *Top Gun* flight school) airspace. When I landed at Palomar Airport, I got an earful from the local Federal Aviation Administration (FAA) and was lucky to retain my pilot's licence. We were also fortunate not to get blown away by an F14 fighter jet. I told David about the incident later: he was unphased.[m]

Sean McCarthy joined SGX in business development, and with his help, we formed additional relationships with Aventis, Anadys, Boehringer Ingelheim, Celera, and Millennium. They were not especially lucrative deals. The Millennium deal, for example, was a stock deal.[n] Michael (Mike) Grey,

---

[l]As it happened this time, I was not wrong about the commercial potential of a specific inhibitor of the mutant enzyme. The drug ponatinib (Iclusig) developed by ARIAD Pharmaceuticals is a testament to that view.

[m]He has recently completed a book about the late Frederick (Fred) Frank—the legendary "biotech" banker at Lehman Brothers. Throughout the 1980s and 1990s and early noughties, Fred and his team, including his wife brokered many pharma-biotech deals including the Roche–Genentech transaction.[21,22]

[n]In the Millennium deal, they bought SGX stock and we determined structures with the money they bought it with.

the SGX CBO, did broker a more lucrative "double digit, millions of dollars" deal with Eli Lilly. They were impressed with the gene-to-structure platform we had put together, so SGX undertook to replicate the platform in a next-door building solely for the use of Lilly.[23]

One of the important structures that SGX solved was the protease of SARS-CoV-1 (a coronavirus related to the COVID-19-causing SARS-CoV-2), which had emerged in 2002. Whether this contributed to the commercial success of SGX is an open question, but we did the structure very quickly and were one of the first groups—if not the first—to get the structure.[24]

SGX Pharmaceuticals Initial Public Offering

SGX Pharmaceuticals (as the company name had become) filed to go public, in early 2006 shortly after I left the company to become CEO of Novasite.[25] The deal was led by CIBC World Markets, Piper Jaffray, and JMP securities. Mike Grey had taken over as CEO, and the company looked much more like the drug discovery company it wanted to be. They had in-licensed a drug called troxacitabine (Troxatyl) from Shire. Although not discovered by structural techniques, Troxatyl had shown some limited efficacy in acute myeloid leukaemia (AML), and it was available, which helped move the SGX story on to one of having drugs in the clinic. SGX also had drug discovery programmes in BCR-ABL, Aurora, Met, and Ron kinases and had done more recent collaborative deals with Roche, Amgen, and others. Peter Myers, a well-respected medicinal chemist and ex-Glaxo and -Onyx drug hunter, was now looking after their medicinal chemistry. Chris Henney was chairman of the Board, and they had some other new faces around the board table.

The initial public offering (IPO) was a slog, but it got done. They sold four million shares at $6 per share, raising nearly $25 million.[26] The valuation was much lower than anticipated, so given the quality of the technology platform and Lilly's investment, it was no surprise when Lilly stepped in and bought the company in 2008—consolidating some of their research activities in San Diego. The only surprise was the price—$64 million.°

° I am a tad biased, but I called that a cheap deal. Especially given that Takeda bought Syrrx for more than $200 million less than three years before, and SGX were at least as good as Syrrx were. That was the deal on the table though and, in that financing environment, probably the right decision for shareholders.

GPCRs, MEMBRANE PROTEINS, AND NEW COMPANIES

One of the classes of proteins that the original structural genomics companies were not able to deal with easily was large integral membrane proteins, including GPCRs and ion channels. GPCRs (G protein–coupled receptors), the protein targets for many drugs, had not been crystallised and were more difficult to express in sufficient quantities, as well as to purify. Selectivity and tissue specificity is very important for this drug class, and only a few of the large GPCR family of proteins had been successfully exploited as drug targets. This is still true today.

Robert (Bob) Lefkowitz and Brian Kobilka (see Box 2) expressed various GPCRs fused with stable crystallisable protein domains (such as lysozyme and/or immunoglobulin domains) and were able to make stable derivatives. Some of these formed crystals, which diffracted and enabled the structures to be determined by reference to the known domains.

GPCR Companies

Novasite, the company I joined after SGX, were trying to develop allosteric modulators of members of the GPCR family.[25] We spent 18 months, funded by Advent International, trying to do this without much success. We had a small lab at the other end of Roselle Street from SGX. In 2006/7, it was almost impossible to raise money for early-stage companies such as Novasite.

BOX 2. G PROTEIN–COUPLED RECEPTORS

GPCRs have seven transmembrane domains with the effector binding sites generally on the external side of the cytoplasmic membrane or just inside it. Signals are transmitted by conformational changes of the protein, connecting with the G proteins attached to the inside of the membrane.[27]

The 2012 Nobel Prize in Chemistry was awarded to Bob Lefkowitz and Brian Kobilka for developing methods to express and purify GPCRs and to get diffracting crystals from which three-dimensional structures could be defined. Kobilka cloned the adrenergic receptor genes in the Lefkowitz lab in the 1980s. At that point, they realized (as did others) that this receptor was part of a very large protein family, involved in many physiological processes from reactions to adrenalin, histamine, dopamine, and serotonin as well as to some peptide hormones. It also included a large subfamily of receptors that are involved in the detection of smell and the light-detecting opsins in the retina.

What we were doing was a bit too academic in any case. Before long, I had the unenviable task of firing the 20 staff and closing the doors on Novasite forever. I sold all the equipment and lab supplies to a local Chinese company who were moving into our space for $25,000. It helped to pay off some of the debts. Some people call these experiences "character building." Maybe so, but they are at the same time rather depressing. The science we were doing was quite novel, the team were talented, and we published some papers on the structure of GPCRs like rhodopsin.[28]

Heptares Therapeutics

In 2007, Malcolm Weir and Fiona Marshall (both of whom I worked with at Glaxo) co-founded a company in the United Kingdom artfully named Heptares Therapeutics, specifically to focus on the structure determination of GPCRs. The company was based on the work of Richard Henderson and Christopher Tate at the LMB. It was financed initially by MVM Life Science partners, with additional capital from Clarus Ventures, Novartis Venture Fund, Takeda Ventures, and the Stanley Family Foundation. Fiona Marshall is now President of the Novartis Institutes for BioMedical Research—the first woman to hold a top R&D position in pharma.

Heptares were the first company to apply structure-based drug design to the GPCR family of proteins. About 40% of the small-molecule drugs used clinically target members of this family of proteins, so it was an important space to go after. Heptares determined the structure of many important receptors, including members of the muscarinic acetylcholine receptor class and selected peptide-liganded receptors like GLP-1 and orexin. Sosei, a Japanese Pharma company, bought Heptares for $400 million in 2015, to accelerate their pharma R&D ambitions.[P]

Before the acquisition, Heptares had worked with many of the top pharma companies including Takeda, AstraZeneca, and Novartis, receiving millions of research and development dollars. More recently, Sosei Heptares (as they were rebranded) announced a $2.6 billion deal with Neurocrine Biosciences in San Diego to work on subtype-specific, muscarinic receptor agonists active in the brain, for various neuropsychiatric indications.

[P] Another steal, given what the Heptares platform had accomplished.

CRYO-ELECTRON MICROSCOPY AND NEWER GPCR COMPANIES

Richard Henderson, one of the Heptares founders, was awarded the Nobel Prize in Chemistry in 2017 with Jacques Dubochet and Joachim Frank for cryo-electron microscopy (cryo-EM).[29] This technique, which uses frozen proteins and electron beams to "see" 3D proteins and protein complexes at high resolution, has now taken over from X-ray crystallography as the method of choice. As you might expect, several start-ups have now emerged in the cryo-EM space. Some of them are just service providers but others are attempting to make new drugs.

ConfometRx and Septerna

ConfometRx were set up in 2009 by Brian Kobilka to develop further GPCR structure-based drug design, by looking at the proteins in both their active and inactive states. Initially building on their X-ray crystallography expertise, they recently have turned to cryo-EM techniques, as many GPCRs that would not yield to analysis by X-ray diffraction are giving in to cryo-EM methods, with many structures with and without ligands emerging. They report that they have the ability "to solve a series of iterative structures of GPCR-G protein complexes bound to various candidate compounds (both orthosteric and allosteric; see Chapter 19) reaching as high as 2.8Å resolution," in short, high resolution and multiple relevant forms of the proteins of interest.

Not to be outdone by his Nobel co-laureate, and as a recognition of the importance of both GPCRs and cryo-EM, Bob Lefkowitz also set up a drug discovery company called artfully (again) Septerna, Inc., a Third Rock Ventures company. The VC firm invested $100 million in the Septerna series A round; other investors included Samsara BioCapital and BVF Partners. Jeffrey (Jeff) Finer, a Third Rock Venture Partner, is the present CEO. The size of series A reflects, to some extent, the size of the problem, the capital-intensive nature of cryo-EM platforms, the fact that the company will be working on multiple programs in parallel, and that they will need 40–50 research and development scientists to get going effectively.

Septerna, which should not be taken lightly, is based in South San Francisco. It has a proprietary "Native Complex Platform" for structure-based drug design that includes "compound library screening and structure determination using cryo-EM to tackle native G protein–coupled receptors (GPCRs) to develop novel medicines in multiple therapeutic areas"—a now familiar-sounding story.

In addition to Lefkowitz, Septerna's founders are Arthur (Art) Christopoulos and Patrick (Pat) Sexton, two Australians from the Victorian College of Pharmacy at Monash University in Parkville, Melbourne. Art and I have a little (good) history. I was on the SAB of Bionomics in Adelaide for several years. I visited every year in November usually (but not deliberately) when the Melbourne Cup horse race was on. While at SGX and at a Bionomics SAB meeting, I arranged to meet Art and Pat while they were still at the University of Melbourne. They are very careful, very well-published, and very precise structural pharmacologists. I met them again with my SV Health Investors hat on in November 2018, as I wanted to see if they were up for founding a GPCR X-ray crystallography–based company in Australia. This would probably not have been very easy to set up. Now, post-COVID-19, they have instead co-founded Septerna in what will probably become one of the premier companies in this space: good for them.[q]

There are other biotechnology companies, such as Generate Biomedicines, Gandeeva Therapeutics, and MOMA Therapeutics, vying to use cryo-EM as a platform for target validation and the development of new drugs. An Arena Pharmaceuticals spinout in San Diego, called Longboard Pharmaceuticals, is using the technology to discover novel molecules: they went public in 2021 raising $80 million, so there is certainly investor interest.

UNFINISHED BUSINESS

San Diego has had a focus for companies in the structure-based drug design space. Ray Stevens, the Scripps-based structural biologist and founder of Syrxx, founded Receptos in 2007 with Bill Rastetter, partner at Venrock and former Chairman and CEO of Idec before it merged with Biogen. The founding idea of Receptos was to use structural biology to focus on diseases in which small molecules might outcompete biologics like antibodies.

One of the targets Receptos homed in on was the sphingosine-1-phosphate receptors, a small GPCR family with five members. Sphingosine 1 phosphate regulates endothelial cell function and lymphocyte maturation

---

[q] It will be interesting to see how the two Nobel Prize winners, now with two competing companies, exploit their awe-inspiring technology and utilize new technologies like Cryo-EM for drug discovery.

and trafficking.[r] Using their platform, Receptos developed a potent dual S1P1/S1P5 agonist called ozanimod (RPC 1063), which demonstrated disease-modifying activity in ulcerative colitis (UC) and multiple sclerosis (MS). Celgene bought Receptos in 2015 for $7.2 billion, primarily to obtain this drug in late-stage development. Receptos had reported positive data in UC and had Phase 3 trials underway in MS.

In the same vein, former Receptos CEO Faheem Hasnain and CMO Sheila Gujrathi launched San Diego–based Gossamer Bio in 2015, with a $100 million series A round led by Omega and ARCH Venture Partners. Their most advanced program is seralutinib (GB002), a small-molecule inhibitor of PDGFR, CSF1R, and c-KIT kinases for pulmonary arterial hypertension in phase 2 trials. Some of the former Receptos technology has also found its way into another Stevens start-up, Structural Therapeutics, where he is the CEO. Since 2017, the company, based in Shanghai and San Francisco, have raised more than $150 million to convert targets for biologicals into orally bioavailable small-molecule targets (the same shtick as Receptos), using structural tools that include cryo-EM. Structural Therapeutics raised $150 million in an oversubscribed IPO in the first quarter of 2023, based on their GLP-1 small-molecule agonists. This is one of only a handful of biotech IPOs done in 2023.

STRUCTURE PREDICTION USING AI

Contrast Genex, who were a company formed at the beginning of biotech (Chapter 1), with the very recent AI-driven structure prediction tools that are now available, including DeepMind's AlphaFold (DeepMind is a subsidiary of Alphabet based in London). AlphaFold and the program RoseTTAFold (out of David Baker's lab—Box 3—at the Institute for Protein Design at the University of Washington) have revolutionised the ability to predict computationally protein three-dimensional structure from the primary amino acid sequence.

DeepMind AlphaFold

In 2021, DeepMind announced that they have predicted the structure of 350,000 proteins. This number has now expanded to more than 200 million

---

[r] Lymphocyte trafficking is the target for Tysabri, a successful antibody that Biogen obtained from Elan and developed for multiple sclerosis.

BOX 3. DAVID BAKER

David has come a long way from when I first met him in Seattle in 2002. I was trying to get him involved in SGX. He was then an engaging bicycle-riding, dyed-in-the-wool academic at UW, where my second daughter was at school, and was not that interested in things commercial as far as I could tell. Not anymore, it appears. Baker, at the Institute for Protein Design, often working with ARCH Ventures, is now one of the most prolific company founders out there. Most recently, David launched CHARM Therapeutics in London to find compounds for cancer and other diseases and has raised $50 million in a Series A round led by F-Prime Capital and OrbiMed. CHARM has a program called DragonFold, which will find small molecules that are predicted to bind in a particular way to different protein targets. According to the website and *GeekWire*, Baker's lab have been the source of nine companies to date, including Neoleukin, Cyrus Biotechnology, and Icosavax, which raised $180 million in an IPO in 2021.[30]

---

proteins from all the phyla of life, representing virtually all known proteins.[31] The AlphaFold developers at DeepMind, John Jumper and Demis Hassabis, won the $3 million Breakthrough Prize for their prediction work (AlphaFold2). This is obviously useful for drug discovery. Despite the fact that the AlphaFold protein 3D predictions are pretty accurate, predicting how small molecules and large molecules interact with proteins is more difficult than just modelling the protein by itself. There are many fewer complexes of this type available from which to drive the predictions. Cryo-EM and X-ray crystallography of complexes will continue to add experimental data, from which more accurate *in silico* predictions will be derived. One area of interest that will be particularly beneficial is to see how peptides are presented to the immune system by the class I and class II MHC proteins (see Chapter 14). It is sure to come.

Artificial Intelligence and Drug Discovery

As mentioned in Chapter 6, there is a growing list of biotech companies leveraging computational tools for drug discovery. Recursion Pharma in Salt Lake City, founded in 2013, recently completed an astonishing >$400 million IPO to develop compounds coming from their novel computational drug screening platform for rare genetic diseases. Insitro, in South San Francisco, have brought in $400 million in venture funding to "integrate the languages of biology and machine learning." The price of entry into that structure prediction marketplace has just gone up by an order of magnitude. We will have to wait to see how long it takes to provide us with the promised new drugs.

REFERENCES AND NOTES

1. Shapiro L, Lima CD. 1998. The Argonne Structural Genomics Workshop: Lamaze class for the birth of a new science. *Structure* **6**: 265–267. http://biomednet.com/elec-ref/0969212600600265

2. Larry Shapiro is presently a Professor of Biochemistry at Columbia University and Chris Lima is Chair of the Structural Biology Program at The Memorial Sloan Kettering Cancer Center.

3. Werth B. 1995. *The billion dollar molecule: one company's quest for the perfect drug.* Simon & Schuster, New York.

4. 1999. Series A financing. *BioWorld Today* **10**: 238.

5. The SGX SAB: In addition to Wayne and Barry, we assembled a first-rate SAB consisting of prominent scientists in the field: Steve Benkovic (an enzymologist from Penn State), Fred Blattner (*E. coli* geneticist, University of Wisconsin), Stephen Burley (X-ray crystallographer from the Rockefeller University), Geoff Duyk (from Exelixis), Sean Eddy (computational biologist from Washington University), David Eisenberg (a structural biochemist from UCLA), Andrej Šali (computational biologist from the Rockefeller University), and Peter Wright (from the Scripps Institute). Sung-Hou Kim, a structural biologist, was also on the original SAB until he decided to set up a competing company called Plexxikon. I fired him.

6. 2000. Emerging company profile: Structural GenomiX: the importance of being 3-D. *BioCentury*, March 6, 2000.

7. 1999. *Start-Up*, December 1999, Vol. 4, number 11.

8. 2000. The next wave of the genomics business. *MIT Technology Review*, July/August 2000.

9. Declan B. 2000. Wellcome discusses structural genomics effort with Industry. *Nature* **406**: 923–924. doi:10.1038/35023298

10. 2000. Series B financing: Structural GenomiX raises $32 M in series B financing. *BioWorld Today*, April 5, 2000.

11. Pollack A. 2000. The next chapter in the book of life is written in the proteins. *New York Times*, July 4, 2000.

12. Gershon D. 2000. Structural genomics—from cottage industry to industrial revolution. *Nature* **408**: 273–274.

13. 2000. Newspaper report on building the beamline. *Chicago Tribune*, July 26, 2000.

14. 2000. Biotech raises $45 m from private sources. *San Diego Union Tribune*, September 21, 2000.

15. 2001. Proteins—not genes—could be clue to human complexity, disease. *Wall Street Journal*, February 13, 2001.

16. 2001. Searching for the real stuff of life. *Financial Times*, February 21, 2001.

17. Erikson D. 2021. Selling structure. *Start-Up*, March 2001. Vol. 6, p. 18.

18. The story of Gleevec: Vasella D, Robert Slater R. 2003. *Magic cancer bullet: how a tiny orange pill is rewriting medical history*. Harper Business Books, New York.

19. 2006. Novartis/SGX pair up to develop leukemia drugs. *Wall Street Journal*, March 28, 2006.

20. Duncan DE. *Wired Magazine*, April 2001.

21. Duncan DE. 2021. *A philosopher on Wall Street: how Fred Frank forged the future*. Greenleaf Book Group Press, Austin, TX.

22. See also Fred Frank obituary: Cranmer J. 2021. Fred Frank, co-founder of the biotech industry, is dead at 89. *BioCentury*, September 2021.

23. 2003. SGX installing platform as part of Lilly alliance. *BioWorld Today* 14: 77, April 22, 2003.

24. 2003. Company says it mapped part of the SARS virus. *New York Times*, July 30, 2003.

25. 2005. Emerging company profile: Novasite: conceptually quick. *BioCentury*, June 27, 2005.

26. SGX Pharmaceuticals IPO S1, filed with the Securities and Exchange Commission on October 14, 2005.

27. Rosenbaum DM, Rasmussen SGF, Kobilka BK. 2009. The structure and function of G protein coupled receptors. *Nature* **459**: 356–363. doi:10.1038/nature08144.

28. Salom D, Lodowski DT, Stenkamp RE, Le Trong I, Golczak M, Jastrzebska B, Harris T, Ballesteros JA, Palczewski K. 2006. Crystal structure of a photoactivated deprotonated intermediate of rhodopsin. *Proc Natl Acad Sci* **103**: 16123–16128. doi:10.1073/pnas.0608022103

29. Baker M. 2018. Cryo-electron microscopy shapes up. *Nature* **561**: 565–567. doi:10.1038/d41586-018-06791-6

30. Schubert C. 2022. David Baker, head of Seattle's Institute for Protein Design, launches London biotech startup. *GeekWire*, June 9, 2022.

31. 2022. News at a glance: AI reveals protein structures, racism's impact on memory, and rising tiger numbers. *Science* **377**: 560.

CHAPTER 11

Drugs for Rare Diseases

Drug development for rare diseases is a bit different than for common diseases. For perspective, there are about 40 million people in the United States who suffer from migraine, for which Glaxo developed sumatriptan, and there are many millions more who suffer from asthma or diabetes. When you consider rare diseases—especially Mendelian inherited diseases—the number of patients is very much smaller. For example, there are about 30,000 people with cystic fibrosis (CF) in the United States today with around 1000 new patients per year. There are about 100,000 people, mostly African–Americans, with sickle cell disease in the United States, and there are nearly a million people with multiple sclerosis.[1] By definition, the number of patients with rare diseases is low, making access to the patients more difficult and pricing of new drugs more contentious. Nevertheless, many of new technologies we discuss later (e.g., antisense and RNAi, gene therapy, and gene editing) and the companies that applied them have emerged from a focus on rare Mendelian inherited disorders.

AURORA BIOSCIENCE AND VERTEX

While I was at Sequana, Avalon Ventures (Kevin Kinsella) set up a screening company called Aurora Biosciences, just down the road from us in La Jolla. Frank Craig and Harry Stylli from the Glaxo screening group, who worked for me for several years, were hired to help set up Aurora. I visited the company a few times but was never convinced that a screening company was the right thing to invest in.[a]

Aurora built a highly automated screening system not at all like what we had at Glaxo. The screening room looked a little like a large-scale miniature gauge railway. The well plates moved along tracks to various stations, where different parts of the screening process took place (e.g., adding reagents or

---

[a] Once again, I was proved wrong, this time by Aurora Biosciences.

compounds to the wells). Aurora's setup took the screening process to a whole new level: they affectionately called it "The Beast."

The company also developed a proprietary set of fluorescence assay technologies through a collaboration with the late Roger Tsien at UCSD (who won the Nobel Prize in Chemistry in 2008). Tsien had made many proteins that fluoresced at different visible light wavelengths, and Aurora were able to use these reagents to screen many different types of drug targets, including G protein–coupled receptors (GPCRs) and ion channels. The quality and novelty of Aurora's capabilities were validated by the number of life science companies they collaborated with including many of the top pharma companies.

Cystic Fibrosis Transmembrane Regulator

One of the targets that Aurora chose to work on, in collaboration with the Cystic Fibrosis Foundation (CFF), was the cystic fibrosis transmembrane conductance regulator (CFTR) protein. They were trying to answer two questions: (1) was it possible to find compounds that increased the activity of the mutant protein on the cell surface and (2) could they find compounds that helped to move a functional, but "stuck," mutant protein from the interior of the cell to the cell surface? Both types of mutation had been found in the *CFTR* gene, leading to these characteristic phenotypes of the protein. Eventually, both types of compounds were found. The first ones had the relative disadvantage of not working on the most common CFTR mutation (ΔF508—see Chapters 7 and 10) but were effective with other mutant versions of the protein.

Aurora were bought by Vertex in 2001. According to the press release announcing the acquisition, "The merger enables Vertex to integrate Aurora's industry-leading capabilities in the development of cell-based assays and screening instrumentation for use in drug discovery directed at ion channels, G-protein coupled receptors (GPCRs), kinases, proteases, and phosphatases, and for use in target validation in a wide range of gene families. Vertex's ongoing drug discovery efforts will also benefit from Aurora's predictive pharmacology and proteomics technologies, which use high-throughput assessments of toxicology and metabolic marker, to establish therapeutic proof-of-concept and safety of drug candidates in early clinical testing."

Nowhere in the press release was the CF project mentioned. But Vertex inherited the CFF collaboration and the CF drug discovery programme as part of the acquisition, and they had the foresight to recognise the importance of the CF compounds that Aurora were attempting to make.[b] The story of Vertex Pharmaceuticals has already been very well told in two books by Barry Werth, so it will not be re-iterated here. Suffice it to say that Josh Boger, Richard (Rich) Aldrich, and Vicki Sato, as well as others at Vertex, have been important contributors to the biotech industry.[2,3]

Drugs for CFTR Modulation

The two types of CFTR-directed compounds that Aurora discovered were turned into three very important drugs for CF patients, for Vertex, and for the Cystic Fibrosis Foundation. One cannot overemphasize the role of the Foundation, which had taken the important initiative to fund the drug discovery programme in the first place: a first for a nonprofit charity.[c]

The two types of drugs developed are now referred to as CFTR channel potentiators and CFTR correctors, respectively. Ivacaftor is a potentiator, which increases the chloride ion channel activity of the mutant CFTR protein, especially those with what are called class III gating mutations, like G551D. This was the first CF drug that was developed (Kalydeco). CFTR correctors such as Elexacaftor and Tezacaftor work by helping to move a mutant CFTR (e.g., the ΔF508 mutant version) to the cell surface.[d] These were the second and third drugs (Orkambi and Symdeko) developed for CF by Vertex for use in combination with Kalydeco. Vertex has now developed an all-in-one combination of the three compounds in a drug called Trikafta. The combination of increased (mutant) CFTR protein on the cell surface, with potentiation of chloride channel activity, allows increased transport of chloride, resulting in thinned mucus secretions and substantial increases in the ability of CF patients to breathe.

As of today, Vertex *completely* dominates the CF market. They generated an astonishing $7.6 billion in revenue in 2021, $5.7 billion of which came from Trikafta, patents for which do not expire until 2037. This programme contributes to the company's $100 billion market cap (December 2023).

---

[b] Paul Negulescu, the leader of Aurora Discovery Biology, takes much of the credit for making this happen. He is now an SVP at Vertex running their Nav1.8 pain programme, which also had its origins in San Diego (see later in the chapter).

[c] This was a $47 million decision driven by Robert (Bob) Beall at the CF Foundation in the year 2000. It made a lot of money for the charity to re-invest.

[d] The common mutation (ΔF508) that occurs in 90% of affected individuals is affected positively by these two corrector compounds.

With the benefit of hindsight, the Aurora acquisition for $600 million was another biotech steal. Indeed, without the CF portfolio, there might possibly be no Vertex at all by now, especially because other drugs they had or were developing at the time were neither first-in-class nor best-in-class.[e]

GENZYME

Genzyme, another company specialising in making drugs for rare diseases, also had a notable singular event. But it led to a very different outcome than the experience of Vertex.

In-depth stories of Genzyme and their distinctive CEO Henri Termeer have been written recently, providing a great source of additional information on the company, the products, and their leader.[4,5] Genzyme started in 1981 from an association between Roscoe Brady at the New England Enzyme Center at Tufts Medical Center in Boston and Henry Blair, and enzymologist at the Tufts University School of Medicine. Brady had discovered that Gaucher's disease, a rare lysosomal storage disease affecting less than 6000 patients in the United States,[f] was caused by the lack of an enzyme called glucocerebrosidase. This deficiency leads to an accumulation of cerebrosides in the lysosomes of the cells of the patients, causing the characteristic pathology. Brady thought to use a purified enzyme from normal placenta, which was a rich source of the protein, as an "enzyme replacement therapy," but he needed somewhere to have it made.

Blair took the lead in starting the company, putting it in an old clothing warehouse in the red-light district of Boston between Washington, Boylston, and Kneeland Streets.[g] Oak Investment Partners and Advent International were the initial investors in Blair's start-up. Genzyme Pharmaceuticals and Fine Chemicals was born from the 1982 acquisition of two U.K. companies: Whatman Biochemicals in Maidstone Kent (famous for their filter papers, which I well remember using in the lab) and Koch Light Laboratories (makers of fine chemicals).

Henri Termeer was brought in from Baxter Travenol[h] as President & Director in 1983 and named CEO in 1985, where he remained until the Sanofi acquisition in 2011. Genzyme went public in 1986, raising $28

---

[e] This is a great example of a singular event that can make or break a company. We will return later to the subject of singular events and their importance in biotech.

[f] At the time, the true number of Gaucher's patients was not known.

[g] It is still the centre of the Boston "massage" industry today.

[h] Termeer was part of the original Baxter mafia. He died too young in 2017 in Marblehead, Massachusetts.

million. A further $10 million was raised via a Research and Development Limited Partnership (RDLP) to fund the first Gaucher's trial, with the enzyme purified from placentas (mostly from Bio-Merieux, who were using placenta as a source of gamma globulin for their polyclonal IgG product). Genzyme treated 12 patients with the purified enzyme, all of whom responded. Ceredase was subsequently approved in 1991, at an average annual cost of $150,000 (later $200,000). This was considered to be—and was—expensive, but the Waxman "Orphan Drug Act" of 1983, a law passed to facilitate and ensure the development and sale of orphan drugs, applied here. The price was also justified by the fact that most patients required more than one dose of the original drug at a time anyway. Regardless, the cost–benefit analysis was in Genzyme's favour, and the insurers reimbursed the product.

It was life-changing for the Gaucher's patients to have access to a drug that not only stopped but reversed their pathology. To increase the drug's availability and purity, Termeer acquired Integrated Genetics (IG) in August 1989 (run by one of his ex-Baxter colleagues, Bob Carpenter). IG were essentially a molecular biology and recombinant DNA-based cloning shop working on t-PA and other potential products. IG technology allowed Genzyme to rapidly clone a cDNA for human glucocerebrosidase and make a recombinant version of the enzyme by expressing the cDNA in CHO cells. This product, called Cerezyme, was approved in 1994.

Other Enzyme Replacement Therapies

Genzyme also planned to make the enzymes that were missing in people with other lysosomal storage diseases, such as Fabry's disease or Pompe's disease. There are about 5000 patients with Fabry's disease in the United States, who do not produce enough of the active form of the enzyme α-galactosidase A (α-GAL). This enzyme prevents sphingolipids (related to cerebrosides) from collecting in blood vessels and tissues, thereby affecting the heart, kidneys, brain, central nervous system, and skin of mutation carriers. In developing a drug for Fabry's disease, Genzyme had competition from Transkaryotic Therapies (TKT), who were developing the same potential product. The trials were set up differently, and TKT failed to show an adequate response in their patients.[i] Genzyme prevailed and launched Fabrazyme, the recombinant form of the enzyme, in 2001 (now marketed by Sanofi as Agalsidase beta).

---

[i] TKT were sold to Shire in 2005 for $1.6 billion.

Pompe's disease is another glycogen storage disease but is even rarer than Gaucher's and Fabry's diseases. Pompe's disease, which affects one in every 28,000 people, is caused by mutations in the gene coding for acid α-glucosidase, preventing glycogen breakdown in lysosomes. The resulting buildup of glycogen damages the body's muscles, affecting mobility and the ability to breathe, and affected individuals usually die very young. Genzyme's enzyme replacement therapy for this disease, called Myozyme, was approved in 2006.

Genzyme grew by a combination of selling their products and diverse acquisitions in addition to investing in their own research. All their acquisitions were aligned to the strategy of continuing to build a large rare diseases franchise. Genzyme acquired thymoglobulin from SangStat for treating organ transplant rejection. They also obtained thyrogen, a thyroid-stimulating hormone (TSH) replacement therapy, from Neozyme. Genzyme also ended up serendipitously owning CAMPATH 1 (alemtuzumab), by virtue of buying Ilex Oncology.[j]

Genzyme also formed a relationship with BioMarin (Box 1) to make an enzyme replacement therapy for mucopolysaccharidosis type 1 disease (MPS 1), the result of a buildup of long-chain sugars. Under their agreement, BioMarin undertook to manufacture laronidase (Aldurazyme), which was then commercialized by Genzyme.

BOX 1. BIOMARIN

BioMarin were founded in 1997 with an investment of $1.5 million from Glyko Biomedical, a reagents company (MPM Bioventures, Grosvenor Fund, and Florian Schönharting were other seed investors). The company went public in 1999 on the back of the Genzyme deal to make Aldurazyme, which was developed out of academia by Emil Kakkis and his team in Los Angeles (the history of which can be read in his self-referential book).[6] Like Genzyme, BioMarin grew largely by acquisition of smaller companies with niche products. They also developed the first medicine for patients with phenylketonuria (PKU), a rare disease affecting 17,000 patients worldwide. PKU is caused by mutations in the gene coding for phenylalanine hydroxylase, resulting in the buildup of the amino acid phenylalanine in the blood, organs, and brain of affected individuals. BioMarin's PKU drug is a pegylated version of the enzyme phenylalanine ammonia-lyase (Palynziq), which metabolises phenylalanine. More recently, the company have been developing gene therapy medicines both for PKU and for haemophilia (see Chapter 13).

---

[j] As mentioned, this was the monoclonal antibody (alemtuzumab) that we were growing at Celltech that had been licensed to treat chronic lymphocytic leukaemia (CLL). Sanofi would later repurpose and relaunch it as Lemtrada, a drug for MS.

The Virus Problem

Genzyme were the third-largest independent biotech company before their acquisition by Sanofi for $20.1 billion in 2011. Revenues in 2009 were ~$4 billion. The reason for the acquisition was largely catastrophic management following the discovery of a virus contamination of Genzyme's cell reactors making their product. This quality issue happened in the iconic cathedral-like 180,000-sq-ft Cerezyme production plant next to the Charles River in Allston, Massachusetts, where most of their products were made. The contamination was caused by a calicivirus, vesivirus 2117—a nonhuman pathogen, which probably got into their reactors from raw materials like the foetal bovine serum being used in the process.

The calicivirus contamination problem was compounded by several issues. The first was that their production was based on the perfusion growth of CHO cells rather than a more traditional fed batch system. In essence, this means that the perfusion runs lasted for several months, and thus signs of infection (seen as effects on yield) only become apparent towards the end of a run. Second, both Cerezyme and Fabrazyme were being manufactured in this one facility, meaning that it was running at full capacity 24/7 with limited downtime for maintenance. Third, there was insufficient experience in the company to know how to deal with "consent decree negotiations" with the FDA to manage the problem after the discovery.[k] It took much longer than anticipated to sort out the issues, resulting in many patients being let down by the inability of the company to supply their life-supporting drugs.

Despite off-loading several businesses, including their genetic diagnostic company they acquired from Integrated Genetics, to raise needed cash during this challenge, Genzyme themselves were subjected to a semihostile bid from Sanofi. They soon succumbed. The company was known as Sanofi-Genzyme until 2022, when it became Sanofi again.

The "Activist Investors" who had taken seats on the Genzyme Board waited for no one, wanting as large a return on their investment as they could get, in the shortest amount of time.[l] Sanofi for their part felt that a Cambridge (MA) base for their expanding R&D ambitions was a *sine qua non*, but they had to pay a lot of money (in addition to the acquisition price) for the remediation of the Allston plant.

---

[k] In essence, consent decree negotiations in this context are aimed at reaching agreement about the actions to be taken by a company (Genzyme) to remedy nonconformance issues to avoid government (FDA) litigation.

[l] They probably sold the shareholders short given the Genzyme products.

In early 2021 National Resilience, Inc., a new and fast-growing manufacturing start-up, bought the lease to the Allston building. Resilience were founded with an initial $800 million in late 2020 to "change the manufacturing paradigm." The founding team included pharma executives, manufacturing scientists, venture capitalists (VCs) (including ARCH ventures), and former regulatory officials. The iconic building looks as magnificent now as it always did (see Plate 22 for a picture of the Alston old Genzyme iconic manufacturing facility). Unfortunately, Resilience announced in February 2023 that they were going to close the facility and lay off all the staff, presumably for financial reasons.

Genzyme's surprising and rather disappointing denouement is yet another example of a singular event (in this case, viral contamination) determining the destiny of a biotech company. Several other examples are in Chapter 18.

OTHER COMPANIES, RARE DISEASES, AND APPLICATIONS

Convergence Pharmaceuticals

Nav1.7, Nav1.8, and *Nav1.9* all encode sodium channel proteins that are known to be involved in pain sensation. *Nav1.7,* like the gene *PCSK9* mentioned before,[7] has been validated genetically from both gain-of-function and loss-of-function mutations. Individuals with gain-of-function mutations are pain-sensitive and suffer episodic pain, and individuals with loss-of-function mutations are insensitive to pain.[8] The second might sound like a good thing at first blush, but it can lead to severe burns and other mechanical trauma.[m]

Convergence Pharmaceuticals were a Cambridge (UK)-based start-up founded to exploit compounds that blocked the Nav1.7 channel (in-licensed from GSK). They were financed by SV Health Investors, who led a $35 million series A round in 2010, and Simon Tate left GSK to become the CEO. Vixotrigine, their lead orally bioavailable compound, was in the clinic for sciatica and was being profiled for clinical studies in small fibre neuropathy and trigeminal neuralgia, leading to their acquisition by Biogen (an initiative led by Doug Williams) for $675 million in 2015. Unfortunately, the

---

[m] You may remember the Stieg Larsson book *The Girl Who Played with Fire*, which was turned into a 2009 film. The girl's father's henchman did not feel pain—possibly (I surmise) owing to Nav1.7 mutations.

compound failed to show good efficacy in sciatica but has shown some positive, albeit modest effects in small fibre neuropathy[n] and trigeminal neuralgia. Biogen recently announced that this programme will not go forward.[o]

In contrast, Vertex is developing a novel highly selective Nav1.8 inhibitor (VX-548) for neuropathic pain, for which they have obtained Breakthrough Therapy Designation. It went into phase 3 trials in late 2022 and will be in phase 2 for other moderate and severe pain indications. It has shown some efficacy at the highest dose in a recent phase 2 trial (see Chapter 19).[9]

Praxis

David Goldstein, a professor of genetics at Duke University, went to the Broad Institute of Harvard and MIT to give a seminar in the fall of 2013. He chose to speak on the genetics of epilepsy. There are many genetically defined Mendelian forms of epilepsy in which the gene and the mutations causing the disease had been uncovered, a primary interest of the Goldstein lab at Duke. After the seminar Ed Skolnick, the ex-R&D Chief at Merck, a Core Member of the Broad, and a partner at Clarus Ventures (now part of Blackstone), asked him to come and talk to the partnership about epilepsy genetics and drug discovery opportunities.[10]

Several of the genes that were found to cause these Mendelian forms of epilepsy encoded druggable targets. It was thought that they might be targets for the more common general forms of epilepsy as well, which affect several million people worldwide. David duly met with Nicholas (Nick) Galakatos, Ari Brettman, and Kiran Reddy (who had just joined Clarus from Biogen). Together, they parsed David's complete list of epilepsy genes down to the ones that might be most relevant and potentially useful for the more common forms of epilepsy. They chose KCNT1, a potassium-sodium channel that, when mutated, caused an ultra-rare form of childhood epilepsy, SCNT 2A, a voltage-gated sodium channel, and SCN8A, part of the Nav1.6 sodium channel. Based on these discussions, Clarus decided to invest, starting a company called Praxis Pharmaceutical, with David Goldstein as a founder and Kiran Reddy as founding CEO.

---

[n] Small fibre neuropathy is characterized by severe pain that typically begins in the feet or hands. Symptoms tend to be worse at night and can have a significant impact on overall quality of life.

[o] The disruption at Biogen over aducanumab probably did not help this or other "orphan" programmes there.

There is an interesting personal coincidence here. Not only were David and I starting to work together on the amyotrophic lateral sclerosis (ALS) exome-sequencing project at Biogen (Chapter 8), but I had been on the scientific advisory board (SAB) of an Australian Biotech company called Bionomics in Adelaide since 2004, which started life in 1999 with an interest in epilepsy genes. Bionomics had been collaborating with an epilepsy geneticist, Steven (Steve) Petrou at the University of Melbourne, who later became associated with Praxis.[P]

Praxis kick started their SCN8A/Nav1.6 programme with a licensing collaboration with Gilead Sciences. They had compounds that increased the SCN8A sodium current peripherally but were also brain-penetrant and thus could potentially be used for epilepsy. They also in-licensed a positive allosteric modulator (PAM) of the $GABA_A$ receptor ($GABA_AR$) from Purdue Pharma ($GABA_AR$ is the target of benzodiazepines, used to treat anxiety). The PAM was anticipated to be useful for depression.

Praxis have continued to work on the PAM $GABA_A$ programme. PRAX-114, which acts extrasynaptically, is in clinical trials for postmenopausal depression. In a collaboration with Ionis, Praxis are developing an ASO targeting gain-of-function mutations in SCN2A to reduce the expression of the protein. They are also developing PRAX-944, a novel T-type calcium channel blocker for essential tremor. PRAX-562, their third clinical programme, is a selective persistent sodium channel blocker, in development for the potential treatment of rare adult cephalgias, and developmental and epileptic encephalopathies (DEEs).[11,12]

Praxis were well-financed: they raised $210 million privately in 2020 alone and followed that with a $190 million IPO. This money, and likely more, will certainly be needed, as phase 2 and 3 trials for depression are neither cheap nor fast.

Actio Bio

The OMIM (Online Mendelian Inheritance in Man) database referring to genes causing Mendelian inherited forms of disease and the large whole-genome sequencing sets (like the U.K. Biobank) contain significantly more information about mutations causing rare forms of disease than they used to. When Praxis were formed, there were only very few mutations identified

---

[P] Steve and I had shared some interesting nights out in Adelaide after Bionomics SAB meetings.

in the genes known to cause the rare forms of epilepsy, and there were no systematic ways to test whether the mutations were gain- or loss-of-function. Today, for many rare diseases, most of the mutations in the genes that cause the syndromes have been uncovered, and there are rapid and systematic ways to test their activity *in vitro*.

The computational ability to look at overlaps of phenotypes involved in common and rare diseases is now at a place where shared genetic pathways can be identified. David Goldstein and John McHutchison (ex-Gilead), who worked together when they were both at Duke University, have now formed a company called Actio Biosciences to exploit this genotype–phenotype juxtaposition. It is a little like Praxis in concept only more so, because so much more data can be accessed. The company is in San Diego, financed by DROIA Ventures, EcoR1 Capital, and Deerfield. Canaan Ventures recently led their series A financing. David left Columbia to go and run Actio as CEO.[q] They are focusing on developing drugs for rare diseases and partnering with pharma for those drugs that are relevant for more common disorders, sharing at least some of the pathological molecular pathways.[13]

Bioverativ

Bioverativ was spun out of Biogen in 2016. The thinking was that this business would benefit from a more focused team approach than they were getting within the company and would increase shareholder value more substantially as a separate entity. John Cox, who was head of commercial and technical operations at Biogen, became the CEO.[r] The official separation date of the two companies was February 2017. Biogen investors received a special dividend of one share of Bioverativ stock for every two shares of Biogen stock held as of January 17. Bioverativ began trading on January 12, 2017.

I was working as a venture partner for SV Health Investors when John asked me if I would help with the R&D parts of the haemophilia programmes as a consultant for Bioverativ (I was reasonably familiar with the programmes from Biogen times). He knew that "I spoke it like it was and not how some might like it to be," which he appreciated.

---

[q] I am a consultant to Actio Bio.

[r] I had worked with John at Biogen helping him and his commercial team to become more familiar with some of the R&D programmes.

Bioverativ moved to a new facility in Waltham on Second St. shortly afterwards. It was the best space I have ever worked in. It was modern in design, and spacious enough that the whole company of about 400 people, including R&D, could fit into the one building. It was a great atmosphere: we built a "can-do; science matters because the patients matter" culture. You could feel it when you walked in the door; and nearly everybody who did walk through the door commented on the tangible buzz.[s]

It only took a heartbeat or two to decide to take up John's subsequent offer of being the permanent head of R&D in April 2017. To be able to walk to speak to legal, finance, R&D, HR, and commercial—all in about five minutes—was a privilege that I have not experienced before or since. There were Tuesday breakfasts to be had in the café downstairs, and what we called "stair talks"—where seminars took place in front of the stairs in the large open space. It was a very special place to work.

Bioverativ's Science

The science revolved around making new versions of ELOCTATE (a long-acting Factor VIII) and ALPROLIX, a long-acting version of Factor IX (see Chapter 2). The sales of these products were much better under Bioverativ management than they had even been at Biogen. We also inherited the CD34 sickle cell and β-thalassaemia gene-editing programmes from Biogen in collaboration with Sangamo, as well as their interests in both a Factor VIII and a Factor IX lentivirus gene therapy programme in collaboration with Luigi Naldini in Milan (see Chapter 13 for more details).

The newer versions of Factor VIII and Factor IX were protein pro-drugs that required exquisite processing to become the active molecules. I remember first learning about the new Factor VIII construct with its complicated XTENylated regions (see Chapter 9 for origins of XTEN technology), called BIVV001 from Rob Peters and his team at Biogen. I thought at the time that it was going to be tough for these potential drugs to make it and even tougher to get them to work: once again I was wrong—at least for the Factor VIII derivative.

Bioverativ put BIVV001 into clinical development in 2017: it is only quite recently that the molecule, now called Efanesoctocog alfa, was

---

[s]It was a bit like the "Who are those guys?" line in the movie *Butch Cassidy and the Sundance Kid*.

announced to have met all the primary and secondary endpoints in the phase 3 trial that ended in early 2022. The drug also received breakthrough therapy designation from the FDA, the first time that has happened for a Factor VIII derivative.[t] The drug "Altuviiio," approved in early 2023, is the first once-weekly Factor VIII replacement therapy, providing competition for Hemlibra, the bispecific antibody developed by Chugai and marketed by Roche Genentech (see Chapter 4). Once a week treatment with Altuviiio will allow for almost completely normal activity for type A haemophiliacs and should completely control bleeding episodes.[14,15]

The Complement System

We looked to expand outside of the haemophilia space but still in rare diseases. One place Rob and his team started looking was at the complement system.[16,17] We were obviously aware of the success of Alexion with their antibody for the rare disease paroxysmal nocturnal haemoglobinuria (PNH), a haemolytic anaemia affecting some 5000–6000 patients (Box 2).

The focus on the complement system led us to a company called True North Therapeutics in San Francisco. They had been developing an antibody against complement Factor C1s, further up the complement cascade than C3 or C5, for another rather rare haemolytic anaemia called cold agglutinin disease (this disease leads to extreme sensitivity to cold and drops in haematocrit via complement lysis of red blood cells).

BOX 2. ALEXION

Alexion were selling Eculizumab (Soliris) a monoclonal antibody against complement factors 5a and 5b, with a list price of more than $470,000 per year: one of the most expensive drugs in the world. Sales reached more than $4 billion, and the company enjoyed a market cap north of $20 billion. AstraZeneca bought Alexion for this asset as well as several other backup monoclonal antibodies in 2020 for $39 billion. I have always considered Alexion to be a bit like Shire Pharmaceuticals, where most of the innovation came from the acquisitions that the company made, rather than their own research and development. But that strategy still creates shareholder value! Shire also got acquired, by Takeda, for $62 billion in early 2019.

---

[t] It also had priority review at the FDA for approval, which duly occurred in February 2023.

The data in the first six patients treated with the antibody, in which the haemoglobin went up by two points in as many months with no adverse effects, made us believe that what pertained for six patients would probably pertain for 600 or 6000 patients. Even better, despite the large doses of the antibody (TNT 009) used per patient, the side effects were benign. In May 2017, Bioverativ acquired True North for $825 million. The antibody (now called sutimlimab, sold under the brand name Enjaymo) was approved in the United States in February 2022. We were not wrong about it working in many patients.

Being Acquired

I was at the rather quaint Toronto City Airport in November 2017 on my way back to Boston when I got a call from Brian Posner, the Chairman of the Bioverativ Board. He told me that there had been a bid to buy the company, and roughly what it was. I responded by saying that I did not think I could create that amount of shareholder value that quickly from R&D, but I would be able to do it over the next four to five years.[u]

The Bioverativ executive team, consisting of John Cox (CEO,) John Greene (CFO), Richard Brudnick (CBO), Andrea DiFabio (Chief Legal Officer), Lucia Celona (Chief Human Resources Officer), Rogério Vivaldi (Chief Commercial Officer), and me, were duly dispatched to present to the potential acquirers in New York. We had done plenty of pitches before, but it was difficult to know whether to play it up or play it down because—despite the magnitude of the offer—we were all passionate about the company and we did not really want to be acquired. The acquirers were impressed, the Bioverativ Board and the shareholders recognised the value of the offer, and in January 2018, Sanofi announced that it would acquire Bioverativ for $11.6 billion.

The news led to some interesting conversations with ex–Genzyme Bioverativ employees about what it was like to be acquired by a large French pharma company. I knew Elias Zerhouni, who was president of R&D at Sanofi, from my time at SAIC-Frederick when he was still head of the NIH, and I liked him. I also knew his successor, John Reed from San Diego, where he ran the Burnham Institute before he went to Roche.

---

[u]Given what has happened with Altuviiio and Enjaymo, I was probably right, but that was hardly the point.

At that time, the Burnham Institute, just up the road from Sequana, were one of the best-funded research Institutes in the world, owing to his influence and the good science that was being done there. I was hopeful that they would listen to what Bioverativ had to say, and that it would make for a good combination. The acquisition was certainly good financially for the Bioverativ staff.

But the reality reflected my experience at Glaxo, which was that the further down a large (more than 120,000 people) organization such as Sanofi you go, the more difficult it is to get things done. Independent decision-making by relatively junior people is not encouraged and is unlikely to happen: precisely the opposite behaviour and culture to what we were trying to encourage at Bioverativ. Predictably, one year after the takeover less than 50% of the Bioverativ employees remained. The Bioverativ name also went away, just as it did for Genzyme after their Sanofi takeover.

One of the things that should be retained in an acquisition after the products are the people responsible for them and for the products that will come next. As far as I know, no value was attributed in the deal for the people, and very little in the way of retainers and bonuses were put in place by Sanofi to keep the employees. But it is those people that make the company—and Bioverativ had some of the best people in the business. Maybe try to keep the people and manage the company as a wholly owned subsidiary? The response was "we do not do that with the companies we acquire"—another way of saying that we "Sanofi-ize" them. You know you are toast when you convene "innovation days" to promote innovation: innovative companies innovate *every* day without needing to meet formally to do so.

Bioverativ existed as an independent company for 15 months (February 2017 to June 2018). It was a short but great ride, which created lasting value from the products they made and a great return on investment for their investors. But it also left a frustrated management team wanting to do it all over again.

REFERENCES AND NOTES

1. For multiple sclerosis, the numbers come from the MS Society (https://www.nationalmssociety.org)

2. Werth B. 1995. *The billion dollar molecule: one company's quest for the perfect drug.* Simon & Schuster, New York.

3. Werth B. 2014. *The antidote: inside the world of new pharma.* Simon & Schuster, New York.

4. Hawkins J. 2019. *Conscience and courage: how visionary CEO Henri Termeer built a biotech giant and pioneered the rare disease industry.* Cold Spring Harbor Laboratory Press, Cold Spring Harbor, NY.

5. Geraghty JA. 2022. *Inside the orphan drug revolution: the promise of patient-centered biotechnology.* Cold Spring Harbor Laboratory Press, Cold Spring Harbor, NY.

6. Kakkis E. 2022. *Saving Ryan: the 30-yr journey into saving the life of a child.* Impositivity Media LLC, Burbank, CA.

7. Hall SS. 2013. Genetics: a gene of rare effect. *Nature* **496**: 152–155. doi:10.1038/496152a

8. Dib Haij SD, Binshtok AM, Cummins TR, Jarvis MF, Samad T, Zimmermann K. 2009. Voltage-gated sodium channels in pain states: role in pathophysiology and targets for treatment. *Brain Res Rev* **60**: 65–83. doi:10.1016/j.brainresrev.2008.12.005

9. Jones J, Correll DJ, Lechner SM, Jazic I, Miao X, Shaw D, Simard C, Osteen JD, Hare B, Beaton A, et al. 2023. Selective Inhibition of Nav1.8 with VX548 for acute pain. *New Engl J Med* **389**: 395–405.

10. Conversation with David Goldstein, 2021.

11. Praxis website. https://praxismedicines.com

12. Couzin-Frankel J. 2019. Epilepsy's next frontier. *Science* **366**: 1300–1304. doi:10.1126/science.366.6471.1300

13. Fishman MC, Porter JA. 2005. A new grammar for drug discovery. *Nature* **437**: 491–493. doi:10.1038/437491a

14. von Drygalski A, Chowdary P, Kulkarni R, Susen S, Konkle BA, Oldenburg J, Matino D, Klamroth R, Weyand AC, Jimenez-Yuste V, et al. 2023. Efanesoctocog alfa prophylaxis for patients with severe hemophilia A. *N Engl J Med* **388**: 310–318. doi:10.1056/NEJMoa2209226

15. Leissinger C. 2023. Another victory for patients with hemophilia. *N Engl J Med* **388**: 372–373. doi:10.1056/NEJMe2216176

16. The complement cascade is a set of proteins and enzymes that is part of the immune system that helps antibodies and the innate immune system to remove pathogens and damaged cells.

17. Dunkelberger J, Song W-C. 2010. Complement and its role in innate and adaptive immune responses. *Cell Res* **20**: 34–50. doi:10.1038/cr.2009.139

CHAPTER 12

Antisense and RNAi

M any of the technologies that mark the progression of the biotech industry are suitable for treating both rare and (sometimes) more common diseases. Antisense oligonucleotides and RNA interference (RNAi), however, have proven most relevant for treating rare diseases.

The complementarity between the base pairs that make up DNA (i.e., A–T, C–G) is important not only for accurate DNA replication but also for "transcribing" mRNA copies of the DNA sequence, which in turn are "translated" into proteins (the central dogma of molecular biology: DNA makes mRNA that is translated to protein; see Chapter 1).

Proteins do most of the enzymatic and other reactions in the cells that make up an organism and are usually the targets for drugs. But genomic technologies have demonstrated that DNA and RNA themselves can be both therapeutic targets and drugs themselves.

A paper published in 1978 by Paul Zamecnik in the *Proceedings of the National Academy of Sciences* (*PNAS*)[1] described using a short single-stranded piece of DNA to slow the replication, transcription, and translation of an RNA virus called Rous sarcoma virus (RSV) in cell culture. These single-stranded DNA segments were called "antisense oligonucleotides," or ASOs, because they were complementary to the sense mRNA sequence, binding to it via its matched bases.[a]

The observation led to the beginnings of several "antisense companies" that wanted to develop the prospect that ASOs could be a straightforward way to target the expression of any protein (i.e., via blocking its mRNA). Among the first were Hybridon, started by Zamecnik himself, Genta, Isis (now Ionis), and Gilead Sciences. I focus here on the two leading companies

---

[a]Oddly, in this case, the effect on RSV turned out not to be mediated by hybridisation of the short sequence-specific oligonucleotides to the RNA. It was a seminal paper, nevertheless.

in the space, Ionis and Gilead Sciences, because I know them well and have had connections with them both.

ISIS (LATER TO BECOME IONIS)

Stan Crooke was the founder and prime mover in the genesis of Isis. Crooke was the head of R&D at SmithKline Beecham (SKB) when George Poste was head of Research. Even before the ultimate merger with Glaxo, it was clear that SKB would not be an independent pharmaceutical company for very long, given their less-than-impressive pipeline of new drugs.

Crooke left SKB in 1988, but not before he got his small research group at the company interested in the Zamecknik approach. He was seduced by its simplicity and inherent specificity. Apparently,[2] the interest was further inspired by a seminar that Paul Ts'o, a biologist from Johns Hopkins University working on oligonucleotides (oligos) as antiviral and anticancer agents, delivered at SKB.

The field was nascent at that time from a drug discovery point of view, with nothing much happening. In the early 1980s, Marv Caruthers (a founder of Amgen) and his team at the University of Colorado Boulder had developed the phosphoramidite chemistry that empowered the building of automated DNA synthesizers.[3] These new machines made it relatively simple to construct any oligonucleotide, sense *and* antisense. David Ecker, who later became an Isis cofounder, had access to a DNA synthesiser in his lab at SKB. He started to make some oligos in the summer of 1988 and had time to do some initial experiments. Stan Crooke was put in touch with Christopher (Chris) Gabrieli, a young venture capitalist (VC) at Bessemer Ventures, who was aware (as was the entire VC community) of Genentech's success and wanted to replicate it.

Crooke wanted to build an organization that would focus on R&D à la Genentech, and he wrote a business plan about developing antisense drugs.[b] Chris persuaded Bessemer to commit $50,000 in seed money to start the company. Like other start-ups, the Crooke house in Philadelphia became the company headquarters (HQ), where some of early research plans were hatched and SKB pharmacologist Christopher (Chris) Mirabelli joined the team. The company was originally named Isis after the Egyptian god of healing. This choice of name turned out to be an unfortunate choice, thanks to the Iraq war and the emergence of several violent factions in the Middle East. The name was subsequently changed to Ionis, which I will use from now on.

---

[b] This was around July 1988, before Stan formally left SKB.

Palomar Airport and Carlsbad

The small Ionis team considered many of the usual biotech locations (the Bay Area, Boston, etc.) for the company. The Bay Area was already too expensive, and Boston was panned for its lousy winter weather. In SKB days, Stan had frequently flown to Palomar Airport in Carlsbad, near San Diego.[c] He liked the area, which is much more developed now than it was in the early 1990s. Many of the new streets that were built were named after scientists in the anticipation that it would become a biotech hub, competing with Torrey Pines and Sorrento Valley. That never really happened, but Stan and Ionis were committed to building a lab near Palomar, leasing space on Faraday Avenue. While they were waiting for the labs to be built out and in the interests of "getting going" in the face of burgeoning competition, Ionis secured a temporary space in Sorrento Valley about 15 miles south of Carlsbad, close to the train station and just up the road from where the SGX labs were located 10 years later.

That initial space was rough to say the least.[4] It was at the back of a neurology clinic close to the Interstate 5 Freeway and its multiple storage facilities, most of which are still there. They leased the space from Richard Smith, a physician who ran a clinic focusing on amyotrophic lateral sclerosis (ALS): the connection with Smith turned out to be important later. He was always telling them how important their technology could be in treating ALS, as there were no disease-modifying drugs available. Money was tight, but they had enough to put some makeshift chemistry hoods into the labs and began to make oligos.[d]

The Ionis people were clearly up for a scrap, and they soon appreciated as they went around trying to raise money that they were not the only antisense game in town. This affected not only raising money but also hiring people. January weather in San Diego is not January weather in Boston, which certainly helped. Stan Crooke managed to hire a first-rate chemistry group, and they hired great biologists as well, despite the local competition from Genta. Meanwhile, Gilead had persuaded Venrock, who had shown initial interest in Ionis, to invest in them instead. One of the concerns Venrock

---

[c] This is an airport I know well as I trained for my private pilot's licence there. I could land on runway 24 or 6 even now.

[d] When the lab began, they did not have the necessary licence to hold or use ethanol, so they had to go across the border to Tijuana to buy the solvent needed to precipitate their DNA. A familiar story to me, as I had to buy distilled water (intended for car batteries) from the local petrol station to run gels when we started at Celltech in 1981.

articulated was that Crooke was still encumbered by his ex-relationship with SmithKline and whether he was really a free agent. Knowing that good legal advice in a start-up is essential, Ionis approached the law firm Cooley Godward to see if they could help. Lynne Parshall, a young lawyer from Cooley Godward, was assigned to work with Ionis and Crooke. It turned out to be the beginning of a long relationship between Parshall and the company. Ionis raised $5.2 million in its series A round in March 1989, led by Bessemer Venture Partners and included Accel, Rothschild, and Sutter Hill Ventures.

Pharma Deals

Among other visitors to Ionis in Sorrento Valley were representatives of Big Pharma. Ciba-Geigy, headquartered in Basel, Switzerland (a precursor to Novartis), Rhône-Poulenc Rorer (one of the predecessors of Sanofi), and Eisai from Japan turned out to be the most important. Ciba-Geigy were the first to show serious interest in Ionis and their work. They had the advantage of knowing some nucleic acid chemistry and were prepared to use it in a collaboration and already had 25–30 employees working on antisense chemistries. The Ciba deal called for the Swiss company to buy $8.5 million of Ionis stock and to supply research money for them to investigate oligos against four disease targets, including 5-lipoxygenase in order to find a potential drug for inflammation. They also undertook to help fund work to develop the necessary "large-scale" manufacturing processes. Finally, Ionis would get royalties on sales if an approved drug came out of the deal.

The Ciba-Geigy deal, which lasted for eight years, was critical for the development of Ionis. The company also managed to strike three other corporate deals within a year, which provided strong validation that Ionis knew what they were doing technically and had gained the trust of several large companies. Like other groundbreaking start-ups, obtaining money was always a challenge. By the end of 1989, the company had already spent $2.7 million. Bessemer came up with a $1.5 million bridge loan in May 1990, with Stan himself helping to meet payroll for one month before the money came in. The corporate deals enormously helped the cash flow.

By this time, the company had moved into 15,000 sq ft of new lab space in Carlsbad and were burning through money on their first projects at a rate of knots (i.e., very fast). These were by design looking at diseases where systemic application of the oligonucleotides might not be necessary and where topical or application by injection might work. The first project was to develop ASOs against cytomegalovirus (CMV) for intravitreal injection for CMV retinitis, an opportunistic disease that had become common in AIDS patients. Ionis called the compound ISIS-2922 or fomivirsen. It was a synthetic 21-nucleotide phosphorothioate oligodeoxynucleotide.[e] The oligo was aimed at an mRNA that coded for a protein needed for virus replication. With Eisai paying half the cost, Ionis moved the drug into phase 3 trials in 1995. By the end of 1997, fomivirsen had completed four phase 3 trials, one of which was in combination with other drugs. The package Ionis sent to the FDA had data on more than 300 subjects. The FDA approved its use in the United States to treat CMV retinitis in AIDS patients in 1998, just five months after the application was submitted. The drug was branded Vitravene, and approval in Europe and Brazil soon followed.

Vitravene was the result of nine years of research into ASOs, and it was rightly applauded as the first ASO drug in the world. However, as HIV therapies improved and patients no longer progressed rapidly to AIDS, the need for a drug against CMV retinitis became much less than anticipated, rendering Vitravene commercially unsuccessful.

Ionis IPO

The company began to prepare for their IPO and to write their prospectus in 1990. Stelios Papadopoulos, who ran Biotech investment banking for PaineWebber and had a PhD in Cell Biology (see Chapter 9), helped prepare it. Stelios had met with all the various antisense and small interfering RNA (siRNA) companies, and he knew and understood the technology and the potential as well as anyone did. Story has it that he wrote the prospectus for a barely one-year-old company by cutting and pasting from previous company documents and writing the business section by hand one weekend in San Diego.[5] Not only did Stelios believe in Ionis and Stan Crooke but he also wanted PaineWebber, who were the lead manager for the Genentech IPO, to lead the Ionis IPO.[f]

Crooke wanted to use Lehman Brothers, which before its demise were one of the whitest of the white-shoe banks,[6] to take the company public with PaineWebber as co-manager. At the last minute, Morgan Stanley showed up, and Crooke elected to go with Morgan Stanley *and* Lehman

---

[e] All that means is that it had a sulphur atom in place of a nonbinding oxygen in its phosphate backbone, which helped to stabilize and protect the molecule.

[f] This is in stark contrast to how IPOs are managed now, where huge numbers of people and multiple versions of the prospectus are prepared at enormous costs in time and money.

Brothers. Unfortunately, April 1991 was not a very favourable time to go public. Regeneron, based in Tarrytown, New York, went public in that April at $22 a share, but a month later, their stock was trading at less than half the IPO price. Ionis originally sought to sell three million shares priced between $13 and $15 a share. They could not find much interest at that price or at that volume. Instead, in May 1991, they sold 2.5 million shares at $10 apiece, for gross proceeds of $25 million.

Ionis had filed 22 patents with the USPTO, rapidly becoming one of the leaders in antisense technology. They had synthesized more than 350 novel classes of ASOs and were building a 3000-sq-ft manufacturing facility in Carlsbad. The company had grown to 66 full-time employees, nearly half of whom had a PhD or an MD. Lynne Parshall had become VP of Business Development. In March 1992, Ionis began a clinical trial with their compound against human papillomavirus (or HPV), called ISIS 2105. In culture dishes, the drug inhibited the production of a protein required for HPV to replicate. The clinical trial called for the drug to be injected into the skin after HPV-caused genital warts had been surgically removed and look for a reduction, delay, or elimination of wart recurrence. Ionis moved the compound to phase 2 in December, eventually testing the drug in more than 300 people, at considerable cost. It failed to work. This failure contributed to the growing doubt in the field about whether antisense technology really worked specifically at all, or whether the effects were due simply to the induction of a nonspecific anti-inflammatory effect.

Financing and Applications

Despite the rumblings, the company continued to do corporate deals and raise money. They raised ~$19 million from Japanese investors in 1991, $13.7 million later that year in the United States, another $37 million in two public offerings in 1993, and more in both 1994 and 1995. Ionis also did a deal with Boehringer Ingelheim (BI) for an array of inflammation targets. BI bought $28.5 million worth of Ionis stock, gave them a $40 million line of credit, and would pay them another $10 million if they found a compound active against Crohn's disease or another inflammatory disease.

ISIS 2303 (alicaforsen) was an oligonucleotide designed to inhibit the expression of the *ICAM-1* gene. Ionis had it in phase 2 trials for a handful of indications, the most advanced being Crohn's disease. An interim analysis of 150 patients in late 1998 showed the drug caused a steroid-free complete remission in 29% of patients (the placebo arm had a remission rate of just 14%). Ionis expected the full trial to bear similar results, but when it ended in late 1999, only 12% of drug-treated patients in the second half of the study had a steroid-free remission.

This kind of result is quite typical. Often for unknown reasons, positive phase 2 results fail to replicate in a second study or in phase 3, likely having to do with patient allocation bias and heterogeneity. But it was yet another ASO failure for the company: these were beginning to stack up. Ionis did deals with AstraZeneca and Abbott, which raised additional dollars for R&D to keep them solvent. Targets for cancer included inhibiting the expression of protein kinase C (α) and an ASO designed to inhibit C-Raf kinase mRNA. These programmes were not wildly successful either.

It would not be until 2013 that another ASO drug was approved. The drug known as Kynamro (mipomersen) is indicated for familial hypercholesterolemia targeting the apolipoprotein B-100 mRNA.[7]

Gilead Sciences, whom we will talk about shortly, also started life as an ASO company but exited the field in 1998 to focus on antiviral nucleotide- and nucleoside-derived drugs. Ionis paid Gilead $6 million for the patent estate that Gilead held around antisense chemistry and drug delivery, strengthening the overall patent estate of Ionis considerably.

Some of the more recent successes at Ionis have stemmed from relationships that the company had forged over many years. For example, Richard Smith introduced Frank Bennett, who joined Ionis very early on as a founder, to Don Cleveland so that he could hear about "superoxide dismutase" or SOD for ALS. The *SOD* gene was a target in ALS because toxic gain-of-function mutations in the gene had been shown to cause a familial form of ALS disease. It became the first Ionis neurological disease–focused target.

Spinal Muscular Atrophy (SMA) and Nusinersen

SMA was a second neurology target for Ionis and an important and lifesaving therapy for these patients. Frank Bennett knew Doug Williams, who was by now head of R&D at Biogen, from Immunex days. Ionis formed a collaboration with Biogen for this project, in which ASOs were used to modulate splicing in SMA (see Box 1).

Ionis and Biogen focused on developing ASOs that induced the correct splicing of the *SMN2* gene so that more correct SMN2 mRNA was made (regardless of copy number of the gene), the resulting SMN protein making up for the loss of the protein from the *SMN1* gene.

BOX 1. SPINAL MUSCULAR ATROPHY

SMA is a rare disease but still the most common genetic cause of mortality in infants, with a prevalence of one in 10,000 individuals affected. It is an autosomal recessive neuromuscular disorder (i.e., you need two copies of the faulty gene to get the disease) that leads to degeneration of motor neurons and muscle wasting, resulting in the inability of affected children to stand or even to sit up. The mutant gene frequency is such that about one in 25–50 people are carriers. Spinal muscular atrophy is caused by a mutation in the *SMN1* gene coding for the protein SMN1 (survival of motor neuron protein). In individuals affected by SMA, the *SMN1* gene has a deletion or a point mutation in exon 7 that eliminates or blocks the expression of the SMN1 mRNA, so that the normal mRNA is not produced (nor the protein, either). Diagnosis of SMA is relatively straightforward via a simple genetic test for the mutations.

SMN2 is a gene located next to *SMN1* on chromosome 5 that probably arose via gene duplication during evolution. It is 99% related to *SMN1* and the number of copies of this gene in the affected individual carries influences the severity of SMA disease. Most of the transcripts from the *SMN2* gene are spliced incorrectly, leading to only small amounts of mature mRNA that can be translated to form the SMN2 protein. This protein functions like SMN1, so the more copies of the *SMN2* gene a patient carries (and, consequently, the greater the amount of protein expressed), the milder the symptoms of SMA. Most severe SMA type 1 babies who die rapidly only have one or two copies of the *SMN2* gene.[8]

In December 2016, Biogen introduced Spinraza (nusinersen) to treat type 1 SMA. It was a revolutionary drug for a disease for which no drugs were available, and the results in some patients were quite spectacular. Babies presumed unable to sit up or walk were able to do both and survived much longer than expected. The treatment cost patients $750,000 for the first year of treatment, with additional annual treatment costs of $375,000. It has led to sales of more than $2 billion for Biogen and relieved both the parents and the children with SMA. It is truly a lifesaving drug.[9]

The Ionis–Biogen antisense drug has not been without competition. Novartis acquired and developed a gene therapy product, in which the faulty *SMN* gene is replaced by a nonmutant copy in an adenovirus-associated virus vector (AAV9, see Chapter 13) as a once-only venous infusion. In May 2019, the FDA approved the Novartis product with a very long name (onasemnogene abeparvovec-xioi) as Zolgensma for the treatment of SMA in babies less than two years old. Zolgensma has not taken much of the market for SMA treatment, owing to side effects and pricing issues.

Another competing product comes from Genentech, who have commercialized a small-molecule splicing modifier developed by PTC Therapeutics and the SMA Foundation. Risdiplam (brand name Evrysdi) works on the same concept as the ASO by increasing the amount of the SMN2 protein. The FDA approved Evrysdi for the treatment of SMA in adults and children two months of age or older in 2020.

Overall, Ionis received $180 million through contracted work with pharma and raised more than $250 million in the equity markets. They borrowed $72 million to finance parts of the business and had an accumulated loss of $257 million before having any products at all. Recently, AstraZeneca paid $200 million upfront for the ASO that Ionis is developing for transthyretin-mediated amyloid polyneuropathy (ATTRv-PN). The molecule has met the predetermined end points in an interim analysis of their phase 3 trial. There has also been some promising data published recently on an ASO targeting all hepatitis B mRNAs for the treatment of chronic hepatitis B infection. This oligo, called Bepirovirsen, is part of an Ionis–GSK collaboration.[10]

In 2023, the FDA approved Tofersen, the ASO treatment for the 2% of ALS patients with mutations in the superoxide dismutase gene, which was also part of the Biogen–Ionis collaboration. Qalsody, as it is artfully called, costs nearly $15,000 a vial.

Ionis went through some tough times to get to where it is today. With trial failures and lack of cash, they had to do at least one substantial layoff with all the internal trauma that events like that cause.

Yet they have had some notable successes for both the patients and for the company and the staff. I well remember going to Stan's spectacular home in Sedona Arizona, after a Joint Steering committee meeting on the SMA project. It was only a short trip by plane from Palomar Airport to get there. *Audentes fortuna luvat*—fortune favours the bold.

GILEAD SCIENCES

Kansas native Michael (Mike) Riordan graduated with a medical degree in 1984 from Johns Hopkins University, but he was always more taken by technology than by medicine.[11] The idea that you could switch off gene expression specifically using ASOs particularly captivated him. Mike spoke at some length with a Hopkins group that were using methyl phosphonates to make ASOs, and when he subsequently went to Harvard Business School, he began to put the idea of building an antisense company into practice. But it was not until he joined Menlo Park Ventures as a rookie venture capitalist that Mike's company idea began to take real shape.

Originally founded in 1987 under the name Oligogen, Gilead Sciences (as in the biblical "Balm of Gilead") soon became the preferred name.[12] Three scientific founders worked with Riordan to create the company: Peter Dervan at Caltech, Douglas (Doug) Melton at Harvard, and Harold (Hal) Weintraub at the Fred Hutchinson Cancer Research Center. Riordan served as founding CEO and Chairman from the company's inception until 1996. He recruited several other key scientific advisers, including Harold Varmus, the 1989 Nobel Prize winner with Michael Bishop (for discovering onco-genes), Jack Szostak, recipient of the Nobel Prize in Physiology or Medicine in 2009 (for work on telomeres with Elizabeth Blackburn), and Ralph Hirschmann, a long-serving ex-Merck medicinal chemist. Recruiting a stellar scientific advisory board with as many Nobel Prize winners as possible was, of course, a move right out of the best VCs' playbook.

Gilead's first lab space also had a Nobel Prize association. An old Kevex lab in Foster City, where Glenn Seaborg, the 1951 Nobel Prize in Chemistry laureate who discovered many of the brightly coloured trans-uranium atoms, had worked, was available.[g] With this space as impetus, Gilead had a lab with six people working in it by July 1987.

Riordan stirred up a lot of interest in the company from Venture Investors. Benno Schmidt and John Hay (Jock) Whitney, who formed the first U.S. venture firm J.H. Whitney in 1946, led a pre-series A round of $2 million, followed by a Series A round of $10 million with Tony Evnin from Venrock at attractive valuations. Menlo Ventures, the venture capital firm where Riordan had previously worked, also made an investment of $2 million. Despite being a new company with only a few employees—and Riordan being only 29 years old at the time—they managed to persuade Donald Rumsfeld, the ex-CEO of G.D. Searle, to get involved on the Board. Not only did they now have stellar science advisors but they also built a very impressive Board of Directors, including Gordon Moore of Intel and George Schultz (former U.S. Secretary of State). Stellar casts often, but not always, make a difference to the future of a small biotech company.

Visits to the Cold Spring Harbor Genome meeting led to the hiring of Genentech alumnus Mark Matteucci to run Gilead's medicinal chemis-try department, along with Brian Froehler, another accomplished synthetic

---

[g] Seaborg had set up a new company and moved out of the space.

chemist. Michael Bigham joined as CFO and operations head, moving from H&Q, where he was a biotech analyst.[h] Jeffrey (Jeff) Bird completed his MD/PhD at Stanford while consulting for Gilead, getting involved in their clinical trials.[i] Jay Toole from GI (Chapter 2) was hired as one of the best cloners in the business, following his success with Factor VIII. And, in 1989, Norbert Bischofberger came on board as Director of Chemistry and future head of all R&D.

The Gilead–Glaxo Deal

I first met Mike Riordan in 1990, as part of the Glaxo USA team that had made the decision to "get into" ASOs. Their impressive start and stellar personnel meant that it did not take long to consummate a deal between the companies, signed on July 26, 1990 (see Plate 23 for Gilead deal signing). Under the agreement, we would work together on developing ASOs against several targets. I remember chatting with Mike one afternoon about managing research people, when he asked me what the secret was. I said, "Delegate to good people and let them get on with it."[j]

Gilead went public in 1992, raising nearly $90 million. A lot of their early success can be attributed to the combination of Mike Riordan and Mike Bigham. But the move to the big time happened when John Martin, who had joined the company from Bristol-Myers Squibb (BMS) as VP R&D in 1990, became CEO in 1996. Martin ran the company as CEO and then Executive Chairman until 2018. He died in 2021 at the age of 69.

Gilead focused on ASOs until 1996, but they also always had an interest in developing antiviral nucleotide drugs against both influenza and HIV. Gilead launched Vistide (cidofovir injection) for the treatment of CMV retinitis in patients with AIDS in 1996. Oseltamivir was also discovered by Gilead for the treatment of influenza and was licensed to Roche for late-phase development. It is now marketed as Tamiflu.

Gilead acquired NeXstar Pharmaceuticals in 1999 for access to AmBisome, an injectable antifungal treatment with more than $100 million of sales annually, DaunoXome, a less toxic formulation of doxorubicin, and the sales force to go with both products.

---

[h] Bigham went on after Gilead to be a managing partner at the VC firm Abingworth.
[i] Bird ended up as a VC at Sutter Hill Ventures.
[j] Like I really knew what I was talking about.

Tenofovir (Viread) was approved for the treatment of HIV in 2001, leading to an even greater focus on antivirals. Gilead obtained emtricitabine by acquiring Triangle Pharmaceuticals (a Research Triangle Park–based start-up, as the name suggests). Emtricitabine (Emtriva) was also approved for HIV treatment. Gilead's cancer assets were sold to OSI Pharmaceuticals, who later developed Tarceva (erlotinib) for treating epidermal growth factor receptor (EGFR)-expressing lung cancers.[k] Finally, Gilead launched adefovir (Hepsera) for the treatment of hepatitis C in 2003.

From Biotech to Pharma Company

If the above description of Gilead makes them sound more like a pharma company than a biotech company, then it is because that is what they were becoming. They have continued to become a pharma company in both products and in culture. For example, Gilead was one of the first companies to combine anti-HIV medicines to reduce the pill load that HIV-infected individuals had to take. Atripla, approved in 2006, combined a nonnucleoside reverse transcriptase (RTase)[l] inhibitor with a nucleoside and a nucleotide RTase inhibitor (efavirenz [from BMS], emtricitabine, and tenofovir: three different inhibitors in one drug). Gilead's Truvada (combining emtricitabine and tenofovir) was also approved for pre-exposure prophylaxis (PrEP) against HIV in 2012.

It is all too easy, given their success, to imagine that Gilead's position in the industry is due to their innovation and their own drug discovery. In fact, most of their successful products since the mid-2000s have come from acquisitions. Of these acquisitions, their most successful is probably the purchase of Pharmasset in 2012 for $11.2 billion for their hepatitis C inhibitor sofosbuvir. Gilead was heavily criticised for overpaying for this asset at the time. Pharmasset were founded in 1998 to focus on the development of novel antivirals especially for hepatitis, and their acquisition gave Gilead a leadership position in new drug treatments for hepatitis C virus. A year later, Gilead launched Sovaldi (sofosbuvir), followed by Harvoni (a combination of sofosbuvir and ledipasvir). It is not an exaggeration to say

---

[k] OSI was acquired by Astellas in 2010 for $4 billion.

[l] Reverse transcriptase is an important enzyme in the life cycle of lentiviruses like HIV, and a critical laboratory tool for making cDNA from mRNA. David Baltimore and Howard Temin won the Nobel Prize in Physiology or Medicine in 1975 for its discovery.

that these two drugs have revolutionized the treatment of hepatitis C infection, providing complete cures in more than 90% of patients following a daily treatment regimen over two to three months. There were communication missteps by Gilead over arguments about pricing of their hep C drugs, leading to heavy scrutiny over their business practices. But they certainly did not overpay for Pharmasset!

Another acquisition of note was Kite Pharma, one of the leading companies in the development of engineered T cells to treat leukaemia and lymphoma (see Chapter 14). It was notable not just for the novelty of the field but also the acquisition price of nearly $12 billion. I would argue they did overpay for that asset and will return to that story in Chapter 17.

In 2020, Gilead announced that they were acquiring Forty-Seven Inc. for $4.9 billion.[m] This company was developing a human IgG4 mAb (magrolimab) against CD47, a cell surface protein that is expressed by tumour cells to signal to macrophages that the cells should not be attacked. The antibody is being developed for myelodysplastic syndrome and other leukaemias. It is unclear at present whether this mAb has any efficacy in patients at all.

During the COVID-19 pandemic, Gilead's Remdesivir was shown to have some antiviral activity against SARS-CoV-2 and was approved in that indication, but the drug is limited by not being available orally. More recently, other drugs have shown more activity against the virus such as Paxlovid from Pfizer, which is an orally available mixture of two existing antiviral drugs, nirmatrelvir and ritonavir, marking an uncertain future for Remdesevir.

With consistent annual revenues of around $25 billion/year and a market cap north of $75 billion, Gilead are one of the most successful biotech companies ever started. It is notable that their products, except perhaps the CAR T product acquired from Kite (see later), are mostly traditional pharmaceuticals. Some of these products were not first-in-class, but most frequently they were among the best-in-class, especially as combinations. This is certainly a recipe for success, although it is a far cry from Mike Riordan's initial interest in building an ASO company.

---

[m] Forty-Seven Inc. were a start-up out of Irving (Irv) Weissman's lab at Stanford and were supported by Lightspeed Ventures in 2016 leading a $75 million series A round. Jonathan MacQuitty, a long-time Abingworth-based VC and the original CEO at Genpharm (see Chapter 4), was the lead investor and founding CEO of the company, before their IPO in 2018.

RNAi AND ALNYLAM

Craig Mello and Andrew Fire were awarded the Nobel Prize in Physiology or Medicine in 2006 for their discovery of a naturally occurring form of modulation of gene expression in *Caenorhabditis elegans* that they called "RNA interference" (RNAi)— work they published in 1998.[13,14] RNAi is mediated by small single-stranded RNA molecules (ssRNA), derived from a long double-stranded RNA (dsRNA) by various specific ribonucleases (the dsRNA is known as "siRNA"). The derived ssRNA then binds to and mediates the degradation of specific mRNAs, leading to selective control of the expression of certain genes.

Several companies were set up to exploit this phenomenon, but the most important one without question is Alnylam Pharmaceuticals. John Maraganore, who was at Millennium in 2001 (see Chapter 7), got a call from MIT's Phil Sharp to talk about RNAi. Being familiar with it, John brokered a meeting at Millennium with the internal teams, who quickly became excited about the technology as a way to understand the function of many of the genes that Millennium had cloned.[15] Sharp visited Millennium with Thomas Tuschl from Germany in tow. Tuschl's lab had some important founding intellectual property (IP) in the space where small RNA molecules mediated interference. Given the number of genes that they wanted to investigate, Millennium sought an exclusive license to all the technology. The external team did not want to do that, so a company was set up instead to exploit the technology and license some of it nonexclusively to Millennium for some of their targets.

Alnylam were officially started in June 2002 with Christoph Westphal (Polaris Ventures) and John Clarke (Cardinal Partners) as founding investors. Christoph was also the founding CEO, but they needed to identify a permanent CEO quickly. Phil called Maraganore again and asked him to come in for a chat in his office at the MIT Cancer Center. It was a successful meeting, and John joined Alnylam as the sixth employee in December 2002 as CEO, a job he would stay in for almost 20 years.

Maraganore's first task was to in-licence all the Tuschl IP at the Max Planck Institute in Munich. This was not as easy task, because the Institute insisted that the U.S. company also have a European presence. This led to a 2003 merger with the German company Ribopharma and the consolidation of their IP on short interfering RNAs.

Paul Schimmel, Thomas Tuschl, Phil Zamore, David Bartel, and Phil Sharp were founders and advisors to Alnylam. The name of the company at that time was the uninspiring Precision Therapeutics. It was soon changed thanks to Paul Schimmel's aunt, an Arabic scholar who knew that the central star of Orion's belt was the centre of a string of pearls, called Alnilam from the Arabic "al-nizam."[16]

Alnylam (with a y) got going with the major goal of delivering siRNAs (which are double-stranded) to cells, using lipid nanoparticle carrier formulations of various kinds. Barry Greene, who joined Alnylam as president in 2003, worked with John as the two critical leaders of the company from both a cultural and scientific perspective for many years.[n]

Pax Oligo

While in the process of going public in 2004 with Bank of America Securities, Citigroup, ThinkEquity, and Piper Jaffray as their bankers, Alnylam received a "cease and desist" letter from Ionis, who claimed infringement on their IP, specifically the emotionally special to Crooke "Stan Crooke" claims.[15] John called Stan immediately and, with Vin Miles, flew to Carlsbad to hammer out a cross-license deal with Crooke and Lynne Parshall.[o] In the 28 days it took to put it together, Alnylam secured access to Ionis IP for RNAi. In return, Ionis had access to Alnylam IP for ASOs. This deal included all the Gilead IP that Ionis had bought and included original IP from Hybridon. This was a "Pax Oligo" win–win deal: it allowed competition for targets but in a differentiated way with different types of molecules.

Alnylam were the first preclinical company to go public after the biotech bubble burst in 2000, thanks to the human genome pronouncements from Clinton and Blair (Chapter 8). All the other companies had late-stage assets, and there were no innovative companies able to go public at that time apart from Alnylam. They were planning to sell about 5.5 million shares at $10, but owing to substantial lack of interest, ended up selling them at $6/share, raising only a little more than $30 million. The shares traded flat for quite a while.

However, post-IPO deals with Merck in late 2004 in ophthalmology and Novartis in 2005 for 35 targets in different therapeutic areas helped Alnylam's financial picture. There was much consternation, though, when Steve Friend from Merck phoned John one morning in 2006 and told him that Merck had just bought Sirna Therapeutics, a significant competitor to Alnylam. The Alnylam Board were distinctly underwhelmed by this duplicity and cancelled

---

[n] The company was always scientifically driven without being unfocused, with a product development mindset and always with the patients in mind.[17]

[o] It was the only way for them to be able to go public with the requisite timing.

the Merck deal shortly thereafter. They leveraged the Merck–Sirna acquisition to do a rival deal with Roche in 2007, one of the largest-ever preclinical biotech deals at the time, which included a $300 million up-front payment.

These kinds of deals and the cash that comes with them are company-defining, especially so when the company has mostly mouse data, some limited nonhuman primate experimental data, and a lot more work to do on drug delivery. The first two targets highlighted in the IPO were for the modulation of vascular endothelial growth factor (VEGF) and α-synuclein for Parkinson's disease (PD). Neither worked.

Ironically, Merck fell out of love with siRNA, and in 2014 closed the Sirna facility and laid off most of the staff. Alnylam were able to buy Sirna and their IP for $175 million.[P] Other pharma companies also decided that siRNA and ASO technology was going nowhere. Alnylam were feeling it and trading below cash. Consequently, the executive team (driven by John) initiated a "5 × 15 plan"—meaning they were going to invest in only five drug discovery programmes designed to be successful by the year 2015.

Real Drugs

Their first drug to be approved in 2018 was patisiran (brand name Onpattro), an siRNA acting against hereditary transthyretin-mediated amyloidosis (a disease afflicting about 50,000 people). Patisiran targets the mRNA coding for an abnormal form of transthyretin, which reduces the TTR protein in the liver forming the complexes that cause the characteristic polyneuropathy. Onpattro was not only Alnylam's first drug but also the first siRNA product to be approved by the FDA. A second-generation drug for this indication, called vutrisiran (Amvuttra), is also now available (see Biotechnology Product Time Line [pp. 441–449]).

Givosiran (Givlaari) was approved in 2019 for the treatment of hepatic porphyria by inhibiting the mRNA coding for delta-aminolevulinate synthase 1, an enzyme involved in the synthesis of heme. In 2020, Alnylam received their third approval, this time for an siRNA oligonucleotide targeting a rare metabolic disorder called primary hyperloxuria, caused by mutations in genes controlling the synthesis of oxalic acid and leading to the buildup of oxalate. Lumasiran is an siRNA against glycolate oxidase mRNA coding for an enzyme upstream of the alanine:glyoxylate aminotransferase mutated

---

[P] What goes around comes around: at least sometimes.

in patients with primary hyperoxaluria. The drug reduces the accumulation of oxalate manifesting as kidney stones, which, as those of us who have had them know, are extremely painful.

The other rather important programme that Alnylam started was the development of an siRNA against PCSK9, the protein that Regeneron and Amgen had both targeted with an antibody, to reduce low-density lipoprotein (see Chapter 6 for a fuller discussion of PCSK9 in the context of Regeneron). In this approach, the siRNA targets PCSK9 mRNA and reduces its translation. Alnylam out-licensed this product to The Medicines Company (TMC) to conduct clinical trials in patients with hypercholesterolemia. Positive phase 3 data led to the acquisition of TMC by Novartis for $9.7 billion in 2019. Inclisiran, which is sold under the brand name Leqvio, was approved in 2021, as a competitor for the anti-PCSK9 antibodies. Under Alnylam's deal with TMC, they retained 50% of the rights to the drug and were able to sell the royalty stream to Blackstone for $2 billion in 2020. It was one of the biggest royalty deals of that type ever done in biotech.

Another of the five drug discovery projects they targeted under the 5 × 15 strategic plan was to develop an siRNA for antithrombin III mRNA to down-regulate the major thrombin inhibitor and thus promote clotting in haemophilia patients. The idea of knocking out a protein that controls thrombosis this way (inhibiting an inhibitor) was quite radical at the time, given that most therapeutic approaches were variations on the theme of promoting clotting with Factor VIII or Factor IX derivatives (see Chapters 7, 11, and 13). No one thought that it was going to be safe to inhibit a thrombin inhibitor, owing to the risk of clots and thrombosis as a side effect.

History has shown that, while this concern was not unfounded, the approach does in fact work. Sanofi entered into a license agreement with Alnylam for fitusiran, the siRNA that they developed against antithrombin III mRNA. It is now in phase 3: some thrombosis has been seen, leading to a lower dose being used in the trial and a delay in the filing of the NDA.

Given the rare diseases that these drugs target, Alnylam also had to be at the forefront of making sure that patients get the drugs they need and that they are reimbursed appropriately. Maraganore made this vision an important part of the Alnylam mantra.

With revenues of more than $1 billion in 2022, Alnylam is another biotech story in which the company had to endure rough times with fortitude and persistence before transitioning to a stable and successful company. As usual, it took time, money, grit, stubbornness, and a little good fortune.

siRNAs do have drawbacks as drugs, including the challenge of delivery to the right place in the body and not just to the liver, but Alnylam (and others) continue to work on addressing those issues.

Remember this: Alnylam raised more than $7.5 billion over 20 years, more than half of which came from pharma partners. John Maraganore has now joined ARCH Venture Partners (see Chapter 18) as a partner, so that he can help to finance the next set of Alnylam-like biotech success stories.[17]

OTHER COMPANIES IN OR NEAR THE ANTISENSE/siRNA SPACE

Regulus

Regulus Therapeutics were set up in 2007 in San Diego as a joint venture (JV) between Alnylam and Ionis. The purpose was to target miRNAs: short (micro) RNAs that control gene expression, with a focus on cardiovascular disease, cancer, and infectious disease. The most advanced drug candidate developed to date is now in the clinic for ADPKD (acute dominant polycystic kidney disease). We had a collaboration with Regulus at Biogen, and the company have formed other alliances with GSK and Sanofi. The original CEO was Kleanthis Xanthopoulos, who also founded Anadys, a company who developed some hepatitis C inhibitors subsequently acquired by Roche in 2011 for $230 million.[q]

Dicerna Pharmaceuticals

Dicerna, founded in 2007, were another company in the space with some longevity. They were sold to Novo Nordisk in 2021 for $3.3 billion, based on their ability to make proprietary derivatives of RNAi complexed to forms of N-acetylgalactosamine (GlcNac), called "GalXC" and "GalXC-Plus." These were designed to extend the gene-silencing capability of RNAis and reduce their immunogenicity in liver and other tissues. The acquisition followed on from a strategic alliance formed in 2019 with Novo to test the system with a variety of liver-targeted miRNAs.

---

[q]Kleanthis and I used to trade war stories about being CEOs of biotech start-ups in San Diego among the Hybritech mafia. He left Regulus in 2015 and is now CEO of Shoreline Biosciences located once again on Roselle St. in Sorrento Valley—old familiar stamping ground, making natural killer (NK) cells from iPSCs. Given that NK cells are the focus of Catamaran, an SV Health Investors company that I cofounded, our interests once again overlap.

Molecular Evolution and SELEX

Larry Gold's lab in Molecular, Cellular and Developmental Biology at the University of Colorado, Boulder had been studying transcription and translation in bacteriophage T4-infected *Escherichia coli* for some time. They were especially interested in the translational regulation of mRNAs. Craig Tuerk, a PhD student in the lab, was looking at the control of translation of T4 DNA polymerase. He made a library of ssRNAs that bound to an octonucleotide hairpin loop in the polymerase mRNA. He made a library of 48 RNAs to see which would bind. As expected, the known RNA "antisense" sequence bound but there was also a variant with four nucleotide changes that also bound. This selection process was called "SELEX" (Systematic Evolution of Ligands by EXponential enrichment). SELEX invoked the concept that sequence in small RNAs created a shape that defined binding in addition to the primary sequence.[18] Meanwhile, Jack Szostak and Andrew (Andy) Ellington were doing similar work on the shape-defining oligonucleotides called "aptamers."[19] The Gold lab subsequently refined the process of preparing and screening vast libraries of RNA sequences for the rare members of such libraries that have interesting 3D protein-binding properties. SELEX yielded high-affinity and high-specificity nucleic acid antagonists for a variety of proteins.

NeXagen was set up by Larry in 1991 to exploit the technology. They made both Macugen, the first VEGF antagonist approved for use in age-related macular degeneration (AMD), and an aptamer that bound to PDGF for the same indication. NeXagen merged with Vestar in 1995, a company that had products including AmBisome (see above), to form NeXstar. This merged entity was acquired by Gilead Sciences in 1999 for $550 million. NeXstar scientists also published a remarkable selection experiment aimed at a small molecule—theophylline—which is nearly identical to caffeine. In that work, an RNA aptamer was identified that bound to theophylline about 10^4 times more tightly than to caffeine, a selectivity impossible to achieve with monoclonal antibodies.[20]

Perhaps the important concept to appreciate is that these small RNAs were not like beads on a string all lined up in a straight line, but formed globular structures based on their primary sequence, which could bind to proteins like conventional small molecules (i.e., like drugs). There is now structural information available supporting that concept.[21]

SomaLogic

Some NeXstar scientists were willing to join Larry in 2000 to create SomaLogic, a company primarily aimed at developing diagnostics using aptamer technology. They were able to negotiate the repurchase of the required NeXstar IP back from Gilead to be able to continue to that work.

As Larry has written, SomaLogic were started around a simple idea of quantitating proteins in complex mixtures like blood by aptamer binding.[21] This technology could be used diagnostically as a way of finding biomarkers across a range of concentrations for many diseases and their treatments. SomaLogic created an aptamer-based technology that would allow the relative levels of thousands of proteins to be measured all at once in a single sample. This was not quite as simple as it sounds: there is a lot of inherent noise (i.e., nonspecific interactions) that occur with these multiplex binding assays at scale. SomaLogic scientists engineered a new kind of aptamer, containing protein-like side groups, very similar to those on amino acids. Both the affinity of these modified aptamers for their chosen protein and the kinetics of binding turned out to be important. They also had to be amplifiable to allow for their continued selection and refinement. The name SOMAmers (Slow Off-rate Modified Aptamers) given to the binders reflects that biochemistry. During the last few years, the scientists at SomaLogic have used thousands of proteins (mostly human, some other mammals, and some bacterial proteins) as SELEX targets for specific SOMAmer selection.

The primary use of the collection of SOMAmer reagents has been to quantify proteins in biological samples by a large multiplex platform assay called "SomaScan." Biomarkers have been found robustly, quickly, and reproducibly, using small amounts of precious samples. Once found, the biomarkers can be measured using more conventional assays or by using SOMAmer-based approaches, especially when multiplexing is required. One good example of this is a study looking at serum biomarkers in Duchenne muscular dystrophy published in 2015 in *PNAS*, which Larry refers to as one of his own favourite papers (and he has published hundreds over the years).[22] It clearly demonstrates the power of the SOMAmer technology.[r]

---

[r] I spent many hours with Larry and SomaLogic while at Biogen trying to come to a business relationship that made sense to both parties, so we could use this technology in our biomarker discovery activities. Unfortunately, we could never get to a business arrangement that worked for both parties. SomaLogic were always very clear about their business model, and it did not suit Biogen well.

SomaLogic have built a service business, with a turnover of nearly $70 million in 2021, based on measuring proteins for diagnostics and as biomarkers. They maintained their focus as a private company until April 2021, when they went public via a SPAC (a special purpose acquisition company). This was a non-traditional way to get a public listing, but, for SomaLogic at least, it seems to have worked. They raised $630 million with a valuation of more than $1 billion.[s] The participants in the offering were well-known biotech investor like Casdin, and Illumina and Novartis were strategic investors. The SomaScan platform has been expanded recently to measure 7000 proteins per sample, and they are nearing a 10,000-protein version. Aside from the life science tool sales, the vision is still to be able to quantitate large numbers of proteins in samples from both healthy and diseased people and be able to deliver comprehensive prognostic and diagnostic information from blood, urine, and other relevant human samples. Somalogic has now merged with Standard Biotools, another Casdin Company.

BOX 2. SMALL MOLECULES FOR RNA MODULATION

Various companies have been set up over the years to look for more conventional small molecules that work by binding to mRNA. PTC Therapeutics developed their splicing modifier for SMA by screening for conventional small molecules that affected splicing. Ribometrix, an SV Health Investors–financed company, is using screening and structural methods to find and develop small molecules that affect RNA. Arrakis, led by Michael Gilman, make highly selective RNA-targeted small molecules that are designed to bind directly to RNAs, especially those coding for targets that have been difficult to drug by other means. Like many early start-ups, they have interests in many therapeutic areas. They also have the benefit of strategic alliances with both Roche and Amgen. One of their competitors in this space, Expansion Therapeutics, just completed an $80 million series B financing to continue their work in ALS, frontotemporal dementia (FTD), and other neurodegenerative disorders, using much the same approach as Ribometrix and Arrakis.

If access to capital makes a difference, then perhaps the biggest competition for these three companies comes from Skyhawk Therapeutics. They have raised more than $700 million, with $133 million in September 2021 alone. Skyhawk have partnerships with Merck, Bristol-Myers Squibb, Biogen, Sanofi, Takeda, and Genentech. Vertex did a deal with them in December 2020 with $40 million upfront. Like the other companies, Skyhawk are focusing on

---

[s] The stock value has retreated over the course of 2022–2023, with a change in internal leadership and the Board of Directors.

difficult-to-drug central nervous system (CNS) targets for neurodegeneration, and their lead candidate is in the IND-enabling stage. They have many other candidates in autoimmune, cancer, and neuromuscular targets at various pre-clinical stages. Judging by their partnerships and the money they have raised, people clearly think Skyhawk is the real deal in this space.

ASOs, siRNA, and associated nucleic acid–based technologies have created new modalities for the biotech industry and have led to several new and important drugs for mostly rare diseases, including some small molecules targeting RNA directly (Box 2). As we will see in the next chapter, it is also possible—although hard—to use the genes themselves as drugs.

REFERENCES AND NOTES

1. Zamecnik PC, Stephenson ML. 1978. Inhibition of Rous sarcoma virus replication and cell transformation by a specific oligodeoxynucleotide. *Proc Natl Acad Sci* **75**: 280–284. doi:10.1073/pnas.75.1.280

2. Jarvis LM. 2019. Stanley Crooke on finally making sense out of antisense. *Chem Eng News*, Vol 97, Issue 18.

3. Beaucage SL, Caruthers MH. 1981. Deoxynucleoside phosphoramidites—a new class of key intermediates for deoxypolynucleotide synthesis. *Tetrahedron Lett* **22**: 1859–1862. doi:10.1016/S0040-4039(01)90461-7

4. Conversation with Frank Bennett, CSO at Ionis, 2021.

5. Conversation with Stelios Papadopoulos, 2021, 2022.

6. White-shoe banks are named after the white Oxford shoes that the senior bankers used to wear in the 1950s, a name that emerged in the 1970s. A "white-shoe firm" is an old-fashioned term for the most prestigious, well-established banks. Over the years, several white-shoe firms have been acquired by bigger rivals or have gone out of business.

7. Wong E, Goldberg T. 2014. Mipomersen (Kynamro). A novel antisense oligonucleotide inhibitor for the management of homozygous familial hypercholesterolemia. *Pharm Ther* **39**: 119–122.

8. Tisdale S, Pellizzoni L. 2015. Disease mechanisms and therapeutic approaches to spinal muscular atrophy. *J Neurosci* **35**: 8691–8700. doi:10.1523/JNEUROSCI.0417-15.2015

9. Huggett B. 2021. Podcast *Nature Biotechnology*. Hope lies in dreams. Stan Crooke rose from poverty to start a company that pioneered antisense drugs. After decades of struggle and years of public doubt, the company created a treatment that has saved thousands of children from the brutal disease spinal muscular atrophy. https://www.nature.com/immersive/d41587-021-00015-5/index.html

10. Yuen M-F, Lim S-G, Plesniak R, Tsuji K, Janssen HLA, Pojoga C, Gadano A, Popescu CP, Stepanova T, Asselah T, et al. 2022. Efficacy and safety of Bepirovirsen in chronic hepatitis B infection. *New Engl J Med* **387**: 1957–1968. doi:10.1056/NEJMoa2210027

11. Conversations with Mike Riordan, 2021.

12. Balm of Gilead: The Bible records that in ancient times there came from Gilead, beyond the Jordan, a substance used to heal and soothe. It came, perhaps, from a tree or shrub and was a major commodity of trade in the ancient world. It was known as the Balm of Gilead.

13. Fire A, Xu S, Montgomery MK, Kostas SA, Driver SE, Melo CC. 1998. Potent and specific genetic interference by double-stranded RNA in *Caenorhabditis elegans*. *Nature* **391**: 806–811. doi:10.1038/35888

14. Watts G. 2006. Work on RNA interference brings Nobel triumph. *BMJ* **333**: 717. doi:10.1136/bmj.333.7571.717

15. Conversation with John Maraganore, 2022.

16. Alnylam and Orion's belt. *Alnilam* is the middle star in the famous three-member belt of Orion, the hunter where all three stars are in a straight line.

17. Armstrong A. 2022. With Alnylam in rear view, Maraganore wants to be a biotech multiplier, matchmaker and mentor. *FIERCE Biotech*, January 6, 2022.

18. Tuerk C, Gold L. 1990. Systematic evolution of ligands by exponential enrichment: RNA ligands to bacteriophage T4 DNA polymerase. *Science* **249**: 505–510. doi:10.1126/science.2200121

19. Ellington AD, Szostak JW. 1990. In vitro selection of RNA molecules that bind specific ligands. *Nature* **346**: 818–822. doi:10.1038/346818a0

20. Larry Gold conversation, 2021.

21. Gold L. 2015. SELEX: how it happened and where it will go. *J Mol Evol* **81**: 140–143. doi:10.1007/s00239-015-9705-9

22. Hathout Y, Brody E, Clemens PR, Cripe L, DeLisle RK, Furlong P, Gordish-Dressman H, Hache L, Henricson E, Hoffman EP, et al. 2015. Large-scale serum protein biomarker discovery in Duchenne muscular dystrophy. *Proc Natl Acad Sci* **112**: 7153–7158. doi:10.1073/pnas.1507719112

CHAPTER 13

Gene Therapy

The field of "gene therapy" (i.e., the therapeutic use of genes themselves) is a story played out in two acts: Act I, the early ideas and experiments, followed by a hiatus owing to the deaths of some patients in gene therapy trials; Act II, a renewed enthusiasm for gene therapy leading to some astonishing—but expensive—new treatments for patients.[1]

ACT I

Well ahead of gene therapy's time, Theodore (Ted) Friedmann and Richard (Rich) Roblin published a paper in *Science* in 1972 entitled "Gene therapy for human genetic disease?"[2] In this paper, they outlined the potential for incorporating DNA sequences into patients' cells for treating people with genetic disorders. Although it was not until the use of positional cloning in the 1980s and early 1990s (Chapter 7) that many of the genes causing Mendelian forms of human diseases such as cystic fibrosis (CF) and Duchenne muscular dystrophy (DMD) were discovered, Friedmann's and Roblin's thinking was already directed towards how to replace the "disease genes" with the normal copy so that it gets expressed in the right place in the body at the right time. It took 15 years for Friedmann's and Roblin's ideas to be put into practice.[a]

For a gene to be useful therapeutically, the DNA needs to be transcribed and translated, which requires it to be introduced into nucleus of a recipient cell. The initial plan was to use either naked DNA or DNA complexed with cationic (positively charged) lipids in lipoprotein particles.[3] These DNA/lipid/lipoprotein structures were quite simple to make and easy to scale up, but suffered from poor transfection rates (i.e., they did not get into cells very

---

[a] The gene therapy process and the vectors available in the 1990s are well-described in a special gene therapy report in *Scientific American* in 1997.[3]

well) and low levels of expression of the transferred genes. As we discuss later, a similar process for mRNA has now become very sophisticated.[b]

When I was working on foot-and-mouth disease virus (FMDV) at Pirbright in the 1970s, we used to infect tissue culture cells with the RNA isolated from FMDV using calcium phosphate or DEAE dextran to bind to the RNA and protect it during transfection. New virus particles and infectious virus could be generated with that method in only a few hours. This was transferring an mRNA to the cytoplasm of the cell (where virus replication normally took place), so it was not a stretch to imagine that this experiment with a virus mRNA would work.

It was much more of a stretch to think about getting genes into the nucleus of the cell, integrated into the chromosome preferably at the site where the gene usually resides and to get the gene expressed—the proposition underlying the emerging field of gene therapy.

Ex Vivo and *In Vivo* Gene Therapy

Two approaches have been used for gene therapy to date. In the first, CD34[+] stem cells were taken from the patient, transfected with the gene of interest, and then the cells returned to the patient. This was called "*ex vivo*" gene therapy. Its origin was in stem cell transplantation, where cures could be obtained for some leukaemias and lymphomas by taking stem cells from a human leukocyte antigen (HLA)-matched donor and, after purging the recipient's bone marrow with high-dose chemotherapy or radiation, replacing their bone marrow with the donor's cells.[4,5,c] As Malcolm Brenner had already transfected bacterial DNA into CD34 cells as a way to mark them to see what happened to them after transplantation, it was not a big jump to think about incorporating human genes into CD34 cells in the same way.

In contrast to *ex vivo* gene therapy, "*in vivo*" gene therapy was simpler in concept: the DNA, in an appropriate form, would be directly injected or infused into the patient. *Ex vivo* gene therapy has an advantage because the cells can be analysed before re-infusion, to check, for example, that the appropriate gene has been transferred and is in the right place in the DNA of the cell. This is not possible for *in vivo* methods in which the "vector" goes

---

[b] In fact, it is how the mRNA in the SARS-CoV-2 vaccines is delivered.

[c] This lifesaving therapy earned Don Thomas and Joe Murray in Seattle the Nobel Prize in Physiology or Medicine in 1990.

where it goes, and you find out where that is after it gets there. Most *in vivo* gene therapy vectors end up in the liver with some in the spleen.

Specialized vectors based on viruses, which are natural vehicles for delivering DNA or RNA to cells to infect them, were developed to make gene therapy more effective. Two major types of virus-based vectors have been developed over the years. One set—the γ-retrovirus vectors—was developed from Moloney murine leukaemia virus. The γ-retrovirus constructs have now largely been replaced by viruses based on lentiviruses, which are safe integrating derivatives of human immunodeficiency virus (HIV). Luigi Naldini was one of the prime movers in developing these retrovirus vectors, working with Inder Verma and others at the Salk Institute in San Diego, before he went back to Milan.[6]

The other major type of virus vectors is based on DNA viruses, including adenovirus and adeno-associated viruses (AAVs—Box 1).

BOX 1. AAV VECTORS

Adeno-associated viruses are small DNA viruses (parvoviruses) that infect several primates, including humans. They are replication-defective single-stranded (4.8-kb) DNA viruses that only replicate in cells that are also infected with a "helper" virus, such as adenovirus (hence the name) or, in some cases, herpesvirus—an even larger DNA virus. These larger co-infecting viruses provide some of the proteins needed for the AAV to replicate. The vectors are designed to replicate themselves (and the genes that they carry) in either dividing or nondividing cells, and there are several immunologically distinct serotypes of AAV, an important attribute with respect to where the gene that the AAVs are engineered to carry is expressed. An additional feature that makes AAV an attractive candidate as a gene therapy vector is that they can persist in an extrachromosomal state without integrating into host cell DNA. If the vector and gene do integrate into the genome, it tends to be at a particular site on chromosome 19 a more predictable consequence than the integration of retrovirus-based vectors.[7]

The major disadvantage of AAV vectors is that there is an ~4-kb size limit to the amount of DNA that can be packaged into the vector, limiting the size of the gene that can be transferred to the cells. Until very recently, AAVs were not thought to cause disease in humans, but they have now been implicated in rare forms of hepatitis in children.[8,9] Immune responses to AAVs were initially thought to be mild, but recently the immune responses to AAV-based vectors have led to serious side effects after treatment in some patients.

Preexisting antibodies to the virus limit the usefulness of the common serotype (AAV2) for certain applications because the antibodies prevent the AAV from binding to cells to deliver any gene they carry. Other AAV serotypes are now being used that have different levels of circulating antibodies in humans and, conveniently, different cell tropisms (i.e., affinities for different cell types). There are serotypes such as AAV5 for the vasculature, AAV6 for bronchial epithelial cells, AAV7 with affinity for muscle cells, AAV8 for hepatocytes in the liver, and AAV9 for the brain.

THE FIRST COMPANIES

Gene therapy was first tried in humans in 1989 for a severe combined immunodeficiency, an autosomal recessive disorder caused by mutations in the adenosine deaminase gene. William French Anderson, then at the NIH in Bethesda, Maryland, inserted the *ADA* gene into a retrovirus vector and introduced the virus by infusion into a patient: a four-year-old girl named Ashanthi DeSilva. DeSilva responded positively after several infusions over 12 days.

The success of this trial was seen as a major accomplishment, leading to the enthusiastic launch of numerous other trials throughout the 1990s. French Anderson formed a biotechnology company called Genetic Therapy Inc. (GTI) that, among other things, manufactured the ADA vector with the appropriate safety features under strict good manufacturing practice (GMP) FDA requirements. Both the research lab and the development group were located on Clopper Road in Gaithersburg, Maryland, just up the road from the NIH. Although French Anderson was employed at the NIH, he spent a lot of time at GTI, where he was often seen walking the halls usually, for some reason, in his slippers.[10] GTI's major clinical project was to insert the herpesvirus thymidine kinase (*HSV-TK*) gene into tumour cells using a vector and then to expose the patient to ganciclovir (which will kill cells expressing the HSV enzyme). The initial trial was done in 12 patients at the NIH Clinical Center. Sandoz (part of Novartis after merging with Ciba-Geigy) formed a strategic alliance with GTI in 1991 and subsequently acquired the company in 1995 for about $300 million.

Being Big Pharma, Sandoz behaved as one might expect. Their idea of the clinical trial and what was needed was very conventional: they decided that the project needed proper project management tools, although a Microsoft Excel spreadsheet would probably have sufficed for such a small number of subjects. Some clinical effects were seen, but they were thought to be due to the inflammation induced by the procedure rather than the effect of the transferred gene. The experiments were discontinued.[10]

At Glaxo, just before I left to go to Sequana, we initiated a collaboration on cystic fibrosis transmembrane conductance regulator (CFTR) gene therapy with a lipid nanoparticle company in the Bay Area called Megabios. The idea was to use cationic lipid complexes to transfect the lung epithelium of CF patients with a normal copy of the CFTR gene. We were not alone in doing this: Genzyme had started a similar project, but we were one of the first groups to try the approach. This programme went all the way to treating the nasal epithelium and lungs of CFTR patients with the gene complexes by inhalation. As others also found, there was not enough expression of the transferred DNA to make any difference to symptoms, and this programme was terminated.

Another approach that was taken for cancer was to introduce lipoplexes containing the *HLA-B7* gene, coding for an important histocompatibility antigen,[6] into the tumour cells. The idea was to "light up" the cancer cells for the immune system by expressing B7 in them to stimulate an immune response. The company Vical, started in San Diego in 1988 by Philip (Phil) Felgner (formerly of Syntex), specialised in making and using DNA-cationic lipid complexes. They did a trial of the B7 approach in 90 patients by injecting tumours with lipoplexes of the *HLA-B7* gene. In some of the melanoma patients, the tumours shrank, but this also strangely occurred in the controls, where there was no transgene—perhaps, a result of inflammation induced by the process.

Setback for the Field

The death of 18-year-old Jesse Gelsinger in a University of Pennsylvania gene therapy clinical trial in 1999 led to moratorium on gene therapy until more was known about the safety of the vectors and several other parts of the process.

Gelsinger had a genetic condition known as ornithine transcarbamylase deficiency, in which a mutation in this gene compromised his ability to deal with ammonia, which then accumulated in his system. He was part of a trial in which an adenovirus vector containing the normal gene was infused. Unexpectedly, Gelsinger suffered a severe immune reaction resulting in the release of large amounts of toxic cytokines, called a cytokine storm or CRS (cytokine release syndrome), and he died four days later. A subsequent FDA investigation uncovered several issues. They concluded that Gelsinger should never had been in the trial in the first place and discovered that other patients had experienced serious side effects from the treatment. The deaths of monkeys given a similar treatment had also apparently not been disclosed.[11,12,d]

Unfortunately for the field and for the patients, there were also deaths recorded in a γ-retrovirus-based gene therapy trial for a severe combined immunodeficiency (a disease caused by mutations in the gene for the IL-2 receptor γ chain). CD34 cells were collected from the bone marrow of several boys with this X-linked disorder. These cells were then transduced with a retrovirus vector containing the IL-2 receptor γ chain gene and the modified cells reinfused. Some patients started to make T cells after the infusion, indicating that the gene transfer had worked. Unfortunately, three years after the therapy several of the boys developed leukaemia, assumed to be caused by the vector integrating into certain locations in the genome in the transfected cells, causing activation of oncogenes.[9]

ACT II

It took a long time for the gene therapy field to begin to recover from these setbacks. Most of the gene therapy treatments that have been developed since, whether *ex vivo* or *in vivo*, have been directed to severe Mendelian diseases (where there are few treatment options available) using both modified AAV and lentivirus vectors.

Disorders of the Eye

One of the most popular indications for the reemergent field of gene therapy has been inherited blindness disorders, as the eye is an "immune privileged site," meaning that there are no B and T cells present to cause immune responses. It is also, obviously, an anatomically accessible site. Mutations in many different genes can cause inherited blindness (Box 2), two of the best-known forms being retinitis pigmentosa and Leber's congenital amaurosis (LCA).

LCA is a group of hereditary retinal diseases, with various phenotypes and at least 29 genotypes (with more to uncover). One form of LCA caused by biallelic mutations in the *RPE65* gene affects between 1000 and 2000 patients in the United States. When light hits photosensitive pigments in the

---

[d] *The Washington Post* minced no words: "Gelsinger's death is the latest in a series of setbacks for a promising approach that has so far failed to deliver its first cure and that has been criticized as moving too quickly from the laboratory bench to the bedside."[12]

BOX 2. OTHER GENE THERAPIES FOR THE EYE

China's Neurophth Therapeutics' leading candidate, NR082 (rAAV2 -ND4), was granted an orphan drug designation by the U.S. FDA in 2020 for the treatment of Leber's hereditary optic neuropathy associated with mutations in the *mt-ND4* gene, and early clinical studies are promising.

   Normal copies of the GUCY2D gene code for guanylate cyclase 2D, an enzyme in the pathway that rod and cone cells in the retina use to convert light into neural signals. A lack of this protein blocks the recovery of this pathway, preventing further signalling from rod and cone cells, leading to vision loss (another form of LCA). Gene therapy clinical trials for blindness caused by GUCY2D gene mutations are also going ahead.

   In Europe, companies such as Nightstar Therapeutics, Horama, Eyevensys, and MeiraGTx are developing multiple gene therapies targeting the different genetic mutations that cause blindness.

---

retina, it changes a molecule called 11-*cis* retinal (a form of vitamin A) to all-*trans* retinal to initiate the visual cycle. The RPE65 protein helps to convert the all-*trans* retinal back to 11-*cis* retinal completing the cycle. The lack of this protein causes progressive vision loss, leading eventually to blindness.[13]

   The results of several clinical trials in children showed that the delivery of AAVs carrying the *RPE65* gene by subretinal injection into the back of the eye yielded positive results: patients recovered functional vision without any apparent side effects. In 2017, the FDA approved voretigene neparvovec-rzyl[e] (Luxturna), the first *in vivo* gene therapy, for the treatment of blindness caused by LCA. The price of this treatment is US $850,000 for both eyes (see Biotechnology Product Time Line [pp. 441–449]).

   Luxturna was the first directly administered gene therapy approved in the United States targeting a disease caused by mutations in a specific gene. The safety and efficacy of Luxturna were established in a clinical development programme with a total of 41 patients between the ages of 4 and 44. All participants in the clinical study had to have confirmed biallelic *RPE65* mutations. The primary evidence of efficacy of Luxturna was in only 31 participants, based on the ability of treated and nontreated patients to complete an obstacle course at low light levels.

   For several reasons, the FDA granted Priority Review and Breakthrough Therapy designations for this drug application to the developer Spark

---

[e]The awkward name does actually mean something: *voretigene*, gene for the retina, and neparvovec, a new parvovirus-based vector (i.e., an AAV derivative).

Therapeutics. Spark had been founded in 2013 by Katherine (Kathy) High, Jeffrey (Jeff) Marrazzo, and Steven (Steve) Altschuler primarily to develop gene therapy for haemophilia, the focus of their studies at the Children's Hospital of Philadelphia, although they also had interest in LCA and choroideremia. Spark went public in 2015, raising $161 million. They were subsequently acquired by Roche for $4.3 billion in 2019, following a protracted takeover process.

GSK AND STRIMVELIS

GSK obtained approval in Europe for Strimvelis, a gene therapy product for adenosine deaminase deficiency (ADA SCID), in 2016. It has not been approved in the United States. This was an *ex vivo* product, in which a γ-retrovirus vector carrying an ADA cDNA gene construct was transfected into CD34 cells taken from the patient. After transduction, the cells were returned to the patient, who had been treated with busulfan as a pretransplant conditioning regimen. More than two dozen patients were treated, and the results were encouraging, although there was some concern about the retrovirus integration sites and the risk of leukaemia.

In 2017, GSK took the curious decision to out-license all their rare cell and gene therapy assets, including Strimvelis. Maybe it was not so curious, if you think that this kind of project was more the province of biotech anyway, and GSK perhaps should not have initiated the study at all.

Being at Bioverativ at the time, we were one of several companies invited to review the GSK programmes and, if interested, make a bid for them. GSK had a long-standing collaboration with Fondazione Telethon and Ospedale San Raffaele, and Luigi Naldini in Milan, with whom Bioverativ were also working on gene therapy for Factor VIII and Factor IX. They also had arrangements with MolMed (a San Raffaele spinout) as a product manufacturer. Several rare disease programmes were on offer by GSK, including both *in vivo* and *ex vivo* gene therapy opportunities. The inherent problem with the deal was that not only would we have to license the products, but we would also be expected to take on some of the staff working on the programmes. Inconveniently for us, these were primarily based in the United Kingdom and Milan.

Orchard Therapeutics

Bioverativ did not get the deal. Instead, GSK out-licensed the assets to a small gene therapy company in the United Kingdom named Orchard Therapeutics. Orchard had raised a $33 million series A round in 2016, led by F-Prime Capital, on the back of their gene therapy programmes in ADA SCID and mucopolysaccharidosis III. The founding scientists had been in gene therapy for a while, so it was not such a bad option for GSK. It also gave a home to at least some of the GSK cell and gene therapy employees, which Bioverativ could not have done.

I was left with the distinct impression, however, that this was a stitchup: the Orchard deal being the plan all along. The bids that we and other companies submitted were simply to legitimize a decision that had already been made to choose Orchard as their partner. Consistent with my impression was that, on the back of the deal that closed in April 2018, Orchard went public on Nasdaq in November, raising $225 million.

Most of the programmes that GSK were out-licensing are now part of the Orchard pipeline, and they continue to support Strimvelis.[14] We also thought it was notable that the GSK β thalassaemia gene therapy programme, in which Bioverativ had a particular interest, was not included in the original portfolio of products that we looked at. This was a third-generation lentivirus CD34 product—much like that from bluebird bio (see below)—with encouraging data. Curiously, it did get out-licensed to Orchard as part of their deal with GSK.

The second approved product (and another GSK programme) for Orchard in Europe is Libmeldy, an *ex vivo* lentivirus-based CD34 cell product to treat aryl sulphatase deficiency. The loss of two copies of this gene causes metachromatic leukodystrophy in juveniles. Gene therapy, early in the disease course, prevents some of the cognitive decline that follows after the inability of the affected children to walk independently. Orchard recently announced that a toddler named Teddi Shaw was the first child in the United Kingdom to receive Libmeldy, at a cost of £2.8 million to the National Health Service (NHS).

On reflection, I cannot help wondering why GSK wanted to rid themselves of the very assets that would help them to develop cell and gene therapy for more common diseases, but it is consistent with their strategy of not entertaining rare disease therapies. GSK's loss was very much Orchard and Orchard's shareholders' (including GSK) gain. In December 2023, Orchard had a market cap just shy of $400 million. I suspect that is more a reflection of the prevailing environment than of the value of their programmes.

NEUROMUSCULAR AND OTHER DISEASES

X-linked myotubular myopathy (XLMTM) is a rare and serious form of neuromuscular disease characterized by extreme muscle weakness, respiratory failure, and early death. The disease is caused by mutations in the *MTM1* gene that prevent expression of a protein called myotubularin, required for normal muscle function. Mortality rates are at least 50% in the first 18 months of life, but some less-affected patients live a few years longer. The disease affects approximately one in 40,000 to 50,000 males. In most patients, normal developmental motor milestones are not achieved at all or are delayed. Astellas Gene Therapies have been developing a novel gene therapy product AT132, an AAV8 vector containing a functional copy of the *MTM1* gene, for the treatment of XLMTM, a programme they obtained by acquiring Audentes for $3 billion in 2019.

It has been far from problem-free, largely because the copy number of the virus particles they used in the trial was much higher than anyone had ever used before. Despite some early efficacy, severe side effects were seen on dose escalation in the phase 1/2 trial. Three patients from the highest dose (3.5×10^{14} vector genomes/kilogram) died of liver complications within four weeks of treatment. Dosage was dependent on weight, and those patients who died were among the heaviest treated. Astellas subsequently reported that a fourth boy had also died in the trial, which is now on clinical hold. These tragic deaths speak to the importance of "relative safety," which in this case is clearly related to the amount of virus vector infused.

uniQure, a Dutch company formed in 1998, were the first company to obtain approval for a gene therapy medicine in Europe. The European Commission granted approval in 2012 for uniQure's alipogene tiparvovec for treating a rare lipoprotein lipase (LPL) deficiency. Their gene therapy product was an AAV1 vector encoding the LPL S447X gain-of-function mutation in the *LPL* gene, injected into muscle. Lack of demand coupled with a high price ($1,000,000) led to commercial failure, and the drug was withdrawn in 2017. uniQure have continued to develop other AAV-based gene therapy products, including ones in clinical trials for Huntington's disease and haemophilia B. They have recently encountered toxicity in their Huntington's disease gene therapy trial, again a result of high vector copy numbers.

FACTOR VIII AND FACTOR IX

The ability to transfer the Factor VIII gene or the Factor IX gene for haemophilia A and B gene therapy, respectively (Chapter 2), was clearly of importance for Bioverativ, as our core business was based on Factor VIII and Factor IX replacement therapy for these two diseases. It is fair to say that gene therapy for haemophilia B (loss of Factor IX) has proceeded faster than that for haemophilia A.

Many companies have declared interest in gene therapy for haemophilia.[15] Spark Therapeutics (now part of Roche) are developing an AAV-based gene therapy for Factor IX with Pfizer, as well as their own Factor VIII programme. BioMarin in California have an AAV5/Factor VIII programme, and Pfizer hooked up with Sangamo (see Chapter 15) for their AAV6-based Factor VIII programme. Dimension Therapeutics also had a Factor VIII programme in collaboration with Bayer, but they were acquired by Ultragenyx in 2017 for $151 million and are somewhat behind the competition.

Heme B

In addition to the uniQure and Spark programmes, Freeline Therapeutics (working with Ted Tuddenham, a very experienced haemophilia expert in the United Kingdom) have reported clinical results using a Factor IX gene in a modified AAV3 vector. These Heme B studies use the Padua version of the Factor IX gene, encoding a protein with a higher specific activity than the normal protein. The uniQure product Hemgenix was approved in November 2022 and is marketed by CSL Behring at a one-time price for a one-time treatment of $3.5 million. Hemgenix is the fifth gene therapy product approved in the United States (see Biotechnology Product Time Line [pp.441–449]). The Heme B gene therapy product fidanacogene elaparvovec that Pfizer licensed from Spark is also making progress beating standard of care in their clinical trial of 45 patients and is showing good control of bleed rates. Following their phase 3 data, the FDA accepted their Biologics License Application (BLA) in June 2023.

Heme A

BioMarin received conditional marketing approval in the European Union (EU) in 2022 for Roctavia (valoctogene roxaparvovac), their Factor VIII AAV-based gene therapy for patients with no antibodies to AAV5 or to Factor VIII (i.e., antibodies that inhibit either the virus or the activity of Factor VIII). The therapy is given as a one-time infusion and costs $1.5 million in Europe. There was a sustained positive treatment effect for up to two years following a single infusion in more than 100 patients, and safety data for up to five years in further patients in a supportive trial. The therapy significantly increased Factor VIII activity levels in most patients.[16] Bleeding rates were reduced by 85% and many of the patients no longer needed Factor VIII replacement therapy. Common side effects were raised transaminase levels, indicative of immune-related hepatotoxicity. Patients will have to be followed for 15 years to monitor for long-term efficacy and safety.

BioMarin recently provided more patient data in the United States as requested by the FDA ahead of a March 2023 decision deadline. They reported, from the trial of 134 patients with severe Heme A who received the one-time treatment, that after three years the average number of bleeds per year was down by 80% and that 92% of patients were no longer getting regular infusions of Factor VIII. Roctavia finally got the green light from the FDA in June 2023, some seven months after Hemgenyx. BioMarin is pricing the drug at $2.9 million.

The major issue with both the Heme A and Heme B gene therapy approaches is the same as that bedevilling most gene therapies: obtaining stable and consistent levels of the protein produced by the transferred gene posttherapy. In the haemophilia gene therapy approaches, the lack of control over the expression of the transgene once delivered to the liver cells can result in considerable overexpression of the gene, leading to much more than 100% of the normal levels of the protein. Even if expression is at a reasonable level to begin with (say 50%), the amount of factor produced tends to go down over time. In haemophilia A for example, Pfizer and partner Sangamo Therapeutics paused their phase 3 trial of Factor VIII gene therapy to reduce dose, because some patients had Factor VIII activity of 150% or more than normal. Freeline also reported very high levels of Factor IX in some of their Heme B trial participants.

The levels are almost certainly dependent on the vector serotype and the amount of vector (in terms of virus copy number) injected initially. Another consistent—and related—issue is the immunogenicity of the vector. In a phase 2 clinical trial of the Spark Factor VIII therapy, two of seven patients receiving the highest dose of the drug had immune reactions.

We will have to wait and see what the appetite from the haemophilia community will be to gene therapy, especially after the HIV infections that occurred in patients in the 1980s associated with contaminated cryoprecipitate. The idea of a "one-time infusion and no more factor replacement" is more likely to appeal to younger patients. It is going to be difficult to thread the needle to get the appropriate dose of virus for the right expression level for a reasonable duration for each patient. It will not be well accepted if patients who have undergone gene therapy for Heme A or Heme B continue to have to stock exogenous factor in the fridge in case they bleed. The availability of weekly factor replacement therapies like Altuviiio, (see Bioverativ in Chapter 11) will also impact the decisions to undertake gene therapy. Heme B patients with inhibitors where there are severely limited clinical options are more likely to benefit from gene therapy than Heme A patients.

BLUEBIRDS, β-THALASSAEMIA, AND SICKLE CELL DISEASE

Among the first gene regions to be characterised when gene cloning methods were developed were the β-globin gene region on chromosome 11 and the α-globin gene cluster on chromosome 16. The structure of the β-globin gene "locus" was published in *Cell* in 1980 by Ed Fritsch, Richard Lawn, and Tom Maniatis[17] (see Plate 24 for the β-globin locus).

The chromosome 11 genes were cloned from the lambda genomic library that Lawn had made (Chapter 2).[f] Understanding these two regions has proven fundamental to working out how the switch from foetal haemoglobin (Hb), consisting of two γ-globin chains and two α-globin chains, is made to the two β-globin chains complexed with the two α chains that make up adult haemoglobin.[g]

Having in hand the genomic sequence of the β-globin gene locus (and the clones containing it) has also allowed scientists to work out the molecular nature of various inherited haemoglobinopathies such as β-thalassaemia and sickle cell disease (SCD) (see Box 3).[18]

BOX 3. β-THALASSAEMIA AND SICKLE CELL DISEASE

β-thalassaemia comes in several forms but is usually caused by the absence of β chains. There are nearly 350 different changes in the β-globin gene cluster that can cause β-thalassaemia and there are quite remarkable phenotypic variations in the disease, owing to all sorts of mutations in and around the β-globin gene locus or in "modifier" genes. β^0-thalassaemia occurs when no β-globin is produced. β$^+$ and β^E are forms in which a reduced amount of β-globin is produced. People can have a mix of "genotypes" (i.e., a β^0/β^E genotype, or one copy or

---

[f] The α-globin gene region on chromosome 16 was isolated and characterised a little later.

[g] Well before the gene studies, Max Perutz had determined the structure of the haemoglobin protein (for which he was awarded the 1962 Nobel Prize in Chemistry). Foetal haemoglobin has a higher oxygen affinity than normal adult Hb, as the foetus needs haemoglobin that binds oxygen more tightly to overcome the relative scarcity of oxygen in the womb.

no copies of β^0), which are less severe in phenotype. β^0-thalassaemia patients are almost always transfusion-dependent to keep enough functional red blood cells in their blood. A predominant side effect from frequent transfusions is iron overload, which can be treated but is a consistent issue for the patients.

Sickle Cell Disease

Before the globin genes were cloned, Vernon Ingram at the Cavendish Laboratory in Cambridge (UK)[h] in the early 1950s used protein chemistry and chromatography techniques to discover the protein mutation causing SCD was a glutamate-to-valine amino acid change at residue six of the β-globin chain. This change results in an altered haemoglobin, in which the abnormal β chain forms a complex with the α chains ($\alpha2\beta^{s2}$). Under conditions of low oxygen tension (hypoxia), red blood cells containing the altered Hb form a characteristic sickle cell phenotype, which gives rise to the name of the disease.[19]

SCD is of significant human importance. It is more common than CF, affecting ~100,000 people in the United States, most of whom are black African–Americans. In sub-Saharan Africa, SCD leads to high mortality in children under five, usually caused by bacterial infection. It has high morbidity, even in more affluent areas, affecting nearly three hundred million people worldwide. The sickle cell trait is under selective pressure because humans who are carriers of one copy of the HbS allele ($\beta\beta^s$) are more resistant to malaria than their normal counterparts ($\beta\beta$). The level of α-globin chains is one of the variations that modulate the disease phenotype: if fewer are produced (e.g., because of α-globin thalassaemia), then the SCD is less severe. Another influence on disease severity is the residual level of foetal haemoglobin in the patient. Conventional drugs like hydroxyurea increase foetal haemoglobin levels and are used to treat SCD (although they have only a modest effect on the disease).

Gene Therapy for β-Thal and SCD

Genetix was a gene therapy company formed in 1992 out of MIT, focused on gene therapy for SCD and β-thalassaemia, and pioneering *ex vivo* gene therapy with retrovirus vectors.[i] CD34$^+$ stem cells taken from the bone marrow of sickle cell or β-thalassaemia patients were transfected with the normal β-globin gene, in the expectation that the cells would repopulate a patient's bone marrow on re-infusion with functional erythrocytes containing the normal β-globin gene instead of the faulty copy (for SCD), and thus form a functional haemoglobin with the available α chains.

---

[h]Where the structure of DNA was determined by Watson and Crick.

[i]The company had licensed some of its vector technology from Cell Genesys (see Chapter 14).

Genetix had a checkered financing history and were ultimately refinanced by Third Rock Ventures in 2010, re-emerging as bluebird bio. Nick Leschly, the son of Jan Leschly (the professional tennis player, ex-SKB CEO) and partner at Third Rock, became bluebird bio CEO. bluebird bio continued the sickle cell and β thalassaemia studies but also embraced CAR T therapy by making a B cell–directed CAR T-cell product for multiple myeloma (see Cell Therapy, Chapter 14).

bluebird bio made and developed for each patient a genetically modified autologous CD34$^+$ cell-enriched population containing haematopoietic stem cells transduced with a lentiviral vector encoding the $\beta^{A\text{-}T87Q}$-globin gene[j] (stored in cryopreserved bags for intravenous infusion). The product, called Zynteglo, was approved in 2022 to treat transfusion-dependent β-thalassaemia (TDT). The same product is also being developed for SCD. As with other gene therapy products designed to be a "one-and-done" treatment, pricing has been set at the rather eye-popping amount of $2.8 million per infusion—although it is unlikely that every patient will have to pay this amount. The pricing of this product with respect to the recently approved gene editing approach to SCD will also be of some interest (see Chapter 15).

This number was not plucked out of the air: it is based on the estimated lifetime costs of managing a patient with TDT β-thalassaemia. However, bluebird bio have failed to get marketing access in Europe for this medicine so far despite being flexible about payment schedules and have withdrawn it. This brings into sharp relief the economics of these kinds of treatments for poorer patients who need access to these life-changing medicines.

SCD is much more common in African–Americans than in Caucasians—precisely the opposite of cystic fibrosis. The way the disease has been managed in the United States is an unfortunate reflection of that difference, although it is changing (see Box 4). Oxbryta and Adakveo may both help the change that, but at a price of more than $100,000/year for each drug (not discounted), the patients will have to rely heavily on insurance or Medicaid to cover the cost. Hydroxyurea is *underutilized* at about $16,000/year, but it is still much cheaper than the new drugs. The argument being made to support the price of these new drugs is that these medicines may reduce the cost of hospitalizations of patients with SCD, owing to anaemia or the vaso-occlusive crises that are extremely painful and debilitating, necessitating hospitalisation. The epidemiology of SCD and the affected patient

---

[j] This version of the β-globin gene codes for a β-globin protein better at oxygen binding.

BOX 4. NEW DRUGS FOR SCD AND ANAEMIA

Fortunately, as gene therapy for SCD moves forward, several new drugs for the disease have been approved. Global Blood Therapeutics (GBT), a 2011 start-up out of Third Rock Ventures, screened for compounds that increased the affinity of haemoglobin for oxygen and would therefore be likely to reverse the tendency of mutant haemoglobin cells to sickle under hypoxic conditions. One compound they identified, GBT440, binds to the valine amino acid at the N terminus of the α-globin chain and increases haemoglobin's oxygen affinity. To be effective, however, most of the patient's α chains need to be modified, meaning that high doses of drug are needed. The results of a recent phase 3 trial showed that 1500 mg daily of GBT440 (voxelotor) increased haemoglobin levels in sickle cell patients. The drug also reduced the number, but did not eliminate entirely, the vaso-occlusive crises that some SCD patients endure. The FDA approved the drug as Oxbryta in late 2019, and Pfizer announced in August 2022 that they would acquire GBT for nearly $5.4 billion. The preclinical programmes that Rob Peters and his team were pursuing at Bioverativ for SCD were licensed to GBT by Sanofi in 2021, which means that Pfizer now have them as well in their portfolio.

Novartis developed crizanlizumab, an anti-P-selectin (a kind of clotting protein) mAb. In clinical trials, the antibody reduced vaso-occlusive episodes in SCD patients by 50% after monthly infusions. Crizanlizumab (Adakveo) was also approved in 2019 although its efficacy is modest.

Novo-Nordisk acquired Forma for $1.1 billion in cash, mainly for their clinical stage pyruvate kinase R activator etavopivat, an antisickling agent.[k] Agios is also developing a compound for the same target: the EU recently approved Pyrukynd (mitapivat) as the first disease-modifying therapy for haemolytic anaemia in adults with pyruvate kinase deficiency.

population and the price of these new drugs are important to bear in mind as well when considering gene therapy as an alternative treatment. SCD is the polar opposite of CF, so what pertains to CF and contributes almost entirely to the success of Vertex will be difficult to reproduce or justify for SCD (or other diseases reflecting socioeconomic disconnects).[21,l]

GENE THERAPY FOR THE BRAIN

One of the original targets for gene therapy was congenital aromatic L-amino acid decarboxylase deficiency (AADC deficiency). This is an extremely rare

---

[k] Forma went public in 2020, raising $278 million in the fifth largest IPO of that year.[20]

[l] I am unconvinced, as are others, that the companies involved in SCD today completely appreciate that reality, much less that they know how to address it.

condition in which the absence of the enzyme, or mutations in the gene coding for it, causes a lack of dopamine and other neurotransmitters in the brain with devastating neurological consequences for the children who inherit it. The results of a small gene therapy phase 1 study replacing the gene were published in 2012, reporting the observation that the neurotransmitter dopamine was restored in seven patients between 4 and 9 years old affected by AADC deficiency. In May 2022, PTC Therapeutics obtained European approval for eladocagene exuparvovec (Upstaza), an AAV vector carrying the AADC gene injected into the putamen of the brain. The subsequent dopamine synthesis by the restored enzyme improved motor symptoms of the disease.

Voyager Therapeutics

Voyager Therapeutics were set up by Third Rock Ventures in 2013 to exploit the use of AAV subtypes and variants specifically for gene delivery to the brain. Bernard Ravina, a Parkinson's disease (PD) expert with whom I worked closely at Biogen, was hired as their CMO and Steve Paul, a neurobiologist who ran R&D at Eli Lilly, was their original CEO. One of their programmes was to use this same enzyme replacement approach that PTC used for AADC deficiency, for the treatment of PD, in which dopaminergic neurons are lost and dopamine levels become low[m] (the enzyme AADC converts L-DOPA to dopamine, thus increasing the therapeutic benefit of the drug for PD patients). This programme had started at Avigen, one of the early AAV-based gene therapy companies focusing on AAV vectors, before they were acquired for $12 million by Genzyme in 2005.

In 2018, the VY ADCC AAV2-based programme was Voyager's most advanced clinical candidate, and the FDA had granted fast-track approval for it. It was being evaluated in an open-label, phase 1b clinical trial. The programme was licensed initially to Sanofi as part of a broad gene therapy deal. Sanofi subsequently returned the programme to Voyager, who then licensed this PD programme and other programmes in Friedreich's ataxia to Neurocrine Biosciences (see Chapter 6). Unfortunately, a clinical hold was put in place by the FDA in late 2020 over some disturbing magnetic resonance imaging (MRI) findings in treated patients in the phase 2 trial. Voyager subsequently announced that Neurocrine had terminated the PD portion of their agreement.

---

[m] The reason why L-DOPA is traditionally used to treat early-stage PD.

Voyager is still in business and is developing gene therapies for Huntington's disease by attempting to knock down the expression of the expanded repeat allele that causes the disease. They are focusing in part on constructing novel AAV vectors to direct genes to the central nervous system (CNS). Alfred (Al) Sandrock, the former Biogen R&D chief, is now Voyager's CEO.

GENE THERAPY FOR MUSCLE

Sarepta: DMD Gene Therapy

As discussed in Chapter 7, Duchenne muscular dystrophy (DMD) is caused by mutations in a very large gene coding for the muscle protein dystrophin. The gene was cloned by positional cloning and mutations in the gene affect about one in 3500 to 5000 males. Dystrophin plays a key structural role in muscle fibre function: reduced protein levels cause excessive damage to muscle cells and, after a short time, they cease to function. There are several novel approaches being taken to try to treat DMD by reestablishing dystrophin production in the muscles of affected individuals. Exon skipping using antisense oligonucleotides or small molecules is one interesting approach we have already discussed (see Chapter 12).

A gene therapy approach to DMD would provide an active working copy of the dystrophin gene to the muscle cells, rather than trying to correct the consequences of the mutant version. But there are two major problems with that idea. The first is that both the gene and cDNA derived from the mRNA are very large. The DMD gene is one of the largest in the genome, so even a complete cDNA (much less the whole gene) will not fit into any of the vectors that have been developed for gene therapy so far. Secondly, getting the gene to the relevant cells in all the large and small muscles throughout the human body is a daunting prospect for any gene therapy. To mitigate the issues at least partially, Sarepta Therapeutics and other companies have made smaller genes coding for the parts of the dystrophin protein that can still function the same way as the natural protein. Sarepta have placed these "minigenes" (they call them "micro-genes") into special muscle targeting AAV vectors (AAVrh74). The micro-genes are under the expression control of a transcriptional promoter from a muscle gene (*MHCK7*) so that the micro-gene is only expressed in muscle cells. Sarepta is working with the FDA to be able to use dystrophin levels as a surrogate biomarker for their trials, rather than waiting for longitudinal disability data from the patients. The FDA granted accelerated approval for Elevidys (delandistrogene moxeparvovec-rokl) and their AAV vector for "micro-dystrophin" in June 2023 for the treatment of boys aged four to five with a confirmed mutation in the DMD gene: the gratifying result for a great deal of time and hard work.

Gene Therapy for Heart and Other Diseases

I met people from Rocket Pharma at the J.P. Morgan conference in January 2018. I was quite taken by their disease focus, which was differentiated from any of the other gene therapy companies I had met. I also liked their pragmatic style and their leadership. Rocket have both AAV- and lentivirus-based programmes.

Danon disease is a rare inherited hypertrophic cardiomyopathy caused by mutations in the *LAMP2* gene, which codes for a lysosome-associated membrane protein. Rocket's Danon programme is an AAV9 vector containing the LAMP2B transgene administered as a single i.v. infusion. The expectation is that this will restore lysosomal function in the cardiac muscle of Danon patients. Their second programme is in Fanconi anaemia, a genetic disorder of DNA repair. RP-L102 is a CD34$^+$ cell-derived product in which the *FANCA* gene is delivered by lentivirus vector into autologous stem cells, which are then returned to the immune-depleted patient.

Leukocyte adhesion deficiency type I (LAD I) is characterised by life-threatening, recurrent bacterial infections, caused by mutations in the *ITGB2* gene and preventing proper neutrophil function. The condition is almost always fatal owing to neonatal bacterial infection. The disease is usually treated by bone marrow transplantation (with all the concordant morbidity that goes with that procedure). Rocket have developed a lentivirus vector containing the *ITGB2* gene (RP-L201) for autologous stem cell therapy with the patient's own gene-treated cells. In an early-stage clinical trial, RP-L201 increased levels of CD18 correcting the deficiency that is a hallmark of the disease. Recently, all seven patients in a phase 2 pivotal trial met the preset overall survival criterion allowing the company to file for expedited approval. I think the rocket might just have taken off for this rare disease.

NEW VECTORS

Many novel AAV vectors have been engineered to facilitate manufacturing or to try to increase the size of the genes being packaged, to alter the capsid protein to avoid antibodies, or to alter tissue tropism. It was discovered

somewhat serendipitously that the AAV genome exists in curious concate-
nated forms, going to the nucleus and replicating as exogenous circles of
DNA. Engineering these DNA forms might allow gene therapy without
needing the AAV coat proteins and with the ability to deliver larger gene
loads. Atlas Ventures formed Generation Bio in 2016 to try to exploit this
observation. Geoff McDonough was recruited from SOBI to be CEO, with
Douglas (Doug) Kerr as CMO.[n] The Generation Bio platform uses these
closed-ended DNAs in cell-targeted lipid nanoparticles as vectors. They are
targeting several diseases, including Heme A, phenylketonuria (PKU), and
Gaucher's disease. This approach should allow redosing (not possible with
conventional AAV vectors) as well as better control of overdosing and gene
expression, all with a scalable manufacturing process. It is all yet to be proven.
Generation Bio went public in June 2021 raising ~$230 million in an offering
led by Wedbush PacGrow and including J.P. Morgan, Jefferies, and Cowen.

PHARMA OPTIMISM FOR GENE THERAPY

Despite recent setbacks, many in pharma and biotech remain optimistic
about current gene therapy and AAV-based vectors. For example, AskBio
(Asklepios BioPharmaceutical) is a wholly owned subsidiary of Bayer headquartered in Research Triangle Park (RTP), North Carolina, with a broad-
based AAV gene therapy platform as the nucleus of Bayer's cell and gene
therapy interests. Much of gene therapy's recent success can be attributed to
considerable advances in viral vector technologies used to deliver the genetic
material. Dyno Therapeutics, for example, are focused on making novel safer
AAV vectors and have formed multiple relationships with pharmaceutical
companies in the space. The same is true of Affinia with their interests in CF
and DMD in a partnership with Vertex, who have a CF franchise to protect.

There is so much current activity in gene therapy that manufacturing
capacity to make vectors is being exceeded. Wait times for start-ups to get
their vectors made can be 12 months or longer, despite new capacity being
built all the time. The vector production process is still largely based on the
original multi-plasmid method, which is costly, time-consuming, and inef-
ficient. ARCH Ventures recently set up a company called Resilience in San
Diego, mentioned in Chapter 11, who have raised more than $800 million

---

[n] I worked with Doug at Biogen, where he led the development of the ASO Spinraza for
SMA (Chapter 12).

supposedly to revolutionise the production of not only vectors for gene therapy but also for cell therapy.

According to Statista,° there were 341 gene therapy clinical trials in phase 1 and 1b, 733 in phase 2 and 2b, and 147 in phase 3 in 2022. Many of these are being done by companies that I have not mentioned in these pages. Several of the trials are designed to test variations of the vectors or the genes being used to treat many of the same diseases that have already been mentioned above. Of the Big Pharma players, Eli Lilly, have demonstrated a very strong commitment to gene therapy. In 2021, Lilly spent nearly $1 billion to acquire Prevail and their three AAV9-based gene therapies for neurodegeneration. They have now upped the ante by acquiring Akouos, an AAV-based company focusing on hearing loss for close to $500 million. Akouos' leading candidate targets the oterferlin (*OTOF*) gene, mutations in which cause hereditary deafness (see Chapter 19).

As you have seen part of the success of gene therapy has to do with vector design and the ability to get the vectors (and their payload) into the appropriate cells. Much of the groundwork for this capability was set as part of the development of cell therapy in which the cells themselves are the drug (e.g., the CD34 (*ex vivo*) gene therapy approach is itself a form of cell therapy). We look at the history of this technology more comprehensively in the next chapter.

REFERENCES AND NOTES

1. 2022. The trials of gene therapy. *The Economist*, August 27, 2022.

2. Friedmann T, Roblin R. 1972. Gene therapy for human genetic disease? *Science* **175**: 949–955. doi:10.1126/science.175.4025.949

3. Special report on Gene Therapy, *Scientific American*, June 1997. See progress at the time picture on pages 97 and 98.

4. From the Be the Match website (https://bethematch.org): Human leukocyte antigen (HLA) typing is used to match patients and donors for bone marrow or cord blood transplants. HLA are proteins—or markers—found on most cells in your body. Your immune system uses these markers to recognize which cells belong in your body and which do not.

5. Davis DM. 2013. *The compatibility gene: how our bodies fight disease, attract others, and define our selves.* Oxford University Press, Oxford.

6. Matsukova M, Durinkova E. 2016. Retroviral vectors in gene therapy. In *Advances in molecular retrovirology* (ed. Saxena SK), Chap 5, pp. 143–166. InTech Open, London.

°A leading provider of market and consumer data.

7. Keeler AM, Flotte TR. 2019. Recombinant adeno-associated virus gene therapy in light of Luxturna (and Zolgensma and Glybera): where are we, and how did we get here? *Annu Rev Virol* **6**: 601–621. doi:10.1146/annurev-virology-092818-015530

8. Lewis T. 2021. The quest to overcome gene therapy's failures. *Scientific American*, November 1, 2021. Tragic side effects plagued the field's early years, but researchers are finding ways to minimize the risk.

9. Ledford H. 2022. Gene therapy's comeback: how scientists are trying to make it safer. *Nature* **606**: 443–444. doi:10.1038/d41586-022-01518-0

10. Conversation with Elizabeth Ashforth, an ex-Glaxo colleague who was at GTI at the time, 2021.

11. Verma I. 2000. A tumultuous year for gene therapy. *Mol Ther* **2**: 415–416. doi:10.1006/mthe.2000.0213

12. Weiss R, Nelson D. 1999. Teen dies undergoing experimental gene therapy. *Washington Post*, September 29, 1999, page A01.

13. Yu-Wai-Man P, Chinnery PF. 2000. Leber hereditary optic neuropathy. 2000 Oct 26 [Updated 2021 Mar 11]. In *GeneReviews* (ed. Adam MP, Feldman J, Mirzaa GM, et al.) University of Washington, Seattle. https://www.ncbi.nlm.nih.gov/books/NBK1174/

14. Orchard Therapeutics website. https://www.orchard-tx.com

15. See Table 4 from Peters R, Harris T. 2018. Advances and innovations in haemophilia treatment. *Nat Rev Drug Discov* **7**: 493–508. doi:10.1038/nrd.2018.70

16. Ozelo MC, Mahlangu J, Pasi KJ, Giermasz A, Leavitt AD, Laffan M, Symington E, Quon DV, Wang JD, Peerlinck K, et al. 2022. Valoctocogene roxaparvovec gene therapy for hemophilia A. *N Engl J Med* **386**: 1013–1025. doi:10.1056/NEJM oa2113708

17. Fritsch EF, Lawn RM, Maniatis T. 1980. Molecular cloning and characterization of the human β-like globin gene cluster. *Cell* **4**: 959–972. doi:10.1016/0092-8674(80)90087-2

18. Weatherall DJ. 2001. Phenotype–genotype relationships in monogenic disease: lessons from the thalassaemias. *Nat Rev Genet* **2**: 245–255. doi:10.1038/35066048

19. Orkin SH, Bauer DE. 2019. Emerging genetic therapy for sickle cell disease. *Annu Rev Med* **70**: 257–271. doi:10.1146/annurev-med-041817-125507

20. Cameron T, Chris Morrison C. 2021. 2020 biotech IPOs shatter all records. *Nat Rev Drug Discov* **20**: 93–94. doi:10.1038/d41573-021-00019-5

21. Geraghty JA. 2022. *Inside the orphan drug revolution: the promise of patient-centered biotechnology*. Cold Spring Harbor Laboratory Press, Cold Spring Harbor, NY.

CHAPTER 14

Cell Therapy

Today, there is a great deal of excitement (and some hype) around "activated cell therapy" (ACT), particularly in cancer treatment. Most of that excitement is around chimeric antigen receptor (CAR) T-cell therapies and the amazing difference that these cell treatments have had on the lives of patients with haematological malignancies, including acute lymphocytic leukaemia (ALL), chronic lymphocytic leukaemia (CLL), and non-Hodgkin's lymphoma (NHL). CAR T cells are made from T cells taken from a patient, modified in a special way, and returned to the patient in a single procedure—designed to be a one-time treatment.[1] It is also a form of gene therapy because the T cells are modified by adding certain genes to them using lentivirus-based vectors[2] (Chapter 13). All the presently approved CAR T products are directed to haematological malignancies, like leukaemia, lymphoma, and multiple myeloma.[a]

Before discussing how T cells are being used therapeutically as "ACT," a short primer on T cells may help with understanding this exciting new therapeutic approach (see Box 1). We will then touch on the early days of cell therapy,[b] as that period sets the scene for what is happening now.

THE EARLY DAYS OF CELL THERAPY

The early cell therapy pioneers targeted solid tumours, not haematological malignancies. Investigators like Drew Pardoll and Elizabeth Jaffee at Johns Hopkins and Glenn Dranoff at the Dana-Farber Cancer Institute pioneered the idea of taking tumour cells from a patient, transfecting them with a

---

[a] It is important to remember that most cancer occurs in organs and tissues ("solid tumours"), most commonly in the breast, lung, colon, kidney, and prostate gland. CAR T-cell therapy does not address these cancers—at least not yet.

[b] I was fortunate to be on the Board of Dendreon, one of the early cell therapy companies, so I got to learn some of the science involved firsthand.

BOX 1. THE DISCOVERY OF T CELLS

T cells were discovered in the 1960s by Jacques Miller while working at the Walter and Eliza Hall Institute of Medical Research (WEHI) with Gustav (Gus) Nossal in Melbourne after obtaining his PhD in London. While in the United Kingdom, Miller worked at the Chester Beatty Institute on the Fulham Road, not far from the Chelsea football ground and Battersea Power station.[c] Miller discovered that the thymus was an important immunological organ: when it was removed from a mouse, the animal could not mount a proper immune response to an infectious agent nor reject a skin transplant from a nongenetically identical mouse. His work at WEHI with Graham Mitchell determined that there were two major classes of lymphocytes: "B cells" that made antibodies and "T cells" that carried out other immune functions like organ or skin rejection. Mitchell's lab went on to clone many of the important cytokines in the early 1980s (including GM-CSF). Both scientists helped to make WEHI an immunological research powerhouse.[d]

---

vector expressing granulocyte-macrophage colony-stimulating factor (GM-CSF) to make them immunogenic, irradiating them so that they could not grow, and then returning the altered cells to the patient. The idea was that these tumour cells would act like vaccines: the patients would mount an immune response to the altered tumour cells and to their own tumour, helping their immune system to remove it.

The researchers formed a collaboration with a company called Somatix (not to be confused with another cell therapy company SyStemix set up in 1988 by Irv Weissman), founded by Richard Mulligan from the Whitehead Institute for Biomedical Research, in the Bay Area in 1991.[e] Somatix had developed many of the tools needed to engineer cells for therapy, including different vectors to move genes around, but they were burning through large amounts of cash—more than $100 million in five years—without much to show for it except to find that GM-CSF could stimulate dendritic cells to present antigens to T cells. Cell Genesys acquired Somatix Therapy in 1997

---

[c] The same organisation where my father worked at that time. They knew each other, although not well.

[d] It is quite surprising that neither of them was awarded the Nobel Prize for their contributions to T-cell biology. Instead, the 1996 Nobel Prize in Physiology or Medicine was awarded to Peter C. Doherty and Rolf M. Zinkernagel for their discoveries concerning the specificity of cell-mediated immune defence.

[e] Mulligan trained in Paul Berg's lab and later became a Biogen Board member.

for about $83 million to create what was at the time the largest cell therapy–based biotech company. Steven (Steve) Sherwin, a former Genentech MD, was a founder of Cell Genesys in the late 1980s and the CEO, and Mark Levin from Mayfield provided some of the initial capital.[3,f]

The above process, referred to as "GVAX," was initially used on renal cell carcinoma by Pardoll and colleagues. The process was difficult to scale up, expensive, and time-consuming. Although Cell Genesys were committed to the GVAX platform, they decided to develop it for prostate cancer rather than for renal cell carcinoma or pancreatic cancer (the second being difficult to manage clinically, owing to the rapid mortality of patients). Prostate cancer phase 2 and phase 3 trials failed to drive a response in most people, although some responses were seen infrequently. Like many initially promising therapies, and despite optimism and much hard work, GVAX finally failed in 2008.

Cell Genesys lost 90% of their value after several deaths were reported in the GVAX arm of the prostate cancer trial, and Takeda backed out of a deal they had signed with Cell Genesys only six months before.[g] It was not all doom and gloom: Cell Genesys formed the subsidiary Abgenix to make human monoclonal antibodies in transgenic mice (see Chapter 4), which included panitumumab (Vectibix), a monoclonal antibody (mAb) to the epidermal growth factor (EGF) receptor that became an important oncology drug. Abgenix were bought by Amgen for $2.2 billion in 2006. Cell Genesys sputtered on a bit and quietly merged with BioSante Pharmaceuticals in 2009 in a $38 million deal.

DENDREON

GVAX was not the only prostate cancer cell "vaccine" being developed. Dendreon, based in Seattle and led by Chris Henney, the CEO (a veteran of Immunex and Icos), had also set their sights on prostate cancer.[h] The company started life as ACT (Activated Cell Therapy) in Mountain View, California, in 1992, founded by Bill Haseltine, together with Edgar (Ed) Engelman and Samuel (Sam) Strober from work they were doing at

---

[f] Sherwin founded several companies and was also on the Board of Biogen when I worked there, so I met him often.

[g] Rest assured that Takeda did not get their $50 million up-front money back![3]

[h] I joined the Board of Dendreon in late 1999 and got some firsthand experience of how difficult it is to make cell-derived products.

Stanford. The company began by trying to make a sterilisable separation device for stem cells for the Stanford blood bank, but that turned out to be a rather limited business opportunity.[4]

Two decades prior to the founding of ACT, Ralph Steinman and colleagues at Rockefeller discovered that dendritic cells were the cells that present antigens to the immune system.[i] The company, now renamed Dendreon, decided to develop methods to make a dendritic cell–based product that would present tumour antigens to the patient's immune system and invoke an immune response: a challenge much easier in principle than in practice. This would be an autologous product—that is, the cells came from the patient who was going to be treated after the cells were modified.

After Chris became CEO, Dendreon moved from Mountain View to Seattle, to labs on 1st Street with views of the water. The original investment was from HealthCare Ventures, who had started HGS among other companies. Vulcan Ventures (Paul Allen money), Kummel Investments, and Sanderling Ventures were all investors in the Dendreon Series C round. Chris was hired as CEO after a 20-minute interview with one of the HealthCare Ventures people in their tennis whites, after being flown all the way from Seattle to New Jersey. That approach turned out to be a somewhat typical way of operating for that investor group. Recall the Wally Steinberg HGS tennis story related in Chapter 8. This tennis game was not local either: it was in Boca Raton.[4]

Money and Deals

Not only did Chris raise money from venture funds but he also did a deal with Kirin Company (the Japanese beer and beverage company). He knew them from working with George Rathmann at Icos. In 2000, shortly after I joined the Board, Dendreon went public and raised $45 million in a deal led by Prudential Vector Healthcare, SG Cowen, and Pacific Group Equities.[5,j] Their pipeline consisted of the prostate vaccine Provenge and a similar product for myeloma called Mylovenge. Provenge, also called sipuleucel-T, was a

---

[i] Steinman was one of only two scientists to receive a Nobel Prize (in this case for Physiology or Medicine) posthumously: he died from pancreatic cancer between being nominated in early 2011 and the award ceremony in December 2011.

[j] I left the Dendreon Board after Chris was replaced by Mitchell (Mitch) Gold as CEO in 2003. Hans Bishop, who went on to be the CEO of Juno and is now Chairman of Sana Biotechnology and a founder and President of Altos Labs, became Dendreon's Chief Operating Officer.

product made by stimulating dendritic cells taken from a prostate cancer patient with an antigen "cassette" consisting of a fusion protein of prostate acidic phosphatase (PAP) and GM-CSF.[k] This dual protein was designed to stimulate an immune response to the endogenous PAP found on the tumour cells, with the GM-CSF providing stimulation to the dendritic cells. The fusion protein was designed to be taken up by dendritic cells using a specific receptor signal.[l] The initial clinical results and the phase 3 trials of Provenge showed a survival benefit in late-stage prostate cancer.[6] Provenge was approved by the FDA in 2010, the *first* therapeutic cancer vaccine to be approved.

Provenge was priced at $31,000 for each of three infusions over a month, making a total cost of $93,000 per course of treatment. As a result of this pricing and overly optimistic sales forecasts, Dendreon stock went on a roller-coaster ride. After rising above $50 based on the initial expected sales of Provenge, the stock then plummeted on revised and more accurate forecasts of the likely number of people that could be treated.

The relative success of Provenge cannot really be attributed to Dendreon itself. The "vein-to-vein" turnaround time of three days for Provenge was a lot faster than most of the T-cell therapies used today but, owing to logistical issues with drug supply, cost of goods, and lack of sales, the drug has not been a huge commercial success. This forced Dendreon to file for Chapter 11 bankruptcy in 2014 and restructure the business.

Valeant bought the Dendreon assets in 2015 for almost $500 million, a number mostly reflecting the cost of two manufacturing facilities in Seal Beach, California, and Union City, Georgia, near Atlanta, ending Dendreon's plan to become a contract medical organization that supplied complex cell-based products.[m] Those assets were then sold on to a Chinese conglomerate for $819.9 million in 2017.

Dendreon is still going today with modest but steady income from Provenge. They cannot be thought of as a failure. Dendreon is a notable example of the trials and tribulations that face a biotech company developing cell-based products ahead of anyone else and should be remembered for that reason. It should also not be forgotten that by 2022 more than 40,000

---

[k] Dendreon were awarded a U.S. patent for the composition of sipuleucel-T in 2001.

[l] Dendreon were planning to attack other tumours using the same general approach, by varying the antigen in the cassette.

[m] To manufacture their initial product Provenge, Dendreon had built a facility in Morris Plains, New Jersey: it was sold to Novartis for $40 million in 2012 to raise some much-needed cash.

prostate cancer patients had been treated with Provenge, an important and historic drug for the field and for the patients who have benefitted from it.

CAR T CELLS

2018's *A Cure Within* by Neil Canavan describes some of the history of T cells and T-cell therapies, including the people involved with the development of CAR T cells.[7] As Canavan recounts, it took the usual mixture of vision, persistence, fortitude, courage, and optimism to design a living T-cell product. Of course, it took much longer to become a reality than was ever imagined.

The idea was developed directly from fundamental advances in immunology (see Box 2) but also depended on other discoveries. One was the development of stem cell transplants for leukaemia and inherited immunodeficiencies (see Chapter 13), as that work demonstrated that it was possible to use live cells to reconstitute the immune system of patients after lymphoablation. Steven (Steve) Rosenberg's pioneering work at the National Cancer Institute (NCI) of taking tumours from patients, isolating T cells that had

---

BOX 2. SOME T-CELL BASICS

T cells come in several different flavours. For T-cell therapy, the distinction between $\alpha\beta$ T cells and $\gamma\delta$ T cells is important. They differ from each other depending on whether an $\alpha\beta$ T-cell receptor (TCR$\alpha\beta$) or a $\gamma\delta$ T-cell receptor (TCR$\gamma\delta$) is paired on the surface of the cell in a complex with another protein called CD3.[n] Although there are similarities in the structure of TCR$\alpha\beta$ and TCR$\gamma\delta$, they differ in several ways, the most important being that $\gamma\delta$ T cells can recognise antigens independently of the major histocompatibility complex (MHC; another set of immune system proteins), but $\alpha\beta$ T cells are "restricted" to recognizing only peptides derived from antigens that are presented in a complex with the MHC. This difference has increased interest in using $\gamma\delta$ T cells therapeutically. $\gamma\delta$ T cells tend to congregate in the gut and are thus important for gut mucosal immunity. There is also a whole different class of T cells called "T regulatory cells" or Tregs, which control the function of other T cells. The balance between T-cell activation and Tregs is vital for proper immune homeostasis, a factor in successful T-cell therapies.

---

[n] The "holy grail of immunology," as the elusive T-cell receptor was described, was cloned independently by Tak Mak at the University of Toronto and Mark Davis at Stanford University. The structures were published in seminal papers in *Nature* in 1984.[8]

infiltrated the tumours, growing them *in vitro* in the presence of a cytokine called interleukin-2, and then returning them to the patient (the so-called tumour-infiltrating lymphocyte or TIL therapy), with sometimes remarkable therapeutic effects was a critical part of the story.[1]

How CAR T Cells Began

The work of two people in the early 1990s has been recognised as the origin of CAR T-cell therapy. Michel Sadelain, a postdoc student at the Whitehead Institute in Cambridge (MA) in Richard Mulligan's lab, began to engineer T cells using retrovirus vectors. Zelig Eshhar from the Weizmann Institute in Israel also engineered T cells but with a portion of an antibody fused to part of a TCR. Zelig pursued this idea because he appreciated that the newly discovered TCRs were made up of immunoglobulin-like domains that could be adapted to bind to antigens on tumour cells and potentially kill them.

Although Zelig published the initial CAR concept in 1993,[9] the concept of engineering T cells to kill tumour cells was developed in a series of steps. Normal T cells need a co-stimulatory signal to activate properly and to transfer activation signals into the cell to drive proliferation. Sadelain introduced the CD 28 gene, which codes for a T-cell stimulator discovered by Carl June at the University of Pennsylvania, into his gene constructs to provide that co-stimulation. 4-1BB (CD137), a member of the tumor necrosis factor (TNF) receptor superfamily of genes, is another co-stimulatory gene used in CAR T constructs.

The first effective CAR T cells were developed to treat B-cell neoplasms by binding CD19, an antigen on the B-cell surface. Sadelain's group showed that a CAR engineered to bind to CD19, using an antibody fragment as the binder and CD28 as the co-stimulatory domain, could kill CD19-positive tumour cells in mice. The CAR T structure was further elaborated by including an additional signalling processing domain—shown in Figure 1 (Box 3) as signalling domain 2. It is conventionally a CD3ζ domain, one of the components of the CD3 complex, found on the surface of T cells.

Meanwhile, Phil Greenberg and Stan Riddell at the Fred Hutchinson Cancer Center in Seattle, Malcolm Brenner at MD Anderson Cancer Center, and other investigators at Memorial Sloan Kettering Cancer Center (MSKCC) were expanding autologous T cells directed against viruses such as cytomegalovirus (CMV) and Epstein–Barr virus (EBV), to help stem cell

BOX 3. CAR T CELLS

The development of CAR T cells relied on the understanding of antigen presentation by dendritic cells and the response of T cells to antigen stimulation. A CAR is a composite engineered gene that codes for a protein that binds to an antigen on a tumour cell and passes that signal to the inside of the T cell via its other domains to activate it.[10]

In principle, when those CAR T cells are put back into the patient from which they were derived (an autologous transplant), they home in on the antigen on the tumour cells in the patients and kill them without killing the normal cells (or the patient).

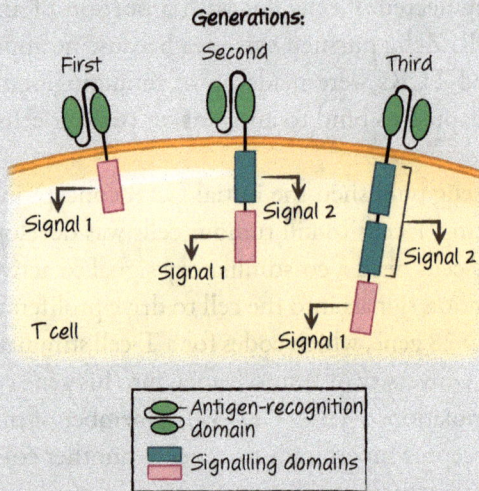

Figure 1. Chimeric antigen receptors on T cells (CAR T). (Figure adapted from Jackson et al., *Nat Rev Clin Oncol* **13**: 370 [2016], with permission from Springer Nature @ 2016.)

transplantation patients infected with these viruses as the lympho-ablation step in the transplant procedure wiped out their immunity to these viruses. These groups developed "T-cell expansion protocols" that allowed specific sets of antiviral T cells to be multiplied sufficiently *in vitro* so that they could be used to augment immunity in transplant patients.°

° This work was obviously highly relevant to the manufacturing of engineered tumour-specific T cells: a manufacturing protocol for CAR T cells was published in 2009.[11]

First-in-Human Studies

The first CAR T-cell trials in humans were performed by Carl June and his colleagues while he was at the National Naval Medical Center in Bethesda in HIV patients where CD8 T cells (the "effector" T cells that kill other cells) were engineered to attack HIV-infected CD4 cells in the host. These "first-generation" CARs, which lacked the co-stimulatory domains, were directed at gp120, an HIV protein expressed on the surface of a patient's infected CD4 cells. These clinical trials were done in collaboration with Cell Genesys and showed some efficacy, but the development of effective antiviral drugs such as protease and reverse transcriptase inhibitors led Cell Genesys to lose interest in the cell therapy approach to HIV.

"EXPLOITING" CAR T

Two major companies were set up early to exploit CAR T technology: Kite Pharma and Juno Therapeutics. A comparison of the two both scientifically and financially is of some interest here. It is also an entertaining story about what happens when investors get slightly overenthusiastic about the prospects of innovative science.

To summarise the business outcome first: Kite were acquired by Gilead Sciences in August 2017 for $11.9 billion in cash. Following suit, Juno were bought by Celgene for $9 billion in early 2018. Celgene in turn were acquired by BMS in 2019 for $74 billion primarily for the cash flow from Revlimid but also for Juno themselves.[P]

Kite Pharma

Kite Pharma were founded in 2009 by Joshua Kazam and Arie Belldegrun, an Israeli surgical oncologist. David Chang became R&D chief and CMO.[12] By this time, neither Cell Genesys nor Somatix nor Systemix—the first cell therapy companies—existed. Some of the old Cell Genesys gene vector intellectual property (IP) was folded into Kite from the University of California San Francisco (UCSF).[3] The first clinical data from Kite emerged in 2013, when the company announced that their CD19 CAR produced robust and durable responses in patients with diffuse large B-cell lymphoma

---

[P] These purchases will be covered more in Chapter 17, because these pricy transactions have set an "unfortunate" precedent for the industry.

(DLBCL), a form of non-Hodgkin's lymphoma (NHL), the most common haematological cancer in which cells are positive for B-cell markers such as CD19 and CD20.[q] Kite subsequently formed a *Cooperative Research and Development Agreement* with the NCI to help them develop their CAR-T CD19 directed products, largely to facilitate patient access.

On the commercial side, Kite did several deals to increase their global footprint. They formed a relationship with Amgen in 2015 for their next-generation products. That same year, they acquired the "T-Cell Factory B.V.," a privately held Netherlands biotechnology company, for €20 million, to form Kite Pharma EU and to get access to additional T-cell receptor (TCR)-related IP. This acquisition also gave Kite access to European manufacturing facilities. In January 2017, the company signed a lucrative strategic partnership with Daiichi Sankyo to develop and commercialize their CAR T-cell therapy in Japan. That same month they formed a joint venture with Fosun Pharmaceuticals (Shanghai) to do the same in China.

The FDA approved Kite Pharma's therapy, Yescarta (axicabtagene ciloleucel), in October 2017, making it the first CAR T therapy approved by the FDA for the treatment of adult patients with relapsed or refractory large B-cell lymphoma (LBCL). The FDA further approved Yescarta's use in adult patients with LBCL refractory to other chemoimmunotherapy in April 2022.[r]

In 2020, the FDA also approved Tecartus (brexucabtagene autoleucel, a T cell–enriched version of Yescarta) for the treatment of adult patients diagnosed with mantle cell lymphoma (MCL), another B-cell malignancy. Tecartus was given accelerated approval and was granted priority review, breakthrough therapy status, and orphan drug designation: the whole nine yards for a rare disease therapy.

Juno Therapeutics

Juno were set up in 2013, a few years after Kite Pharma. They were based on the work done at MSKCC by Michel Sadelain and Isabelle Rivière (Michel's

---

[q]DLBCL accounts for some 30%–40% of cases of NHL. There are two major biologically distinct molecular subtypes of DLBCL: germinal centre B-cell (GCB) and activated B-cell (ABC) types. ABC DLBCL is associated with substantially worse outcomes when treated with standard chemoimmunotherapy called CHOP or R CHOP.[13]

[r]Although initial responses to Yescarta are high and a few patients demonstrate long-term survival, most patients eventually relapse.

wife, collaborator, and co-conspirator, and an excellent molecular immunologist in her own right). Renier Brentjens, who did the original mouse work with CD19-directed CAR and persuaded MSKCC to be interested in it therapeutically for humans, was a founder.[s] Phil Greenberg and Stan Riddell from the Fred Hutch and Michael Jensen from the Seattle Children's Hospital were also co-founders.

Apart from a gene-editing deal signed in 2015 with Editas and the acquisition of AbVitro in 2016 to get single-cell sequencing capability, by far the most important deal for Juno and CEO Hans Bishop was a 10-year deal with Celgene in 2015 worth a billion dollars. Not only did this provide some security of funding for their R&D over several years, but it also led directly to the acquisition of the company by Celgene in early 2018.

The first product to come out of Juno was lisocabtagene maraleucel (liso-cel, Breyanzi), the company's autologous anti-CD19 CAR T therapy. FDA-approved in 2021, the product had a defined composition of purified $CD8^+$ and $CD4^+$ CAR T cells and could be used for the treatment of adult patients with relapsed or refractory DLBCL, following at least two previous other therapies.

A Third Player

There was a third player in this CAR T space that was not a new cell therapy company. Carl June had set up a substantial cell therapy lab at the University of Pennsylvania. That group developed a CD19-directed CAR like Kite's and Juno's but included the *4-1BB* gene rather than the *CD28* gene as the co-stimulatory domain. The University of Pennsylvania licensed the technology to Novartis in 2012 in an R&D agreement that founded Penn's Center for Advanced Cellular Therapies (CACT).[t] In April 2017, tisagenlecleucel (Kymriah) received breakthrough therapy designation from the FDA for the treatment of relapsed or refractory DLBCL, like the Kite and Juno products. Kymriah was FDA-approved for use in ALL and DLBCL: the first CAR T product to be approved by the FDA for both diseases.

The treatment of patients with these three CD19-directed CAR T products in the real world is revealing that cytokine release syndrome (CRS)—a

---

[s]Brentjens led the first ALL trial with a CD19-directed CAR T-cell product.[10]

[t]The agreement was partly driven at Novartis by John Delyani, now the CBO at Repertoire Immune Medicines.

dramatic side effect that can be fatal—is seen more often with Kymriah than the other two products, possibly because of the different co-stimulatory domains.[u]

Another differentiator of the products that makes a difference in the marketplace is the ability to manufacture the product. It is generally a two-to-three-week process to take the cells from the patient, engineer, and grow them before infusing them back into the patients. The process is time-consuming and expensive. Yescarta and Breyanzi seem to behave much the same clinically but the turnaround time to go from patient to cells and back to patient is 17 days for Yescarta versus 24 days for Breyanzi. Although that does not seem like a big difference, this could lead to patient differentiation depending on the status of their disease, especially for more aggressive forms of DLBCL and ALL that can kill patients quickly.

In general, Yescarta and the related drug Tecartus should be preferred to Breyanzi and Kymriah, but we will see what happens in clinical practice. Being able to deliver the product will be of paramount importance. The cost of these medicines is not a differentiator: each is close to $400,000/year without discounts. Put into the context of potentially lifesaving medicines, this number has been accepted as being quite reasonable by the various drug pricing organizations.

2seventy bio Inc: CAR T for Myeloma

The gene therapy company bluebird bio, which is developing therapies for β-thalassaemia and sickle cell disease (Chapter 13), spun out their oncology interests into a new company called "2seventy bio" in early 2021. bluebird bio had a reputation for being slightly edgy. The origin of the new company name illustrates this: "2seventy bio" is derived apparently from the fact that 270 mph is the speed of turning a human thought into action.[v] It is an awkward name to either remember or to write down.

Nick Leschly is the CEO of 2seventy bio, and Andrew Obenshain now leads bluebird bio in continuing to progress that company's sickle cell and β-thalassaemia products (Chapter 13). 2seventy bio is further developing Abecma (idecabtagene vicleucel), their bluebird bio–derived B-cell

---

[u]CRS is increasingly well-managed with anti-IL6 antibodies and steroids, but it is still complicated and even dangerous for some patients.

[v]I am not an expert in branding, but that choice seems rather specious to me. It depends on the thought, I guess. It is more like the title of a record album like Eric Clapton's *461 Ocean Boulevard* than the name of a cell therapy company. I remember that album rather better.

maturation antigen (BCMA) CAR T therapy for multiple myeloma, directed at BCMA found on B cells making antibodies. In the clinical trial of more than 100 patients, there was an overall response rate of 72%, and at least 65% of patients had a complete response lasting 12 months or more. As with other CAR T-cell therapies mentioned above, CRS was a notable side effect.[w]

Bristol Myers Squibb, who were collaborating with bluebird bio on Abecma, received the first FDA approval for a cell therapy product in multiple myeloma in 2021. Abecma was approved for the treatment of adult patients with relapsed or refractory multiple myeloma after four or more prior lines of therapy.

BMS and 2seventy bio are not alone in this therapeutic area. Legend Biotech, a Chinese company, licensed their anti-BCMA CAR T to J&J for $350 million in 2017. They did a huge IPO in the summer of 2020 raising $424 million. The FDA approved their product CARVYKTI (ciltacabtagene autoleucel) for the same indication as Abecma in 2022, giving patients with relapsed/refractory multiple myeloma two CAR-T therapies to choose from for treating end-stage disease.

CAR T CELLS AND SOLID TUMOURS

CAR T cells have been used primarily for haematological cancers, but there is a good deal of work being done to adapt the approach to solid tumours (where there is far more biological heterogeneity than in blood cancers, and the tumour microenvironment is not conducive for T-cell killing of tumour cells). "Armored CARS" are so-called because they are engineered not just to bind to an antigen on the surface of a tumour cell but also to release a protein, such as a cytokine, which will increase the proliferation of adjacent T cells or prevent the suppression of T-cell function (a hallmark of many solid tumours).[10,14]

The choice of antigen is crucial. There are many cell-surface antigens on tumour cells that are not seen on normal cells, so it is obvious to choose one or more of these as the target of CAR T cells. That choice can still be fraught with difficulty: Tmunity Therapeutics stopped a phase 1 clinical trial of a prostate-specific membrane antigen (PSMA)-directed CAR T in prostate cancer because two patients died from immune effector cell–associated neurotoxicity syndrome (ICANS), a serious side effect that is also sometimes seen in CAR T treatments for DLBCL.

---

[w] 2seventy bio plans to bring forward other CAR T candidates in lymphoma and leukaemia, as well as several solid tumours. Their R&D activities have been taken over by Regeneron.

Another challenge is increasing the persistence of the CAR T cells and avoiding "exhaustion" (i.e., cessation of killing activity), especially when CD19 B cells, for example, continue to be made as disease progresses.

There are simply too many companies to mention here that are pursuing these approaches. And there was, at least until very recently, a huge appetite amongst investors to invest in pretty much any company that claimed to be in the immunotherapy and ACT space (see Chapter 17).[15]

CAR T "OFF THE SHELF"

Among many approaches (see Box 4), one of the most exciting being taken is to engineer CAR T cells that can be used "off the shelf" for any patient, instead of being taken from a particular patient, engineered, and put back into the same patient. These so-called allogeneic cells require some quite sophisticated gene editing technologies for them to work. We will get into some of these editing approaches in the next chapter. An interesting example of one such company is the aptly named Allogene.[12] This company was started by the founders of Kite in 2018, to create allogeneic CAR T products. Pfizer had a 25% stake in the company at their inception, in exchange for some

---

BOX 4. THE LATEST CAR T-CELL COMPANIES

An interesting new CAR T company that emerged in 2022 is Capstan Therapeutics, based on work from Jon Epstein's lab at the University of Pennsylvania. In a kind of *in vivo* gene therapy approach, they directed a CAR T mRNA-containing lipid nanoparticle to CD5 on T cells. Those T cells took up the mRNA and expressed the hybrid protein it coded for—effectively creating CAR T cells *in vivo*. Capstan raised an initial $63 million from venture funds, including Novartis Ventures, OrbiMed, Vida Ventures, and RA Capital. More than $100 million was raised in their series A round led by Pfizer Ventures, Leaps by Bayer, and others. Founders include some by-now-familiar University of Pennsylvania/ Johns Hopkins names including Carl June, Bruce Levine, Drew Weissman, and Hamideh Parhiz, experts in cell therapy and mRNA technology. Laura Shawver, an experienced biotech CEO who was President of SUGEN Inc. and CEO of Synthorx (sold to Sanofi), is their CEO. Adrian Bot, the former CSO at Kite, is now the CSO of Capstan, and the chief technology officer is Priya Karmali. You could consider Capstan the new "*in vivo* CAR T paradigm." The key, as in gene therapy, will be getting enough of the mRNA you want to express into the appropriate T cells. As T cells come in many flavours, some antibody or other targeting molecules may be required.[16]

CAR T rights they owned. Owing largely (it seems to me) to the success of Kite financially, the company raised $300 million in their series A round led by Texas Pacific Group (also a Kite investor) and Gilead Sciences, who had bought Kite. Allogene's IPO in October of 2019 raised $324 million, and the company, even at such an early stage, had a unicorn market cap of $3 billion. We talk a little more about these kinds of valuations for very early-stage companies in Chapter 17.

T-CELL "PRODUCTS" OTHER THAN CAR T

TCR T Cells

Another approach being taken by many companies[1] is to engineer T cells to express TCRs that recognise peptides from tumour-specific antigens, presented in the context of that patient's specific human leukocyte antigen (HLA) protein (the MHC). T cells normally "see" an antigen on the surface of a cell (e.g., a virus-infected cell) not as a defined protein, but as peptides derived from the protein, bound to one of the patient's HLA proteins (see Box 2). This is a robust way of amounting a cellular immune response to pathogens because the foreign proteins that are recognised can be from the inside of the cell as well as from the outside.

In short, engineered TCR T (or TCRT) cells use naturally occurring receptors to recognise proteins found inside tumour cells. Just as seen with CAR T cells, there are many companies trying to develop TCRT cells, including Kite and Juno.[1] Like many others, Kite is targeting tumour antigens called NY ESO and MAGE, expressed on many tumour types. TCRT cells are not without their issues, the main ones being persistence, efficacy, and safety. As seen before, choosing the wrong antigen can have devastating effects, as even small amounts of a peptide recognised by a high-affinity TCR in an engineered T cell can kill normal cells expressing the peptide[1]. A variation on this theme, introduced early in this chapter, is to use γδ T cells to kill tumours where the MHC restriction does not come into play (see Boxes 2 and 5).

BOX 5. γδ T-CELL COMPANIES

GammaDelta Therapeutics were set up by Adrian Hayday in London in 2016 and supported by Abingworth Ventures (now part of the Carlyle Group, a huge financial powerhouse). The objective of GammaDelta is in their name: kill tumour cells by directing γδ T cells to cell-surface tumour antigens. Takeda Pharmaceuticals agreed to join Abingworth in investing, to the tune of $100 million, giving Takeda an exclusive option to acquire the company later at a predetermined price, if certain milestones were met.[x] Takeda exercised their option in 2021.

As part of their R&D activities prior to acquisition, GammaDelta spun out a separate company called Adaptate Biotherapeutics in 2019, focusing on developing bispecific antibodies (T-cell engagers), designed to modulate the activity of variable delta 1 (Vδ1) gamma delta ($\gamma\delta$) T cells and to activate them in tumours. As with GammaDelta, this was another Takeda "build-to-buy" deal. Takeda exercised their option to acquire Adaptate at about the same time as they announced that they were buying GammaDelta.

The combination of GammaDelta's cell therapy–based platform and Adaptate's antibody-based $\gamma\delta$ T-cell engager platform puts Takeda at the forefront of $\gamma\delta$ T-cell therapy. Yet it remains to be seen whether these cells have an advantage over other T-cell approaches that employ the more conventional $\alpha\beta$ TCR pairs (Box 2).

Tumour-Infiltrating Lymphocytes (TILs)

As mentioned earlier in this chapter, "nonengineered" approaches to T-cell therapy are also being explored. There has been considerable interest in the use of TILs from melanoma and other tumours—ongoing work from the Rosenberg lab at the NCI. TILs are taken from the patient, the cells are expanded with different cytokine cocktails, and then the cells are returned to the patient. It is still unknown what are the characteristics of the TILS that have the antitumour effect, but it is known that they do not usually make up a big proportion of the TIL population—so billions of cells are needed to see any effects.

Multiclonal T Cells

Multiclonal T cells (MTCs) are also nonengineered cells that are being used to treat cancer. Repertoire Immune Medicines was formed from a combination of Torque Therapeutics and Cogen Immune Medicines by Flagship Pioneering in 2020 to pursue this approach. I worked there for 18 months as Chief Scientific Advisor with John Cox as CEO and several others from the old Bioverativ team. They were one of only a very few specialist MTC companies.[1] In this process, patients are apheresed and their T cells and the dendritic, antigen-presenting cells separated. Peptides from representative

---

[x] In venture capital parlance, a "build-to-buy" deal.

tumour antigens are mixed with the dendritic cells, which are then added back to the T-cell fraction to stimulate the proliferation of T cells specific to the antigens from which the peptides were derived. After a period of growth and further stimulation, the cells are returned to the patient.

A good example of this approach was a study in HPV 16–positive head and neck cancer, where peptides from the key oncogenic proteins E6 and E7 were used to stimulate the proliferation of T cells that recognise those proteins that cause the tumour cells to proliferate. Unfortunately, Repertoire and others have seen only modest clinical responses targeting these antigens in this cancer type using this approach. One interpretation of the existing modest clinical activity is that it is a "cell numbers" game—that is, the number of the appropriate and rare killing T cells that are infused is very important, and the more that can be infused without causing side effects the better. Repertoire were probably not using enough of the right cells.

Natural Killer (NK) Cells

There is also interest in using cell types other than T cells to kill tumour cells. Natural killer cells, or NK cells, do that for a living. In 2019, Kevin Pojasek and I with backing from SV Health Investors started a company called Catamaran Bio to exploit NK cells for solid tumour therapy.[17] We were neither the first nor the last company set up to exploit NK cells, but we believed we had the most effective approach to engineer them and to grow them (see Box 6). NK cells can be obtained from peripheral blood from donors, from stem cell banks, or from induced pluripotent stem cells (iPSCs). Donor NK cells can be used to treat many patients, as they are not recognised as foreign by the immune system: they are naturally "allogeneic."

BOX 6. OTHER NK CELL COMPANIES

CAR NK cells have been used in the clinic successfully for leukaemia.[18,19,20] There are several other emerging companies taking the same approach as Catamaran to focus on CAR T and NK cells or just NK cells, including the now-public companies Fate Therapeutics (set up by John Mendlein several years ago) and Nkarta in South San Francisco. Shoreline Biosciences on Roselle St. in Sorrento Valley (right next to where SGX were based) is another private NK-based company of the same vintage as Catamaran, at which Kleanthis Xanthopoulos is the CEO.

Kevin and I had a memorable trip to Minnesota in the early days of incubating the company. We persuaded Branden Moriarity, an Associate Professor at the University, to become involved, and on that trip, the company was named Catamaran Bio. The technology platform Catamaran have built is called "Tailwind" to go with the nautical name and logo. The differentiating feature of Catamaran is the use of transposons rather than virus vectors to genetically engineer the cells and to use a dominant-negative protein to affect the tumour microenvironment. They have some very interesting solid tumour programmes, and high-quality VC and corporate VC investors. Unfortunately this company failed owing to being unable to finance its expensive R&D.

The success of CAR T and other modern cell therapies relies on the ability to engineer the cells to direct them to an antigen or to improve the way they behave *in vivo*. In the next chapter, we will see that engineering cells using gene editing techniques, especially those that have become available recently, is now extremely sophisticated.

REFERENCES AND NOTES

1. Harris T, Fitzgerald J, Coyle AJ, Reiss T, Sauer K. 2022. T cell therapies for cancer. *GEN Biotechnol* 1: 262–289. doi:10.1089/genbio.2022.0011

2. Sadelain M. 2017. CD19 CAR T cells. *Cell* 171: 1471. doi:10.1016/j.cell.2017.12.002

3. Conversation with Steve Sherwin, 2021.

4. Conversation with Chris Henney, 2021.

5. Dendreon Form S-1, as filed with the Securities and Exchange Commission on November 17, 2014.

6. Kantoff PW, Higano CS, Shore ND, Berger ER, Small EJ, Penson DF, Redfern CH, Ferrari AC, Drelcei R, Sims RB, et al. 2010. Sipuleucel-T immunotherapy for castration-resistant prostate cancer. *N Engl J Med* 363: 411–422. doi:10.1056/NEJMoa1001294

7. Canavan N. 2018. *A cure within: scientists unleashing the immune system to kill cancer.* Cold Spring Harbor Laboratory Press, Cold Spring Harbor, NY.

8. Williams AF. 1984. The T-lymphocyte antigen receptor—elusive no more. *Nature* 308: 108–109. doi:10.1038/308108a0

9. Hwu P. 2015. CCR 20th anniversary commentary: chimeric antigen receptors—from model T to the Tesla. *Clin Cancer Res* 21: 3099–3101. doi:10.1158/1078-0432.CCR-14-2560

10. Jackson HJ, Rafiq S, Brentjens RJ. 2016. Driving CAR T-cells forward. *Nat Rev Clin Oncol* 13: 370–383. doi:10.1038/nrclinonc.2016.36

11. Sadelain M, Rivière I, Riddell S. 2017. Therapeutic T cell engineering. *Nature* **545:** 423–431.

12. Wright R. 2020. Allogene Therapeutics—the juggernaut built on Kite's flight. *Life Science Leader*, January 31, 2020.

13. CHOP and R CHOP: Immunochemotherapy regimens consisting of <u>C</u>yclophospha-mide, <u>H</u>ydroxydaunorubicin hydrochloride (doxorubicin hydrochloride), Vincristine (<u>O</u>ncovin), and <u>P</u>rednisone with and without <u>R</u>ituxan used to treat both indolent and aggressive forms of non-Hodgkin lymphoma.

14. Majzner RG, Mackall CL. 2019. Clinical lessons learned from the first leg of the CAR T cell journey. *Nat Med* **25:** 1341–1355. doi:10.1038/s41591-019-0564-6

15. Ledford H. 2023. The race to supercharge cancer fighting T cells. *Nature* **613:** 626–628. doi:10.1038/d41586-023-00177-z

16. 2022. Starry syndicate powers a $165 million bet on in vivo CAR -T therapies enabling Penn spinout Capstan to set sail. *FIERCE Biotech*, September 22, 2022.

17. 2020. SV Health–incubated Catamaran debuts with $42M to develop CAR NK pipe-line. Emerging company profile. *Biocentury*, November 23, 2020.

18. Siegler EL, Zhu Y, Wang P, Yang L. 2018. Off-the-shelf CAR-NK cells for cancer immu-notherapy. *Cell Stem Cell* **23:** 160–161. doi:10.1016/j.stem.2018.07.007

19. Leslie M. 2018. New cancer fighting cells enter trials. *Science* **361:** 1056–1057. doi:10.1126/science.361.6407.1056

20. Liu E, Marin D, Banerjee P, Macapinlac HA, Thompson P, Basar R, Kerbauy LN, Overman B, Thall P, Kaplan M, et al. 2020. Use of CAR-transduced natural killer cells in CD19-positive lymphoid tumors. *New Engl J Med* **382:** 545–553. doi:10.1056/NEJMoa1910607

CHAPTER 15

Gene Editing

As previously discussed (Chapter 13), the human β-globin genes were among the first human genes to be cloned.[a] The regulation of this gene locus has subsequently been the subject of much wonderful work, particularly by Stuart Orkin's lab at the Boston Children's Hospital. Among Orkin's and others' studies was the discovery of a region (called an "enhancer") in the second intron of the *BCL11A* gene, specifically controlling the switch from γ-globin to β-globin synthesis.[1] This finding led directly to the idea that "editing out" this intronic region might be a way to engineer CD34 stem cells to produce erythrocytes where γ-globin synthesis remained on and possibly lead to a treatment for diseases in which β-globin gene function is compromised.

Various gene-editing tools, including zinc fingers, transcription activator-like effector (TALE) nucleases (TALENs), and CRISPR-Cas9, are available to do this kind of "gene editing." We will look at each of these in turn, along with the companies that have developed and exploited them. I hope it will become apparent how different technologies are being used to make gene edits, not only for sickle cell disease (SCD) and β-thalassaemia but also for several other diseases. The concept of base editing (i.e., the exchange of single nucleotides in genomic DNA rather than making less precise edits with earlier generation gene-editing tools) naturally follows on.

---

[a] In the β-globin gene cluster on chromosome 11, the embryonic, foetal, and adult globin genes are switched on and transcribed into mRNA sequentially through development. Foetal haemoglobin (HbF) consists of two γ-globin chains coupled with two α chains. Neonates switch from the synthesis of γ globin to β globin, and the adult form of Hb (two β chains and two α chains) replaces the foetal form. Hereditary persistence of foetal haemoglobin (HPFH) is a rare hemoglobinopathy that results in patients having up to 30% of HbF in their blood.

SANGAMO BIOSCIENCES—ZINC FINGER NUCLEASES

A "zinc finger" is a section (domain) of a protein that can bind DNA. It is characterised by containing a zinc ion coordinated to a pair of cysteine amino acids and a pair of histidines, with an inner hydrophobic core. Zinc fingers can be linked in tandem linearly and recognise nucleic acid sequences of different lengths and sequences. Specific recognition of DNA or RNA sequences is defined by different zinc finger proteins that target specific genes. Zinc finger–based "gene-editing proteins" are made by fusing tandem zinc finger domains with effector domains, such as nucleases or integrases, to form chimeric effector proteins.[2]

Two major players were developing zinc finger technology in the mid-1990s: Aaron Klug's lab at LMB in Cambridge (UK)[b] and Sangamo Biosciences on the west coast. Edward (Ed) Lanphier formed Sangamo BioSciences in 1995 to focus on the regulation of gene expression by using zinc finger technology. Ed had been at Eli Lilly, Synergen (before they got bought by Amgen), and Somatix. While at the latter, he became aware of the work of Chandrasegaran (Chandra) Srinivisan, who had been fusing zinc fingers to the restriction enzyme DNA nuclease Fok1, derived from a *Flavobacterium* that infected plants. This was an unusual but not unique restriction enzyme, as it recognises a pentameric sequence (GGATG) but then cleaves the DNA 10 base pairs further downstream of that recognition sequence. Chandra's fusion created a zinc finger nuclease (ZFN) that could bind to DNA and then cut it at a specific sequence (Box 1).

Lanphier licensed Chandra's work and, with $750,000 from friends and family, started Sangamo Biosciences. Given that this was a gene-editing technology platform still in search of valuable applications, this was a ballsy move. Sangamo started in Richmond, California, in a lacklustre building shared with Pixar. Chandra joined the Board of Sangamo, as did Herb Boyer and Bill Rutter. Other investors included JAFCO and Lombard Odier.

The company went public in 2000 in an IPO led by Lehman Brothers, Chase H&Q, ING Barings, and William Blair & Co, raising about $50 million. The prospectus was very broad, citing potential applications of the technology for plants, animals, humans, and industrial applications, using what was called a "universal gene recognition" platform for controlling gene

---

[b]Klug was awarded the Nobel Prize in Chemistry in 1982 for his structural studies of protein and nucleic acid complexes.

BOX 1. ZFNs AND TALENs

Each zinc finger domain binds to three nucleotides, so specific ZFNs can be made by linking in tandem five or six ZFNs. To make a specific cut two different ZFN constructs are made: one targets nucleotides on one DNA strand and the other binds to nucleotides on the complementary strand. Each of the ZFNs contains one half of the FokI nuclease, so DNA is cleaved only when the two ZFNs are bound at the same time to the target double-stranded DNA (dsDNA) sequence.[3]

TALEs were originally found in *Xanthomonas*, a pathogenic bacterium that infects plants. On entering the nucleus of plant cells, TALEs bind to target promoters and induce gene expression. TALE proteins consist of three functional domains: a bacterial secretion signal, a nonspecific DNA-binding domain, and a domain that interacts with plant transcription Factor IIA. It is in the DNA-binding region where the programmability lies. It consists of a variable number of 33–35 amino acid tandem repeats, with the two amino acids located at positions 12 and 13 of each repeat defining its specificity (the so-called repeat-variable di-residues [RVDs]; see Fig. 1 for details). The specificities of all possible RVD combinations have been decoded, allowing TALEs specific for single nucleotides or several nucleotides to be made. By re-arranging the repeats, the DNA-binding specificities of TALEs can be changed at will. As with zinc fingers, attaching the catalytic domain of FokI nuclease to the TALE protein nucleotide recognition sequences creates proteins (TALENs) that can bind in a sequence-specific manner to DNA and cleave it.[4]

One of the advantages of TALENs is that they are very precise. The nucleotide recognition sequence is long (30 nucleotides) so that there is much less likelihood of "off-target" gene modifications occurring than with ZFNs or CRISPR-Cas9-directed editing.

---

expression. Based on this breadth, Sangamo signed up 17 biotech and pharma companies as collaborators, and the company received several government grants.

Acquiring Gendaq

The IPO allowed Sangamo to acquire Gendaq, a company set up by Aaron Klug and Yen Choo in London in 1999, doing much the same work as Sangamo. They had recently shown, for example, that the effects of an oncogene in a transformed cell line could be overcome by blocking the gene with a zinc finger protein. By buying Gendaq in a stock-for-stock deal, Sangamo acquired a research team of 16 scientists, 24 patent applications in the ZFN field, and more than $6 million in cash.

The Only Gene-Editing Game in Town

The question for the company became: which gene-editing game do you want to play to win? When I first met them, it was clear that the technology was versatile and relatively straightforward, but it was unclear what applications were their focus. They had collaborations with Dow AgroSciences for plant applications, with Sigma-Aldrich for reagent supply, and they had their own therapeutic programmes.

Despite ZFNs being used to generate both recombinant organisms and plants, the early 2000s was a lean time for Sangamo Biosciences. Many good papers were published and there were prominent members of the gene-editing community in the Sangamo team. It included Fyodor Urnov, a Russian expatriate, now something of a doyen of gene editing generally, Michael Holmes from Bob Tjian's lab at UC Berkeley, and Philip Gregory, an English postdoc who is now the CSO at 2seventy bio after being in the same position at bluebird bio. He is now Head of Cell Therapy at Regeneron.

Sangamo acquired the AAV gene therapy firm Ceregene in 2013, signalling that gene therapy and gene editing were going to be the therapeutic focus of the company. They had an impressive docket of intellectual property (IP) covering their technology including that from Ceregene, consisting of more than 500 patents or patent applications. One of their clinical programmes was in HIV in which they designed a ZFN to edit out the *CCR5* gene from CD4 lymphocytes (Box 2).

This *CCR5* deletion experiment was, contrary to popular belief and despite the view of the CRISPR-Cas9 fraternity (see later), the first-ever gene-editing study done in humans. The paper published in the *New England Journal of*

BOX 2. HIV GENE EDITING

CCR5 (C-C chemokine receptor 5) is the human receptor protein that allows HIV to infect CD4 T cells. There are people who have a mutation in the gene (*CCR5* Δ32) that removes part of the gene, so the normal receptor is not made, which can compromise HIV infection. About 100 million people across the world carry the deletion mutation in the *CCR5* gene, and those rare individuals with two copies of the deleted gene are resistant to HIV.

An autologous T-cell product was made by using zinc finger technology to engineer patient cells to remove the *CCR5* gene, and then return the cells to the patient. In a collaboration with Carl June at the University of Pennsylvania, Sangamo showed that the infusion of autologous CD4 T cells in which the *CCR5* gene was rendered permanently dysfunctional by ZFNs was safe.

Medicine in 2014[5] showed further that the *CCR5* gene knockout recapitulated the natural *CCR5* gene mutation, providing resistance to HIV infection. It was clear, though, that the market for this kind of cell product for treating HIV was going to be redundant, owing to the success of antiretroviral therapies. It was nevertheless an important first for the technology and for Sangamo.

On the back of this trial and the promise of the many other therapeutic opportunities that could be derived by gene therapy and gene editing, Sangamo raised $100 million in a follow-on offering led by J.P. Morgan in 2014, to support clinical development and manufacturing.

Sangamo also formed two important haematology alliances, one in haemophilia with Shire (which subsequently became the Pfizer Factor VIII gene therapy deal; Chapter 13) and one with Biogen-Idec in 2014 to use ZFN technology to treat SCD and transfusion-dependent β-thalassaemia (TDT).

SANGAMO AND BIOVERATIV

Based on the work of Daniel Bauer and Stuart Orkin, the focus of the Biogen deal (which Bioverativ inherited) was to knock out the expression of the *BCL11A* gene either by targeting the coding sequence (exon 2) of the gene or by targeting the intronic region of the gene where the erythroid-specific enhancer was located. I was aware of the project while I was at Biogen, but obviously became much more familiar with it when the programme was rolled into the Bioverativ spinout. It was very exciting to be involved in a groundbreaking CD34 stem cell therapy programme.

Ed Lanphier retired in 2016, handing the CEO reins over to Alexander (Sandy) Macrae, ex-GSK and Takeda, and a Scottish physician. I met Sandy during our Bioverativ interactions and made two big mistakes. The first one was to suggest that I was an honorary Scotsman, by virtue of having two Scottish grandchildren via my eldest daughter, and, second, to tell him that they lived in Milngavie (pronounced Mulguy) near Glasgow. The response was rapid and unequivocal: "You cannot be an honorary Scotsman!" and "Milngavie is way too posh."[c] "I was born and brought up on the River Clyde

---

[c] For those who do not know the word "posh," it has an interesting nautical connotation. In colonial days on ships to India, the most expensive cabins were on the left-hand side of the ship (port) going out to India and on the right-hand side (starboard) coming home. This was because they were mostly on the shady side of the ship and the cabins were much cooler. Port Out-Starboard Home was the origin of the word "posh." Only posh people could afford shady cabins.

in Govan," Sandy told me, a rough part of town where the shipyards used to be and where the comedian Billy Connolly originally worked when he left school. I was firmly put in my place.

Macrae immediately focused the company on therapeutic applications, changing Sangamo Biosciences to Sangamo Therapeutics. They focused on the haemophilia A gene therapy programme with Pfizer, the SCD/TDT collaboration with Bioverativ, and an AAV gene therapy for the mucopolysaccharidosis 1 (MPS1) and MPS2 (Hunter syndrome; Chapter 11). They performed the first ever *in vivo* gene-editing experiment in a human, by altering the gene causing Hunter syndrome with an AAV vector containing a ZFN and a copy of the normal gene coding for iduronate-2-sulfatase (IDS): a construct designed to edit out the flawed gene that had been inherited and replace it with the normal gene.

Intronic *BCL11A* Editing

The gene-editing programme with Bioverativ (BIVV003) focused on the intronic *BCL11A* deletion, because there was better engraftment of the modified CD34 cells compared to those with exon edits.[6] Two ZFN mRNAs on each side of the site to be cleaved were delivered to CD34 cells via electroporation. When the two ZFN proteins came together at the cleavage site the FokI nuclease was reconstituted, and the DNA was cut.

Once the Investigational New Drug (IND) application was filed, we could start treating patients. It was slow-going and very frustrating, owing to manufacturing delays and to the prolonged recruitment of appropriate TDT patients (the first indication we were going after). Although we were not ahead of the bluebird bio gene therapy approach (replacing the β chain with β-T87Q-globin; see Chapter 13), we were competitive with the David Williams *BCL11A* shRNA gene therapy approach being pursued at the Boston Children's Hospital[7] (shRNA is a form of siRNA in which the RNA is single stranded but takes on a partially double-stranded hairpin structure) and we were (in 2016) well ahead of any of the CRISPR-Cas9-based gene-editing competition.[8]

The Sanofi acquisition of Bioverativ in 2018 did nothing to help to speed up the programme. Just the opposite, in fact. Sanofi were thoroughly equivocal about the commercial viability of the whole approach and, despite my best efforts, could never really get behind what we were trying to do. Although the approach clearly works in patients, the lack of commitment from Sanofi to continue has been confirmed by the return of the programme to Sangamo.[d]

I remember speaking at the 2018 American Society for Hematology (ASH) meeting in San Diego with Julie Makani, a very well-respected haematologist with an SCD clinic in Dar es Salaam in Tanzania. An SCD clinic here in the United States might have 100–200 patients registered. Her practice in Tanzania had 2000 patients. They had a huge need for antibiotics to treat *Pseudomonas* infection, which kills young kids with the disease. But there was also overwhelming interest and enthusiasm in totally new approaches to cures—including gene therapy or gene editing of CD34 cells and stem cell transplantation. This, in my opinion, is just the sort of opportunity that large pharma should embrace. I am not naïve: it would certainly require new commercial models, but it is far from impossible. If Bioverativ were still going and we still had the programme and it worked, I would be advocating for making our treatment available in Africa, in addition to treating patients in Detroit and other U.S. locations where the incidence of SCD is high in the African–American population.

TALENs

To quote Jens Boch, one of the discoverers and pioneers of TALEN technology (Box 1), "'TALE nucleases' (TALENs) followed in the footsteps of ZFNs and ignited the genome editing revolution."[9] TALENs were the first gene-editing technology that could be designed and built with relative ease[e] to target any specific genomic sequence with high precision and efficiency. TALENs were applied to genome editing in crops and livestock, and they were, like ZFNs, one of the first genome editing technologies to be used clinically.

When I asked Jens why they did not set up a company to exploit TALEN technology, he told me that setting up a biotech company in Germany was not that easy, and he added that the scientists felt that a new company was not the best way to get the technology to be used widely. So, they filed the IP and licensed it to several companies to use.

It is interesting that very few companies are using TALENs to modify cells for therapeutic purposes, although many large pharma companies have

---

[d] An unfortunate and frustrating but predictable outcome.
[e] They are much easier to design and use than zinc finger proteins.

BOX 3. CELLECTIS—A TALEN COMPANY

Based in Paris, France, Cellectis are the only biotechnology company that are dedicated to using TALEN technology to engineer T cells. Formed in 1999, as their website claims, they have more than 20 years of gene-editing experience, which is more than can be said for most of the CRISPR-Cas9 companies. Cellectis were one of the first companies to use edited cells therapeutically. They made the product UCART22 (universal CAR T) that targets CD22, a protein found on the surface of most B cells and B-cell malignancies. Cellectis scientists are also using TALEN technology to generate "off-the-shelf" allogeneic T cells to treat leukaemia, and the company have collaborations with various companies including Servier and Allogene Therapeutics.

For applications in plants, Cellectis spun out a company called Calyxt, which is based in Minneapolis. They have used the technology to engineer soybeans to make oil with better properties and to synthesise new plant-derived compounds.

---

taken licenses so that they can use TALENs unencumbered for many other purposes (Box 3). It may yet turn out that the rather complex IP situation that pertains with CRISPR-Cas9 will lead to more use of TALEN technology therapeutically.

Boch also told me that he found it sometimes puzzling why everyone is so hyped up about CRISPR-Cas9 but disregarded TALENs. For simple mutations, both are equally efficient and easy to use. He noted how quickly TALEN enabled genome editing happened in plants (the first genome-edited product brought to market) and in animals, where even complicated gene replacement was performed easily in cattle. Key breakthrough achievements in the genome editing field (e.g., editing wheat to be resistant to mildew fungus, or mutating 100 genes in one shot in sugarcane) are often accredited to CRISPR-Cas9 technology; they were in fact done using TALENs. More recent TALE tools enable the editing of the genome of chloroplasts and mitochondria with high efficiency, something that has not been done using CRISPR-Cas9. In 2011, *Nature Methods* awarded TALENs and ZFNs its "Method of the Year," although both are rather underappreciated now considering the much-ballyhooed arrival of CRISPR-Cas9.

CRISPR-Cas9

For some people, gene editing and CRISPR-Cas9 are synonymous. But CRISPR-Cas9 arrived quite late to the gene-editing party. I was not at the meeting, but I have been reliably informed[9] that the excitement of the first

TALEN-focused gene-editing meeting took the attendees' collective breath away. Many ideas of applications that suddenly became possible were brought forward. But most of these are now the province of CRISPR-Cas9 activity, although they might equally well have been done by TALEN technology.

There is an enormous amount of hype around CRISPR that never happened with either ZFNs or TALENs, including a plethora of literature on the history of the discovery of CRISPR-Cas9, its uses, and a great deal of self-promotion, a proxy for "being competitive scientists" (see below). Basically, you cannot believe everything you may read about CRISPR-Cas9 and its development.[10,11]

CRISPR-Cas9 Discovery

The CRISPR-Cas system was discovered in steps, starting in the early 1990s. Like restriction enzymes, zinc fingers, and TALEs before it, the CRISPR-Cas9 system was discovered in bacteria. CRISPR stands for "clustered regularly interspaced short palindromic repeats," repeating DNA sequences found in many bacterial genomes. They were first identified in *Escherichia coli* in 1987 by a Japanese scientist, Yoshizumi Ishino, but their relevance and function were a complete mystery. Six years later, a group in the Netherlands discovered that the spacer sequences in the clusters differed in the DNA of different strains of *Mycobacterium tuberculosis*.

Francisco Mojica and coworkers at the University of Alicante, Spain, also noticed similarities between CRISPRs and sequences in certain bacteriophages and plasmids. They figured out that bacteria containing CRISPRs that were the same as (i.e., "homologous") to those in a particular bacteriophage could not be infected by that phage. This trait was analogous to the restriction-modification system in bacteria, which provides resistance to plasmid exchange, and which led to the discovery of all the restriction enzymes that gave rise to recombinant DNA technology.[f]

In the CRISPR system, the spacer sequences in the CRISPR arrays are transcribed into short RNA copies (crRNA), which guides a "CRISPR-associated sequence protein" (Cas, a nuclease enzyme) to cleave the complementary nucleic acid of the infecting phage virus, preventing

---

[f]This discovery also supports the view that prokaryotic biology still has many things to teach us and likely is still hiding powerful new tools that can help us to understand human biology and human disease.

it from replicating. The third component of the system was recognised by Emmanuelle Charpentier, who discovered a longer RNA called Tracr, without which the editing would not take place. Tracr RNA directs the synthesis of the crRNA that guides the nuclease to its target (see Box 4, Figure 1).

There are many contributors to the history of CRISPR-Cas9 and its growing utility as a gene-editing tool in mammalian cells. Mojica was the first to name the bacterial sequences "CRISPR" and identify that they were copies of sequences found in bacteriophages, but it took a long time to get the then-obscure observation published.[12]

Yogurt

Bacteriophages are the bane of companies that ferment milk to make cheese or yogurt because they infect and kill the *Lactobacillus* (or other fermenting bacteria) in the culture. Philippe Horvath, who worked for Rhodia Food (soon acquired by Danisco), learned of the CRISPR repeats at a scientific meeting. In collaboration with Rodolphe Barrangou in Wisconsin, Horvath and his colleagues started to use the sequences to examine *Streptococcus thermophilus*, another key bacterium used in yogurt making, to see what "phage sequences" they contained. The more repeats that the bacteria contained, the more viruses they were resistant to, which was good for the yogurt fermenting process. They also identified the Cas genes adjacent to the repeats and showed that without them the bacteria were no longer immune to infection. Danisco filed a patent on the discovery in 2005 that issued in 2006, followed by a *Science* paper in 2007.[13] Kira Makarova and Eugene Koonin at the National Library of Medicine at NIH applied sophisticated bioinformatics tools to catalogue the Cas proteins that cleave specific sites in RNA or DNA (e.g., Cas 9, cleaves a site in DNA, whereas Cas 13 cleaves another site in RNA). Meanwhile, Virginijus Šikšnys, at Vilnius University in Lithuania, discovered that the Cas proteins looked and acted like nucleases and was the first to suggest that they could be used for gene-editing purposes.

Protospacer-Adjacent Motifs

The discovery of the protospacer-adjacent motifs (called PAMs) is also an important part of the story. These short 2–6 base pair palindromic sequences lie adjacent to the Cas cleavage sites and were shown to be necessary for Cas editing. PAM sequences are only found in the *phage* nucleic acid sequence (i.e., not the bacterial sequences), ensuring only viral sequences are cleaved and not the host DNA—a kind of primitive but effective "immunity" against phage infection.

It was the foresight and hard work of four people and their labs that really ushered CRISPR-Cas9 into the big time: George Church at Harvard, Feng Zhang (a former student of Church) at the Broad Institute, Emmanuelle Charpentier at the time at the University of Vienna before going to Sweden, and Jennifer Doudna at UC Berkeley. It was appreciated by all of them that the system could be adapted to cleave DNA (and RNA) in a sequence-specific fashion, and the ability to add DNA at the cleavage site would create a versatile genome editing tool (Box 4).

Charpentier and Doudna were the first to show that a three-component system of a synthetic guide RNA, the PAM recognition motif, and the

---

BOX 4. CRISPR-Cas9 AND OTHER GENOME EDITING TECHNOLOGIES (INCLUDING BASE EDITING)

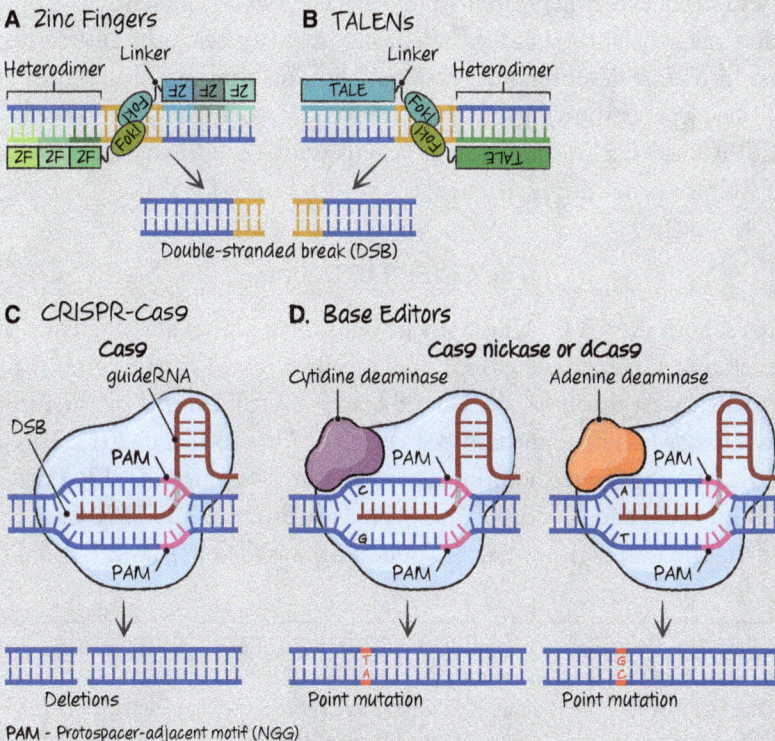

Figure 1. Current DNA-editing technologies.

Cas nuclease from *Streptococcus pyogenes* could work *in vitro*. They further demonstrated that the system could be adapted to cleave DNA with a single-guide RNA (sgRNA). Zhang and Church's labs showed that a similar system worked in eukaryotic cells.

The Story Evolves

The sequence of events behind who discovered what and when they did in the CRISPR-Cas9 story has been documented many times,[10,11] but it is just as important to remember what went before. As has been pointed out by several people, the fact that CRISPR-Cas9 works *in vitro* to cut DNA sequences specifically does not necessarily mean that it would work on chromatin inside eukaryotic cells. But it certainly made sense to try, which is what Zhang (working with Luciano Marraffini, a long-time CRISPR scientist at the Rockefeller, and George Church's lab) duly did.[14,15,g]

Doudna openly admitted at the time that her lab was not able to do those kinds of experiments in mammalian cells, but subsequently and quickly after the others, they did show editing in eukaryotic cells and rushed a paper into *eLife* describing the results.[16] Jin-Soo Kim in South Korea also published a paper showing that the system could work in eukaryotic cells.[17,h] CRISPR-Cas9 is a lot simpler to use in eukaryotic cells than either ZNFs or TALENs for gene editing, particularly at a genome-wide scale.

The CRISPR IP Debate

In 2012, both the CVC (University of California, University of Vienna, and Emmanuelle Charpentier group) and the Broad/MIT/Zhang groups filed their patents. At this time, the United States was still covered by "first-to-invent" rules, switching to the European norm of "first-to-file" in 2013. This unfortunate timing created additional confusion in assigning CRISPR-Cas9 IP.

Both CVC and Broad/MIT claim to have invented CRISPR-Cas9 editing first. The UC Berkeley patent application was filed in May 2012 and the

---

[g] There was, of course, the precedent of ZFNs and TALENs derived from bacteria that had been shown to work in eukaryotic cells.

[h] The fact that four groups could get CRISPR-Cas9 to work in mammalian cells by including a nuclear localisation sequence and optimising the CRISPR mRNA to be translated better in human cells (both obvious things to do) is one of the arguments why the real invention was the system itself rather than the fact that it worked in eukaryotic cells.

Broad patent application was filed in December 2012. There is a huge amount of work involved in laboratory notebook review to find out who is or is not the first to invent, and what they invented. If you have ever gone back and looked at your lab notebook several years after you did the experiment (and I have), you will know just how hard it is to remember exactly what you were doing, let alone someone else attempting to interpret what you noted down.[i]

To cut a very long and important—but ongoing—IP story short, the United States Patent and Trademark Office (USPTO) has ruled in favour of the Broad/MIT/Zhang team. The CVC group are appealing the decision. There are thousands of other CRISPR-related patents in addition to these foundational ones that cover gene editing in human and plant cells, many of them held by the Broad and the CVC (especially new ones around base editing; see below). Europe is, not surprisingly, more favourable to the CVC patent. One of the reasons for this is that during the patent application review, the Broad/MIT patent application was narrowed in scope and Marraffini, who was originally a co-inventor, was removed from the application.[j] A good article in *Nature* summarised the situation on the CRISPR IP. The IP situation continues to be unresolved.[18]

CRISPR Awards

Numerous awards have been made to various people in the CRISPR story, including the much-deserved Nobel Prize in Chemistry in 2020 to Emmanuelle Charpentier and Jennifer Doudna (the first time the award has been made solely to two female scientists). The prestigious Breakthrough prize worth $3 million and the equally prestigious Kavli Prize worth $1 million were awarded to Doudna and Charpentier as well (although the Kavli Prize was also, not unreasonably, shared with Virginijus Šikšnys[k]).

Another more obscure prize, the CRISPR Chutzpah Award (CCA), should go to Eric Lander, former director at the Broad. He wrote a very

---

[i] Lab notebook sign-off (important for IP reasons) was patchy when I was still in the lab, so it is just as well that there is now harmonisation of the United States and Europe on a first-to-file approach.

[j] For those of you used to thinking about names on patent applications being like names on a publication, think again. Only those people who contributed to the *invention* described in the claims and the supporting data may be on the application. Having the wrong people on the application can invalidate a subsequent patent. Whether this was or was not true for Dr. Marraffini in this case is not for me to say.

[k] Some feel Šikšnys should also have shared the Nobel Prize.

eloquent but largely inaccurate history of the discovery of CRISPR-Cas9, which was published in *Cell* in March 2016.[19] It promoted Zhang's and Broad's role in the discovery (and, generously, that of Šikšnys) but played down what Charpentier and Doudna did, without the appropriate credit being given to either of them. It may have been a backhanded attempt to support the Broad patents. Lander has since apologised for the oversight. Cell Press/Elsevier have hopefully learned that conflicts of interest need to be disclosed and that their editors need to do their job as far as ensuring accuracy of content, especially of lead articles/editorials.[l]

I met Emmanuelle Charpentier at the Cold Spring Harbor Genome Meeting in May 2017. She arrived by taxi at the same hotel as I was staying in. I asked her, rather tongue-in-cheek frankly, whether she was hungry and wanted to come and have something to eat that evening in the little town of Cold Spring Harbor. We had a pleasant dinner and talked about biotech and the industry and some of the places she had worked (this turned out to be 10 institutions in seven cities in five countries). But we did not talk much about any detailed CRISPR stuff: she probably thought I would not understand it.[m]

I did talk a bit about the zinc finger experiments we (Bioverativ) were doing with Sangamo, but she did not give much away about what CRISPR Therapeutics, the company she founded, was doing. I am quite sure that I remember the dinner way better than she does, given everything that has happened to her since then. I am equally sure she has dined with many more important and interesting people than me, both before and since. But I was pleased to have met her in the light of her undisputed and pivotal role in the development of the CRISPR technology.

CRISPR-Cas9 has revolutionised life sciences research and the biotech industry in much the same way that recombinant DNA and monoclonal antibodies did decades before. Those two earlier technologies spawned an entire industry. So, what does the commercial exploitation of this fundamental new technology look like? In some ways, it is a microcosm of how the biotech industry developed from recombinant DNA: but speeded up a bit.

---

[l] The journal *Cell* has come a long way since Benjamin Lewin started it in 1974. Some of you may remember fondly the spoof "*Cool*" edition of *Cell* that was produced at the time. I still have a photocopy of it, and it is available in all its glory at https://cell.com/pb/assets/raw/journals/research/cell/cell-timeline-40/spoof.pdf. It is still remarkably current!

[m] I am sure I was suitably banal. I do remember commenting on the University in Umeå, where she worked. Umeå is on the same latitude as Reykjavik in Iceland, where it is almost always dark in winter and correspondingly light in summer. Scintillating stuff.

There were four major companies set up early on to exploit CRISPR technology: Caribou Biosciences, CRISPR Therapeutics (CRISPR TX), Editas Medicine, and Intellia Therapeutics. These were followed by the "base editing" companies Beam Therapeutics, Prime Medicine, and Verve Therapeutics. Each company has a good combination of great scientists and people, intellectual property, and political nous.

CARIBOU BIOSCIENCES

The first company founded to exploit CRISPR-Cas9 was Caribou Biosciences, set up in Berkeley in 2011 by Doudna and Rachel Haurwitz, a scientist with commercial inclinations from the Doudna lab, who would become President and CEO.[n] The primary purpose was to use CRISPR technology in diagnostics and agriculture, but not in human therapeutics. Other co-founders included Martin Jínek, the former postdoctoral fellow in Doudna's lab, now at the University of Zürich, who made the first sgRNAs and led the 2012 study leading to the important *Science* paper, and James Berger, a professor at the Johns Hopkins University School of Medicine in Baltimore.

In the process of building the company, they licensed the CVC IP. At that time, the terms "gene editing," "CRISPR," or "Cas9" did not trigger an "automatic cheque writer, bank teller machine, or VC wallet-opening," and the company—believe it or not—struggled to find money.

Incubating Caribou

Caribou were initially located in the basement of Doudna's UC Berkeley lab building, but subsequently moved to 7th Street in downtown Berkeley. Despite their initial intent, the company have now firmly pivoted to the human cell therapy game, with lead programmes making allogeneic (edited) CAR T cells for enhanced cell killing and persistence, and in deriving edited CAR natural killer (NK) cells from induced pluripotent stem cells (see Chapter 14). Their editing technology has some novel features, including using DNA/RNA hybrids as sequence guides and novel or modified Cas enzymes. These developments were important from an IP point of view for freedom to operate against Intellia Therapeutics (see below).

---

[n]Doudna spent a short time at Genentech, leaving Berkeley briefly for industry (about two months). That brief experience may have helped her to appreciate that a company based on the CRISPR technology was worth forming.

Caribou now have a phase 1 clinical trial in progress for their allogeneic CAR T-cell product derived from healthy donor T cells. It is directed to CD19 and the gene coding for PD1 (a check point protein; see Chapter 5) edited out to potentially increase the persistence of the T cells: basically, the equivalent to treating patients with an anti-PD1 antibody like Keytruda.

The first results of this CAR T product reported in June 2023 are encouraging, producing complete remissions in nearly half of the small number of lymphoma patients treated. Their second CAR T programme is directed against B-cell maturation antigen (BCMA) for multiple myeloma. Neither of these programmes are particularly novel but nevertheless, after a series of successful financings including a $115 million series C raise, the company went public in 2021. Although admittedly at the height of the COVID IPO bubble, they still raised more than $300 million in an upsized offering managed by Bank of America Securities, Citigroup, and SVB Leerink.°

CRISPR THERAPEUTICS

Emmanuelle Charpentier and Rodger Novak, her friend and one-time colleague who was at Sanofi heading up anti-infectives research, formed CRISPR Therapeutics in 2012. Unlike Caribou, they planned to treat patients with the technology from the company's inception. The CFA Shaun Foy was also a founder of the company and served as CBO early in the company's history. Novak became the CEO until Samarth Kulkarni, who had been at CRISPR Therapeutics almost from the beginning, stepped into the role in 2017. CRISPR Therapeutics investors included Versant Ventures, SR One, Bayer, and Celgene.

Although CRISPR Therapeutics are targeting SCD and β-thalassaemia and editing CAR T cells much like the other gene-editing companies, they deserve credit for the single-minded way they have gone about it and the business decisions that they have made. One of the most important moves was to form a multibillion-dollar partnership with Vertex for their sickle cell and TDT programme. This collaboration has since been amended and extended. The original programme is almost an exact replica of what we were doing at Sangamo/Bioverativ, targeting the erythroid lineage enhancer in intron 2 of the *BCL11A* gene, except that they are using CRISPR-Cas9 for editing rather than ZFNs.

°Given the compelling pull of their by then Nobel laureate founder, maybe it is not so surprising after all.

Good Execution

CRISPR Therapeutics have executed very well. Those of us at Bioverativ heard their drumbeat, for sure. They seemed to have much less trouble accessing patients than we did and, although they were well behind us to start with, they passed us quickly. They may well win the gene-editing race for SCD and TDT, base editing notwithstanding (see below). In January 2021, CRISPR Therapeutics scientists published the first paper in the *New England Journal of Medicine* showing results of the transplantation of edited autologous CD34 cells for two patients, one with TDT and the other with SCD.[20] They have done further clinical trials with this approach, and with Vertex are awaiting FDA approval of this first CRISPR-based cell therapy.[21] In November 2023, CRISPR Therapeutics and Vertex announced that their gene-editing product for SCD and TDT was approved in the United Kingdom. It is called Casgevy (exagamglogene autotemcel [exa-cel]).

CRISPR Therapeutics went public in the fall of 2016, raising $56 million. Citigroup, Piper Jaffray, and Barclays acted as the book runners. The amount raised was less than either Editas or Intellia, both of which went public at the same time (all were less than Caribou), although Bayer bought 2.5 million shares at $14, increasing the amount raised by $35 million. Goldman Sachs led a follow-on financing in 2018, raising more than $100 million. In 2019, CRISPR Therapeutics had net income of some $67 million and had some 300 employees compared to the 77 they had at IPO: an aggressive but sensible growth rate.

EDITAS MEDICINE

Gengine Inc. were incorporated in September 2013, but wisely changed their name to Editas Medicine. All of the CRISPR protagonists apart from Emmanuelle Charpentier, who formed CRISPR TX, were involved in Editas at the beginning, including Feng Zhang, George Church, Jennifer Doudna, Keith Joung, and David Liu from Harvard University. The company was financed by top-tier Boston-based VCs including Polaris Venture Partners, Flagship Ventures (now Flagship Pioneering), and Third Rock Ventures. Kevin Bitterman from Polaris, now at Atlas Venture, was the initial CEO. Doudna quit Editas in mid-2014, owing to the Broad versus CVC patent dispute (despite filing their application later than the CVC, the Broad patent issued in May 2014 following an accelerated review; see above). This assignment of IP focused the attention of Editas away from the

CVC patent estate to such an extent that Doudna felt the company could no longer be trusted, and she could no longer be involved with it. She resigned from the SAB and as a founder. The CVC patent was issued later in the year.

Editas and Biogen Connections

Biogen are intertwined in the history of Editas. Former Biogen employee Katrine Bosley was CEO from 2014, until her unexpected resignation in 2019. Cynthia Collins ran the company for a short time, followed by James Mullen in 2021, who had been CEO of Biogen and Board Chairman of Editas. Gilmore O'Neill,[P] former CMO of Sarepta Therapeutics and a former SVP at Biogen, became CEO on June 1, 2022, with Mullen becoming Executive Chairman of the Board.

Editas raised $120 million in their series B funding from Bill Gates and other investors. The company went public in February 2016, raising more than $90 million. Morgan Stanley and J.P. Morgan acted as joint book-running managers for the offering, with Cowen and Company as lead manager, and JMP Securities as co-manager—a suitably stellar banking team. Given what some of their competitors were doing, Editas sensibly formed a strategic alliance with Juno to use CRISPR editing for CAR T cells for cancer therapy, instead of trying to do it all themselves.

The lead clinical programme for Editas is to use CRISPR for correcting Leber congenital amaurosis, an eye disease (see Chapter 13). This programme, which targets the CEP 290 gene, entered the clinic in 2018 and is currently in proof-of-concept trials. Following a clinical hold, the company announced that their proprietary Cas12a enzyme–based CRISPR approach to SCD and TDT (EDT301) had started recruiting patients for the phase 1/2 RUBY trial. This gene editing targets the promoter of the γ globin gene (where mutations that control foetal globin expression are found), rather than the *BCL11A* gene that others are targeting. This programme is not without considerable direct gene-editing and base-editing competition, from both CRISPR Therapeutics and Beam Therapeutics (see below).

INTELLIA THERAPEUTICS

Once they learned that Doudna was resigning from Editas, Jean-François Formela at Atlas Venture (whom I knew well from the SGX days) and

---

[P] Gilmore is one of the most compassionate doctors you would care to meet. I enjoyed working with him on multiple sclerosis while we were both at Biogen.

Nessan Bermingham, recently recruited as an entrepreneur in residence at Atlas, wasted no time in licensing the CVC IP for use in human therapeutics.[22] They were well aware of the developing CRISPR technology but had failed for a variety of reasons to become involved as investors in either Editas or with Versant Ventures in the formation of CRISPR Therapeutics. Atlas persuaded Mark Fishman at Novartis to invest in and work with Intellia, as the company was now called. Bermingham became the founding CEO and John Leonard from AbbVie, a seasoned pharma R&D guy, became CSO. The academic scientists involved in founding the company were all long-term contributors to the CRISPR scene. They included Barrangou, Haurwitz, Marraffini, Erik Sontheimer, and Derrick Rossi, a stem cell scientist and founder of Moderna and Magenta. Intellia raised $15 million in their series A round at the end of 2014.

Novartis—Intellia Deal

In January 2015, Novartis and Intellia formed a deal, by which Novartis obtained rights to use CRISPR for their CAR T programme and the companies agreed to collaborate on ways to use CRISPR for haematological indications, such as SCD and β-thalassaemia. Intellia were able to come to an arrangement with Berkeley to license the CVC patents for human therapeutics, partly because of their relationship with Novartis as a series A investor.[q]

With Caribou already on the scene as a licensee to the CVC patents, one may well ask how Intellia managed to do this. It was a bold move but recall that Caribou at that time said they were not interested in human therapeutics, so Intellia were able to license the IP for use in that field.[22] Paradoxically, now that Caribou are deep into human therapeutics, they had to negotiate a license back from Intellia to get the CVC IP for that use.[r]

Intellia focused not just on CRISPR but also on delivery systems and obtained a license from Novartis to use their lipid nanoparticles to deliver the gene-editing molecules so that they could use them *in vivo* and not just for editing CD34 cells *ex vivo*. They formed an alliance in 2016 with Regeneron, a company not well known for doing deals with small companies, to use CRISPR-Cas9 to treat transthyretin amyloidosis, one of the

---

[q]It is always good to have a Big Pharma partner in the wings of a licensee team, especially if you are a university licensing group.

[r]In fact, the reason Caribou are using novel hybrid guides is to develop IP independent of the original CVC IP.

targets of Alnylam and their siRNA approach (Chapter 12). Regeneron paid $75 million upfront to Intellia with milestones and royalties attached.

On the strength of their alliances, Intellia raised $70 million in a series B round in September of 2017 led by OrbiMed. They went public in May 2016, raising $112 million in an IPO led by Credit Suisse, Jefferies, and Leerink.

In Vivo Editing

The Intellia and Regeneron teams caused quite a splash when they were the first to show *in vivo* gene editing in humans using CRISPR-Cas9. They published their study of gene editing in transthyretin amyloidosis in the *New England Journal of Medicine* in August 2021.[23] Using lipid nanoparticles to deliver a sgRNA targeting the *TTR* gene and the Cas9 endonuclease, they showed that editing of the gene resulted in reduction of transthyretin in the serum of the six patients in the trial. They have since done the same thing for hereditary angioedema using CRISPR-Cas9 to target kallikrein, a serine protease made in the liver.

Novartis cancelled their *ex vivo* CD34 cell deal with Intellia for SCD and TDT in March 2023. Whether this was a strategic move by Novartis concerning *ex vivo* medicines or simply a reflection of the fierce competition in this indication (or something else) is unclear.

It is interesting to compare the relative market capitalisations of the four original CRISPR companies as a measure of value. This does not mean to say that the one with the highest market cap is necessarily the most successful or that the market cap means that they made the most money for their shareholders—but it does tell a story. On July 24, 2023, Caribou Biosciences stood at $550 million, Editas Medicine at $700 million, Intellia Therapeutics at $3.8 billion, and CRISPR Therapeutics at $4.5 billion. Compare those numbers to Cellectis (TALENs) and Sangamo (ZFNs), with market caps of 134 million euros[s] and $232 million, respectively (both are probably undervalued actually).

The story that the market caps tell is that CRISPR technology is preferred to others, and that products count (the closer you are to them, the higher your valuation). Technology alone is not enough: some argue that Editas spent way too much time refining the technology and not enough time developing products, whereas CRISPR Therapeutics focused on targets

---

[s]Cellectis have suffered from an FDA clinical hold applied to their CAR T programme, a result of nervousness about the fidelity of the multiplexed gene edits they were incorporating.

and got on with it. A market cap of $5 billion allows you a lot more flexibility than one of less than $1 billion.

Intellia should be congratulated for going after *in vivo* applications rather than taking an *ex vivo* cell therapy approach and differentiating themselves from the rest of the CRISPR pack. It is a long road to a commercial product and lots of water yet to run under the mythical gene-editing bridge (perhaps one that crosses the Charles River). Furthermore, there is also fierce competition arising from elsewhere, coming in the form of a more precise gene-editing technology called "base editing."

BASE EDITING

In its simplest form, CRISPR-Cas9 editing just removes a bit of DNA at a sequence specifically guided by the sgRNA. The cuts that are made are not uniform, and the repair mechanism used is called NHEJ (or nonhomologous end joining). The result is a set of small deletions that all have the same effect (removing the function of the gene), but not every edited cell will have the same sequence at the edited site. If you want to make identical alterations in every cell precisely, then you need to engage a process called homology-directed repair (HDR) rather than NHEJ. However, HDR requires a sequence template to execute precise alterations in a double-stranded DNA. Also, HDR does not really work very well either in nonreplicating cells because you need cells to be dividing for enough of the machinery that makes HDR work to be present. Homologous recombination is much more precise than NHEJ and relies on different enzymes to carry out the function. For us, the details are not as important as the nature of the change.

Precise gene editing by CRISPR is thus limited by NHEJ. For example, it is not possible to use CRISPR-Cas9 to precisely change the GAG→GTG mutation that causes SCD back to GAG at codon six of the β-globin gene. Base-editing methods were developed—and companies founded—to get around at least some of these limitations.

BEAM THERAPEUTICS

David Liu's lab at Harvard has played a central role in developing base-editing tools. He has founded two companies to develop different aspects of base-editing technology, Beam Therapeutics and Prime Medicine, and he is also involved in Verve Therapeutics.

I met David several years ago after I left SGX Pharmaceuticals. He had founded a Flagship Ventures funded company exploiting DNA-coded combinatorial chemistry called Ensemble Discovery Corp. They were developing a breakthrough system: DNA Programmed Chemistry, which could synthesise molecules that were difficult to make by conventional means. They were looking for a CEO, and I was interested in the technology and the fact that they had a good deal of cash in the bank. But I could not wrap my head around how to apply the technology in any rapid way to make new drugs, so I was disinclined to join them. Probably this was a mistake, given what happened to Novasite (where I chose to go instead; see Chapter 9), but I was impressed even then in 2006 by the creativity and quality of the science and innovation in the Liu lab at Harvard.

Beam Therapeutics were conceived in 2016 and were well-financed from the start. It was apparently called Beam because it stood for "base editing and more." David Liu, Keith Joung, Feng Zhang, and John Evans were founders, and Evans became the CEO. They are located on Main Street in Cambridge (MA). Beam raised $87 million in their series A round from ARCH and F-Prime Capital in 2018, followed by $135 million in a series B round in 2019 led by the original investors but also including Eight Roads Ventures, Omega Funds, and Google Ventures, among others. At that time, the company had 10 active base-editing programmes and more than 70 employees. The company went public in early 2020 raising another $180 million. J.P. Morgan, Jefferies, and Barclays were the joint book runners, with Wedbush PacGrow acting as lead manager.

The Beam base-editing platform evolved directly from the Liu lab. Their base editors consist of all the recognition components (guide RNAs) of the CRISPR system, attached to a modified inactivated Cas nuclease that only nicks one strand of the DNA at the recognition site but does not cleave the DNA.

The editing complex contains either a cytidine deaminase enzyme that changes the single-nucleotide cytidine to uridine, or a synthetic enzyme that the company evolved *in vitro* that changes an adenine to a guanine. These changes convert the base pairs from C-G to A-T and A-T to G-C, depending on the editing enzyme used. The efficiency is reasonably high for both enzymes, but the cytidine deaminase is rather better at editing than the ABE (adenine base editor) (see Box 4). One of the problems, at least initially, was to ensure that only the base that was targeted was changed and not the ones nearby. Therefore, the presence of a PAM recognition sequence is important because it restricts sites that can be edited.

The base-editing technology and in particular the use of the different base-editing enzymes were originally developed in the Liu lab by Alexis Komor and Nicole Gaudelli. The IP was filed by the Broad and Harvard and licensed to Beam Therapeutics. Beam also accesses all the other necessary gene-editing technologies, such as electroporation, lipid nanoparticles, and various virus vectors (including AAV). Like many other gene-editing companies, the editing technology was reduced to practice using the SCD model in mice (the data were published in *Nature* in June 2021[24]). In this case, they edited the GTG codon (coding for valine) to a GCG codon (coding for alanine). Although the normal protein has a glutamate at position six, alanine is found at that position in a naturally occurring nonpathogenic variant of β globin called HBB Makassar. The alanine does not cause the haemoglobin to sickle under low oxygen conditions, as seen with the valine mutation. Sixteen weeks after transplantation of the edited mouse cells into the SCD mouse model, 68% of the donor-derived stem cells were found to be edited.

Beam Pipeline

The Beam pipeline includes not only the correction of SCD but also the induction of foetal haemoglobin in TDT (BEAM-101) and many other opportunities, most of which are being worked on by other companies and have been mentioned previously. A market cap of $2.5 billion in July 2023 would indicate that investors view the Beam programmes to be strongly competitive with these other gene therapy and editing companies.

BEAM-101 is a patient-specific (autologous) CD34⁺ human stem cell product that incorporates base edits that change bases in the β-globin cluster, to those found in HPFH individuals, which are known to increase levels of γ globin. The clinical trials are recruiting both SCD and TDT patients, and the company planned to enrol subjects in the second half of 2022. Beam Therapeutics reported in May 2023 that patient recruitment was on target.

BEAM-102 is the human derivative of the mouse experiment described above, in which the valine codon in the sickle β chain is changed to alanine. Beam also have a multiplex base-edited anti-CD7 CAR T-cell therapy designed to treat acute lymphoblastic leukaemia (ALL), submitting an IND for that programme in 2022. Unfortunately, the FDA placed the programme (Beam 201) on clinical hold pending more information on the genetic changes potentially occurring in the multiplex edited T cells. They are expecting to enrol the first patients into a clinical trial in 2023. This was not a surprise, given that other companies making multiplexed allogeneic T-cell therapies have also faced clinical holds, including Allogene and Cellectis.

BEAM-301 is a liver-targeting liquid nanoparticle (LNP) formulation of base-editing reagents designed to correct the R83C mutation in glucose 6 phosphatase, which causes a form of glycogen storage disease for which there are no gene-altering treatments available today.

At a recent CRISPR conference, Beam highlighted their creation of an improved class of cytosine base editors (CBEs), and they disclosed that they had made an additional editing deaminase enzyme, CABE-T, that can conduct both C-to-T and A-to-G edits. Given the clinical hold for Beam 201, the specificity and selectivity of these base editors could be important in the future.

PRIME MEDICINE

Prime Medicine are a more recent spinout from the David Liu lab. They are based on the ideas and experiments of co-founder Andrew Anzalone, who was a postdoc in the Liu lab. Instead of attaching the CRISPR-editing apparatus to RNA coding for a base-editing enzyme, they attach it to reverse transcriptase. Included in the guide RNA is the sequence that is to be replaced. The guide RNA shuttles the complex to a specific sequence in the DNA, the Cas enzyme nicks the DNA, and the reverse transcriptase copies an RNA extension of the guide RNA (called a "pegRNA" or a "prime-editing guide RNA) into the nicked site, replacing the sequence at the site with the correct sequence of nucleotides: a permanent edit. This simple "search-and-replace" approach to gene editing can handle many different sorts of mutation. It is reported to be very specific, and it works in dividing *and* nondividing cells. It creates very few double-strand breaks, reducing any off-site activity.

This approach does have one significant disadvantage: the CRISPR RTase complex is large, so getting it into cells is not trivial. For that reason, Prime Medicine are advancing targets for the eye, the liver, and CD34 *ex vivo* applications first. I find this a little disappointing, given the power of the technology and its potential.

The investors clearly do not view it that way. Many of them are the same as those that invested in Beam Therapeutics, including ARCH Ventures, F-Prime Capital, Google Ventures (GV), and Newpath Partners. The refreshingly diverse and experienced Board[t] is highly experienced in

---

[t] It is perhaps what a modern biotech Board should look like (i.e., not full of people who have passed their sell-by date): a clear problem for some other biotech Boards that shall remain nameless (they know who they are).

building companies, including people like Bob Nelsen from ARCH and David Scheinkein from GV. Interestingly, the CEO at Beam Therapeutics, John Evans, is also on the Board of Prime Medicine, with whom Beam have a collaboration around IP, delivery, and manufacturing.

Prime Editing

After publishing the prime-editing method in *Nature* in October 2019[25] where details of the rather complex constructs can be seen, and licensing the technology from the Broad, Prime Medicine launched in the summer of 2020. They emerged from stealth mode in June 2021, raising more than $115 million in a Series A and $200 million in series B, presumably at an increased valuation. Not only have they recruited a stellar Board but they also have hired a great executive team—I think they will have a lot of fun. Keith Gottesdiener, who has clinical development experience at Big Pharma and was the CEO at Rhythm Pharmaceuticals, is CEO. Jeremy Duffield, a British MD who was at Biogen with me and at Vertex, is CSO, and Richard Brudnick is CBO (I worked very closely with Richard at both Biogen and Bioverativ, and at Codiak where he was CBO). Despite the unforgiving market conditions at the time, Prime Medicine filed to go public in late September 2022 and raised $175 million the next month. Recent studies have shown that it is possible to use an adenine prime-editing approach to correct the sickle cell mutation in mice,[26] and it is an efficient way to re-induce foetal haemoglobin expression.[27]

VERVE THERAPEUTICS

Beam and Prime are not the only players in the editing space.[28] Verve Therapeutics were set up in 2019 specifically to target cardiovascular disease using base-editing techniques. The founders include Keith Young, a founder of Editas Medicine, Sek Kathiresan (from MGH/Harvard/Broad), and Kiran Musunuru at the University of Pennsylvania. Their nearly $60 million series A round was led by GV and F-Prime Capital, and included Biomatics Capital and ARCH. This investor group must be beginning to sound familiar to you: they are clearly VCs committed to gene editing. Again, rather unsurprisingly, they have a collaboration with Beam Therapeutics and have licensed the IP they need from the Broad. They have a collaboration with Verily for gene delivery methods.

Verve are targeting high cholesterol via lipoprotein A modulation through the enzyme PCSK9 (this enzyme was discussed previously with respect to Regeneron mAbs and as a target for antisense technology—people with loss-of-function mutations in this gene have low cholesterol and are protected against heart disease). Verve's approach is to introduce one of these mutations by gene editing into the *PCSK9* gene in the liver. Verve 101 is a lipid nanoparticle containing a guide RNA and an mRNA that codes for an adenine to guanine base editor enzyme, directed to one specific base pair to inactivate the gene. Unlike other therapeutic approaches to *PCSK9*, this would be a one-time treatment. They are initially targeting patients with familial hypercholesterolemia (FH). They have demonstrated in nonhuman primates that one dose is sufficient to turn off the *PCSK9* gene, with virtually no off-target modification. In July 2022, Verve dosed their first patient with VERVE 101 in their phase 1b clinical trial in FH. There are clearly some concerns about VERVE 101 as the programme has been put on clinical hold by the FDA, although patients are being treated in the United Kingdom and New Zealand. Verve signed a deal with Lilly in mid-2022 for another editing approach to lowering the levels of lipoprotein a (Lpa) by editing the gene directly, rather than controlling levels via *PCSK9*. Lilly paid Verve $60 million up-front in cash and for Verve stock and will pay for the R&D on this target. There should be more dollars to come in milestone payments and royalties.

Verve also went public in the mid-2021 "bubble," raising more than $300 million in a J.P. Morgan–led deal. Their market cap in July 2023 was $1.3 billion, considerably lower than it was before the clinical hold.

Provided they can see some products on the horizon, investors just *love* gene editing (see Box 5).

BOX 5. NEW PLAYERS IN THE GENE-EDITING GAME

Metagenomi

The well-funded Emeryville-based (where Chiron and Cetus used to be) company Metagenomi was founded in 2018 and has made some noise recently owing to an oversubscribed $175 million series B round led by PFM Health Sciences and Farallon Capital (following their $75 million series A round). Additional new investors included Eventide Asset Management, Deep Track Capital, and Frazier Life Sciences among others, plus existing investors Leaps by Bayer and RA Capital. Moderna, who formed a collaboration with Metagenomi in 2021, also invested.

As their name suggests, Metagenomi are looking in microorganisms for other gene-editing systems and enzymes that might be adapted for gene editing in humans. Finding smaller enzymes is one of the objectives, especially because current systems are sometimes too large for effective transfer. They are also building more flexible systems in which the PAM restriction (a necessary part of the CRISPR sequence recognition system) can be altered or eliminated. Metagenomi are also pursuing therapeutic opportunities in the CAR T and TCR T space, with a focus on making edited allogeneic off-the-shelf T-cell therapies. They are planning to do an IPO in late 2023.

Aera Therapeutics

Nearly $200 million has been invested in Aera Therapeutics, a new (2022) Feng Zhang Venture with Akin Akinc from Alnylam as CEO and John Maraganore (ex-Alnylam) as Board Chair. Aera are developing a new virus-like delivery method for getting nucleic acids into cells in different organs and tissues of the body, a primary limitation of gene therapy methods to date. The company have attracted the usual cast of gene-editing investors, including Bob Nelsen at ARCH and Issi Rozen at GV. They folded the Lux capital funded VNV (Virus Not Virus), a Utah company founded by John Shepherd, into Aera to give them a running start. Akinc, Nelsen, ex-Vertex president Vicki Sato, GV general partner David Schenkein, and Lux Capital managing partner Josh Wolfe are on Aera's Board. They may have competition from another David Liu-Keith Young start-up called Nvelop Therapeutics aimed at the same goal. Nvelop have secured $100 million from Newpath Partners, Atlas Ventures, F Prime Capital, and 5AM Ventures. They are currently at 100 Technology Square, in Cambridge (MA).

GENE WRITING

Despite the exuberance of investors, gene deletions and precise base editing do have limitations. To attempt to overcome these, the Flagship Pioneering company Tessera Therapeutics are "pioneering" gene writing, a technique with apparently—and I quote—the "potential to cure practically any genetic disease." Tessera's technology is based on a class of mobile genetic elements called "transposons."[u]

Barbara McClintock at Cold Spring Harbor was awarded the 1983 Nobel Prize in Physiology or Medicine for the discovery of transposons in maize. Transposons use an enzyme called "transposase" to transfer pieces of DNA into the genome of a cell. Catamaran Bio (see NK cells in Chapter 14) are using this system to transfer CAR elements into NK cells, rather than using virus vectors like AAV or lentiviruses.

---

[u]Plasmids, which were part of the beginnings of biotechnology (see Chapter 1), are technically mobile genetic elements in that they can be transferred between bacteria. They formed the basis of the first cloning "vectors" for that reason. The viruses used in gene therapy could also be considered as mobile genetic elements, as they contain genes that they can transport into (infect) cells.

In addition to transposons, there are also "retrotransposons," which are transcribed into RNA that then gets fixed into a DNA site selectively by reverse transcriptase. These retrotransposons are regulated by small RNAs called piRNAs. Tessera have spent time building a catalogue and database of retrotransposon mobile genetic elements and identifying their specificity. These retrotransposons can be adapted to make large and small changes in cellular DNA, now referred to as "gene writing" (see Box 6).

Tessera's founders are chemical engineer and entrepreneur Geoffrey von Maltzahn and Jacob (Jake) Rubens, who incubated the company for some time at Flagship Pioneering. They announced a $230 million series B round in early 2022 and closed a series C round of $300 million in April 2022, where several leading venture funds participated. They have hired an impressive group of people to run the company, including Michael Severino, an ex-Amgen and ex-AbbVie person as the CEO, Howard Liang (an experienced CFO), and Michael Holmes (ex-Sangamo) as CSO. Anne-Virginie (AV) Eggimann was hired as Chief Regulatory Officer from bluebird bio.

Since their formation in 2018 and "coming out" in 2020, Tessera have created multiple platforms to enable the transfer of short or long sequences site-specifically into a genome, without the double-strand breaks that can cause off-target effects. The technology allows them to create both large and small deletions or insertions in the genome without using viral vectors. The more than $500 million they have raised to date should enable them to expand

---

BOX 6. GENE WRITING AND GENE ARITHMETIC: ORBITAL THERAPEUTICS

The fact that most VCs seem to already be invested in RNA technologies, gene editing, or base editing does not seem to stop other such companies from forming. Orbital Therapeutics were launched in October 2022, focusing on RNA therapeutics that are not RNAi-based. ARCH Venture Partners and Newpath Ventures are leading the investment charge, and John Maraganore, now at ARCH Venture Partners after leading Alnylam for some 20 years, is a founder. Giuseppe Ciaramella, an ex–Moderna employee and a founder of Beam and the current Beam President, is serving as CEO. Like Prime Medicine, Orbital will work closely with Beam Therapeutics on technology development and application. Drew Weissman, who was a co-discoverer of the use of modified nucleosides for building mRNA therapeutics while at the University of Pennsylvania and the 2023 Nobel Laureate in Physiology or Medicine, is also involved with this company.

the 200-plus person team that they have assembled and drive gene writing to the clinic for the treatment of liver diseases and rare genetic disease.[v]

MYTHICAL COMPANY HUMGEN: EDITING HUMAN EMBRYOS

One wonders if one day a company called "HumGen" will be set up to edit human embryos on demand. I jest not. The technology, based on CRISPR-Cas9 gene and base editing, *in vitro* fertilisation (IVF), and re-implantation, is certainly available and the experiments are possible; so setting up such a company is not as far-fetched as it might seem. The methods used and the ethical debates surrounding this new capability are all considered in three books of note: Kevin Davies' *Editing Humanity*, Walter Isaacson's *The Code Breaker*, and David Goldstein's thoroughly engaging book *The End of Genetics: Designing Humanity's DNA.*[10,11,29]

Doudna organised a mini-Asilomar in early 2015 to discuss the potential of CRISPR-Cas9 for editing humans. David Baltimore and Paul Berg, who attended the original Asilomar more than 40 years before, were both there and helped organise a follow-up "summit" meeting in DC the following December. The conclusions of this meeting were all about the safety and societal implications of human genome editing. *The New York Times'* Nicholas Wade, who wrote the original commentaries on Asilomar, published a piece that reported that the scientists at this latest meeting had "sought a moratorium on edits to the human genome"[30]: but what they actually said (and ultimately recommended) was that it is "irresponsible to proceed" until the risk was better understood.

It is now possible to use genome editing via CRISPR-Cas9 with IVF to edit the DNA in pre-implantation human embryos. Indeed, it has already been done: He Jiankui's ill-considered effort[w] in China in 2018 to confer resistance to infection with HIV by manipulating the *CCR5* gene resulted in the birth of two apparently normal girls. So, we know that the power of genome editing to make heritable edits to the human genome is all too real. It is quite likely that the children have other changes in the genomes apart from the designed one, making their future uncertain.[x]

---

[v] They do not say yet "all diseases," which probably reflects the fact that delivery of the writing machinery to cells is still a limiting factor, just as it is for other gene-editing technologies.

[w] He Jiankui was even trying quite hard to launch a human gene-editing company in China in 2017.

[x] I hope they experience normality, but by definition, they are not normal now.

The Gen(i)e is out of the Bottle

The problem we face today is not *how* to do the engineering, but *what* genes to engineer, a complex issue both biologically and ethically. It is easier to change the likelihood of inheriting a recessive Mendelian disease by altering the mutated gene, but this can already be done by genetic screening and mate selection. It is much more difficult to understand the outcome if you interfere with other genes in the hope of designing "better" babies, which I am assuming for the sake of this thought experiment is what HumGen, the mythical human gene-editing company, would be set up to exploit.

David Goldstein suggests that the drive for parents to ensure that their babies could be free of undesirable mutations (or changed for desirable traits) will fuel further adoption of the technology before we are ready for it. I agree with him and think that it will. But how far away from that precision technology are we? I would wager, based on all that has gone before in this chapter, not very far. But the technical ability is not nearly as important as understanding what genes we can safely change to improve someone's lot.[31] We are a long way from that. Most traits you can think of improving (intellect, height, body weight, and looks) are controlled by hundreds if not thousands of genes working together.

But heritable human genome engineering (HHGE) is coming whether we like it or not, and there is probably a HumGen or similar company launch in our relatively near future. But who would be the venture investors? Who would buy at the probably oversubscribed IPO? I wonder how it would feel to be a genetically engineered human being.

REFERENCES AND NOTES

1. Liu N, Hargreaves VV, Zhu Q, Kurland JV, Hong J, Ki W, Sher F, Macias-Trevino C, Rogers JM, Kurita R, et al. 2018. Direct promoter repression by BCL11A controls the fetal to adult hemoglobin switch. *Cell* **173**: 430–442. doi:10.1016/j.cell.2018.03.016

2. Klug A. 2010. The discovery of zinc fingers and their applications in gene regulation and genome manipulation. *Ann Rev Biochem* **79**: 213–231. doi:10.1146/annurev-biochem-010909-095056

3. To understand how zinc fingers and TALENS work, see Box 1 and Figure 1 in Box 4. In Figure 1, zinc fingers recognize triplets and the Fok1 nuclease operates as a dimer, cutting in the spacer region between two distinct ZF target sites. A TAL effector nuclease (TALEN) is similar in principle to the ZF nuclease, with the components of the array recognizing individual nucleotides. See Chandrasegeran S, Carroll D. 2016. Origins of programmable nucleases for genome engineering. *J Mol Biol* **428**: 963–989. doi:10.1016/j.jmb.2015.10.014

4. 2011. Addgene's guide to TALEN technologies, Newsletter, December 2011. http://www.addgene.org/talen/guide

5. Tebas P, Stein D, Tang WW, Frank I, Wang SQ, Lee G, Spratt SK, Surosky RT, Giedlin MA, Nichol G, et al. 2014. Gene editing of *CCR5* in autologous CD4 T cells of persons infected with HIV. *New Engl J Med* **370**: 901–910. doi:10.1056/NEJMoa1300662

6. Chang K-H, Smith SE, Sullivan T, Chen K, Zhou Q, West JA, Liu M, Liu Y, Vieira BF, Sun C. 2017. Long-term engraftment and fetal globin induction upon *BCL11A* gene editing in bone-marrow-derived CD34⁺ hematopoetic stem and progenitor cells. *Mol Ther Methods Clin Dev* **4**: 137–148. doi:10.1016/j.omtm.2016.12.009

7. Esrick EB, Lehmann LE, Biffi A, Achebe M, Brendel C, Ciuculescu MF, Daley H, MacKinnon B, Morris E, Federico A, et al. 2021. Post-transcriptional genetic silencing of *BCL11A* to treat sickle cell disease. *New Engl J Med* **384**: 205–215. doi:10.1056/NEJMoa2029392

8. Orkin SH, Bauer DE. 2019. Emerging genetic therapy for sickle cell disease. *Ann Rev Med* **70**: 257–271. doi:10.1146/annurev-med-041817-125507

9. Conversation with Jens Boch, 2022.

10. Davies K. 2021. *Editing humanity: the CRISPR revolution and the new era of genome editing*. Pegasus Books, New York.

11. Isaacson W. 2021. *The code breaker: Jennifer Doudna, gene editing, and the future of the human race*. Simon & Schuster, New York.

12. Mojica FJM, Díez-Villaseñor C, García-Martínez J, Soria E. 2005. Intervening sequences of regularly spaced prokaryotic repeats derive from foreign genetic elements. *J Mol Evolution* **60**: 174–182. doi:10.1007/s00239-004-0046-3

13. Barrangou R, Fremaux C, Deveau H, Richards M, Boyaval P, Moineau S, Romero DA, Horvath P. 2007. CRISPR provides acquired resistance against viruses in prokaryotes. *Science* **315**: 1709–1712. doi:10.1126/science.1138140

14. Cong L, Ran FA, Cox D, Lin S, Barretto R, Habib N, Hsu PD, Wu X, Jiang W, Marraffini LA, et al. 2013. Multiplex genome engineering using CRISPR/Cas systems. *Science* **39**: 819–823. doi:10.1126/science.1231143

15. Mali P, Yang L, Esvelt KM, Aach J, Guell M, DiCarlo JE, Norville JE, Church GM. 2013. RNA-guided human genome engineering via Cas9. *Science* **339**: 823–826. doi:10.1126/science.123203

16. Jinek M, East A, Cheng A, Lin S, Ma E, Doudna J. 2013. RNA-programmed genome editing in human cells. *eLife* **2**: e00471. doi:10.7554/eLife.00471

17. Cho SW, Kim S, Kim JM, Kim J-S. 2013. Targeted genome engineering in human cells with the Cas9 RNA-guided endonuclease. *Nature Biotechnol* **31**: 230–232. doi:10.1038/nbt.2507

18. Ledford H. 2022. Major CRISPR patent decision won't end tangled dispute. *Nature* **603**: 373–374. doi:10.1038/d41586-022-00629-y

19. Lander ES. 2016. The heroes of CRISPR. *Cell* **164**: 18–28. doi:10.1016/j.cell.2015.12.041

20. Frangoul H, Altshuler D, Cappellini MD, Chen YS, Domm J, Eustace BK, Foell J, de la Fuente J, Grupp S, Handgretinger R, et al. 2021. CRISPR-Cas9 gene editing for

sickle cell disease and β-thalassemia. *N Engl J Med* **384:** 252–260. doi:10.1056/NEJ-Moa2031054

21. Kingwell K. 2023. First CRISPR therapy seeks landmark approval. *Nat Rev Drug Discov* **22:** 339–341.

22. Conversation with Jean-François Formela, 2022.

23. Gillmore JD, Gane E, Taubel J, Kao J, Fontana M, Maitland ML, Seitzer J, O'Connell D, Walsh KR, Wood K, et al. 2021. CRISPR-Cas9 in vivo gene editing for transthyretin amyloidosis. *New Engl J Med* **385:** 493–502. doi:10.1056/NEJMoa2107454

24. Newby GA, Yen JS, Woodard KJ, Mayuranathan T, Lazzarotto CR, Li Y, Sheppard-Tillman H, Porter SN, Yao Y, Mayberry K, et al. 2021. Base editing of haematopoietic stem cells rescues sickle cell disease in mice. *Nature* **595:** 295–302. doi:10.1038/s41586-021-0360

25. Anzalone AV, Randolph PB, Davis JR, Sousa AA, Koblan LW, Levy JM, Chen PJ, Wilson C, Newby GA, Raguram A, Liu DA. 2019. Search-and-replace genome editing without double-strand breaks or donor DNA. *Nature* **576:** 149–157. doi:10.1038/s41586-019-1711-4

26. Everette KE, Newby GA, Levine RM, Mayberry K, Jang Y, Mayuranathan T, Nimmagadda N, Dempsey E, Li Y, Bhoopalan SV, et al. 2023. Ex vivo prime editing of patient haematopoetic stem cells rescues sickle-cell disease phenotypes after engraftment in mice. *Nat Biomed Eng* **7:** 616–628.

27. Mayuranathan T, Newby GA, Feng R, Yao Y, Mayberry KD, Lazzarotto CR, Li Y, Levine RM, Nimmagadda N, Dempsey E, et al. 2023. Potent and uniform hemoglobin induction via base editing. *Nat Genet* **55:** 1210–1220. doi:10.1038/s41588-023-01434-7

28. Ravindran S. 2019. Got mutation? 'Base editors' fix genomes one nucleotide at a time. *Nature* **575:** 553–555.

29. Goldstein D. 2022. *The end of genetics: designing humanity's DNA.* Yale University Press, New Haven, CT.

30. Wade N. 2015. Scientists seek moratorium on edits to human genome that could be inherited. *New York Times,* December 3, 2015.

31. Silver LM. 2007. *Remaking Eden: how genetic engineering and cloning will transform the American family.* Ecco Books, New York.

CHAPTER 16

Vaccines

I did not know that the technologies I grew up with in the biotechnology industry would be so important in helping to control COVID-19. Vaccines tend to be thought of as a relatively low-tech and low-margin business, despite pharma bringing in multiple billions of dollars in revenue each year from their vaccine franchises. Even if the individual vaccines are not expensive, vaccinating millions of people makes for good revenue.

This chapter begins with some vaccine history, exploring virus vaccines as illustrations of the kinds of vaccines that are available, and the technologies used to make them. We will then turn to COVID-19 and try to summarise the extraordinary story of the development of the SARS-CoV-2 vaccines: a story that includes both biotech and pharma.

THE BEGINNING OF VACCINATION

Most people are aware of Edward Jenner and the history of smallpox (variola) vaccination in the late 1700s. The practice of "variolation," in which material was taken from smallpox lesions in patients and used to "vaccinate" other people against the disease, had been going on for some time before Jenner made his observations about cowpox (i.e., that bovine lesions caused by cowpox (vaccinia) could be used to variolate people against the related disease smallpox with less morbidity than using human smallpox lesions). It was also well known at the time that dairy maids were resistant to smallpox, owing to their regular exposure to cowpox. Based on that knowledge, Benjamin Jesty was the first person to vaccinate humans (his family) with vaccinia taken from lesions on a cow's udder.[1]

I was vaccinated against smallpox as a child and then again at the University of Birmingham. Henry Bedson, from whose lab the Birmingham smallpox infection occurred that killed a research photographer in 1976, was in the Virology department where I was doing my PhD. We all had to be revaccinated with vaccinia.

Owing to the eradication of smallpox, routine vaccination against the disease was stopped in 1972. This is one reason why the recent monkey-pox outbreak, primarily in gay and bisexual men (reminiscent of HIV), happened. Fortunately, there are vaccines available that protect against monkeypox.[a]

Pasteur

Louis Pasteur made a significant contribution to the development of the vaccines we use today. He along with Émile Roux was instrumental in developing an attenuated (living but weakened) rabies virus vaccine from dried spinal cords from infected animals, introducing the concept of attenuation to vaccine making. These days, attenuated human tissue culture cell-derived rabies vaccines are available, made by Bavarian Nordic. Fred Brown's lab was working on rabies virus at Pirbright in the mid-1970s, so we were all vaccinated against this virus as well (in this case with an inactivated virus vaccine, made from virus-infected diploid monkey cells). Pasteur was also responsible for several bacterial vaccine developments, including cholera and anthrax.

YELLOW FEVER

I learned about yellow fever during my masters Virology degree course in 1972 before I did my PhD. Max Theiler developed a yellow fever vaccine called 17D in 1937 and was awarded the Nobel Prize in Physiology or Medicine in 1951 for doing so. Working at Rockefeller University, Theiler showed that the yellow fever virus could be transmitted to mice, speeding up the infectivity studies that previously could only be done in monkeys. Theiler made an assay for the detection of neutralising virus antibodies and noted that "passaging" the virus multiple times through mice produced a weakened virus that did not kill mice or monkeys.[2] This attenuation led to the development of the 17D live attenuated vaccine in 1938. This vaccine was launched in Brazil, and more than 400 million doses of the 17D virus vaccine were administered from Rockefeller to people in countries where yellow fever was endemic, especially South America

---

[a] Imvamune is a live-virus vaccine and Imvanex is a nonreplicating live vaccinia virus vaccine. Both are licensed for vaccination against smallpox, and Imvanex is licensed specifically to prevent monkeypox. They are made by a company called Bavarian Nordic in Denmark. Their headquarters is in Hellerup, where the Biogen antibody manufacturing facility (no longer owned by Biogen) that I once visited is located, north of Copenhagen.

BOX 1. YELLOW FEVER INFECTIONS

The various infection patterns that the virus uses are interesting, as they depend on geographical location and the transmitting mosquito vector. Yellow fever is caused by a flavivirus, a small RNA virus transmitted via the bite of an infected mosquito of the genus *Aedes* or *Haemagogus*. Yellow fever is related to other mosquito-borne diseases like West Nile, St. Louis encephalitis, dengue fever, and Japanese encephalitis viruses. Mosquitoes acquire the virus by biting infected humans or primates. The disease has three transmission cycles[3]: sylvatic (jungle), savannah, and urban. In the jungle cycle, the virus is transmitted by mosquitoes between monkeys in the forest (though humans can become infected by travelling into the jungle and getting bitten by an infected mosquito). The savannah cycle happens in Africa, where the virus is transmitted by mosquitoes to both humans and monkeys when they are living close to one another. There can also be human-to-human or monkey-to-human infections transmitted by the mosquito vector. The urban cycle primarily involves transmission of the virus to humans by *Aedes aegypti* mosquitoes, occurring mostly in Africa and South America.

(Box 1). Some of my relatives on my mother's side of the family succumbed to yellow fever in Buenos Aires at that time.

It is extraordinary that Theiler's safe and effective vaccine is still produced today by essentially the same methods that he and his colleagues used more than 80 years ago.[b]

DENGUE FEVER

Contrast yellow fever to dengue (break-bone) fever, another mosquito-borne disease caused by another flavivirus. Dengue is endemic in many parts of the tropical and subtropical world and has significant mortality associated with it. It usually causes a disease with aches and pains like an influenza infection, but the morbidity from it is still a major public health concern. There are four serotypes of the virus and infection with one does not protect against the others. Dengue infections are not only a health problem but also a substantial economic burden. Development of dengue vaccines began in the 1920s but has always been hindered by the need to create immunity against all four dengue serotypes (Box 2).

---

[b] Something to remember in these days of COVID-19 and the very sophisticated technologies that are being used to make vaccines against coronaviruses.

BOX 2. DENGUE FEVER VACCINES

The problem with the current vaccination strategy for dengue fever is that for naïve individuals (i.e., with no dengue antibodies), vaccination with all four serotypes results in making antibodies to all four serotypes. Subsequent infection with one of the dengue serotypes, particularly serotype 2, can lead to the production of antigen–antibody complexes. These can cause serious haemorrhagic complications and death—the so-called dengue haemorrhagic fever—in some individuals. This is also true for natural infections, in which a second infection can lead to the same severe disease. The new vaccine has therefore been restricted to people who have been infected with dengue previously, significantly complicating the vaccination process. Only one vaccine is available (Dengvaxia, a live attenuated virus vaccine made by Sanofi-Pasteur) presently, although several others of various kinds are in the works. The Sanofi vaccine is a recombinant version of the 17D yellow fever vaccine, in which the premembrane and envelope genes of the yellow fever virus are replaced by those from the four dengue virus serotypes.

Dengvaxia was approved in the United States in 2019 for the prevention of dengue disease in people aged 9–16, who have had laboratory-confirmed previous dengue infection and live in endemic areas. Takeda are developing a recombinant vaccine (Qdenga) based on the dengue type 2 serotype rather than the yellow fever virus template.[4] It has been reported that Indonesia, where there is a high incidence of dengue fever, will be introducing Qdenga in 2023 for all comers—not just those who were previously infected. The safety of this vaccine has been shown in the trials, but we will have to see what happens when it is used at a large scale in the general population. Takeda recently withdrew the Biologics License Application (BLA) with the FDA in the United States for this vaccine for "data collection" reasons, whatever that means.

POLIOVIRUS

Another virus back in the headlines after a long absence is poliovirus. My first encounter with a polio vaccine was in the early 1950s, taking the attenuated Sabin oral composite polio vaccine as a drop on a sugar lump. The Salk killed vaccine is used now in most countries, owing to the occasional reactivation of virulent polio type 2 from the attenuated vaccine. One of the reasons I remember it so well, apart from the sugar cube, is because my cousin Alison Brading suffered from paralytic poliomyelitis in Nigeria in the early 1950s when out there with her family (poliovirus was endemic in

West Africa at that time). Her father Brigadier Norman Brading was an eminent British soldier and became a senior hospital administrator in colonial Nigeria after World War II. The infection nearly killed her, but her father was able to obtain an iron lung in which she was placed. I remember visiting Alison when she came home from Africa at the Wingfield Hospital in Oxford (now the Nuffield Orthopaedic Centre), not far from where we lived. She was still in an iron lung to help her breathe and in the hospital for more than a year and paralysed from the mid chest down. Subsequently, she became an Oxford anatomy professor, drove a modified car, and lived alone in the outskirts of Oxford for many years despite being wheelchair-bound: an amazing achievement and testament to Alison's strength.

Neither Sabin nor Salk was awarded the Nobel Prize for their work. Instead, the Nobel Prize in Physiology or Medicine went to John Enders, Thomas Weller, and Frederick Robbins in 1954 for their studies on growing poliovirus in tissues and tissue culture. Their work led to the attenuated serotype vaccines, made by multiple tissue culture passages, as well as the inactivated virus vaccines, made by killing the virus with formalin.

If biotech had been around then, you can bet that Jonas Salk would have set up a company to make the killed poliovirus vaccine, and later Albert Sabin would have formed a company to develop the attenuated viruses. As it was, the March of Dimes Foundation funded the development of the killed vaccine, which was never subjected to patent protection. To their credit, the pharmaceutical companies helped to develop both vaccines to make them available widely.

There is an effort under way to fully eradicate poliovirus globally, but there are still "hot spots" in Pakistan and in Afghanistan. These outbreaks are not caused by wild-type virus but by revertants of the attenuated version of type 2 poliovirus, which have regained pathogenicity by regaining the ability to infect neuronal tissue and cause permanent paralysis. Sequencing analysis has shown that it only takes the substitution of very few nucleotides to turn the type 2 attenuated virus back into a wild-type virus able to infect neuronal tissue.[5,6] These revertant viruses have been found recently in sewage water in both London and New York, indicating that they are already circulating in the population: the vaccination levels in parts of those populations are clearly too low.

Sanofi-Pasteur, which makes the inactivated poliovirus vaccine, would probably have acquired Salk's company if there had been one. Instead, Salk's legacy is the famous Salk Institute, near the glider port and across the road from UCSD in San Diego, overlooking the Pacific Ocean. The Salk Institute is where Francis Crick and many other eminent scientists chose to work later in their careers.

HEPATITIS B VIRUS

At about the same time as we were unsuccessfully trying to make a subunit vaccine against foot and mouth disease virus (FMDV) by using cloned VP1 (see Chapter 2), several groups were cloning the DNA from hepatitis B virus. Baruch Blumberg, who won the Nobel Prize in Physiology or Medicine in 1976 for his discovery of the hepatitis B virus, first discovered a novel antigen (called the Australia antigen or AuAg) in the serum of individuals with viral hepatitis. David Dane and June Almeida working in London in 1970 also saw virus-like particles in infected serum after mild detergent treatment, which they modestly called "Dane" particles. These particles were subsequently found to consist of the hepatitis B virus core antigen (HBcAg) and the virus nucleic acid, and AuAg was identified as a polymeric form of the Hep B surface antigen (HBsAg).[7]

Cloning of the Hepatitis B Surface Antigen

In 1979, three groups reported almost simultaneously the cloning and expression in *Escherichia coli* of the hepatitis B virus proteins: Pierre Tiollais at the Pasteur Institute in Paris (an Institute with a deep history in infectious disease research); Pablo Valenzuela, Pat Gray, Howard Goodman, and Bill Rutter at UCSF in California (before Chiron was formed); and Ken Murray's team working in Edinburgh in association with Biogen.[8,9,10,11]

Ken Murray and Peter Hofschneider drove the Biogen cloning effort in Edinburgh for both professional and personal reasons: Murray had been involved in the 1969 viral hepatitis outbreak in the dialysis unit at the Edinburgh Royal Infirmary that killed several people, so he deeply appreciated both the need and the opportunity for a recombinant vaccine. The starting materials for their cloning experiments came from the bacteriology department in Edinburgh. In December 1978, the Edinburgh team filed their intellectual property (IP) and a year later, the patent application issued under a very long title: "Recombinant DNA molecules and hosts transformed with them which produce polypeptides displaying HBV antigenicity and genes therefore and methods of making and using these molecules, hosts, genes and polypeptides." The patent was granted in 1987, giving Biogen exclusivity over the subject matter. The group at UCSF and Chiron started their cloning with DNA from Maurice Hilleman at Merck. Their paper was published in *Nature* a week after the second Murray paper had been submitted, so it was a competitive race. The UCSF group expressed the 220-amino acid sequence of the surface antigen in a yeast expression system in collaboration with yeast geneticist Ben Hall at the University of Washington in Seattle, as the protein failed to be expressed at high levels in either *E. coli* or mammalian cells. Expression in yeast cells was very successful in making large amounts of protein (>800 mg/L) that was immunogenic in rabbits. The IP for the expression of the antigen in yeast was assigned to the Regents of the University of California.

Murray and the Biogen team did not win the race to make the vaccine: Biogen were not convinced that this was a commercially viable project. It was left to Ken Murray to broker a deal with John Beale at Burroughs Wellcome to support the expression studies in yeast. Biogen subsequently licensed the technology to SmithKline Beecham, which then commercialised it. Both the Chiron group (Merck) and the SmithKline vaccine group at Rixensart in Belgium were able to make the protein at large scale with good yields.

Hilleman's team at Merck had originally produced an HBsAg-based vaccine using inactivated protein from infected individuals, but this was removed from the market once the yeast-derived vaccines were available. Recombivax HB, the Merck vaccine, was the first recombinant hepatitis B vaccine to be approved in 1986, emanating from the work of the Hilleman group at Merck and their UCSF, UW, and Chiron collaborators.

A Milestone Vaccine

The Hep B vaccine was the *first* recombinant DNA-derived vaccine and only the *third* recombinant product ever to be approved: a somewhat forgotten milestone. Engerix-B, a similar product from SmithKline Beecham, was approved in Belgium in 1986 and in the United States in 1989. GenHevac from the Pasteur Institute group, which was made in mammalian cells, was approved in France in 1989 but was not licensed anywhere else.

Merck and SmithKline had to cross-license the IP that was filed by the three competing teams for them to commercialise their vaccines. It seems that this was done following normal business procedures, and the vaccines were much more successful commercially than anticipated. The two main

BOX 3. TOWARDS NEWER HEPATITIS B VACCINES

There have been attempts to make HBsAg derivatives that contain the preS sequence, which encodes the preS domain of the surface protein. PreS is also immunogenic and can illicit neutralising antibodies. The thinking was that this additional peptide sequence may increase the immunogenicity of the sAg. Despite there being some people who do not respond to the yeast-derived HBsAg vaccines, these preS studies have not resulted in new hepatitis B vaccines. It remains to be seen whether an mRNA- or DNA-vectored version of the surface antigen work with or without preS will work better than the yeast-derived protein. Given the conformation of the virus-like particles consisting of HBsAg made in yeast, I somehow doubt the wisdom of trying to develop nucleic acid–derived versions.

hepatitis B vaccines today are still those made by Merck and GSK from technology that is more than 40 years old. They are both used as part of childhood vaccination schedules and for at-risk individuals (three shots of these vaccines are sufficient to protect 90% of the people inoculated against hepatitis B) (see Box 3).

The hepatitis B surface antigen has been used as a "particle expression donor" for other vaccines. For example, the new circumsporozoite protein (CSP) malaria vaccines are based on the C terminus and central amino acid repeats of the CSP (the immunodominant protein found on the surface of the malaria parasite) fused to the HBsAg (Box 4). This portion of the fusion protein causes it to self-assemble into virus-like particles in yeast, which can be easily purified and formulated into a vaccine.

BOX 4. MALARIA VACCINES

The new R21/Matrix-M malaria vaccine from the Jenner Institute in Oxford (which developed the ChAdOx COVID-19 vaccine) contains fewer amino acids from the HbSAg protein than the original GSK (RTS,S) construct, resulting in a vaccine with fewer HBsAg monomers and more CSP repeats on the surface of the virus-like particles. Both these vaccines are administered with adjuvants and are immunogenic and protective. The latest results from the R21 trial in Africa are very exciting, with 80% protection in children.[12,13] This bodes well for the much sought-after malaria vaccines being available more widely by 2025. It is a little surprising that some of the biotech companies have not joined the fray here. It would not be at all surprising to learn that their commercial assessment of malaria vaccination is unfavourable: and, of course, wrong.

MOLECULAR GENETICS INC.

Biogen and Chiron were not the only early 1980s recombinant DNA companies with interests in making vaccines. Molecular Genetics Inc. (MGI) of Minnetonka, Minnesota, were once mentioned in the same breath as Genentech and Chiron, but for many of today's biotech people, they are a forgotten company: they are not a household name for having cloned anything. However, they did have a long-standing interest in cloning and expressing genes from herpes simplex virus in *E. coli*, with a view to making HSV vaccines. They filed some robust IP on expressing HSV proteins in *E. coli*, raised lots of money, made many contracts with other companies, and developed some new products. MGI went public in 1982 and formed a collaboration with Lederle Laboratories (American Cyanamid) around their HSV vaccine interests. They eventually got sold to Eisai in 2008 for $3.9 billion. It is unclear if that was a good outcome for their founders and shareholders.

HUMAN PAPILLOMA VIRUS (HPV) VACCINES

Another recombinant virus vaccine made in yeast is for human papilloma virus (HPV), which has similarities in origin to the hepatitis B surface antigen vaccines. The recently deceased Harald zur Hausen won the Nobel Prize in Physiology or Medicine in 2008, for his discovery of the family of HPVs, a prize he shared with Françoise Barré-Sinoussi and Luc Montagnier for their discovery of HIV. Only a small number of serotypes of the large HPV family of viruses create serious health problems. Some of them cause genital warts, and others cause cervical and other squamous cell carcinomas over time (specifically HPV serotypes 16 and 18). zur Hausen was the first to isolate HPV from cervical cancer tumours, but a vaccine to prevent infection against these viruses did not materialise for several more years.[14]

Work by several groups, including the University of Queensland in Brisbane, Australia; Georgetown University; The State University of New York Rochester; and the NCI, led to the vaccines that are available today. Ian Frazer and Jian Zhou in Australia were the first to show that the outer capsid proteins of HPV (L1 and L2) could form virus-like particles when expressed in yeast, like the hepatitis B sAg (see above). MedImmune obtained this virus-like protein technology from the University of Queensland and partnered with GSK to develop their HPV vaccine. The Georgetown team were also investigating the properties of these capsid proteins and filed patents. Both the Georgetown and the Queensland sets of IP were subsequently licensed to Merck and CSL (Commonwealth Serum Laboratories in Australia, now CSL Seqirus), as well as to GSK. The studies were published in 1991, but it took several more years to turn the findings into vaccines. Credit must also be given to the NCI, in particular Douglas (Doug) Lowy and John Schiller, who were given the Lasker Award for their work showing that the virus-like particles produced a protective immune response against HPV in animals, and for driving the vaccine studies.

Gardasil

The first HPV vaccine on the market (Gardasil) was produced by Merck. It protects against four major HPV serotypes including HPV 16 and 18 (which cause cervical and head and neck cancer) and HPV 6 and 11 (which cause genital warts). Gardasil was approved in 2006 in the United States and Australia following extensive clinical trials. By 2007, the vaccine was approved in 80 countries. The equivalent vaccine from GSK (Cervarix) was approved in 2009 in the United States and several other countries. It is marketed primarily outside the United States, protects against HPV 16 and 18, and is effective in preventing at least 70% of the cancers caused by HPV. In 2014, a new vaccine against nine HPV strains (Gardasil 9) was approved, providing protection against HPV 16, 18, 6, and 11 like the original, but adding additional HPV serotypes 31, 33, 45, 52, and 58, all of which have been implicated in cancer causation. Owing to the safety and efficacy of these recombinant vaccines, more than 100 countries worldwide now vaccinate against HPV as part of their vaccine schedules.[c]

INFLUENZA VIRUS

I wonder how many people have really had influenza rather than just a "touch of flu" caused by a different virus.[d] If you get influenza, you know it.[e] Influenza is an unpleasant disease with high morbidity that infects millions

---

[c] Given the current right-wing resistance to vaccines, one wonders whether this vaccine— given what it protects against—will not come under pressure as "undesirable."

[d] Despite the off-used phrase, there is no such thing as a "touch of flu."

[e] My brother-in-law was at our house in Surrey in the 1970s with an infection of the "real" flu strain circulating at the time (H3N2). The illness lasted at least two weeks, and for half of that time, he had a temperature of greater than 102°F.

worldwide and kills more than 25,000 people a year in the United States alone. It is not surprising that there have been large efforts over the years to produce effective flu virus vaccines.[15]

Types of Influenza

There are many challenges to making vaccines against influenza. First, the virus comes in two major serotypes: influenza A and B (although minor serotypes C and D also exist; see Box 5). Influenza A viruses are divided into subtypes based on two proteins on the surface of the virus; haemagglutinin (H or HA—called that because it agglutinates red blood cells) and the neuraminidase (N or NA—an enzyme which removes sialic acid residues from proteins and is involved in the release of virus from infected cells). There are 18 subtypes of HA and 11 subtypes of NA that can be paired in influenza A. Influenza B also has two major lineages but fewer subtypes.

The influenza virus that caused the Spanish flu pandemic that began in 1918 was an H1N1 influenza A strain. This viral outbreak infected more than 500 million people across the world and probably killed more than 50 million of them.[f] Unusually, it was young adults that bore the brunt of the 2018 flu mortality. The older generation may have had some protection from the flu outbreaks that had occurred at the end of the nineteenth century.

BOX 5. DISCOVERY OF INFLUENZA VIRUS

Influenza virus was first isolated and characterised in the United Kingdom in the early 1930s by Wilson Smith, Sir Christopher Andrewes, and Patrick Laidlaw at the MRC laboratories in Mill Hill. They isolated influenza A from the nasal secretions of infected patients. Similar work was done at Vanderbilt in the United States, where it was shown that flu virus could be grown in the allantois of chicken eggs. Much of the work to develop flu vaccines relies on this method of production, as does all the very elegant structural work done on the HA and NA proteins. I remember seeing this process firsthand when I visited John Skehel's lab at the National Institute for Medical Research at Mill Hill, where hundreds of eggs were inoculated with flu virus every day.

---

[f]This was the deadliest pandemic infection to affect humans since the Black Death plague in 1346–1353 caused by the bacterium *Yersinia pestis*, transmitted to humans by infected rat fleas.

The Influenza Genome

The flu virus genome is made up of separate RNA segments. The RNAs coding for the HA and NA proteins can reassort in the wild by interchanging their RNA coding segments with other flu viruses infecting wild birds, pigs, or other animals. This gives rise to the subtypes such as H1N1 and H3N2 that routinely circulate in human populations (differences between the two types are referred to as "antigenic shifts"). The vaccine development problem is exacerbated further by genetic variation (mutation) that occurs in the HA and NA genes as the viruses replicate (called "antigenic drift"). These changes can overcome prior immunity that has been derived by either previous infection or vaccination.

Given that the great flu pandemic was still recent history when the virus was first characterised, there was considerable activity directed at making a vaccine.

Original Flu Vaccines

During the Second World War, scientists in both the United Kingdom and the United States derived the first inactivated flu vaccines, using various methods (e.g., formalin) to inactivate the virus. The vaccines were tested primarily on military personnel. Influenza B was discovered at this time, leading to bivalent inactivated vaccines. Antibody titres were determined by inhibition of red blood cell haemagglutination, mediated by the HA protein. The first commercial vaccine production methods were developed after 1944, following Stanley's description of the process for growing the virus in eggs, purifying it by differential centrifugation, followed by inactivation and formulation.

The original flu A and B vaccine strains, which were passaged multiple times in eggs and grew to high titres, are the basis for all egg-derived flu vaccines. Simply replacing the HA or NA genes of the vaccine strain with the RNA-coding segments from circulating strains enables vaccines to be made against viruses of different subtypes like H3N2 (replacing the H1N1 segments in the PR8 strain). This was fortunate because several new influenza viruses arose subsequently. The H2N2 epidemic virus in the late 1950s circulated for about 10 years, until it was replaced by the so-called Hong Kong flu, an H3N2 serotype. Bivalent vaccines containing the predominant A and B types were produced in response to the emergence of these new flu viruses.

Trivalent and Quadrivalent Vaccines

When two A subtypes circulated at the same time in 1978, a trivalent vaccine was made (H1N1/H3N2/B). Further understanding of variation in influenza B, which is not as extensive as for A, has now led to the development of quadrivalent vaccines. These protect against two subtype A viruses (H1N1 and H3N2) and two subtype B viruses. The precise viruses used in the vaccine change from year to year, based on the antigenic drift that is seen in the HA and NA genes of the flu viruses circulating in the northern and southern hemisphere of every year, and are reviewed constantly by the World Health Organization.

There are several quadrivalent flu vaccines. Flublok from Sanofi was originally licensed as the first trivalent influenza vaccine to be made using recombinant DNA technology. It is now quadrivalent, with the HA and NA proteins of the viruses expressed in an insect cell/baculovirus vector system, allowing higher amounts of the HA protein to be made and incorporated into the vaccine. This vaccine (and the baculovirus system) was obtained in Sanofi's acquisition of a company called Protein Sciences (founded by Dan Adams, a Biogen founder) for $650 million in 2017. Flucelvax Quadrivalent made by CSL Seqirus, containing virus grown in cell culture, is approved for people six months and older. Both these vaccines are egg-free, a trait that makes them more desirable for those people who are allergic to egg-derived proteins. Sanofi's Fluzone is a high-dose quadrivalent vaccine made in eggs for use in people aged 65 and older, and Fluad is an adjuvanted quadrivalent vaccine made in eggs, sold by CSL Seqirus.[15]

Fluad is worth a little more description. It has its origins at Chiron and was launched as an adjuvanted influenza vaccine in the late 1990s. Chiron expanded their vaccine portfolio by acquiring the vaccine company Sclavo in Italy and the European vaccine business of Behring.[g] Chiron also manufactured an MMR vaccine (measles, mumps, and rubella) at their Liverpool plant in the United Kingdom, in addition to Fluad. The acquisition of PowderJect, a United Kingdom–based vaccines company, made Chiron into the second-largest flu vaccine provider in the world and one of the largest vaccines businesses. PowderJect, a needle-free protein delivery company, had been founded by Professor Brian Bellhouse as

---

[g]Behring were a part of Hoechst known as Behringwerke, located in Marburg, Germany. They were started in 1904 by Emil von Behring, the 1901 Nobel Prize winner in Physiology or Medicine for work on bacteria like diphtheria.

BOX 6. AN ASIDE: RACING AT THE 24 HOURS OF LE MANS

I was always a bit jealous of Drayson. Not for his money, but for the fact that he was able to start Drayson Racing Technology and compete at Le Mans in 2009 in an Aston Martin and in 2010 in an LMP 1 car. His team ran in many long-distance sports car races. I overlapped with Lindsay Owen-Jones in my house at school, who later became the CEO of L'Oréal who also raced at Le Mans in a McLaren. Two connections for an unrealised (and somewhat unrealistic) ambition of mine to race at Le Mans. I have been to the Le Mans race several times, most notably the 1970 Le Mans, which was not only the backdrop for the Steve McQueen movie, but it was also a straight fight between the long-tailed 917 Porsches and the 512S Ferraris (see Plate 25 for a photograph of the cover of the 1970 24 Hours of Le Mans Race programme). There were eight 917s and 11 512Ps participating initially. There was no chicane on the Mulsanne Straight then, and speeds of close to 260 mph were recorded. The noise of the cars changing gear going through Tertre Rouge, the corner before the Mulsanne Straight, was just mind-blowing.

---

an Oxford University spinout, financed by ISIS innovation. Paul Drayson (later Lord Drayson; see Box 6), who was the son-in-law of the founder, became CEO and Clive Dix, from Glaxo Group Research, became CSO.[h] The company was bought by Chiron for £542 million in 2003, making the founders and senior management a good return. Chiron were subsequently bought by Novartis, who divested their vaccines business to GSK for $7.1 billion in 2014.

PowderMed

Drayson and Dix spun some of the Powderject technology into PowderMed following the Chiron acquisition. PowderMed had developed a unique technology to deliver DNA directly to the immune system, using very small gold particles coated in DNA. The particles were delivered under the skin using pressurised helium. SV Life Sciences invested in PowderMed, which had a pipeline of vaccine candidates for the H5N1 strain of influenza and also for hepatitis B. PowderMed were sold to Pfizer in 2006 for what was reported by the *Financial Times* to be around $400 million. Clive Dix subsequently became the Chairman of the pain company Convergence, another SV Life sciences investment and a spinout of GSK (see Chapter 11). Clive also

---

[h]I worked with Clive, as well as playing squash with him (he nearly always beat me).

worked closely with Dame Kate Bingham on the U.K. Vaccine Taskforce during the COVID-19 pandemic.[16]

Live Attenuated Flu Vaccines—Aviron

Live attenuated flu vaccines remain available years after they were first developed in Russia. FluMist quadrivalent is a live attenuated flu nasal spray approved for people between the ages of 2 and 49. FluMist has a very definite and unique biotech connection. It was developed by a company called Aviron in Mountain View California. Aviron were co-founded in 1992 by Chairman and CEO Leighton Read, a partner at Interhealth Partners.[i] Peter Palese, a flu virologist from Mt. Sinai in New York, Bernard Roizman, a herpes virologist at the University of Chicago, and John Skehel in London were all also involved. Accel and Institutional Venture Partners (IVP) were investors in the company.

Aviron focused on developing live attenuated vaccines on the basis that they provided more durable immunity than killed vaccines. Their major focus was on a nasal spray administration of a cold-adapted attenuated influenza virus. This virus form had been created at the University of Michigan by John Maassab, who had worked on attenuated flu viruses for a long time.[j] The company had interests in attenuated parainfluenza virus 3 (a virus pathogen important in immune-deficient patients) and EBV. Clinical trials were already underway when Aviron went public in 1996, in an offering that raised $18 million led by Robertson Stephens, Bear Stearns, and H&Q. They also had alliances with several companies, including Sang-A Frontec in South Korea and SmithKline.

Aviron was acquired by MedImmune in 2001 for about $1.5 billion, before their FluMist nasal spray flu vaccine—a trivalent vaccine at that time— was approved in 2003. MedImmune's strategy behind the purchase was to increase their presence in respiratory virus diseases by adding to their lead product Synagis, a mAb against respiratory syncytial virus (RSV). Aviron had signed a co-promotion deal with AHP/Wyeth, but AstraZeneca now

---

[i] Reid had worked with Alejandro Zaffaroni at Affymax. He was also on the Board of Axys Pharmaceuticals, the company that was formed when Arris bought Sequana (Chapter 7). The late Alan Mendelson, a much respected and well-known biotech lawyer at Latham & Watkins, who worked with Amgen in the early days, was also on the Board.[17]

[j] Maassab was on the Aviron Scientific Advisory Board.

markets the vaccine after they acquired MedImmune. It has had patchy success owing to the development of the quadrivalent vaccines, but it is still a legitimate alternative for younger people.

MedImmune

MedImmune were an important biotech company started as "Molecular Vaccines" by Wayne Hockmeyer and Franklin Top in 1988 located in Gaithersburg, Maryland, to focus on therapies for RSV infection. Hockberg had previous experience with conjugated paediatric vaccines at Praxis Biologics. MedImmune were financed initially by Wally Steinberg at HealthCare Ventures with an investment of $3.5 million. Their major claim to fame was the development of Synagis, a mAb treatment for RSV infection in young children. MedImmune were acquired by AstraZeneca in 2007 for $15.6 billion.[18]

Fourth-generation flu vaccines are currently in development. Both Moderna and Pfizer, for example, have quadrivalent mRNA-based vaccines in clinical trials. The modified mRNA coding for the HA and NA from the prevalent type A and two type B viruses are made and incorporated into lipid nanoparticles (LNPs), just like the current successful SARS-CoV-2 mRNA vaccines.[19]

CORONAVIRUSES

When the coronavirus diseases SARS-CoV-1 and MERS first arose, making vaccines against them became a large effort. However, as these coronavirus diseases declined more rapidly in the population than expected, that effort lost some steam, and the vaccines never went all the way through development and to the market. The SARS-CoV-1 infections did lead to several important observations including the understanding that although protective antibodies to the viral spike protein were produced after infection, antibodies and T cells recognising additional virus proteins were also found. These additional immune responses were thought then to be an important additional component of protection against the virus.

After the onset of the SARS-CoV-2 pandemic in late 2019/early 2020, there were more than 50 different approaches taken to make vaccines against the deadly disease, not only to protect individuals against the illness but also to control its transmission. All of them were variations on the themes outlined in Box 7.

BOX 7. VACCINE TECHNOLOGIES

The technologies to make virus vaccine candidates can be broadly categorised into four groups.

Whole-Virus Vaccines

Whole-virus vaccines are comprised of inactivated or attenuated forms of live virus to promote protection, mimicking what happens in a natural infection. This is a well-trodden path for the development of vaccines against many viruses and has been the province of large pharma. The polio vaccine is probably the best-known example of this group.

Protein Subunit Vaccines

Subunit vaccines consist of one or more of the most important immunogenic proteins of the virus, including the most important external proteins. These are usually made by recombinant DNA methods and formulated into a vaccine, often with an adjuvant that increases the immunogenicity of the vaccine. The hepatitis B and HPV vaccines are subunit vaccines.

Vector Vaccines

Virus vector vaccines are composed of a DNA or RNA vector, combined with a DNA or RNA copy of a virus gene or several virus genes. The delivery of the recombinant virus vector to a patient results in the expression of the virus protein(s) and the induction of immunity. As in gene therapy with AAV vectors, previous exposure of people to the vector can reduce efficacy of the vaccine.

Vaccines Based on Messenger RNA (mRNA)

The fourth category is the now infamous mRNA-based vaccines. They consist of a modified mRNA coding for a virus protein delivered in a lipid nanoparticle/polyethylene glycol formulation. The best-known examples of this are the SARS-CoV-2 vaccines produced by BioNTech/Pfizer and Moderna.

Sars-CoV-2

Killed Virus Vaccines

Several Chinese companies focused on making killed whole SARS-CoV-2 virus vaccines. Sinovac Biotech rapidly developed such a vaccine, called CoronaVac, that was tested in clinical trials in several countries. CoronaVac is used widely not only in Asia but also in South America and the Caribbean. It provides some protection (<80% efficacy) against COVID-19 disease and death, a so-so response that is offset partially by the convenience of the vaccine not requiring frozen storage.

Valneva are a specialty European vaccine company based in France that had a manufacturing facility in Livingston near Edinburgh in Scotland. Valneva and Dynavax together developed an inactivated whole virus vaccine against SARS-CoV-2 (VLA 2001), which was rapidly tested in the United Kingdom as part of the government's COVID-19 Vaccine Procurement Taskforce. Valneva made an agreement with the U.K. government to provide up to 100 million doses of the vaccine to be made in Scotland. The vaccine clearly worked, providing very good immunity to infection. Despite that outcome, the U.K. government cancelled the vaccine order in September 2021, presumably on the grounds that they had a more than adequate supply of other types of vaccines.[k] The European commission also cancelled an advance purchase agreement for 60 million doses.

These cancellations are unfortunate for several reasons. The most important of these is that using a whole virus vaccine in combination with a subunit or mRNA vaccine may give better booster protection against virus variants in the long term. In this way mimicking to some extent an actual infection, there will be T cells and antibodies recognising multiple virus proteins developed in the infected person and not just those targeting the spike protein. One might not expect many government bureaucrats to understand that though, and, in this case, they met expectations.[16]

Subunit Vaccines

Novavax are a 1987 vintage biotech vaccine company based in Gaithersburg, Maryland, and focused on making a variety of subunit vaccines. Before the COVID-19 pandemic, Novavax were known for developing subunit vaccines against influenza and RSV. They were also working with the NIH on a vaccine for Ebola virus, an exotic and lethal haemolytic virus found in equatorial Africa.

The company had been through some hard times with their development of ResVax, a recombinant F-protein (fusion protein, which helps the virus enter a host cell) vaccine for RSV. This subunit vaccine had failed to provide efficacy against RSV infection in several trials, even though it was shown to be immunogenic (see Box 8). Novavax were financed by a mixture of private and public investments, including more than $89 million from The Bill and Melinda Gates Foundation to support their paediatric RSV programme.

---

[k]There may have been a political component as well. Valneva employed many people in Livingston, Scotland, and the vaccine was approved by the U.K. MHRA (Medicines and Healthcare Products Regulatory Agency) in the spring of 2022, so it is hard to understand why the contract was terminated on purely "supply and demand" grounds.

BOX 8. RSV SUBUNIT VACCINES

GSK have been very successful recently with Arexvy, their new RSV vaccine approved in May 2023. This vaccine contains a recombinant subunit pre-fusion RSV F-glycoprotein antigen (RSVPreF3) combined with GSK's proprietary AS01 adjuvant (Stimulon), which they obtained from Agenus. This adjuvant–antigen combination is thought to help to overcome the decline in immunity seen in older people, for whom the vaccine is primarily being made. In October 2022, GSK announced data indicating that the vaccine was very efficacious with high levels of protection against RSV A and B subtypes seen in older adults and this vaccine has been approved. Pfizer have followed a similar approach and their vaccine (called Abrysvo) is also available now. Moderna are rapidly closing the gap in this race with an mRNA-based prefusion F-protein vaccine for adults more than 60 years old.[20]

When COVID-19 emerged, the Coalition for Epidemic Preparedness Initiative (CEPI) and Operation Warp Speed contributed $384 million and $1.6 billion, respectively, to Novavax to speed up the development and manufacturing of their NVCoV2373 subunit vaccine, based on the SARS-CoV-2 spike protein plus adjuvant, with the aim of making 100 million doses available by January 2021. The phase 3 trials of the vaccine showed an 89% efficacy, including against at least some of the newer variants. At present, this Novavax COVID-19 vaccine (Nuvaxovid) has been approved by the FDA in a two-dose primary series for people more than 12 years old, becoming the fourth COVID-19 vaccine available in the United States. The development of Nuvaxovid shows what a relatively small biotech company can do in what is traditionally the domain of the Big Pharma vaccine companies, such as GSK, Merck, and Sanofi, none of whom have yet developed a subunit vaccine against SARS-CoV-2.

DNA Virus Vectors

Vaccitech were founded by Sarah Gilbert and Adrian Hill in 2016, as a spin-out from the University of Oxford's prestigious Jenner Institute. To get started, they raised money from Oxford Sciences Innovation (OSI), GV, Sequoia Capital, and some Asian investors. The company has built a platform for immunising animals (including humans) with different DNA virus vectors. One vaccine is based on the modified Ankara vaccinia virus, and another on a chimpanzee adenovirus vector.[1] In various inoculations with various antigens, these vaccines were shown to elicit both antibody and T-cell responses.

---

[1] The primate adenovirus vector DNA backbone was chosen because there are few antibodies to this virus in the human population, compared to a human adenovirus.

Vaccitech were quick off the mark in early 2020 to adapt their adenovirus vector to express the spike antigen of SARS-CoV-2. They had some previous experience with coronaviruses as they had already been developing an adenovirus-based vaccine for the MERS virus spike protein, a coronavirus relative of SARS-CoV-2 that caused a deadly coronavirus disease outbreak in the Middle East derived from infected camels.

Vaccitech's direct competitor in this virus vector vaccine space was Johnson & Johnson (J&J), who were developing a similar (serotype 26 adenovirus) vector construct, a recombinant, replication-incompetent virus encoding a full-length and stabilised SARS-CoV-2 spike protein. The Chinese company, CanSino Biologics, have also developed an adenovirus-vectored SARS-CoV-2 vaccine based on serotype 5 adenovirus. This is an inhaled mucosal vaccine: its efficacy remains to be seen. Other countries are developing similar Ad5-nCoV-S recombinant vaccines against COVID-19, but scant data are available on their efficacy.

Vaccitech worked closely with both AstraZeneca and Oxford Biomedica, a 1996 spinout from Oxford University, to scale up their ChAdOx1 vaccine, as it was later named (Box 9). This vaccine, now called Vaxzevria, formed the backbone of the vaccine supply for the United Kingdom. Vaxzevria clearly produces excellent immunity against SARS-CoV-2.[16,21]

mRNA Vaccines

Derrick Rossi, a Canadian educated at the University of Toronto and at the University of Helsinki, will go down in history as the principal founder of Moderna. Rossi was a stem cell biologist, interested in induced pluripotent stem cells (iPSCs) made from differentiated cells using transfected transcription factors.[m]

BOX 9. OTHER VACCITECH VACCINES

Vaccitech are not a one-trick pony. They have vaccines in the works for HBV (VPT-300), HPV (VPT-200), and shingles (VPT-400), plus two cancer vaccines aimed at prostate cancer and non-small cell lung cancer, respectively. They went public in April 2021, raising more than $100 million. With revenues from the COVID-19/AZ vaccine, they should continue to be a well-funded and successful operation.

---

[m]The 2012 Nobel Prize in Physiology or Medicine was awarded jointly to Sir John Gurdon and Shinya Yamanaka "for the discovery that mature cells can be reprogrammed to become pluripotent."

Instead of using virus vectors and DNA, Rossi wondered whether you could simply reprogram cells by transducing mRNA coding for the transcription factor genes into the cells. He was aware of the work done by Jon Wolff in Madison at the University of Wisconsin, who had injected mice with luciferase mRNA and showed that they made the protein. He was also aware of the work of Katalin Karikó and Drew Weissman at the University of Pennsylvania (UPenn) in Philadelphia, who had synthesised RNA *in vitro* containing modified bases such as pseudouridine, which did not change the coding of the RNA but reduced the induction of innate immunity signalling and thus increased the half-life of the mRNA *in vivo*.[n]

ModeRNA (AKA MODERNA)

In 2007, Rossi was working at the Harvard-affiliated Immune Disease Institute (IDI) in Cambridge (MA). He was sufficiently excited by the results he obtained with his mRNA stem cell transduction experiments to think about forming a company.[o]

As related in the 2022 book *The Messenger* by Peter Loftus,[22] Rossi was encouraged to go and meet Timothy (Tim) Springer by Ryan Dietz, the IDI lawyer Rossi was working with to make sure the requisite IP was filed. So, Rossi duly presented his results and his idea of a company in the spring of 2010. Springer was something of an entrepreneur as well as a structural biologist of some repute at Boston Children's Hospital. He had made considerable money from selling LeukoSite (a company he founded) to Millennium in October 1999 in a stock-for-stock deal that closed at close to $1 billion.

Rossi's presentation went well, and Springer agreed to invest. They met with Robert (Bob) Langer, a biochemical engineer and another serial entrepreneur and MIT professor, who became a founder. Kenneth (Ken) Chien, an MD and stem cell research scientist that Rossi knew, also became a founder. The group took the story to Flagship Ventures, where they met not only Flagship's leader Noubar Afeyan but also Doug Cole, an MD and managing partner. Flagship invested alongside Springer.[p] ModeRNA, as

---

[n] Karikó and Weissman have subsequently received many awards, including the prestigious Lasker Award and the 2023 Nobel Prize in Physiology or Medicine.

[o] Weissman and Karikó had already started a company in this arena called RNARx, based in Philadelphia. Their institution (UPenn) had filed IP, but the company never got traction financially.

[p] Tim Springer is now a Harvard Professor, an investor in start-ups, a serial entrepreneur, and very wealthy. He started the successful companies Scholar Rock and Morphic Therapeutic, in addition to being a Moderna investor.

they were known then, were born. The initial plan was to use mRNA rather than the proteins themselves for "enzyme replacement therapy."

Bancel Becomes CEO of Moderna

Afeyan persuaded Stéphane Bancel, the young CEO of bioMérieux, a multibillion-dollar French diagnostics company, to become Moderna CEO in early 2011. They had worked together before on the Board of BG Medicine, a diagnostics company based at the time in Waltham, Massachusetts.[q] The company went public in early 2011 (shortly before Bancel became the CEO of Moderna) in a deal led by Lazard Capital Markets and Cowen and Company, raising a little more than $30 million.[23] Bancel and Afeyan are a pair of very smart individuals, but once their minds are made up about something, it is difficult to get them to be changed. This can be either a good thing or a bad thing, depending on the decision that is made. As is often the case with strong-minded individuals, they prefer to surround themselves with people who do not say "no" very often. However, if you look at Moderna carefully, it seems that many more good decisions got made than bad ones.

As might be anticipated given the strong personalities involved, not everything went smoothly. Derrick Rossi, the founder of Moderna, cut most of his ties with the company and stepped off the SAB in 2013 for that reason.

First Labs

Like most start-ups, Moderna existed in a hole-in-the-wall lab for many months, with the first employee (Jason Schrum) synthesising mRNAs of various kinds to see what kinds of activities they had. Bancel and the team raised an additional $40 million in December 2012, supporting the growing company focus on injecting *in vitro*–synthesised RNA encapsulated in proprietary LNPs to make proteins of known therapeutic utility, like EPO and VEGF. The data they generated helped to persuade AstraZeneca, led by Bancel's fellow Frenchman Pascal Soriot, to do a $240 million upfront, five-year deal with milestones and royalties, in 2013. The deal was to

---

[q] Stelios Papadopoulos and I were BG Medicine Board members, which gave me a chance to watch Afeyan and Bancel work together. It comes as no surprise to me to read what has been written about them in the Loftus book.[22]

develop mRNA therapeutics for cardiovascular disease, metabolic disease, kidney disease, and cancer. AZD 8601, a VEGF mRNA product, has since been in phase 1 clinical trials, but little else of note emerged from this original deal.

The money from this deal, coupled with a $25 million grant from DARPA (Defense Advanced Research Projects Agency) and an additional raise of $110 million from investors, gave Moderna much more money in the bank and allowed them to move to more appropriate and larger space in Technology Square in Cambridge. Alexion Pharmaceuticals also paid Moderna $100 million in January 2014 for product options to develop mRNA treatments for various rare diseases, not one of which has made it to the clinic almost 10 years later.

Moving into Vaccines

Moderna's move into vaccines was fortuitous but had some logic to it. One significant problem with using mRNA as a template *in vivo* to make a protein is that none of the normal controls over synthesis and degradation of the mRNA are incorporated into the construct. In effect, you get whatever expression you get, depending on where the mRNA ends up and how long it can hang around. This is probably not such a problem for a vaccine, for which the purpose is to make enough protein antigen and peptides from it to generate an immune response without any other physiological effect. Using mRNA to deliver a biologically active molecule, like a cytokine or a growth factor, in an uncontrolled way may not be so forgiving when there is biological heterogeneity in expression and location.

Henri Termeer, the CEO of Genzyme and a Moderna Board member, was very supportive of a move into vaccines. Such a move had already been predicated in 2016, by a $50 million deal with Merck. Roger Perlmutter, the head of R&D at Merck at the time, was a believer in mRNA approaches to vaccines: Merck had been a vaccine powerhouse for many years with a stable of vaccines, including MMR, chicken pox (varicella zoster), HPV, and rotavirus.

First Moderna Vaccine for Influenza Virus

Moderna chose H10N8 influenza virus as their exemplar project. This largely avian flu virus was seen as *potentially* an important human pathogen. mRNA 1440 was developed as a vaccine and 200 volunteers in Germany were treated with it. Antibodies to the proteins were made and the vaccine was generally safe. At the 2016 J.P. Morgan conference, Bancel announced that Moderna were now a clinical stage company on the back of these results.[r] On the basis of that one mRNA vaccine trial, one might have expected Merck to tie up with Moderna for the development of the COVID-19 vaccine, but that never happened.

In the meantime, The Bill and Melinda Gates Foundation invested in Moderna, and the company were able to obtain more money from BARDA (Biomedical Advanced Research and Development Authority) for work on Zika virus, an emerging virus pathogen threat, and chikungunya virus, a not very common mosquito-borne disease. These grants were important for Moderna, because they strengthened their ties with government research agencies and developed further virus vaccine expertise.

Two Important Events for Moderna

Two events in 2018 were fundamentally important for what transpired later for Moderna. Both showed the belief and fortitude that the company and the executive team had engendered.[s] The first event of note was the opening of their 200,000-sq-ft manufacturing facility for their mRNA products in Norwood, Massachusetts, where all of the COVID-19 "Spikevax" vaccine would be made (until Lonza got involved and started making it under contract in Portsmouth, New Hampshire, and Visp in Switzerland). The second event was their IPO in December 2018, the *largest IPO in biotech history to date*. The company sold 27 million shares at $23, raising $621 million. More on this in Chapter 17.

BioNTech

In Mainz, Germany, a small company called BioNTech were also developing *in vitro* RNA technology. BioNTech were founded in 2008 by Uğur Şahin, Özlem Türeci, and Christoph Huber, with a focus on using mRNA to make

---

[r] Being in the clinic is a value-creating event for any small biotech company.

[s] Moderna were almost universally perceived to be a company without any real prospects. They were very secretive and little of what they were doing was published, so it was very difficult for scientists outside the company to judge the quality of the science that was being done. The prevailing view at the time was that the science was not that great and that the whole thing was mostly smoke and mirrors. No one doubted the fact, though, that the company had been very successful in raising money and doing deals.

cancer immunotherapies as vaccines. They also had interests in developing vaccines for infectious disease including coronaviruses and in using mRNA to replace the enzymes that are missing or dysfunctional in some rare genetic diseases (for which enzyme replacement is used therapeutically; see Chapter 11). Their first round of financing was €150 million.

For cancer vaccines, the idea was to use mRNA to make antigens to elicit the patient's T-cell responses to kill the tumour cells. In stark contrast to Moderna, many of the RNA studies that were done by BioNTech were published, and several patent applications were filed before the publications came out. Katalin Karikó of UPenn (see above) went to work for BioNTech in 2013 and became head of R&D.[t] The Weissman/Karikó IP from UPenn has been licensed to both BioNTech and to Moderna, and the technology was used by both companies to make their respective SARS-CoV-2 mRNA vaccines.

BioNTech Vaccines

BioNTech's foray into vaccines started in 2018 with a deal with Pfizer, focused on influenza virus. Under the deal, BioNTech would make the mRNA vaccine and Pfizer would be responsible for clinical development and commercialisation. Pfizer already had an important vaccine franchise having inherited Prevnar, a pneumococcus vaccine from Wyeth after their acquisition, as well as what they got from PowderMed. BioNTech created their potential flu vaccine after testing different doses and LNPs. This was a precursor deal to the very important 2020 partnership BioNTech did with Pfizer, for the development of BNT162b2, their RNA vaccine for SARS-CoV-2. Sanofi took an €80 million stake in BioNTech in 2019, extending their 2015 cancer vaccine deal (they previously paid $60 million up-front for the rights to five discovery-stage immunotherapies). The Bill and Melinda Gates Foundation also invested $55 million in BioNTech in 2019.

After a $150 million IPO in mid-2019, BioNTech acquired Neon Therapeutics, a cancer immunotherapeutics company in Cambridge (MA) that was running out of money, for $67 million. This acquisition gave BioNTech a presence in Cambridge (MA) and access to Neon's adoptive T-cell and TCR T-cell therapies, which were promising but needed greater investment to make them competitive with other activities in that field. These

---

[t] Kariko K. 2023. *Breaking through: my life in science*. Crown, New York.

projects augmented BioNTech's own activities in melanoma, head and neck squamous cell carcinoma (HNSCC), and breast cancer (see Chapter14).

CUREVAC

Another German mRNA company in the RNA vaccines game but much less well known is CureVac. This company was founded in 2000 by Ingmar Hoerr (a doctoral student), Georg Rammensee, and several others in Tübingen, Germany. Like other companies, they were focused on vaccines for infectious disease and for cancer. Hoerr had discovered that when unmodified RNA was administered directly into tissue, the mRNA was translated into protein. He thought that it might therefore be used directly as a vaccine or to make a therapeutic protein.

CureVac have never invested in the modified nucleotide technology that both BioNTech and Moderna used. Consequently, owing to the induction of innate immunity by the unmodified RNA, they have always been restricted by the dose they can give. Over the years, CureVac formed alliances with several European pharma companies, and by 2017 (before COVID-19), the company had raised €305 million from various investors and was worth more than €1 billion. CureVac obtained a further $34 million from CEPI to help their technology development for COVID-19 and other vaccines.[u] They also raised $640 million in a mezzanine financing in July 2020 with Dietmar Hopp, a German billionaire participating, along with The Bill and Melinda Gates Foundation, the German government, and GSK. Only a month later, they raised more than $200 million in their IPO. So, despite the technological shortcomings, CureVac have been well-financed.

THE PANDEMIC

Much has been written about the SARS-CoV-2 pandemic (including the vaccine aspects I have touched on above). It should be acknowledged that the rapidity with which the DNA and RNA vaccines were developed had as much to do with the work done and the investment made in the respective small biotech companies *before* the pandemic, as it does with the extraordinary work they did once the virus hit and with such high urgency.

---

[u] Their vaccine interests apart from COVID-19 include RSV, Lassa fever, rabies virus, flu, and yellow fever. Some of these efforts are in an alliance with GSK.

As most of us will never forget, the U.S. Centers for Disease Control and Prevention (CDC) announced the COVID-19 pandemic in January 2020. I can well remember talking to Doug Williams in early February, over two cans of Guinness in his office at Codiak BioSciences in Alewife, and us both deciding not to travel anywhere. Indeed, I had brought the Guinness with me, so we did not have to go to the pub and mingle with other people. Almost the next thing we knew was that Biogen had their senior management off-site meeting late in February 2020, which turned out to be a COVID-19 "super-spreader" event. No one knew then what to do if they were infected, except to isolate at home. Several people got quite sick and went to the emergency room at MGH only to wait for hours. Others did not really get sick at all, a pattern that is still true today. Each person's reaction to infection is certainly not defined completely by genetics, as the genome-wide association studies (GWAS) that have been done show only moderate effects of any gene on infection rates. Recently, though, it has been shown that people who have two copies of the HLA allele HLA-B*015:01 are relatively protected from severe disease.[24]

Another thing we did not know was that Moderna, BioNTech, Vaccitech, CureVac, and other vaccine company hopefuls, without missing a beat, had obtained the published sequence of the spike protein on the surface of the coronavirus (which makes up the characteristic "corona" on the surface of the virus as seen by electron microscopy) and were already making mRNA constructs that expressed the protein. The sequences had been posted online early in January 2020, after Chinese investigators had sequenced the complete 35-kb RNA of this new coronavirus. Genetically, the virus was related both to the bat-derived SARS-CoV-1, which infected and killed people in the 2003 outbreak, and to MERS, the lethal coronavirus found in camels in the Middle East.

One of the subtleties that was not lost on Barney Graham and Kizzmekia Corbett at the Vaccine Research Center in Bethesda, Maryland, with whom Moderna were collaborating, was that the conformation of the spike protein (i.e., the three-dimensional shape of it) was important for its immunogenicity.[v]

---

[v] The SARS-CoV-2 spike glycoprotein is a glycosylated type I trimeric membrane protein anchored in the virus membrane by a transmembrane segment and made up of 1273 amino acids. The spike protein binds to the angiotensin converting enzyme 2 receptor on a susceptible cell and undergoes a large structural conformational change after binding. This is quite similar to what happens with both the HA protein of flu and the F-protein of RSV. The prefusion conformation (the so-called arms up) has the receptor binding domain pointing towards the cell before the protein is cleaved into the S1 and S2 regions after binding to the membrane.[25]

To be appropriately immunogenic, the mRNA needed to encode a modified spike protein that was fixed in the "prefusion" conformation. This was done by making a couple of amino acid substitutions in the protein.

Warp Speed

The speed with which all the vaccines were developed and distributed is nothing short of amazing. My experience over many years in the lab is that experiments very rarely work first time and have to be repeated: maybe many times. The speed of progress of the development of these vaccines would indicate that there were very few experimental failures. It also showed a level of collaboration between the government agencies (both in the United States and elsewhere) and the vaccine developers that had never been seen before. Life-and-death decisions require moves to be made quickly. Everybody involved seemed to appreciate the seriousness of the situation and made decisions accordingly.[16]

Three critical collaborations were in place for the DNA and RNA vaccines that involved small biotech companies: Moderna and NIAID (National Institute for Allergy and Infectious Diseases); BioNTech and Pfizer; and AstraZeneca and Vaccitech in the United Kingdom. All three managed to develop and test their vaccines in about the same time frame. It certainly helped Moderna to receive $483 million from Operation Warp Speed, set up by the Trump Administration to accelerate the development of COVID-19 vaccines.[26] Moncef Slaoui, an ex-GSK vaccine head and Moderna Board member, was selected to run the Warp Speed Operation: consequently, he stepped off the Moderna Board. It also helped BioNTech to have a deep-pocketed pharma partner working with them. AstraZeneca and Vaccitech had some issues with rare neurological side effects with their vaccine, slowing their activities in the United States, but not in the United Kingdom and Europe.[w]

Pfizer/BioNTech and Moderna ran almost neck-and-neck the whole way through testing in mice in February of 2020, to phase 1, 2, and 3 clinical trials. One significant difference between the two RNA vaccines was stability. It was found that the Moderna vaccine (mRNA 1273) could be stored

---

[w]The results of the phase 1 study of the Vaccitech/AZ vaccine were reported in *The Lancet* in August 2020.[27]

at −20 °C, whereas the BioNTech/Pfizer vaccine (BNT 162b2) had to be stored at −70 °C to maintain activity.

Vaccinating Millions of People

The Moderna and BioNTech/Pfizer vaccines first went into humans in late March 2020, less than four months after the virus was isolated and the sequence determined: an absurdly short time by any standards. The results of the mRNA 1273 testing in nonhuman primates were published in the *New England Journal of Medicine* in July 2020.[28] The vaccine showed good antibody and T-cell responses and safety at both 10- and 100-microgram doses. It was shown to raise neutralising antibodies in the small phase 1 (human) trial shortly thereafter. From the initial dosing in adults, Moderna settled on a dose of 100 micrograms for mRNA 1273. The BioNTech vaccine, now called Comirnaty, would be used at a dose of 30 micrograms in adults and 3 micrograms in children.

You can remember (or imagine) the relief and the excitement when the results of both the BioNTech and the Moderna phase 3 trials were announced. The Moderna trial had enrolled 30,000 people of various ethnicities. Out of 30,000 patients in this placebo-controlled trial, there were 95 infections, 90 in the placebo group, and only 5 in the vaccinated group, representing a >90% response rate, which correlated with antibody levels to the spike protein. Stunningly, by November 2020—less than a year since the pandemic started—we knew that the Moderna vaccine candidate was 94.1% effective in preventing SARS-CoV-2 infection with only minor side effects. Similar results were reported by the BioNTech/Pfizer team. On December 18, 2020, Moderna's mRNA 1273 was granted emergency use–only authorisation (EUA) by the FDA, as was the Pfizer vaccine.

In December 2020, temporary authorisation was granted in the United Kingdom and Europe for the Pfizer/BioNTech vaccine, followed by the Moderna vaccine in January 2021. It was not just the stronger efficacy of the mRNA vaccines that counted, but there was a slightly higher incidence of serious side effects from both the AZ vaccine (known as Covishield or Vaxzevria) and the J&J adenovirus vaccines—the most serious of which was a low incidence of thrombosis in some younger subjects. Four effective COVID-19 vaccines were available at the beginning of 2021, only a year after the pandemic started.

Vaccine Rollout

The vaccine rollout began in early 2021. I well remember waiting for my number to come up, which it did after the home-care and hospital/clinic health workers were vaccinated. I got my shot of the Moderna vaccine at a gymnasium in Lynn, Massachusetts, the day after our age group authorisation on February 18, 2021 and the requisite booster vaccine a month later. I heaved a huge sigh of relief, and at the same time a thanks under my breath to all the employees in all the biotech and other companies that had made this happen (see Box 10). It will not be forgotten.

BOX 10. SARS-CoV-2 mAbs

In addition to the vaccine companies, the monoclonal antibody companies also took very little time to develop mAbs that neutralised the virus. If the vaccines had not worked as well as they did, these would have been even more important in controlling the death and disruption that the pandemic caused. VIR Biotechnology developed sotrovimab, derived from B cells of patients infected with SARS-CoV-1 that made antibodies that cross-reacted with SARS-CoV-2. VIR had been set up in 2016 with an enormous series A round of $600 million by ARCH Venture Partners, with George Scangos (ex-Biogen) as CEO. Regeneron similarly developed a combination of mAbs, casirivimab/imdevimab, that neutralised the original Wuhan strain of the coronavirus.[x]

Eli Lilly also developed a cocktail of antibodies called bamlanivimab/etesevimab from similar sources. In clinical trials, these antibodies were shown to prevent infections and to reduce hospitalisation from SARS-CoV-2 infection in unvaccinated patients. Unfortunately, as new mutant versions of the virus arose (like the omicron strain), these original mAbs were much less effective at neutralising the new viruses. These initial versions were given emergency use authorisation by the FDA but are now no longer in use. Newer mAbs are being made and tested against the new strains of SARS-CoV-2. AstraZeneca, for example, have developed Evusheld (tixagevimab/cilgavimab). Adimab, a premier antibody engineering company in New Hampshire, set up a company called Adagio (now Invivyd) specifically to make new mAbs against SARS-CoV-2.

As with mRNA vaccines, each of the companies that delivered anti SARS-CoV-2 mAbs had developed very sophisticated platforms for making antibodies and antibody derivatives for other diseases, enabling them to move very fast to the clinic. It was helped by the FDA and its EUA. These companies deserve as much credit as the vaccine companies for the rapid development of these often lifesaving therapeutics.

---

[x]This was the antibody cocktail that then–U.S. President Trump was given, rather than the hydrochloroquine or bleach he recommended for others.

Are We Done with COVID Yet?

Having had several Moderna booster shots including the omicron version, the short answer to that question is "no," or at least "not yet." The ability for the SARS-CoV-2 virus to mutate to increase transmissibility and to avoid immune responses is now even more evident than was originally expected. Dominant strains have been taking over in waves from the original Wuhan strain as the virus passes through people, including the α, β, and δ variants and more recently the omicron variants.[y] There now seems to be a bit of stalemate between the emergence of new strains and the ability of the population to deal with them—either through herd immunity induction (most people having been infected) or a mixture of infection and vaccination with boosters, against the prevailing new strain: similar to what happens annually with flu. This is another way of saying that SARS-CoV-2 will probably be managed in the future by regular vaccinations that reflect prevailing and emerging strains—very much like flu, which still causes mortality and morbidity in the winter months. There are at least two outstanding issues that require resolution concerning SARS-CoV-2—the nature of the Long COVID syndrome and the origin of the virus (see Boxes 11 and 12).

Intellectual Property and COVID-19

It is only now that the IP battle between Pfizer and Moderna has started in earnest. While the pandemic was in full force, all litigation was sensibly put

---

BOX 11. LONG COVID

What causes "Long COVID," and how do the vaccines confer long-term protection from serious disease and death (as the statistics clearly show)? Most of the answers to both these questions lie in the other part of the immune response (i.e., the T cells). Having had the benefit of working for Repertoire Immune Medicines, a T-cell-focused company in Cambridge, during the pandemic, I read the literature in detail about what role the T cells play during the infection. Repertoire scientists published several papers on the subject, well before it became a topic of more popular discussion.[29] The ultimate vaccine needs to induce antibody and T-cell responses. Short-term protection is probably provided by the induction of antibodies, but longer-term protection and freedom from "Long COVID" will probably require a broad T-cell response, enabling the T cells to kill all virus-infected cells.

---

[y] With more than half of the Greek alphabet to go!

BOX 12. ORIGIN OF SARS-CoV-2

The other question that needs an answer is where the virus came from. The centre of the pandemic initiation was the animal market in Wuhan, China. Despite the conspiracy theorists, this is an unequivocal fact. The real question is how did the virus get there? The nearest relatives to SARS-CoV-2 in RNA sequence are viruses found in bats in caves in southern China and Laos. There are some other close relatives found in pangolins, animals that were frequently brought to the live animal market. But it is all-too easy to invoke the scenario that it was the introduction of live animals infected by bat viruses into the market that caused the infection in humans.

One big conundrum is the fact that SARS-CoV-2 isolates that infect humans have a characteristic furin (a cell protease) cleavage site between the S1 and S2 subunits of the spike protein. It is unclear how that furin cleavage site came to be encoded in the SARS-CoV-2 genome.[30] Most publications on the origins of the virus seem to ignore this fact, but there are a couple of books on the subject that are worth a read.[31,32] Given that it was known that a furin cleavage site increases the human transmissibility of the virus and that such gene insertion studies were being done at the Institute of Virology in Wuhan, it is perfectly plausible to think that a lab leak was the origin of the animal market outbreak and that a live human animal in the market was responsible for initiating the outbreak.[33]

In this scenario, an infected Institute of Virology worker (or workers) went to the market and infected other humans, rather than the virus being transmitted by another animal species already at the market. But that is not the only explanation, and it is probably not as simple as that. Recent publications argue that the furin cleavage site sequence arose via a 12-base pair nucleotide insertion by recombination with a very closely related virus in a bat.[30] I do not find that explanation terribly convincing, but, of course, it is possible.

I am not a conspiracy theorist, but I have worked for several years in a very high security virus lab. I do not think that the virus was released deliberately, but several facts still bother me. The SARS-CoV-2 outbreak started in Wuhan in November 2019, where there is an Institute of Virology studying coronaviruses from bats that are the most similar in sequence to the SARS-CoV-2 sequence than any other coronavirus virus found so far. No one has yet reported the sequence of a bat (or any animal) virus that is close enough to the SARS-CoV-2 virus to be the plausible cause. The closest relatives to SARS-CoV-2 are viruses that were isolated more than 1000 km away in Laotian caves and brought to the Institute. It was also known that the Institute were actively doing "gain-of-function" experiments with coronaviruses, such as introducing changes known to increase pathogenicity (like the furin cleavage site).[34]

We may never get to an answer to the question concerning the origin of the pandemic. It is so important to understand it to prevent further pandemics. For some people, not knowing the origin may be a "convenient" situation for various reasons,[35,36,37] but it is not for me. Even if it is fraught with politics, we should not hide from trying to pin down the origin of the pandemic virus in order to learn how to prevent, or at least detect earlier, another almost inevitable one arising in the future.

on hold. Moderna are now suing Pfizer for alleged infringement on some of their COVID-19 IP. BioNTech were following much the same approaches to mRNA vaccines as Moderna, at much the same time from 2012 onwards. They also filed IP, so it will be interesting to see how this pans out. It will also be interesting from a public relations perspective, especially as Moderna have already looked to dismiss a suit from another company named Arbutus that claims that the Moderna vaccine infringes one of their patents.

GET VACCINATED

Vaccines are part of our lives from very early on. They are also a very cost-effective way to limit infectious disease. The total cost of the recommended vaccines in the United States for babies and children is only $1200. In the United Kingdom, vaccinations start at eight weeks, followed by shots at three months and four months after birth. The 6-in-1 vaccine used in many countries (although not in the United States), consists of vaccines against diphtheria, tetanus, pertussis (whooping cough), polio, haemophilus B, and hepatitis B. The first MMR vaccine is recommended at the age of two years, followed by haemophilus B and meningococcus C.

The more familiar vaccines, including flu shots, come later, and there are vaccines available for yellow fever and dengue, if travelling to countries where these viruses are endemic. The vaccine schedule recommended by the CDC in the United States is very similar to that in the United Kingdom, except that a varicella zoster (chicken pox) vaccine is recommended in the United States.

As we (and our immune systems) get older, further vaccinations are important. The HPV vaccine given to teenagers protects not only against HPV infection but also the longer-term cancer consequences of that infection. Even in our 60s and 70s, more vaccines are in order. A varicella shot to protect against shingles, caused by a re-emergence of latent chicken pox virus, is recommended, as well as a vaccine against pneumococcus-induced pneumonia.

Access to COVID-19 vaccines in late 2020 and early 2021 was experienced as a release for many of us: a release from self-imposed lockup and a release from the concern that catching the virus could kill you as it did (and still does) millions of other unfortunate—and mostly unvaccinated— people, both old and young.

It must be clear by now that I am a vaccine advocate. Vaccines are one of the most cost-effective ways to control infection and disease. I am very tired of people who, for no particularly good reason, do not get themselves (or, worse still, their children) vaccinated.[z]

The argument that the COVID vaccines—or any other vaccine for that matter—are either unsafe or not tested enough is just specious and wrong. Vaccines do have side effects and sometimes unfortunately they are serious, but the good done by a vaccine far outweighs the risk. It does not just protect you from the disease: it protects others too. We run the risk of there being a resurgence in avoidable diseases causing childhood mortality such as whooping cough and measles.

Sadly, humans are not very good at assessing either absolute risk or indeed relative risk. Being vaccinated or not being vaccinated is an example of an everyday risk assessment you must do. If you do that assessment even superficially, you will conclude that being vaccinated with a vaccine (like COVID-19) is a really a no-brainer (unless there is a good medical reason why not). Skipping vaccination just because one cannot be bothered, or does not "believe" in it, is a form of human behaviour that is *profoundly* antisocial. It is akin to drunk driving or running red lights and should not be tolerated by a rational and well-informed society but unfortunately we do not live in one.

REFERENCES AND NOTES

1. Riedel S. 2005. Edward Jenner and the history of smallpox and vaccination. *Bayl Univ Med Cent Proc (BUMC)* **18**: 21–25. doi:10.1080/08998280.2005.11928028

2. Tan SY, Pettigrew K. 2017. Max Theiler (1899–1972): creator of the yellow fever vaccine. *Singapore Med J* **58**(4): 223–224. doi:10.11622/smedj.2017029

3. Transmission of yellow fever virus. Centers for Disease Control and Prevention website. https://www.cdc.gov/yellowfever/transmission/index.html

4. Mallapaty S. 2022. Dengue vaccine poised for roll-out but safety concerns linger. *Nature* **611**: 434–435. doi:10.1038/d41586-022-03546-2

5. Famulare M, Chang S, Iber J, Zhao K, Adeniji JA, Bukbuk D, Baba M, Behrend M, Burns CC, Oberste MS. 2016. Sabin vaccine reversion in the field: a comprehensive analysis of Sabin-like poliovirus isolates in Nigeria. *J Virol* **90**: 317–331. doi:10.1128/JVI.01532-15

---

[z]Unfortunately, there are some politics at play here on both sides of the aisle. Influential people who do not get vaccinated or vaccinate their children should remain quiet on the subject so that people can make up their own minds free of unnecessary and factually incorrect information.

6. *Nature* Editorial. 2023. Vaccine-derived polio is undermining the fight to eradicate the virus. *Nature* **618**: 434. doi:10.1038/d41586-023-01953-7

7. Huzair F, Sturdy S. 2017. Biotechnology and the transformation of vaccine innovation: the case of the hepatitis B vaccines 1968–2000. *Stud Hist Philos Biol Biomed Sci* **64**: 11–21. doi:10.1016/j.shpsc.2017.05.004

8. Burrell CJ, Mackay P, Greenaway PJ, Hofschneider PH, Murray K. 1979. Expression in *Escherichia coli* of hepatitis B virus DNA sequences cloned in plasmid pBR322. *Nature* **279**: 43–47. doi:10.1038/279043a0

9. Charnay P, Pourcel C, Louise A, Fritsch A, Tiollais P. 1979. Cloning in *Escherichia coli* and physical structure of hepatitis B virion DNA. *Proc Natl Acad Sci* **76**: 2222–2226. doi:10.1073/pnas.76.5.2222

10. Valenzuela P, Gray P, Quiroga M, Zaldivar J, Goodman HM, Rutter WJ. 1979. Nucleotide sequence of the gene coding for the major protein of hepatitis B virus surface antigen. *Nature* **280**: 815–819. doi:10.1038/280815a0

11. Pasek M, Goto T, Gilbert W, Zink B, Schaller H, MacKay P, Leadbetter G, Murray K. 1979. Hepatitis B virus genes and their expression in *E. coli*. *Nature* **282**: 575–579. doi:10.1038/282575a0

12. Datoo MS, Natama HM, Somé A, Bellamy D, Traoré O, Rouamba T, Tahita MC, Ido NFA, Yameogo P, Valia D, et al. 2022. Efficacy and immunogenicity of R21/Matrix-M vaccine against clinical malaria after 2 years' follow-up in children in Burkina Faso: a phase 1/2b randomised controlled trial. *Lancet* **12**: 1728–1736. doi:10.1016/S1473-3099(22)00442-X

13. Willyard C. 2023. The next frontier for malaria vaccination. *Nature* **618**: S20–S22. doi:10.1038/d41586-023-02048-z

14. Frazer IH. 2019. The HPV vaccine story. *ACS Pharmacol Transl Sci* **2**: 210–212. doi:10.1021/acsptsci.9b00032

15. Barberis I, Myles P, Ault SK, Bragazzi NL, Martini M. 2016. History and evolution of influenza control through vaccination: from the first monovalent vaccine to universal vaccines. *J Prev Med Hyg* **57**: E115–E120. doi:10.15167/2421-4248/jpmh2016.57.3.642

16. Bingham K, Hames T. 2022. *The long shot: the inside story of the race to vaccinate Britain*. Oneworld, London.

17. Cranmer J. 2021. Remembering Alan Mendelson, mentor to biotech lawyers, CEOs. *BioCentury*, October 2021.

18. Kinch M. 2016. *'A prescription for change': the looming crisis in drug development*. UNC Press, Chapel Hill, NC.

19. Neuzil KM. 2023. An mRNA influenza vaccine: could it deliver? *New Engl J Med* **388**: 1139–1141. doi:10.1056/NEJMcibr2215281

20. Wilson E, Goswami JU, Baqui AH, Doreski PA, Perez-Mar G, Zaman K, Monroy J, Duncan CJA, Ujiie M, Rämet M, et al. 2023. Efficacy and safety of an mRNA-based RSV PreF vaccine in older adults. *N Engl J Med* **389**: 2233–2244. doi:10.1056/NEJMoa2307079

21. Clemens SAC, Folegatti PM, Emary KRW, Weckx LY, Ratcliff J, Bibi S, De Almeida Mendes AV, Milan EP, Pittella A, Schwarzbold AV, et al. 2021. Efficacy of ChAdOx

1nCoV-19 (AZD1222) vaccine against SARS-CoV-2 lineages circulating in Brazil. *Nat Commun* **12**: 5861. doi:10.1038/s41467-021-25982

22. Loftus P. 2022. *The messenger: Moderna, the vaccine, and the business gamble that changed the world*. Harvard Business Review, Brighton, MA.

23. BG Medicine S1, as filed with the Securities and Exchange Commission, December 11, 2007.

24. Augusto DG, Murdolo LD, Chatzileontiadou DSM, Sabatino Jr JJ, Yusufali T, Peyser ND, Butcher X, Kizer K, Guthrie K, Murray VW, et al. 2023. A common allele of *HLA* is associated with asymptomatic SARS-CoV-2 infection. *Nature* **620**: 128–136. doi:10.1038/s41586-023-06331-x

25. Zhang J, Xiao T, Cai Y, Chen B. 2021. Structure of SARS CoV-2 spike protein. *Curr Opin Virol* **50**: 173–182. doi:10.1016/j.coviro.2021.08.010

26. For a Republican administration view of the pandemic handling, see Giroir B. 2023. *Memoir of a pandemic—fighting COVID from the front lines to the White House*. Texas A&M University Press, College Station, TX.

27. Folegatti PM, Ewer KJ, Aley PK, Angus B, Becker S, Belij-Rammerstorfer S, Bellamy D, Bibi S, Bittaye M, Clutterbuck EA, et al. 2020. Safety and immunogenicity of the ChAdOx 1nCOV-19 vaccine against SARS COV-2: a preliminary report of a Phase1/2 single-blind, randomised controlled trial. *Lancet* **396**: 467–478. doi:10.1016/S0140-6736(20)31604-4

28. Moderna trial: Jackson LA, Anderson EJ, Rouphael NG, Roberts PC, Makhehe M, Coler RN, McCullough MP, Chappell JD, Denison MR, Stevens LJ, et al. 2020. An mRNA vaccine against SARS-CoV-2—preliminary report. *N Engl J Med* **383**: 1920–1931. doi:10.1056/NEJMoa2022483

29. Sauer K, Harris T. 2020. An effective COVID-19 vaccine needs to engage T cells. *Front Immunol* **11**: 581807. doi:10.3389/fimmu.2020.581807

30. Garry RF. 2022. SARS-CoV-2 Furin cleavage site was not engineered. *Proc Nat Acad Sci* **119**: e2211107119. doi:10.1073/pnas.2211107119

31. Ridley M, Chan A. 2021. *Viral: the search for the origin of COVID-19*. Harper, New York.

32. Farrar J, Ahuja A. 2021. *Spike: the virus vs the people*. Profile Books, London.

33. Quammen D. 2022. *Breathless: the scientific race to defeat a deadly virus*. Simon & Schuster, New York.

34. Select committee on Health Education, Labor and Pensions. Minority Overnight staff. Interim report October 2022. An analysis of the origins of the COVID-19 pandemic.

35. Keusch GT, Amuasi JH, Anderson DE, Daszak P, Eckerle I, Field H, Koopmans M, Lam SK, Das Neves CG, Peiris M, et al. 2022. Pandemic origins and a One Health approach to preparedness and prevention: solutions based on SARS-CoV-2 and other RNA viruses. *Proc Natl Acad Sci* **119**: e2202871119. doi:10.1073/pnas.2202871119

36. Ridley M. 2022. COVID origin case re-opened: a lab leak is a legitimate question. *Health*, November 13, 2022.

37. Quammen D. 2023. Ongoing mystery of COVID'S origin. *New York Times*, July 25, 2023.

CHAPTER 17

Fortunes and Unicorns

There are several examples over the last 400 years of people becoming so obsessed with assets or probable company performance that the commodity becomes overvalued and causes a financial bubble. The definition of an asset bubble is that the value of the asset rises extraordinarily quickly and soon becomes higher than the intrinsic value of the asset. Two of the best-known examples are tulip mania in Holland in the seventeenth century and the South Sea Bubble occurring in the United Kingdom some 80 years later (Box 1).[a]

BOX 1. TULIP MANIA

Tulip mania lasted from 1634 to 1637 in Holland. The Dutch Republic and their rich merchant class led the world in economics and finance at that time. Tulip bulbs were a particularly fascinating asset to the Dutch people because of the brightly coloured flowers that they produced. But tulips came to have value that far exceeded their real worth. Introduced to Holland in 1593, tulip bulbs were apparently traded on "exchanges" in many Dutch cities leading, in 1636, to uncontrolled speculation on their value.[1] It appears that people thought that the value of tulips would be high forever, so they just had to have one or maybe several—especially if you were a rich trader. The price continued to rise until February 1637, when the bubble spectacularly burst, and no one wanted to buy tulip bulbs anymore.

The South Sea Bubble[2]

In 1711, the South Sea Company was set up to try to control the British national debt and to help trade with North American and other colonies. The British government granted the company a trading monopoly, part of which was to

---

[a] I am not suggesting for one moment that biotech stocks are like tulips or the run-up in stock value that happens is like a South Sea Bubble, but it is true that the valuations attributed to some companies over the last 50 years might lead you to think, if you were observing from afar, that there was something a bit "flowery" about it. Remember that most biotech companies formed lose all their money and never make any products at all!

increase profits from the slave trade from Africa to Spain and Portugal. The trading was not as brisk as expected. King George I took ownership of the company, which led to the stock in the company returning inflated interest and encouraging many people to buy it, increasing its value well over what it was worth, based on the trading profits of the company. It got even worse when the government allowed the company to take over the National Debt, causing the interest to soar even further. By the end of 1720, the bubble had burst, the stock returned to a reasonable level, and a lot of the landed gentry lost a lot of money. The government subsequently uncovered considerable bribery and corruption.

GROUP THINK

The "group think" that led to these bubbles does still occur, even in the biotech sector. Group think, for example, prevails over the technology where several companies are formed around the same or similar technology when one or two at most would be sufficient to exploit the technology effectively.[b]

Sometimes companies should not be formed at all around a technology, owing to the time it takes to get any product launched. There are also times, as we have seen recently, when many companies that should have remained private, manage to go public in a bull market. When the market collapses, as in 2022, these small public companies are left high and dry with no way available to raise additional money. It gets very ugly very quickly under those circumstances, as companies fold up their tents and lay off all their people.

Venture capital investors have been attracted to biotech not just by the science, nor by the fact that the companies they invest in are trying to make new medicines to treat patients with horrible diseases, but because good returns on investment can be made. In 1976 at the beginning of the industry, there were not many venture capitalists (VCs) who either understood or were aware of what recombinant DNA technology was. Nor were they aware of the new companies that had been set up to exploit it. Three of the most venerable early adopters were Venrock, J.H. Whitney, and Kleiner Perkins, which set the scene for the investors that followed them (see Chapter 18).

---

[b]As in most things these days, social media just accentuates certain points of view and adds to the feeding frenzy.

FINANCING THE FLEDGLING INDUSTRY

The financing of the biotech industry has changed from the beginning to how it is today. Biotechnology came of age financially when Genentech went public in the fall of 1980, followed by Cetus Corporation in early 1981. Few biotechnology companies had existed before these original biotech IPOs. In 1981, owing to the development of both recombinant DNA technology and the ability to make monoclonal antibodies and the Genentech/Cetus success, many small companies exploiting these technologies were set up. Most of them were financed with relatively small amounts of capital by the rather few venture capital firms around at the time. More money was invested as they met their milestones. Many of them also obtained strategic alliances with pharmaceutical companies. The pattern was to go public after a couple of rounds of venture funding plus a strategic alliance with the possibility of building a lasting company or getting acquired. This was the financial template and the road map for the industry for several decades (see Plate 26 for early IPOs).[3]

Fast forward to today and we see that different investment philosophies have emerged. There are still the traditional start-ups in biotech in which founders with good ideas and good technology will raise small amounts of seed money from certain VCs or angel investors and develop their company from there, doing a seed round followed by series A and series B financings, before an exit by either IPO or acquisition. Success is dependent not only on the science that is done and the meeting of goals but also on the quality of the people and the management teams that they can assemble. Adhering to proposed time lines and budgets is table stakes for the continuation of these companies. They generate data, raise money to generate more data, and raise more money to develop their products.

LARGE FUNDS MAKE BIGGER BETS

A new model has developed over recent years. It is now the province of some of the larger VCs investing in biotech such as ARCH Venture Partners in Seattle, Flagship Pioneering in Cambridge (MA), and Third Rock Ventures in Boston. To some extent, the situation reflects the much larger pools of capital available today as compared to 40 years ago. These VC firms have large enough funds to be able to make big bets on companies right from the start[4] (Chapter 18).

Whether this is a successful strategy or not depends on much the same things as the small stepwise approach: experienced management teams, great science, great technology, and a slice of good fortune thrown in for good measure. The external market for IPOs and the sentiment of nonspecialist investors for the biotech space, which is not very high currently, markedly affects the outcome.

It is obviously better to have more money rather than less. However, the discipline that goes with not having immediate access to very much money may help the smaller companies in the longer run. The Boards of Directors of these companies have an obligation to help management view risk in the appropriate way, to make the right decisions, and to manage the company budget accordingly. The ability to raise money even in a down-round (i.e., at a lower valuation than the previous round of financing), which is dilutive for the original shareholders, is better than going out of business.[c]

Given what biotech has done for health care over the last 50 years, one hopes that the prevailing market sentiment changes soon and that those companies financed with small amounts of money in small steps over time can continue to coexist alongside those set up with large amounts of money from the beginning.

IPO SCALE

Concordant with the early financings, the scale of the IPO process has also changed over time.[4]Before the year 2000, the average IPO brought in around $29 million. In 2000, 58 companies went public, raising just over $30 billion in total. By 2019, the dollar amount had grown to around $120 million per IPO. Argenx, for example, with a potential product in late-stage clinical development, raised more than $750 million in 2020 in their IPO.

These differences reflect, in part, a reality that is referred to in biotech as "IPO windows" (i.e., periods of time when the market is either open to IPOs or not). There was an important open IPO window in late 1999/2000 that coincided with the genome sequencing revolution, and another window that remained open from 2002 to 2007 (until the financial crash of 2008). New windows opened off and on from 2012 to 2021. The IPO window was flung wide open in 2020 and 2021, which were record years for biotech IPOs with proceeds of $5 billion in 2019 to $12 billion the next year. In 2020, 130 companies went public, with the top 10 raising more than $3 billion between them. There were 154 IPOs for biotech companies of various kinds

---

[c]An event that should be called "terminally dilutive."

in 2021, raising an average of $150 million each, some of which were only preclinical stories with nothing in the clinic.[5] In 2022, the window more or less closed and remains virtually so in 2023.

Some have said that the summer of 2022 was the worst financing environment that there has ever been for biotech, with only a handful of IPOs across the whole year. Stelios Papadopoulos, the "Godfather of Biotech," often reminds the biotech community that there is *always* a market for the stock of good companies and that the windows concept is a bit of a myth.[6] The 2022 situation reflected in large part the fact that the less-sophisticated, nonspecialist investors, who thought they saw an opportunity to make a quick buck in 2020–2021, instead lost a lot of money in biotech and ran for the hills.[d]

ORIGINS OF EXUBERANCE

Some of the 2020–2021 investment "overexuberance" undoubtedly comes from the unusual returns that were made on the two cell therapy companies Kite and Juno that I covered in Chapter 14 and from the story of the vaccine company Moderna (and to a lesser extent BioNTech) in Chapter 16. It is worth a more granular look at the financial story of Kite, Juno, and Moderna, as they seem to have biased recent thinking about how to finance biotech companies.

Kite Pharma raised convertible seed rounds in 2011, which converted at a 15% discount in the spring of 2013 into their series A—amounting to a total of $35 million at $1.85/share. A pre-IPO mezzanine round of $50 million was raised as a convertible note into the IPO price, coming in at a 10% discount to the $17.00/share IPO ($128 million) in 2014. In short, Kite raised $85 million as a private company and raised $830 million in a series of follow-on public financings at increased prices after their IPO: a total of $1 billion in equity capital over their lifetime. They were sold to Gilead for $11.9 billion in 2017.

Juno Therapeutics raised $314 million as a private company. They did a large $180 million series A and a $130 million series B round in 2013/2014. Juno went public in December 2014, raising $264 million at $24/share.

---

[d] They did not really understand what they were investing in in the first place. I think it is a good idea before making any investment that you understand what the company you are investing in do and how they are differentiated from the competition in the same space. The most successful hedge funds buying biotech stocks in the public market rely heavily on that good intelligence and have the skilled people to practice it (see Chapter 18).

Juno then struck a deal with Celgene, who purchased 9% of the company for $93/share, raising $850 million in further equity financing. Juno's public follow-on financing raised an additional $250 million. Celgene's 2018 acquisition of Juno valued the business at $11 billion, or $9 billion net of cash. In total, inclusive of the 2015 strategic investment from Celgene, Juno brought in $1.7 billion in equity capital over their short lifetime.

Both the Kite and the Juno deals led to *massive* returns to both their private and public investors, placing these deals in the top 2% of venture capital returns on companies *ever*. Early investors like David Bonderman of TPG and Alta Partners in Kite Pharma, and ARCH in Juno, made huge multiples and enormous absolute returns.[7]

In 2019, Moderna were a company that most people had never heard of, who had a niche technology for making vaccines and potential therapeutics using mRNA. Moderna are a good example of a biotech "unicorn" (see below): until COVID-19 emerged, it was a company with a technology looking for a disease to cure. Cancer vaccines did not have a good track record, and Zika virus and other exotic pathogens did not have much commercial value. Nevertheless, even before COVID-19, Moderna were already a unicorn, having a valuation that was hardly in line with the potential product revenue. Fortunately for all of us, the mRNA technology they and others (notably BioNTech) developed proved very effective in delivering an mRNA coding for the spike protein of the SARS-COV-2 virus to induce immunity as a vaccine (see Chapter 16). Big kudos to the management team at Moderna for execution. With COVID-19, the valuation of the company took off—given what they had achieved. Whether they are truly worth more than some of the pharma companies with multiple products and franchises is, however, a matter of some debate.[e]

Kite, Juno, and Moderna investments have set an unfortunate precedent for the biotech industry. Several new companies have been formed and gone public with valuations way north of their real potential, but not matching

---

[e] It is unlikely that Moderna will be able to maintain such a high valuation based on singular revenues from COVID-19 vaccines, and the flu and RSV vaccine marketplace that they want to enter is much more competitive than COVID-19. They will probably need to acquire one or more small companies who have potential technology and products but little money, to grow their business. Though it may be a challenge for current Moderna leadership, it takes a certain humility and pragmatism to do those kinds of deals: it acknowledges that the little company invented something that you did not invent internally.

the hype with something real. Some investors need to learn to run the numbers again and not be swept up by hyperbole and showmanship. Once those inglorious unicorn valuations are reached, it is almost unacceptable for the companies to fail. This has significant implications for the future of the biotech industry, especially because some of the unicorns have actually already failed and some spectacularly so.

UNICORNS

A unicorn is traditionally a mythical white horse-like animal with a single centred spiral horn on its head. A biotech unicorn is a privately held start-up company worth more than $1 billion. Apparently, the term was first redefined by tech venture investor Aileen Lee of Cowboy Ventures. There was a time when unicorns were as rare in biotech as the mythical creature described above was in the wild. But today there is no shortage of biotech unicorns that have been created, and no shortage of fortunes that have been made (and perhaps lost) by some of these unicorns.

Do the unicorns bias the thinking for future investment in biotech? I think so, based on the examples already mentioned. For example, if you do valuations or net present value calculations based on the returns that the cell therapies were going to make for both Kite and Juno, considering the complexity of the manufacturing process and the number of patients likely to be treated, you will not come up with valuations anywhere close to the numbers the companies sold for.[f]

Who Are the Unicorns and Where Are They?

ARCH and Flagship Pioneering are the VCs leading the charge with this new investment strategy (Chapter 18). Not only does the VC invest, but sometimes the limited partners who invested in the VC firm participate as well. This apparently aligns everyone for the long-term horizon over which the investment is supposed to play out. But recent history suggests that the time frame may be expected to be quite short, which is certainly not in line with the time it takes to build a successful company with many products.

---

[f] And remember that some companies that got sold for around the same price as Kite and Juno and at about the same time had more than a $1 billion/year in *real* revenue.

Consequently, it may be instructive to consider some of the other members of the unicorn herd to see what has happened to them and why.

Sana Biotechnology

Flagship Pioneering and ARCH started Sana Biotechnology, a novel cell therapy company with labs in Seattle, Boston, and South San Francisco, in 2018. They have some Juno heritage, as several of the management team were at that company before it was acquired (including Sana Founders Hans Bishop, who was CEO of Juno, and Steven (Steve) Harr, the CFO of Juno). Sahsen Ventures, F-Prime Capital, and Bradley Horowitz from Google were among the initial investors. The company was to be "taking a long-term view of the cell therapy space." According to their website, they are engineering cells as "medicines to provide new tools to meaningfully change the outcome of many human diseases." Sana's technologies involve cell and gene therapy, stem cell biology, and scalable manufacturing. They raised ~$700 million in their series A round from the founding investors and from an impressive set of other deep-pocketed investors (see Box 2).

Sana's ~$600 million IPO raise in February 2021 was the largest IPO ever for a company with—at that time—no clinical-stage assets. The share price quickly moved from $25 to $40 but has since declined to a little more than $4 (in December 2023). The company have certainly set high expectations for them to deliver clinical candidates and a return on investment in short order. Quite recently, they in-licensed nonexclusively a B-cell maturation antigen (BCMA)-directed CAR T therapy from IASO Bio and Innovent in China, and Doug Williams from Biogen and Codiak joined them as EVP R&D for a short time.

BOX 2. SANA'S TECHNOLOGY

The company in-licensed several different assets to help to create themselves. They have three principal product sets in their pipeline: (1) modified CAR T cells directed to CD19 and CD22 for B-cell malignancies; (2) potential products based on hypoimmunogenic stem cell technology licensed from Harvard; and (3) cell-specific delivery methods based on "fusogen" technologies developed by Geoffrey von Maltzahn and Jacob Rubens in the Flagship lab. This latter technology was derived from the work of James (Jim) Rothman, Randy Schekman, and Thomas Südhof, who won the 2013 Nobel Prize in Physiology or Medicine for studies on vesicle and cell trafficking in cells.

Lyell Immunopharma

Lyell Immunopharma are another so-called unicorn with a noteworthy recent IPO—and an outcome not dissimilar to Sana, at least so far. Lyell was an ARCH company founded and run by Richard (Rick) Klausner, the ex-NCI chief, in 2018. In three years as a private company, Lyell raised close to $1 billion in financing. They did a large $425 million IPO in July 2021 as a preclinical company with a CAR T (LYL 797) directed to ROR 1 (an orphan receptor expressed on the surface of some solid tumours) in development. This CAR T was differentiated by being engineered to resist exhaustion and, by epigenetic manipulation, to have durability and the ability to self-renew. Like Sana's program, this product had not yet reached the clinic, but the company filed the IND for it in 2022. Lyell's share price peaked at $19.84 and now in late 2023 hovers around $2 with a market cap of <$500 million—well out of the unicorn range. Investors clearly believed that Lyell[8] were differentiated from the many other companies pursuing solid tumour CAR T cells with the cells modified to be less exhausted and more durable.[8] I guess it helps that GSK owned nearly 15% of the company, although they have recently axed their collaboration with Lyell.[7] Now they do have (at least) two clinical stage projects (see Chapter 19).

Laronde

Laronde were founded in 2017 to exploit what was referred to as "endless RNA" (eRNA), essentially a large circular mRNA that is resistant to digestion by exonucleases. eRNA technology was developed at the Flagship Pioneering labs by Avak Kahvejian out of work on circular long noncoding RNAs. Normally these circular RNAs cannot be translated, but by engineering conventional start and stop translation signals (like an AUG initiation codon and relevant stop codons) and an internal ribosome entry site (as occurs in some virus RNAs), these eRNAs can code for proteins in their host cell and, with their looped design, protein expression can be sustained.

The company raised $50 million in their series A round from Flagship Pioneering in May 2021 to pursue this programmable RNA platform. The initial findings suggested that eRNA could be injected directly into an animal without a lipid nanoparticle or vector, was nonimmunogenic, and led

---

[8] I remain unconvinced that they have anything that warranted high valuations. There are indeed many companies doing the same thing.

to prolonged expression of the encoded protein. Laronde closed a $440 million series B round in August 2021 on the back of some encouraging data on the expression of the hormone GLP-1. This was one of the largest series B rounds ever recorded and included other investors like T. Rowe Price, Invus, The Canada Pension Plan Investment Board, Fidelity, and BlackRock.

According to their website, Laronde naïvely imagined that they would "create 100 new medicines in the next ten years." Good hyperbole, but simply not going to happen. From various reports in the press, the company was never a very happy place to work.[9] There were many examples of unrepeatable experiments and considerable employee turnover, including at the executive level. Indeed, things seem to have gone from bad to worse following the revelation that the company cannot reproduce some of their original data that purported to show expression of the hormone GLP-1 (as well as GLP-2) from eRNA.

The (Unholy) GRAIL

Another relatively recent unicorn is GRAIL, spun out of Illumina in 2015. GRAIL are focused on cancer diagnostics by deeply sequencing the DNA found in the blood of cancer patients. Among their efforts, GRAIL have developed a liquid biopsy test called the "Galleri test." This test identifies distinct changes in the methylation patterns of DNA that are correlated with different types of cancer, which may allow for their early detection. Other multicancer screening tests using sequencing are also being developed, some in collaboration with the National Health Service in the United Kingdom. For Illumina, GRAIL reflected a strategy that said: "Don't just make money from making and selling sequencing machines: make money from the medical use of the technology as well."

GRAIL were backed by Bill Gates and Jeff Bezos, who both invested in a $100 million ARCH-led series A round. GRAIL also had Juno connections: Hans Bishop, the original GRAIL CEO, was the Juno CEO, and Bezos and ARCH were also Juno investors. Through late 2020, GRAIL raised $1.9 billion in four rounds of financing, the last one being a $390 million series D in May 2020.

GRAIL filed to go public in 2020 to raise another $500 million, but in the fall of that year Illumina decided to acquire them for $7 billion. As we saw in the cases of Kite and Juno, it is a bit difficult to see a justification for this price, given that the company had no commercial products, and the cancer diagnostics market and intellectual property (IP) is significantly fragmented. It cannot possibly be argued that the company was worth that amount of money from potential cancer diagnostics revenue. GRAIL are also spending well over $100 million/year on R&D, collecting very large numbers of patients to look for early-stage cancer markers. Once again ARCH, with almost 10% of the company with the other previous Juno investors, could make a great return on investment someday. But in my opinion, that day is much further off than they anticipated. If it ever arrives.

As a further complication, the Illumina/GRAIL deal has not yet been consummated completely, owing to antitrust issues with U.S. and EU company regulators. In August 2022, the EU finally disallowed the Illumina/GRAIL merger, and in April 2023 the U.S. FTC ordered Illumina to divest their GRAIL holding. It is almost inevitable that Illumina will have to sell the asset at a considerable loss, marking the end to what was one of the most disastrous attempted mergers in biotech history (see Chapter 19).

It gets even worse: Carl Icahn got involved in Illumina as an activist investor and is in a proxy fight with the Illumina Board over the GRAIL deal, which has led so far to new Board members being added and a change in CEO. Between the regulatory bodies and the Icahn-associated investors, any resolution is unlikely to be pretty for Illumina or indeed GRAIL.

OTHER UNICORNS ON THE LOOSE

Like it or not, biotech unicorns do not end with Sana Biotechnology, Laronde, or GRAIL.

Altos Labs

ARCH have also backed Altos Labs, a company set up to advance new therapies for ageing and rejuvenation, who have to date raised more than $3 billion.[10] Anti-ageing is not a new biotech story (see Box 3). The story goes that Rick Klausner, a founder of Juno and of GRAIL, had the idea for Altos while walking in Los Altos with Yuri Milner, a very successful tech investor with the VC firm DST Global, an investor in many of successful tech companies like WhatsApp and Spotify.

Altos Labs were co-founded in January 2022 by Klausner and Hans Bishop, ex-CEO and president of GRAIL. Foresite Ventures and ARCH put in the first money, followed by Bezos and Milner in 2021 (who both—like some of us—have personal interests in anti-ageing, cellular regeneration, and the metabolic reactions that control that process).

BOX 3. EARLIER ANTI-AGEING BIOTECH STORIES: SIRTRIS PHARMACEUTICALS

There is a bit of history of biotech companies interested in anti-ageing that also involves GSK. One such is Sirtris Pharmaceuticals. It is not a very pretty story. David Sinclair, a scientist at Harvard,[h] was working on proteins called sirtuins and their relevance for metabolic conditions such as diabetes and ageing. He set up Sirtris Pharmaceuticals in 2004 with Andrew Perlman and Richard Pops (the CEO of Alkermes) to pursue that relevance. Richard Aldrich, the founder of the Longwood Fund who had been at Vertex Pharmaceuticals, Paul Schimmel, from the Scripps Institute in La Jolla who had founded many companies, and Christoph Westphal, now at the Longwood Fund with Aldrich, were also involved with Sirtris.

Sirtris were principally focused on the molecule resveratrol, a constituent of red wine derived from grape skins, as an anti-ageing molecule. It was purported to be an activator of SIRT1, an NAD-dependent deacetylase enzyme. Sirtris went public in March 2007 raising $60 million. GSK decided to buy the company for $720 million on the stated basis that the company had SIRT1 activators (e.g., SRT 501, a formulation of resveratrol) that worked both *in vitro* and *in vivo*. This claim was subsequently shown not to be true in work done in GSK labs: the *in vitro* activity was shown to be an artefact of the assay and the *in vivo* data in rodents was not reproducible. GSK stopped development of SRT 501 in late 2010 and the development of other clinical stage compounds they had made and shut down the Sirtris subsidiary altogether in 2013. This was not a good way to spend $720 million and not a very good precedent to set for future would-be anti-ageing drug companies. I remain sceptical about the whole area.

Altos Labs have persuaded some very prominent and highly respected scientists to become involved, including Juan Carlos Belmonte (a regenerative medicine expert), Steve Horvath (known for his work on the epigenetics of cell ageing), Hana J. El-Samad (a computational network biologist from UCSF), and Shinya Yamanaka (a 2012 Nobel Prize winner in Physiology or Medicine for his work in reprogramming differentiated cells).

In addition to world class scientists, Altos recently recruited Hal Barron, who was R&D chief at GSK, to be CEO and Co-Chair of the company. The Board of Directors consists of other Nobel laureates, including Jennifer Doudna, David Baltimore, and the chemical engineer Frances Arnold, who won the Nobel Prize in Chemistry in 2018. Altos Labs are actively setting

---

[h] Sinclair apparently always thought he should win (or should have won by now) the Nobel Prize for this work.

up research facilities in San Diego, San Francisco, the United Kingdom, and Japan, and Alexandria Real Estate Equities are building Altos a rather large new research facility in San Diego.

Altos are by no means an average biotech start-up. They are taking a long-term view to the problem, investing heavily in the research necessary to get a better understanding of the ageing and regenerative processes.[11] It could well be that various companies are spun out of Altos as product opportunities emerge from their R&D.

In light of the Sirtris debacle, though, I wonder if any new drugs will emerge from Altos Labs for ageing in the future or from other companies like Rejuvenate Bio, set up by George Church in 2018 to use gene therapy technology to deliver the Yamanaka transcription factors to animals to try to reverse some ageing processes. If I sound like I am embodying the "Risk Factors" section of the S1 of the companies public offering, then you can understand the sentiment. But good fortune to the companies all the same.

Rubius Therapeutics

Unicorns do not even have to be large or even "at large" anymore. One such company was Rubius Therapeutics, who are now formally out of business. Rubius was started in 2013 to use red blood cells as a cell delivery vehicle for cancer vaccines and autoimmunity. The idea was quite unique, but it did not work in humans sufficiently well for them to be able to create a product. The company raised $240 million in their IPO in 2018. Five years later and with a valuation down by 99%, the company shut their doors and liquidated the remaining assets. It is worth recalling the Rubius story because it does represent one of the biggest collapses of any biotech company in the history of the industry. It is also perhaps a lesson for other unicorns. The gene-editing companies in particular, who are all chasing many of the same targets, should take note of this and be careful of what they might wish to be.

Dumping huge amounts of money on the table to create a unicorn (and hopefully a durable company) is not the only way to develop a successful biotech company.

Arvinas

Arvinas had all the necessary ingredients for success. They were founded by people who had made money before for their investors, they were based on new technology with strong IP that was licensed from one of the founder's institutions, and they were invested in by top-tier VCs. They also managed to get an early strategic alliance with Genentech, and they were run by a strong and experienced management team.

Craig Crews at Yale had already formed Proteolix in the Bay Area in 2003, based on technology developed by his lab and by Ray Deshaies at Caltech. Proteolix developed a drug called carfilzomib and were acquired by Onyx in 2009 for $810 million. Crews and his team were also the first to describe formally the concept of PROTACs—proteolysis targeting chimeras (see Box 4).

BOX 4. PROTEOLYSIS TARGETING CHIMERAS (PROTACs)

PROTACs are bifunctional small molecules: one part binds to the protein to be degraded and the other part targets the molecule for degradation. The two parts of the molecule are separated by a flexible linker (see Fig. 1).

This very simple concept means that finding compounds that selectively inhibit an enzyme or receptor target can be replaced by compounds that bind specifically to a target and "tag" it for degradation via a linked E3 ubiquitin ligase (of which there are a variety that can be used). Because the ligase is a catalytically active enzyme, the amount of small molecule needed to induce the degradation is not proportional to the target, so lower concentrations of it can be used. One of the downsides of this approach is that the small molecules tend to be larger than usual, having a molecular weight between 700 and 1100. However, they can be delivered orally, by injection or by infusion, so this may not turn out to be too much of a limitation.

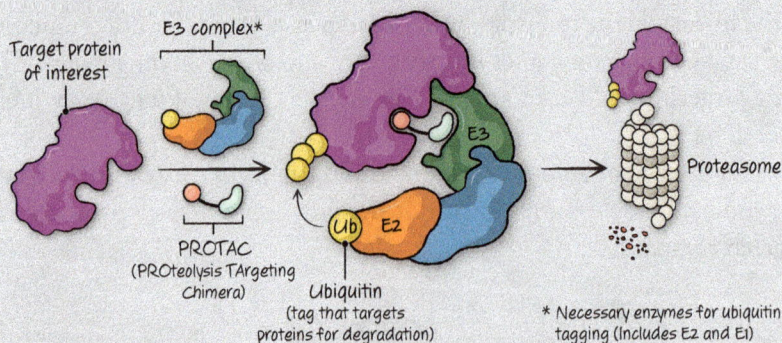

Figure 1. PROteolysis TArgeting Chimeras (PROTACs).

Founding Arvinas

Crews founded Arvinas in 2013, the same year as Rubius Therapeutics, to exploit the PROTACs technology in cancer and other areas, licensing the technology IP from Yale University. They raised $15 million in a small series A round led by 5AM Ventures and Canaan Venture Partners in 2013. A series B in 2015 attracted additional investors, including RA Capital, OrbiMed, and New Leaf Ventures, raising an additional $42 million. Arvinas announced the closing of a $55 million series C financing in 2018 led by Nextech Invest, with participation from Deerfield, Hillhouse Capital, and Sirona Capital, as well as the original investors.

By this time John Houston, a Scottish biochemist I worked closely with at Glaxo and who had been a discovery Senior Vice President at BMS for many years, joined Arvinas and became the CEO. The series C round was essentially a mezzanine financing because in September 2018 they went public, raising $120 million in a Goldman Sachs–led deal. Coincidently, this was the same year that Rubius also went public. The Arvinas financings were made easier by the company forming a deal with Genentech in 2015 for multiple targets which was extended two years later. This was followed by a large $2.4 billion deal with Pfizer for the discovery and development of potential PROTAC clinical candidates. Based on their progress, Arvinas were able to secure yet more money in a secondary Goldman-led public offering that brought in another $400 million.

Arvinas are now a well-funded public biotech company with two clinical programmes, two deals with Big Pharma, and millions of dollars in the bank. Today the company has a strong portfolio with three potential medications: including ARV-471 (Vepdegestrant) for breast cancer, targeting the oestrogen receptor (partnered with Pfizer), and ARV-110 (Bavdegalutamide) for castration-resistant prostate cancer, targeting the androgen receptor.

Given that Arvinas were started only a decade ago, they are a good example of what a successful start-up can look like in that time frame and a good contrast to Rubius. Arvinas will no doubt provide excellent returns for investors without them having to put huge amounts of money into the deal in the first place. Arvinas' market cap stands at $2.8 billion in December 2023: a unicorn that grew its large horn rationally and methodically over time.

Other PROTAC Players

There are other players in this space, partly owing to the Arvinas success. Kymera Therapeutics housed in Watertown, Massachusetts, were founded in 2015. They are targeting some kinases involved in cancer and have an IRAK4 PROTAC in phase 1 for diffuse large B-cell lymphoma (DLBCL) and a STAT degrader in phase 1 for haematological malignancies. Kymera raised $170 million in their IPO in 2020 and have a deal with Sanofi.

C4 Therapeutics, another Boston-based start-up, also have a protein degrader programme in the clinic, specifically directed to multiple myeloma. They also went public in 2020, raising more than $200 million. Other new companies are exploiting the protein tagging approach in different ways including targeting proteins for lysosomal degradation—the so-called LYTACs. Cedilla, Celeris, and Lycea are three such companies, all of which have attracted solid investments to get going. One successful company in a new space tends to spawn the formation of many others.

From a herd that includes Altos, Arvinas, GRAIL, Laronde, and Sana, which unicorn would you choose to ride?

FORTUNES

We are near the end of this chapter without me saying much about "Fortunes" at all. Let us do a little arithmetic.

It is fun to conceive of what a billion dollars looks like. The Bureau of Engraving and Printing states that all U.S. bills weigh a single gram. This means that $1,000,000 in $100 notes, weighs 10 kg and would thus fit in a suitcase, as seen in many bank robbery or kidnapping ransom movies. By simple maths, a billion dollars would be 1000 such suitcases, each one of which weighs 10 kg. So, as 1000 kg is a metric tonne, a tonne of money (as in "they raised a tonne of money," or "they made tonnes of money") is therefore 100 suitcases. You now know that a metric tonne of money is equivalent to 100 million U.S. dollars. Looking back over the companies described in this book so far, it is clear that some biotech principals have indeed made a tonne of money, and their VC investors tonnes more.[12,i]

---

[i] Talking of large amounts of money reminds me of yet another Sydney Brenner story. He was once at a Cabinet meeting in the United Kingdom concerning Science and Technology with then–Prime Minister Margaret Thatcher when she asked him how much a particular piece of science would cost (maybe the Human Genome Project, that part of the story is lost). Sydney answered that it "would be 2 or 3 'FRUs,' Prime Minister." Thatcher shot back "What is an FRU, Brenner?" to which Sydney replied in his typically acerbic way that an FRU was a "Falkland Island Runway Unit." Millions of pounds were spent extending the Port Stanley Airport runway by 6000 metres during and after the Falkland Islands War.[13]

REFERENCES AND NOTES

1. Hayes A. 2022. Tulipmania: about the Dutch tulip bulb market bubble. *Investopedia.* August 22, 2022.

2. Stewart T. *The South Sea Bubble: The 18th century version of the Dot Com boom—and bust.* Historic UK. https://www.historic-uk.com/HistoryUK/HistoryofEngland/South-Sea-Bubble/

3. Murray J. 1984. Initial public offerings of Biotech companies 1980–1983. *Bio/Technology* **2**: 619–622.

4. Drakeman DL, Drakeman LN, Oraiopoulos N. 2022. *From breakthrough to blockbuster: the business of biotechnology.* Oxford University Press, New York.

5. DeFrancesco L, Lähteenmäki R. 2022. Public biotech in 2021—the numbers. *Nat Biotechnol* **40**: 1427.

6. Papadopoulos S. For the 'godfather' of biotech, saving Biogen is the final act of a singular career. *STAT*, August 2022.

7. Kite and Juno: Two CARTs, two charts: dissecting returns from T-cell therapy M&A. Posted January 23rd, 2018 in *Exits IPOs M&As, VC-backed Biotech Returns.* Life Sci VC: 2015.

8. 2022. GSK has axed its NY-ESO-1 T-cell receptor (TCR) pact with Lyell Immunopharma. *FIERCE Biotech*, October 25, 2022.

9. 2023. DeAngellis A, Cross R. The inside story of how data integrity issues roiled a biotech seen as Moderna 2.0. *STAT*, June 12, 2023.

10. 2021. Regalado A. Meet Altos Labs: Silicon Valley's latest wild bet on living forever. *MIT Technology Review*, September 2021.

11. Offord C. 2023. Two research teams reverse the signs of aging in mice. *Science* **379**: 224.

12. Vardi N. 2023. *For blood and money: billionaires, biotech and the quest for a blockbuster drug.* W.W. Norton, New York.

13. A possibly apocryphal story but confirmed to me by George Poste in 2021.

CHAPTER 18

The Essence of Biotech

U p to now, this book has focused more on the new technologies and the companies (and some of the people) exploiting those technologies as the backbone of the industry. But there are plenty of other critical parts that make up the biotech industry: I will try to capture them here.

There are several outcomes that investors and management teams strive for in building a successful biotech company. One of the most important is to develop products that truly help patients by alleviating suffering from common diseases such as autoimmune disease or cancer or from the rare Mendelian diseases. This reason is truly why most people—but not all—are in the biotech business in the first place. A second outcome, which obviously goes with the first, is to be able to provide a good return on investment for the shareholders in your company.

If you are a venture capital–backed entity, there are three main ways to provide a return to your shareholders and investors. The first is to go public via an initial public offering (IPO) so that your shares are traded on the Nasdaq, the NYSE, or on a European or Asian exchange. Your investors (after the customary lockup) can then sell their shares and/or transfer them to their Limited Partners (LPs) for them to sell when they wish to, or simply return the cash to the LPs.[a]

The second option is that your company gets acquired by a larger biotechnology company or pharmaceutical company. This can happen at wildly exaggerated valuations (see Chapter 17). It is often an easier and quicker option than an IPO and provides a more immediate return on investment. It can happen either before or after the company has gone public or developed any products. For example, it could happen because the company has a

---

[a] LPs are generally pension funds, university endowments, sovereign wealth funds, and other organisations who invest some of their money in "high risk assets," usually via one or more venture capital firms.

technology that the acquirer sees as vital to their ongoing business or, more likely, adds value to their products or prospective products.

The third way is to grow the company into an independent commercial entity selling their own products into the medical/pharmaceutical market. This long-term option requires patient investors. Most likely the company will no longer be privately held at this point but will already be a public company. We have met several such companies in the course of this history, such as Biogen, Amgen, Gilead Sciences, and Regeneron. But there are not many such successful independent long-term companies, especially compared to the number of start-ups that get created.

You may justifiably ask what makes the difference between success and failure. What option do you follow, and what is the most likely outcome? Is it the technology, the products, the people, the investors, or the location of the enterprise, or is it just dumb luck?

It is certainly not dumb luck, but fortune favours the brave, and nowhere is that adage more true than in biotech. Chance also plays a role in the sense that chance encounters of people and minds can lead to interesting outcomes. My thesis is that successful outcomes depend on several "phenotypic traits," which define the "essence" of a biotechnology company.[b,c]

Let us look at these traits in turn.

TRAIT 1: TECHNOLOGY

Waves of biotechnology are defined by the emergence of new technologies, as laid out in the preceding chapters. From recombinant DNA to monoclonal antibodies, genetics and genomics, DNA sequencing, antisense and siRNA technology, cell therapy, gene therapy and gene editing, and, most recently, mRNA-derived vaccines. The Bayh–Dole Act that the U.S. Congress passed in 1980 meant that universities got to keep royalties and product licensing

---

[b]By the way, I hate the words "secret sauce" as in "I wonder what their secret sauce is" or "there must be a secret sauce we do not know about." Secret sauces go on exotic food in Michelin star restaurants: they are not attributable to biotech companies. I am also allergic to people saying that something about an inanimate object is "in its DNA" as in "it is in the company's DNA." Companies do not have DNA. People in them have DNA, and it would be much more appropriate to say, "it is in the DNA of the people who work there."

[c]I also apologise to the purists for using the word "trait," which is a term used to describe phenotypic aspects of a plant or an animal rather than aspects of an industry, but I think it works.

fees from research done in their laboratories, which in turn helped the bio-technology industry to get going commercially.[1]

Nobel Prizes

It is quite revealing that several of the people who developed the technologies that have been applied in the business of biotechnology were awarded the Nobel Prize, which I tried to note every time this connection came up throughout this book (see Relevant Nobel Prizes starting on page 451). This is obviously not a coincidence, but I was surprised to see just how often it applies. You can almost track the history of biotechnology by looking at the list of Nobel Prize winners in the life sciences over the last 50 years, though some of them are less directly connected than others. The 1978 Prize to Werner Arber, Dan Nathans, and Ham Smith for the discovery of restriction enzymes, the 1980 Prize to Paul Berg, Fred Sanger, and Wally Gilbert for their work on nucleic acids, and the 1984 Prize to César Milstein, Niels Jerne, and Georges Köhler for developing the hybridoma technology and the production of monoclonal antibodies, all stand out from the early days of biotechnology. More recently, the 2006 Nobel Prize to Craig Mello and Andy Fire for RNA interference, the 2018 Prize to Jim Allison and Tasuku Honjo for cancer therapy by checkpoint inhibition, the 2020 Prize to Jennifer Doudna and Emmanuelle Charpentier for CRISPR-Cas9 gene editing, and the 2023 Prize to Katalin Kariko and Drew Weissman for mRNA-based vaccines are all clearly directly relevant to the more recent developments in the biotechnology industry.[2] On that basis, I expect a Nobel Prize for base editing in the not too distant future.

It is easier to nurture and utilise new technology in a small company than to try to incorporate it into the workings of a large company. This is especially true for "platform" technologies applicable across many therapeutic areas, where large companies find it hard to communicate well enough to build the technology for access by all parts of their R&D organisation. Big Pharma is generally not adept at internal innovation either. That is the province of the VCs and the founders of the technology that build the small platform–based biotech companies, which are only then accessed by Big Pharma through alliances or by acquisition.

One important and often overlooked part of the development of the technology in biotechnology was access to premanufactured reagents to simplify and accelerate the experiments. In the early days, we used to barter restriction enzymes that were made by graduate students or postdocs in various labs. We swapped other things as well: enzymes like DNA ligase, and necessities such as plasmids and cell lines. The opportunity to provide these reagents conveniently led to the birth of companies such as New England Biolabs and Bethesda Research Labs, both of which started out selling the standard enzymes used for cloning. Easy access to reagents and kits provided by companies like Thermo Fisher continues to drive the industry (Box 1). Many activities that used to take a day in the lab can now be done by machine in a couple of hours.[d]

BOX 1. INVITROGEN BECAME THERMO FISHER

Invitrogen were started in 1987 in San Diego (Carlsbad) as a partnership between Lyle Turner, Joseph (Joe) Fernandez, and Bill McConnell. The company started out by manufacturing molecular cloning "kits" for making mRNA and for making cDNA libraries. The amount of time saved in the lab by access to these kits was generous—but it came at a price.

It was not so easy in the beginning for Invitrogen, and without money from the California Export Finance Office, Invitrogen might not have survived. The reason they did make it was probably because of their development of a "TOPO" cloning kit. The key to TOPO cloning, and hence the name, was the enzyme DNA topoisomerase I, which functions both as a restriction enzyme and as a ligase. Its biological role is to cleave and rejoin DNA during replication. TOPO cloning speeded up restriction and ligation of DNA by an order of magnitude, and thus it is not surprising that the kits sold like hotcakes.[e]

By 1998, Invitrogen had sales of more than $33 million, and they went public in 1999, raising almost $50 million. This financing allowed the company to expand their product offerings and their earnings. Invitrogen's real success was via an acquisition strategy, in which several small reagent companies were brought into the Invitrogen fold. They merged with NOVEX, another San Diego reagents company that sold, among other things, precast electrophoresis gels, saving scientists even more time and aggravation by not having to pour their own gels (something I and others of my generation spent many hours doing, as well as cleaning the gel mould plates!).

---

[d]Lab workers today have to be far more computer-savvy than I ever was.

[e]At about this time, Mark Bergseid, one of the first employees at Structural Genomics, introduced me to Lyle Turner who was gracious enough to take me for a flight from Palomar Airport in his recently acquired Pilatus aeroplane—a top-of-the-range single-engine turboprop made in Switzerland. I understand he got a new Pilatus in 2019—such was the success of his start-up. It was certainly a step up from the Cessna 172 or 182 that I used to rent to fly.

A pivotal deal for Invitrogen was the acquisition of Life Technologies in Maryland in 2000. Life Technologies were the result of a 1983 merger of Bethesda Research Laboratories and GIBCO (Grand Island Biological Company, a subsidiary of Dexter Corporation that manufactured cell culture reagents).

Invitrogen's merging with Life Technologies significantly increased the Carlsbad company's presence in the gene cloning, expression, and analysis market, creating a company with many differentiated product lines. By now they had more than $500 million in sales and an annual operating income of $80–$100 million, owning 40% of the expanding "kit" market. One direct connection I had with them was through SAIC-Frederick, where I was running the Advanced Technology Program in 2006–2010. Jim Hartley and his team there had developed the "Gateway" cloning system, which facilitated the rapid cloning of DNA fragments from one vector to another and circumvented the use of restriction digestion and ligation. This technology was developed on behalf of the NCI as part of the FFRDC (Federally Funded Research and Development Centers) and licensed to Life Technologies.

In 2008, Invitrogen purchased Applied Biosystems (part of Perkin-Elmer) the Lee Hood–Mike Hunkapiller company that made DNA sequencing and PCR machines (see Chapter 8), in a $6.7 billion deal. This entity became the new Life Technologies company with about $3.5 billion in annual sales, more than 9500 employees, and more than 3600 licenses and patents. They bought Ion Torrent in 2010 (see Chapter 8). Life Technologies were acquired by Thermo Fisher in 2014 for $13.6 billion, making Thermo Fisher (originally formed from a merger of Thermo Electron with Fisher Scientific) the biggest life sciences reagent supply company in the world (although they divested their cell and cell culture media business to GE for more than $1 billion at that time).[f]

It does not always go that well for reagent companies. For example, Stratagene, a La Jolla–based company that made Lambda arms from which genomic libraries could be constructed, were started in 1984 and were sold to Agilent in 2007 for $250 million.

TRAIT 2: LOCATION, LOCATION, LOCATION

Although Lord Harold Samuel coined this phrase when he founded Land Securities in the United Kingdom in 1944, its origin belongs to a classified Real Estate ad in Chicago in 1926 referring to the desirability of Rogers Park[g]: "Attention salesman, sales managers: location, location, location close to Rogers Park."

---

[f]My youngest daughter Rebecca was once an HR director in Diversity and Inclusion at Thermo Fisher.

[g]A still very desirable and diverse northern suburb of Chicago on the shore of Lake Michigan.

The phrase is very apposite for the biotechnology industry, especially in the United States. It is no surprise that the location of the biotech companies largely reflects the home location of the founders of those companies and the availability of trained scientists for hire locally. The major biotech centres are generally near major universities and university hospitals, the most notable being the hubs in San Diego, South San Francisco, Seattle, and Cambridge (MA).[3]

But location is not just defined by talent access and home turf. There are companies set up to manage and/or buy buildings and turn them into labs and lease them to the fledgling, mostly venture-funded, companies. The major player is Alexandria Real Estate Equities (ARE), which now has more than 75 million square feet of lab and office space leased to many different pharma and biotech companies: fivefold more than the next biggest biotech real estate player.[4]

Alexandria Real Estate Equities

ARE grew up in a symbiotic relationship with the developing biotech industry and, as a result, the company is now much more than just a landlord in the biotech business. The company was founded in 1994 and named after the ancient scientific capital of the world in Egypt, with a logo that pays homage to that namesake. They were started like a garage start-up with $19 million in Series A capital and a vision to create the first real estate company uniquely focused on the life science industry (see Box 2). Jerry Sudarsky, one of the partners with Joe Jacobs of Jacobs Engineering, was presented with a business plan for the Alexandria business model in 1993. Jacobs approached Joel Marcus, a lawyer and CPA, with the idea of representing their interests in this company and overseeing the management team, who intended to provide laboratories and office space to the growing number of biotechnology companies.

Joel became a co-founder with Jerry. Formerly a lawyer at Brobeck, Phleger & Harrison before they filed for Chapter 7 bankruptcy, Joel had worked on many biotech deals in the 1980s, including with VC investors such as KP and with IPO teams like the biotech banking specialists Robertson Stephens. He was also one of the lawyers who worked on the historic Kirin–Amgen erythropoietin (EPO) deal, so he knew the burgeoning biotech scene quite well.

> **BOX 2. ALEXANDRIA REAL ESTATE EQUITIES WERE ONCE A START-UP COMPANY TOO**
>
> Jacobs put $5 million into the Series A round, as did a local San Diego VC called Domain Associates (Jim Blair and Art Klausner were principals). ARE's first purchase was four buildings in San Diego, one of which happened to be 11099 North Torrey Pines Rd. —the building where Sequana was located (see Plate 27 for the old Sequana building). Even though I did not know it at the time, this was the start of a long-term relationship between ARE and me. They were my landlords for several of the companies that I either led or worked for, both in San Diego and in Boston. These included 10505 Roselle Street for SGX, 11025 Roselle St. for Novasite, 225 Binney St.—the new Biogen building, and most recently Repertoire Immune Medicines in One Kendall Square—part of the new Alexandria Center in Cambridge.

Joel and his team followed the "cluster theory" put forward by Michael Porter at Harvard.[5] This theory holds that clusters of companies in the same business generate competitive advantage by being close together. After their initial investments in the nascent San Diego biotech cluster, ARE expanded to Seattle and to the Bay Area. They followed the same cues as the VC investors, being sure that the companies they worked on behalf of had good science, had good management teams, and were well-capitalized. It was not particularly surprising, given Joel's background, that ARE would also often invest in the companies they were leasing space to and Alexandria Venture Investments were formed to formalize this activity, making ARE far more than just a landlord.

ARE became a public company in 1997, raising $155 million, which enabled them to expand from San Diego, the Bay Area, Seattle, and Maryland (Gaithersburg/Rockville) to New York and Research Triangle Park. Step by step the company acquired and built more lab and office space for biotech companies.[3]

Incubator Concept

ARE also initiated the concept of more formal biotech incubator space for "just started" companies. Originally a "hotel" idea, the concept has grown into Alexandria LaunchLabs, with locations in Kendall Square and New York. The LaunchLabs provide very early-stage life science companies with "move-in ready" lab and office space and basic support services: a far cry from the makeshift spaces and support functions that Celltech, Gilead Sciences, Ionis, and many other companies had to scrounge for when they started.

Key Locations and Iconic Buildings

ARE now have a large presence in all the key biotech locations.[h] The company's San Diego properties are primarily in La Jolla at Torrey Pines, Sorento Valley, Sorrento Mesa, and University City. Their largest "campus" by walking proximity is in Cambridge (MA), which includes the 2.4-million-sq-ft Alexandria Center at Kendall Square and the 1.2-million-sq-ft facilities in Technology Square, which they obtained from MIT (who had previously leased it to Novartis and others) in 2006.

ARE now owns a total of 7.5 million square feet in the greater Cambridge/Boston area. Their tenants include Moderna[i] in Cambridge[6] and Bristol-Myers Squibb, their largest client, some of whose buildings were inherited from the companies that BMS acquired.[j]

ARE continued to expand in their key cluster locations throughout the 1990s and into the 2010s. They built the Mission Bay campus in San Francisco, which opened in 2013. In 2002, they acquired the headquarters of ZymoGenetics (Zymo) for $52 million in a lease back deal (Box 3). ZymoGenetics' lab and office space was housed in the old Steam Plant building on Lake Union in Seattle at 1201 Eastlake Avenue, a wonderful building constructed between 1914 and 1921 by Seattle Light and Power. ZymoGenetics did not renew the lease in 2016, and the space was taken over by the Fred Hutchinson Cancer Research Center (see Plates 28 and 29 for pictures of the Zymo building on Lake Union).

ARE's space in New York is also notable. They built a 750,000-sq-ft building on the west side of the city on the Hudson River, not far from Greenwich Village. The Alexandria Center for Life Sciences now houses more than 50 mostly "home-grown" small life science company tenants spun out of New York academia (the centre is New York's only commercial life sciences hub).[k]

---

[h]More than 75% of the more than $2 billion/year in revenue they make comes from the Boston, San Diego, and San Francisco clusters.

[i]Sensibly, ARE participated in Moderna's $100 million-plus Series B financing round.

[j]Moderna have built a 462,000-sq-ft Moderna Science Center at 325 Binney St. in Cambridge, just up the road from Biogen and across the road from Amgen.

[k]Among their notable tenants was ImClone Systems, the infamous Sam Waksal monoclonal antibody company that got bought by Eli Lilly in 2008.

BOX 3. ZYMOGENETICS

ZymoGenetics, one of the early biotech companies, were founded by Earl Davie and Ben Hall at the University of Washington and Michael Smith (University of British Columbia, and winner of the 1993 Nobel Prize in Chemistry). The company had a long-term relationship with Novo Nordisk in Denmark, with a focus on expressing recombinant proteins in yeast and other eukaryotic organisms. They were run by Bruce Carter, a Trinity College Dublin–educated yeast geneticist. Later, before they were sold to BMS in 2010 for $885 million, Doug Williams was the CEO.

The story of how ZymoGenetics got to occupy the Light and Power building is a classic piece of Irish *dánacht* (chutzpah).[7] Apparently, Bruce Carter had always had his eye on the building, even when it was mostly a ruin with broken windows and no chimneys. One day he took senior Novo Nordisk executives on a boat trip that ended up in Lake Union, when the building was lit up by the evening sun. Bruce persuaded Novo to buy the building and refurbish it as the ZymoGenetics headquarters and labs. Much to the consternation of some of the local population, the refurbished chimneys had "ZymoGenetics" emblazoned down each one in big letters. When the complaints came, Bruce simply responded that Seattle Light and Power had used the chimneys to advertise their name in the same way. Case closed.

More Than a Real Estate Company

ARE are more than just a large real estate company in the life sciences sector: they also happen to be good landlords.[1] They were very understanding with me when paying the rent at Novasite became difficult, and there are many stories of them taking the longer-term view, in line with the appropriate "biotech" philosophy. The story of Amylin related in Chapter 6, a San Diego company founded in 1987 and sold to BMS for $5.3 billion, is a case in point.

Biotech Space in the United Kingdom

The life sciences infrastructural support in the United Kingdom pales into near-insignificance in comparison to what has been done in the United States. Granted there are now the Oxford and Cambridge Science Parks,

---

[1]The Alexandria Real Estate Equities philosophy is that patience and opportunity will pay off. They clearly adhere to the principle that those you see in the elevator going up are generally the same ones you see in the elevator going down—so treat people with respect, humility, and humanity.

and the Babraham and Hinxton campuses near Cambridge, but there is no ARE that pulls all of them together in the way that has been done in Boston or San Diego. Part of this problem is historical, owing to the complicated ownership of the land around Oxford and Cambridge. Enara Therapeutics, an Immuno-oncology start-up funded by SV Health Investors, for example, is located in the Oxford Science Park.[m]

The lack of life science company space is even more obvious since the resurgence of interest in biotech after COVID-19, including a particular lack of biotech and pharma manufacturing space. This is an issue the government in the United Kingdom needs to take seriously, as it can severely limit the commercial innovation.[8] It is a particularly pressing problem in the triangle of Oxford, Cambridge, and London. Building lab/office space is expensive and requires significant planning and some specialisation. There are many incubators around the United Kingdom (primarily near the major universities), but these are only useful for small, early-stage companies. Other cities like Edinburgh and Manchester are making bolder moves: for example, the old AstraZeneca site in Alderley Park near Macclesfield has been turned into space for small companies.

I am on the Board of Pheno Therapeutics, a small company working on small-molecule drugs for multiple sclerosis based on the groundbreaking work in the University of Edinburgh labs of Siddharthan Chandran and Neil Carragher. Pheno Therapeutics were founded by the University with Advent Life Sciences and Life Arc as investors and are also located at a very supportive University. Edinburgh is a good place to build a company in the United Kingdom.

TRAIT 3: PRODUCTS

Most of the small biotech companies that are set up fail in two ways: either quite quickly or painfully slowly. In 2015, there were ~2700 Biotech companies of various shapes and sizes. It is more than double that now.[9] Those numbers translate to many failures and a few successes, the outcome for each

---

[m]Their address is Magdalen Centre, Robert Robinson Way, Oxford. Magdalen College Oxford, like many other of the Oxford and Cambridge colleges, owns large amounts of land in the United Kingdom, particularly in their local vicinity. Trinity College Cambridge owns the land on which the Cambridge Science Park is built. There was a time, it is said, when you could walk from one end of England to the other without ever walking off Trinity College Cambridge land. I do not doubt it, but I do not know if it is still true. The same used to be said for the Church of England and that is probably still true.

largely dependent on the products that the companies make rather than the technology used to make them. As I have said, technology takes you only so far. Technology always develops at such a pace that your technology becomes history more quickly than you or your investors anticipate. Look at gene editing: "plain old vanilla" CRISPR-Cas9 is being rapidly superseded by base-editing technologies.

Acquisitions to gain ownership of products happen more often than acquisitions for technology, although sometimes they go together (as seen in the Kite and Juno acquisitions). Several notable product acquisitions include Gilead buying Pharmasset for sofosbuvir in 2011 for $11 billion in a move that anchored Gilead's hepatitis C franchise. Roche similarly acquired InterMune for $8.3 billion in 2014 for pirfenidone, a drug that treats idiopathic pulmonary fibrosis by inhibiting the synthesis of TGF-β-stimulating endothelial cells. This acquisition added InterMune's drug to a Roche respiratory franchise that already included Pulmozyme, a recombinant enzyme cloned and developed by Genentech for cystic fibrosis.

Every so often there have been reviews of the successful early biotech products (e.g., in *Nature Biotechnology* in 2000).[10] It is entertaining to review the top best-selling drugs in 2020 and 2021 and to evaluate whether you call them "biotech" drugs or "pharma" drugs (Boxes 4 and 5), an increasingly more difficult exercise as these industries coalesce. To put the numbers into perspective and provide context, remember that EPO, the best-selling biotech product, had sales of $4 billion in 2004 (see also the Biotechnology Product Time Line [pp. 441–449]).

The new 2022 Inflation Reduction Act passed by the U.S. Congress in 2023 has big implications for drug pricing and development. Medicare will be able to negotiate prices for 10 medicines in 2026. That number grows to 60 by 2029. There is a distinction as well between those drugs (usually

---

BOX 4. ORIGIN OF THE TOP-SELLING DRUGS IN 2020

In 2020,[11] the best-selling drug was the anti-tumor necrosis factor (TNF) mAb Humira, marketed and sold by AbbVie. This drug clearly has its origins in biotech (Chapter 5). Humira has been the world's best-selling medicine for 10 years, making AbbVie more than $20 billion in revenue in 2020. It came off patent finally in 2022, so its days at the top are numbered. As replacements, AbbVie is pushing (to doctors and consumers) both Skyrizi (risankizumab), a humanized monoclonal antibody targeting IL-23a made in collaboration with

Boehringer Ingelheim, and Rinvoq (upadacitinib), a relatively selective JAK1 inhibitor for autoimmune diseases, especially psoriasis.[n]

Coming in at number 2 is Keytruda ($14.4 billion), the Merck anti-PD-1 mAb against cancer. This is also a product of biotech (if by a somewhat circuitous route) by it being a monoclonal antibody humanised by the MRC Technology Centre in the United Kingdom (Chapter 5). Number 3 is Revlimid ($12 billion), the thalidomide derivative that Celgene obtained before they were bought by BMS. If you call Celgene a biotech company (and I think we can), then that is three in a row for biotech.

The number 4 and number 10 medicines in 2020 are both Factor Xa inhibitors, used as anticoagulants in atrial fibrillation patients to reduce stroke risk. Both drugs (Eliquis at $9.2 billion and Xarelto at $6.9 billion) were developed by pharma.

Number 5 is Imbruvica (ibrutinib), a tyrosine kinase inhibitor used to treat B-cell malignancies. This drug was developed by a small chemistry-focused company called Pharmacyclics, founded by Jonathan Sessler, which was bought by AbbVie for $21 billion in 2015. We should count Pharmacyclics as a biotech company despite becoming AbbVie, because the origins of the drug go all the way back to Axys and Celera.

Number 6 is Eylea ($8.3 billion), the vascular endothelial growth factor (VEGF) trap discovered, developed, and commercialised by Regeneron: a clear biotech win. Stelara ($7.9 billion) is number 7, an anti-IL-12/23 mAb marketed by J&J. This drug (and all the mAbs used to treat psoriasis and psoriatic arthritis) we can count as biotech-derived products. At number 8 is Opdivo (nivolumab), BMS's anti-PD1 cancer mAb ($7.9 billion), also a biotech product.

Biktarvy ($7.2 billion), an HIV triple drug combo from Gilead Sciences, is at number 9. It is a combination therapy consisting of a novel integrase inhibitor, plus emtricitabine (which Gilead got from Triangle Therapeutics), and tenofovir alafenamide. Biktarvy is really a typical pharma product but given Gilead's origins in biotech we could place it in the biotech column.

---

[n] The amount of airtime dedicated to direct-to-consumer (DTC) advertising of Skyrizi and Rinvoq medicines on prime-time TV reflects their importance to the AbbVie franchise. TV advertising of drugs is an anathema to me. I think it is a complete waste of money: these enormous >$500 million advertising budgets would be much better spent on R&D. I would be interested to know how they calibrate what the effect of advertising is on the sale of these drugs. It is especially irritating to be told that if you are allergic to a drug, you should not take it: how do you know if you are allergic to it or not if you have not taken it? It is also much easier to remember all the side effects than anything else, even what the drug is supposed to help treat. It seems to me that this whole DTC advertising of drugs is a self-defeating activity. I almost always put the TV on mute when these advertisements interfere with what I am trying to watch. And I suspect I am not the only one. Someone more qualified than I will have to explain to me how the TV advertising of drugs makes money.

> You can do this analysis for the next 11 best-selling drugs yourself. If you do and then total up the score (with a little give and take), it will reveal a biotech origin for between 60% and 70% of the top 20 bestselling drugs in 2020. This feels about right. I know it is to some extent an artificial distinction, but it does put the contribution biotech has made to the medicines that are used and sold today into sharp relief.

biologicals) administered by doctors (or their nurses) reimbursed through Part B of Medicare, and small molecule drugs controlled by Part D pricing. Medicines would be eligible for Medicare negotiations after nine or 13 years on the market, depending on whether they are a small molecule or a biologic, much reducing the time that the company has available to make the kind of money that is listed in Boxes 4 and 5.

In addition to pricing concerns, nearly 70 medications will lose patent coverage by 2030 (the so-called patent cliff). Despite all AbbVie's efforts, Humira sales are already suffering from the drug coming to the end of its patent life. Keytruda's patent expires in 2028 for Merck. Lynparza (Olaparib), the AstraZeneca PARP inhibitor, also comes off patent in 2028, and GSK faces expiration of some key HIV drug patents in 2028 as well.

A review of the top 50 best-selling drugs in 2022 was published recently in *Drug Discovery and Development*.[12] The top 20 are not very different from 2021 with the COVID-19 vaccines and Paxlovid (the antiviral from Pfizer)

BOX 5. CHANGES IN THE LIST OF THE TOP-SELLING DRUGS IN 2021

The 2021 list of top-selling drugs looks a bit different from 2020[13] owing to the contribution of the COVID vaccines and the SARS-CoV-2 mAbs. Comirnaty, the Pfizer/BioNTech mRNA vaccine for SARS-CoV-2 generated $36.8 billion in revenue in 2021, ending the Humira era of being the best-selling drug (even though Humira's revenue was still more than $20 billion in 2021). Moderna made $17.7 billion from Spikevax. The SARS-CoV-2 antibodies generated more than $5 billion and the antiviral remdesivir from Gilead Sciences also made more than $5 billion, even though the drug does not work very well against SARS-CoV-2. The vaccines clearly have biotech origins whereas the antiviral can probably be counted as a pharma product. The other important drug to enter the top 20 in 2021 is Dupixent for atopic dermatitis from Regeneron/Sanofi and Darzalex, an anti-CD38 mAb for multiple myeloma from J&J. Vertex's Trikafta, used for treating cystic fibrosis, also made the top 20 best-selling drugs in 2021. These three medicines all have biotech origins, so the trend continues, with nearly 75% of the top-selling medicines in 2021 having their origins in biotech.

biasing the data somewhat, but with the remainder trending as might be expected given the 2020 to 2021 list changes and the shifting IP situation.

TRAIT 4: PEOPLE

In Chapter 6 of the book *From Breakthrough to Blockbuster*,[9] the authors describe quite well what characteristics mark a biotech entrepreneur. Even though it is an adjective and not a noun, I prefer the word "entrepreneurial" to entrepreneur: it has more of an active flavour. Biotech is full of entrepreneurial people, who are comfortable taking career risks to do extraordinary things (but quite often from a place of financial security!). I used to climb the high conifer trees in my garden when I was a kid. One way of looking at biotech "people" is to imagine a very tall conifer tree with people climbing it. Who is closer to the trunk where the branches are fatter and less likely to break, and who is further out at the ends of the branches, perhaps even swaying about in the breeze a little? Biotech people are generally out on the edges more, whereas those in pharma prefer to be closer to the trunk. You can get high in the tree either way: it just depends on your appetite for risk and how encompassing your preferred view. If you want to change the practice of medicine and create both new drugs and new classes of drugs, you need to feel comfortable at the end of the branches and with no safety net (to mix analogies a bit). There are definitely more of these people in biotech than in pharma.

The People Acquisition Paradox

It is a curious paradox that when a small biotech company is acquired by large pharma or even a bigger biotech company, the one asset that tends to be undervalued is the people. Some of them may even be the founders of the company, who started the whole thing and have intimate knowledge of the technology, the drugs in the pipeline, and even the market for their drugs. Yet when the acquisition happens little to no effort is made to keep them happy or even keep them at all.

To be fair, in some takeovers, the acquirers have tried hard to maintain the culture of the acquired organisation. The 1990 Roche–Genentech deal (finalised in 2009) is an example of that kind of effort, and it even continued after the company was completely acquired. Really good people remained at the company, including the senior leadership team. The consistency of having the same people at the top of the organisation, whom you get to know and understand, is very important for the scientists and other employees. Continuity of a decent leadership team that have worked together is almost a prerequisite for building a successful company. Nevertheless, this asset usually disappears completely in an acquisition, and the acquired company and their products suffer as a result. Genentech's perceived magic has been reflected in their successful products and they still attract very good people to work there. The Wyeth–GI merger in 1992, on the other hand, fully consummated four years later was designed to get the GI innovation philosophy into Wyeth and not transfer Wyeth behaviours into GI. Unfortunately, it turned out to be mostly the other way around (Chapter 2).

Large pharma companies (and large companies in general) like to measure progress by frequent assessment of whether people and projects have "met their goals." This approach turns managing the science into a box-ticking quantitative exercise. How many compounds/mAbs/constructs have you made recently? How many drugs into development? How many will you make by the end of the month/quarter/year? This is a good way to make ensure that employees generate a lot of "stuff," but the "stuff" might not be all that useful. It is not *quantity* that counts, it is *quality*. One good compound/mAb/construct/drug in development is worth much more than many average-at-best ones. As this demonstrates yet again, the drug discovery process is so easy in principle and yet so hard in practice. It is a process of elimination but with application of great judgement at the same time.

Managing R&D

Managing R&D, in biotech and elsewhere, is an art form. Some people have the ability and some people do not. It requires a mixture of skills. A good manager must be clear to the team about what is required, at what cost, and in what time frame: but they should also know when to call a time-out. Setting the goals should be the responsibility primarily of the project teams, not management. Tell me (as R&D head) what you will achieve and in what time frame. They should expect me to hold them to the time lines that *they* set, unless the science and data derived tells them (and me) that it is either not going to work or will take a lot longer. The art comes in allowing the scientists enough space to achieve the goals, without overwhelming them with bureaucracy. It also is about making sure that the scientists set expectations properly. Just because I (as R&D head) might want something to be done in two months, does not mean it can be done in that time frame. Tell me that and then have the team set the expectations together, *which they should then meet*. This is much easier to do in a small company than in a large one. The problems come when the team imagines that the real time line will not be acceptable to management: they then set expectations that can never be achieved because they are not real. As a result, everyone loses.

Employees Are Mobile

One of the most interesting things about biotech people (and often forgotten or ignored by big companies) is that they have legs, and they know how to use them.[o] They can easily move to another company or, more likely if they are the entrepreneurial, start another company. If you want to keep the employees, you need to look after them. The clusters of excellence in Boston, San Diego, and South San Francisco make this mobility easy, as there are many companies close by, doing much the same sort of things scientifically. You literally walk round the corner to get your new job. The new company you go to may even be in the same building with the same investors, so you may not have to walk very far.

A looming danger in Big Pharma and other large companies in my experience is what I have always referred to as the "grey men."[p] You do not need people who, when presented with a new idea, immediately say that because it has not been done before, it is not a good idea for the company to do it. Precisely the reverse attitude is required to drive success. Find a way to give it a go, by any means that will work.[q] The essence of biotech, as Doug Williams once put it to me, is to play offence and not defence.

You may have heard of "skunkworks." These are projects or experiments that are done without senior management's knowledge. If the senior management knows about them, then they are not really skunkworks. This kind of work is the essence of research in biotech. So do not overmanage: hire good people and let them get on with it. The motivation and the reward for the staff comes from being allowed to come up with a new way of doing something or a new project and being given the opportunity (or taking it) to try it out. There is a great feeling of satisfaction when it does work and being responsible for it.

---

[o] With apologies to ZZ Top.

[p] They do not need to be biological men, but they usually are. "Greyness" is an attitude of mind.

[q] Within the bounds of legality, morality, and decency of course.

Good decision-making by the way, like goal setting, is *not* just the province of senior executives. In my opinion, the best person or persons to decide on something are the person(s) that know most about it. Often decisions get made by people "at the top," who know absolutely nothing about what they are being asked to decide about. They in turn ask the wrong people what the right answer might be. As someone once said to me, you are only as smart as you are, and you are unlikely to get smarter over time. What you will be able to become if you watch and listen and learn is wiser, so you can be successful by combining your inherent smarts with experience.[r] Contrary to myth, you are not born with wisdom.

Biotech Culture

As founders of a start-up company, you not only have to raise money, but you also need to do whatever is necessary to make the company work. You can never be too proud or self-important if you want to start and grow a new company. If you need to be the one that sticks stamps on letters, then that is what you do. Many founding CEOs have taken cuts in salaries and moved from plush offices to holes in the wall to make their company happen. The biotech industry is (or at least used to be) full of people like that. It is the passion that drives what you do in the interest of developing technology or new drugs to treat patients, and not the luxurious surroundings in which you do it or the limos you get to travel in. It is called being scrappy. Biotech is a business for scrappy and competitive people. There are many true (and a few apocryphal) stories about senior executives leaving their opulent offices in Big Pharma for a room in a tatty building as a start-up CEO. This was certainly true for Gabe Schmergel at GI, Henri Termeer at Genzyme, and Stan Crooke at Ionis. They moved because they wanted to achieve something and had the passion to do it. Their external environment meant little to them in the short term.

It is also a little-appreciated fact that the pens, mugs, hats, and jackets—the so-called "swag"—that small biotech companies like to buy for their employees can make a difference in building a shared culture. I still have a Biogen umbrella as well as shirts and jackets and mugs and pens from many of the small companies where I have worked. As I mentioned before

---

[r] Put into equation form, $S + E = W$ prevails in biotech, where S is smarts, E is experience, and W is wisdom. S is a constant, so E is very important.

(Chapter 7), I still have a Sequana polo shirt from 1994 that I still wear 30 years later. Though swag items generally have little retail value, the feeling it can give of being in it together for the patients and for the employees and for the company is much easier to find in a small company than in a big one. That does not mean that people working in Big Pharma do not feel they belong and are working on behalf of the patients. It just all feels much closer and real when you are in a stimulating and inspirational small company.

Smaller companies also nearly always invest in weekly and monthly staff get-togethers.[s] Genetic Systems' nibbles, Genentech's Hohos, and the Thursday evening drinks and talks at Biogen are all good examples. At Bioverativ, we used to have breakfast omelettes cooked for us on a Tuesday morning. What a great way to meet people from other parts of the team (albeit that they were in the same building) and talk about prevailing issues. I miss those breakfasts.

This sort of thing does not happen routinely in Big Pharma, although there are isolated pockets here and there that try to capture that spirit without much support from senior leadership.

The Role of the Board

The presence or absence of a competent Board of Directors can make or break a company, especially small and relatively fragile biotech start-ups. How well the CEO works with the Board is equally important.

The investing VCs and sometimes other investors will have a seat on the board of their companies if they have sufficient ownership in the firm. As the company gets bigger, there may be one or two independent Board directors added. Sometimes Board membership is negotiated during the investment process, and a Board seat is not necessarily a given for any investor. The Board members have the responsibility to represent the interests of all the shareholders, unless the company enters the "zone of insolvency," at which point the Board must look after the company's creditors in case of company windup or bankruptcy. It does happen in biotech!

Having been to hundreds of Board meetings over the years, I can tell you that they tend to follow a common format. The Board reviews the

---

[s] Although COVID-19 has changed the way we work forever, Friday "POETS" day ("Push Off Early, Tomorrow's Saturday") still prevails in some larger organisations.

scientific and clinical progress of the company. It makes sure that the management adheres to the budget and advises the company on where their next round of financing may come from. The Board approves the annual goals of the company and can even help management define the desired company culture, although that tends to be created from the bottom up. The trend these days is to review the detailed science separately from the main Board meeting, so that the actual board meeting can focus on financing and other fiscal and governance issues. It is important for the company to prepare the Board papers and presentation well ahead of time. The best CEOs have sidebar conversations with their key Board members before the meeting to make sure everyone is aligned. It is not that easy, as most investors on the board have other companies to look after and they are generally rather busy people.

Being User-Friendly

I was told by the late Alex Barkas at Prospect Venture Partners, who was on the SGX board, that I was not a "user-friendly" CEO. In translation, this meant I did not spend nearly enough time keeping the Board in the picture about what we were doing, especially when it came to bringing in money in the form of business development deals. These deals always take longer to put together than you anticipate and can be a big source of Board frustration, especially when the prospect of the investors having to put more money into the company is on the horizon. Sometimes I felt that the SGX Board would play "intellectual tennis," batting opinions on different issues to and fro, across the table rather than either approving or disapproving management's recommendations for a course of action.

Management recommendations need to be very well thought out and very clearly presented. In retrospect, I focused way too much on the science we were doing, largely because that was what I preferred to do.[t] I think a lot of R&D people find it hard to adapt to life as the CEO of a biotech company for the same reason. To put it bluntly, a successful CEO must (to mix metaphors again) kiss a lot of frogs and rub shoulders with people he or she would not necessarily want to go for a beer with. For me, life is too short for that pretence, whatever the financial upside might be. Give me the science any day of the week.

---

[t] I watch old colleagues of mine who become CEOs make the same mistake all the time.

Where Biotech People Meet

Ritz-Carlton Meeting

The Ritz-Carlton hotel in Laguna Niguel, California, was the venue of an annual biotech CEO meeting sponsored by Burrill & Co. and Kleiner, Perkins, Caufield & Byers (KP). Each October from 1987 onwards more than 200 biotech CEOs gathered to share management ideas and set an agenda for the industry. Although the programme listed a serious set of presentations, there was always time for networking. I only went a few times when I was at SGX and did not have to stay over as it was an easy drive on Interstate 5 from La Jolla. The sessions were held inside, with the meals outside in the pretty courtyard of the hotel.

Various eminent people would be invited to speak and reflect on the past year in the industry. One year, Al Gore, who was a KP partner, came to speak about his health-care agenda: it was that kind of meeting. Fred Frank and Stelios Papadopoulos, both legendary biotech bankers, were frequent attendees. David Ewing Duncan, who I met frequently while in California, wrote a delightful book about (with) Fred Frank entitled *A Philosopher on Wall Street* before Fred died.[14] It does justice to his role in the business.

The annual meeting was organised by Brook Byers from KP and Steven Burrill from Burrill & Co. The original series of meetings at the Ritz finished shortly before Burrill was indicted for tax evasion and securities fraud and sentenced to 30 months in prison. Burrill & Co. had raised a series of funds and had made some investments in companies that made good returns, such as Galapagos and Pharmasset.[u]

Under pressure from the CEOs of biotech and diagnostic companies, Brook restarted the annual meeting, moving it to the Montage Laguna Beach hotel. The meeting takes over the hotel (all 200 rooms) for three days in early October for what is now known as the "Laguna Biotech CEO Forum."[15] True to form, the CEOs like the swag that goes with the meeting, including the polo shirts with the palm tree logo available.[v] The meeting has a similar format as before, with meals outside in the hotel garden in the

---

[u] Pharmasset developed antiviral compounds for HIV, HCV, and HBV, before they were acquired in 2011 by Gilead Sciences for $11.2 billion.

[v] If you look at one of the pictures in the book about Henri Teermer,[16] you can see a picture of Henri on his boat wearing one of the polo shirts.

usually delightful weather. CEOs from the larger companies opine on subjects such as drug pricing or the importance of "breakthrough designation" at the FDA.

J.P. Morgan (H&Q)

One annual biotech meeting that still reflects the "essence of biotech" is the annual J.P. Morgan investment conference, held every year in San Francisco at the Westin St. Francis hotel (and, really, all around Union Square). This annual event was originally called H&Q as it was sponsored by Hambrecht & Quist.[w]

Apart from the recent COVID disruption, the meeting is always held in the second week of January. Both private and public biotech companies present their stories to a mixture of investors, bankers, and scientists, and hype up what their prospects are likely to be. It is the iconic annual meeting of the biotech industry: nothing else comes anywhere close. The most valuable part of the meeting is all the other meetings that take place away from Westin St. Francis, in multiple other hotels, restaurants, and even coffee shops. Pharma companies, big biotech companies, small biotech companies, private companies, start-up ventures, serial entrepreneurs, founders looking to find money for a new company, and venture capitalists all have their meetings around the J.P. Morgan umbrella. Inevitably, although less now than before, there are the parties and drinks with friends. It is the quintessential biotech boondoggle.[x]

The Greek Meeting

On the face of it, being a proficient footballer (soccer player for you in the United States) should be irrelevant to biotech. I can tell you it is not. In 2001, Stelios Papadopoulos and Spyros Artavanis-Tsakonas set up what is now the annual Fondation Santé Greek biotech meeting. Biotech executives and others meet in various locations in Greece for two or three days in the

---

[w] Chase H&Q were formed in 1999, when Chase Manhattan acquired Hambrecht & Quist. The firm changed their name to JPMorgan H&Q in 2000 when Chase Manhattan Bank acquired J.P. Morgan & Co.

[x] I went to my first H&Q boondoggle with Kevin Kinsella in 1995, before the Sequana IPO. Kevin loved being on the stage making the pitch, and his presentations were famously provocative.

early summer to talk about issues facing the industry.[17] Many CEOs, VCs, buy side and sell side folks, and "friends of the family" are invited to discuss pertinent issues at the meeting. It could be the importance of genomics in the morning and interactions with the FDA in the evening. The R&D talks might include new technologies such as gene editing and their long-term implications. The meeting includes a mixture of talks and round tables, followed by a bit of drinking late into the evening. The meeting is different from the VC portfolio meetings because the people who are invited represent different aspects of the industry. It is both educational and fun. The locations in Greece are chosen carefully, not only for their charm and ambience, but also because in the afternoon there are trips to the local Greek ruins or museums and to some well-chosen restaurants.

One of the afternoons is dedicated to the annual football (soccer) match. There are usually about five teams representing the British Isles, the United States, Europe, Greece, and "Others," with different colour T shirts printed for each team. It is taken seriously, but the tournament is always designed so that the Greeks win, often by the organisers inviting someone onto their team who is Greek with limited association with biotech, but who "coincidentally" has played professional football: the classic ringer. Bill Mathews, who was the CEO of Deltagen and a very skilled footballer, was always a star for the British and Irish team (see Plate 30 for a picture of the team in 2003). I think we won once or twice over the years, much to Stelios' disappointment. The results are always reported in *BioCentury*, one of the enduring biotech news sources that is still published (Box 6).

BOX 6. BIOTECH NEWS PUBLICATIONS

BioCentury was set up in 1992 by Karen Bernstein and David Flores. The other original biotech news source was *BioWorld*, a daily publication started in 1990 by Cynthia Robbins-Roth, a former Genentech scientist with David Bunnell. Today, thanks to digital media, there are many different news sources that cover biotech on a daily or weekly basis, including *STAT*, *ENDPOINTS*, and *Fierce Biotech*. They come, like many other things, directly to e-mail and not via a fax machine as they used to.[y]

---

[y] What's a "fax machine"?

A Word about BIO

BIO are the trade association and advocacy (lobbying) group for the biotech industry and its people. The organisation was formed in 1993 by the merger of two smaller trade groups—the Association of Biotechnology Companies (ABC) and the Industrial Biotechnology Association (IBA). BIO were initially led by Washington insider Carl Feldbaum, former chief-of-staff to Arlen Spector, with Kirk Raab, CEO of Genentech, as Chairman of the Board. Over the years, BIO have helped to shape many of the drug and regulatory policies going through congress. For example, they helped to defeat the proposed government controls on drug pricing during the Clinton administration and advocated strongly for the Orphan Drug Act that is so important for those patients with rare diseases, giving seven years of market exclusivity and tax credits for the clinical trials.

When Henri Termeer was Chairman of BIO in 1997, they had some influence over shaping changes to the FDA, assisted by Al Gore. BIO also weighed in heavily on the debate about the use of human stem cells for research during both the Bush and Obama administrations, overturning some of the restrictions that were put in place. James (Jim) Greenwood, a Pennsylvania Congressman, became the second president and CEO of BIO in 2004. BIO supported efforts to obtain a vaccine to curb the 2008 swine flu epidemic, during which an H1N1 virus moved from pigs to birds and then to humans, killing more than 18,000 people. A rather small number compared to COVID-19, but we should have learned more from that "almost" pandemic.

The breakthrough designation for certain drugs that I have mentioned in several chapters was added to the FDA's expedited review in the fifth iteration of the Prescription Drug User Fee Act (PDUFA) act. This was driven by PhRMA—the pharmaceutical industry equivalent of BIO—but BIO also had a say in the matter: it allowed the hiring of additional FDA staff to manage the extra work.

In 2016, BIO were renamed the Biotechnology Innovation Organization. Under Greenwood's leadership, BIO tripled in size, growing to a 176-employee organisation with an $85 million operating budget. There have been some recent changes at the top of the organisation. Michelle McMurry-Heath took over from Jim Greenwood in May 2020, but resigned as CEO in the fall of 2022, eventually replaced by Rachel King. Their lobbying did not appear to do much to affect the new changes that have occurred for biologics and small molecules in the 2022 Drug Pricing bill.

TRAIT 5: INVESTORS

Throughout this book are brief mentions of many of the VC firms who invested in the biotech companies that were formed over the years. It is very important to do a deeper dive into the VC firms that were there at the start of the biotech revolution and specifically call out one or two of the companies I have worked closely with or at least interacted with over the years, including SV Health Investors where (full disclosure) I am an operating partner. Different VC firms have different investment philosophies, and as access to capital has grown there are several flavours of VCs now investing in biotech companies.

There were very few VC firms who invested in biotech in the beginning. One was J.H. Whitney, founded in 1946, which invested in GI where the J.H. Whitney founding partner Benno Schmidt was Chairman for some time. Kleiner Perkins (KP), an investor in Genentech, and Venrock, the venture fund supported by Rockefeller money, were also critical to the early success of the industry.

Kleiner, Perkins, Caufield & Byers (KP)

KP, founded in 1972, have backed both tech and biotech ventures including Genentech, Google, and Amazon.com. Over many years, the firm has raised billions of dollars from their LPs. KP founders are Eugene (Gene) Kleiner (also a founder of Fairchild Semiconductor) and Thomas (Tom) Perkins (who worked at Hewlett Packard in the early days of that company). Brook Byers joined the firm in 1977. They were one of the first VC companies to locate on Sand Hill Rd., the future home of many venture firms investing in biotech.[z]

While KP were deliberating on whether to invest in Genentech, Brook Byers shared a house in San Francisco with Bob Swanson after Bob's original roommate moved back to New York. He remembers copies of *The Double Helix* (Watson's legendary book about deciphering the structure of DNA) and *Nature* on the floor of the apartment.[15] Tom Perkins made the initial investment in Genentech.

---

[z] It was always a fun trip to visit Sand Hill Rd. and imagine some of the conversations that must have gone on in the conference rooms of the various firms located there.

Venrock

Tony Evnin, who joined Venrock in New York in 1974, was one of the first people to become involved in biotech investing.[18] Among his investments were Genetics Institute, Centocor, Gilead Sciences, and Millennium. Venrock always had a reputation for being very supportive of their companies and their management teams and always behaved in an appropriately straightforward manner.[aa] As a result, Venrock were always in high demand as a potential investor. Evnin was on the board of Centocor for 19 years, supporting the company through their many trials and tribulations (see Chapter 4). Other important investments over the years, many of which have already been mentioned, include Adnexus, Arris Pharmaceuticals (who acquired Sequana), Athena Neurosciences, and Fate Therapeutics (a company focusing on natural killer (NK) cell therapy using stem cells as a starting point).

Venrock were an Idec investor with KP, backing Bill Rastetter and his team to develop Rituxan (Chapter 5). They also invested in Illumina, maker of DNA sequencers; Infinity Pharmaceuticals, Steve Holtzman's CEO gig after Millennium; Ligand Pharmaceuticals in San Diego; and Receptos, bought by Celgene for their multiple sclerosis (MS) drug (all companies previously mentioned). They also invested in Sirna, the siRNA company that got acquired by Merck but ended up as part of Alnylam (see Chapter 12).

Syntonix, where Rob Peters and others developed the longer-acting Factor VIII drug Eloctate and the Factor IX drug Alprolix, were also a Venrock-supported company. They invested in Signal Pharmaceuticals, another San Diego company, who were run by ex-Wyeth R&D leader Alan Lewis and had a focus on intracellular signalling and gene and protein regulation with technology based on Tony Hunter's work at the Salk Institute. They were bought for their technology in 2000 by Celgene for nearly $200 million. Sugen in the Bay Area were a similar kind of intracellular signalling company that Venrock also invested in (Chapter 6). They developed sunitinib and were acquired by Pharmacia-Upjohn in 1999 for $650 million.

[aa] The Venrock investing philosophy can be summed up as "doing well by doing good." By adhering to this philosophy, they helped to build some important companies in biotech and made very impressive returns in doing so.

The Larger VC Funds

From the perspective of the sheer size of the funds raised, Flagship Pioneering, ARCH Venture Partners, and to some extent Third Rock Ventures are in a class of their own.[19] Flagship Pioneering closed a $3.4 billion fund (fund VII) in 2021. ARCH Venture Partners announced in June 2022 the closing of ARCH Venture Fund XII, with $3 billion to invest. This new fund followed the January 2021 announcement of the $1.9 billion ARCH Venture Fund XI. Third Rock Ventures announced a new $1.1 billion fund in June of 2022.

Access to capital from these larger funds has resulted in new investment strategies. Companies with different technologies and approaches to new medicines are sometimes incubated internally within the large VC firm in "stealth mode" for up to two years or more. They assemble the team from within, collect the appropriate experts from academia, license some of the technology they need from the home universities of the founders, and then come out of stealth mode—often with all the glamour of a debutante at the ball—with as much as several hundred million dollars already earmarked for the company (see Chapter 17). The money is designed to take them to a large value-creation step such as a drug in the clinic with proof of concept: that is, beyond just defining a drug candidate or an IND filing. It also avoids having the management team spend all their time raising more money, letting them focus on building the company, the science, and the potential products. The companies also have money in the bank for when financial times are tough.

With money like that to invest these VCs need to make big and frequent bets. One way to think about the large investments these VCs do make is to imagine that the first very large round is a consolidated round of Series A, B, and C all rolled together. The valuations of the usual serial "tranches" are pre-determined. There are implications for the preferred shareholders—which the VC investors always are. Liquidity events (or liquidation) reveal some of those implications. Without going into detail, suffice it to say that when push comes to shove in either the positive or negative direction (e.g., an IPO or an acquisition), the preferred shareholders get first dibs on the proceeds (called liquidation preferences). Given that they usually pay way more for their shares than the common shareholders, one could argue that this is fair. But sometimes the liquidation preferences are so egregious that the common shareholders and staff and founders get completely washed away and get nothing.[bb]

[bb] Recall the acquisition of Prospect Genomics by SGX in Chapter 10.

ARCH Venture Partners

ARCH in the 1980s were the commercialisation office of the University of Chicago and the Argonne National Lab, where SGX built the X-ray beam line to allow us to determine protein structures rapidly. Steven (Steve) Lazarus, another Baxter Travenol alumnus and deputy dean of the University of Chicago business school, used his small team to find technologies on which to start companies. To help to do so they created a small venture fund of $9 million, ARCH Venture Fund 1. Based on the success of this fund, ARCH Venture Partners were born. The principals now include Bob Nelsen and Keith Crandell, who were there almost from the start. More recently, Richard Heyman, John Maraganore (of Alnylam), and Vicki Sato (Vertex) have joined the partnership. According to their website, they are not interested in following the crowd: "We are contrarian, bold, and imaginative risk takers. We follow the science to found companies based on revolutionary technologies that can impact people's lives."

I think that this is largely true, and they have been very successful at doing it, particularly recently. Their investment in Juno, for example, made spectacular returns. This tends to make it easy for them to attract new LPs, as well as new companies to invest in.

Flagship Pioneering

Flagship are the other 800-pound gorilla in the biotech VC zoo that have become even bigger and more powerful, based on their early investment in Moderna. They have a wonderful suite of offices overlooking the Charles River in Cambridge (MA), close to the Royal Sonesta Hotel. Flagship were started by Noubar Afeyan, the founder and CEO of PerSeptive Biosystems, an instrument company acquired by Perkin-Elmer in 1998 (Chapter 8).

There are many principals on the Flagship team. Doug Cole is Chairman of the Board of Repertoire Immune Medicines, where I worked most recently. This company was formed in 2020 by combining two Flagship companies: the innovative and proprietary immune decoding platforms of Cogen Immune Medicines and the immuno-oncology platforms of Torque Therapeutics. Flagship likes to populate the Boards of their companies with people that they have worked with before, so there is a lot of overlap in the Board membership of many of their portfolio companies. It is part of a deliberate command and control strategy. John Mendlein, Simba Gill (ex-Celltech and CEO of Evelo, a Flagship company), and Chris Austin are all people I have worked with or interacted with in the past, who are or were part of the Flagship Pioneering organisation. Apart from Moderna and Repertoire, other investments and company creations include Editas, Sana Therapeutics, Laronde, Codiak BioSciences, and Tessera (the gene writing and editing company). As with ARCH Venture Partners, their co-investors tend to be the investors that they have made money for and with previously.

Third Rock Ventures

Third Rock Ventures (TRV) was started in 2007 by three ex-Millennium executives, Mark Levin, Bob Tepper, and Kevin Starr. These three had worked together at Millennium for some time, which brings the trust you need to build successful organisations. Their core values are mentioned clearly on their website, which I find refreshing. They bring the same passion and intensity to their investing as they did to Millennium.

TRV define the starting of companies in four stages. The first, "discovery," stage is where academics and advisors, with expertise in the area in which they want to create a company, are brought together to put the plans in place. This can take two years or more. In the second stage, they tend to launch the companies with more money than the norm to avoid the constant headache of money raising, so the management team can focus on the science and building the company (the third stage). TRV like to put their executives into management roles to begin with until the company has some stability. TRV have had more than 60 companies in their portfolio from which 18 products have been launched: this is the fourth and transformational stage.[20]

We have met some of the companies in which they have invested, including bluebird bio, the gene therapy company that was run by Nick Leschly; Editas, one of the first gene-editing companies; Magenta, a company dedicated to making bone marrow transplantation more successful; and Neon, the Cambridge (MA) T-cell company acquired by BioNtech. They also invested in Septerna, the new G protein–coupled receptor (GPCR)-based structural biology–focused company in the Bay Area (Chapter 10), and Voyager (p. 289), another gene therapy company with Al Sandrock, the ex-Biogen R&D chief, as the CEO. Another 2018 company quite close to home that TRV started was Casma Therapeutics, who were in some respects a direct competitor for Caraway, the lysosome-focused company I founded with SV Health Investors that was acquired by Merck. Other successful companies include Agios, Blueprint Medicines, and Foundation Medicine (sold to Roche).

Other VC Firms of Note

Atlas Venture

Atlas Venture stepped up with Prospect Venture Partners to invest in the Series A for Structural GenomiX (SGX; Chapter 10). Jean-François Formela, who is a managing partner at Atlas with Bruce Booth, Kevin Bitterman, and others, was on the board. Given this connection—and the fact that he and I both like to race Porsches—it is not a surprise that I have a soft spot for this rather successful venture capital company. We have had some fun adventures racing too (Box 7).

I first became aware of Atlas Venture when they were still located on Boylston St. in Boston (they are now in Cambridge after a stint in Waltham). The firm has roots in Europe, as it was initially formed in 1980 as a subsidiary of the Dutch bank NMB—now part of the ING group. The company started investing in biotech in 1990 and formed a separate "biotech only" group in 2014. Atlas raised more than $3.0 billion of investor commitments across nine venture capital funds, adding Atlas XI in 2017 ($350 million) and Atlas XII ($400 million) in 2020. Several of their key investments are in companies already mentioned, including the gene-editing company Intellia, gene therapy company Generation Bio, Translate Bio (a Cambridge mRNA

BOX 7. ATLAS ADVENTURES INSIDE AND OUTSIDE THE BUSINESS

Jean-François, Kevin Bitterman from Atlas, and I flew to Monticello racetrack in New York State from Boston for a day of races at the track with Bruce Ledoux from J.P. Morgan. The pilot of our small plane showed up in shorts and flipflops, which did not inspire much confidence. Not only did he seem to fall asleep on the way there, but he had to do three go-arounds to land the plane, owing to fog at that time of the morning covering the runway. We found out later that he had more than 14,000 hours of flying time in his logbook, flying for FedEx (the average airline pilot does 700 hours/year, not to exceed 1000, so he was a very experienced pilot). Kevin nevertheless thought better of it and decided to hitch a lift back from the venue to Boston in one of the trucks carrying the cars.

It is a well-known fact that VC firms like to have their annual portfolio meetings in salubrious locations such as the south of France, Monte Carlo, or even more exotic places such as Marrakesh. I was coming back from the annual Atlas Venture meeting for their portfolio companies in Cannes in 2001, when 9/11 happened. I got stuck in the United Kingdom for an extra four days because there were no flights back to San Diego from Heathrow until the Thursday of that terrible week.

company acquired by Sanofi for 3.2 billion in 2021), and the PROTAC company Kymera Therapeutics (Chapter 17).

Prior exits of note from Atlas' earlier investments include Alnylam, deCODE Genetics, and Exelixis. Atlas frequently co-invested with Index Ventures, a Swiss venture firm. In 2016, a biotech-only fund called Medicxi Ventures was started by Index principal investor Francesco De Rubertis, who was an investor in SGX and is a managing partner and founder. Medicxi recently announced the closing of a new $400 million investment fund ("Medicxi IV").

SV Health Investors

I have worked with SV Health Investors formally since 2016, when I left Biogen and before I joined Bioverativ. This has given me a front-row seat in a venture capital company focused on biotech and a much better understanding of how it all works. Fundamentally, VC investing is based on recognising good science that can be turned into (or is already) a technology platform from which products will emerge. Alternatively, it could be good science that underpins directly or indirectly, some existing or emerging products.

Good VC work is not just about what you know about the science (or can find out by proper due diligence), it is also about who you know. I do not mean that in a derogatory, name-dropping way. The venture business is dominated by backing entrepreneurs and their teams who have made money with good science before, which informs their future investments. It is a little incestuous but if the people are good at what they do and you trust them, then it makes sense to back them multiple times—with the proviso that the innovative science must also be first class.

SV Health Investors were originally known as Schroder Ventures. I first visited them in 1992 in Covent Garden in London while I was still at Glaxo Group Research. I met Catherine (Kate) Bingham (who took a time-out from SV in 2020 to chair the U.K. government's COVID-19 Vaccine taskforce) around that time to talk about creating a "cell cycle" company in the United Kingdom. It never happened, but we kept in touch. Kate, Mike Ross (a protein chemist who worked at Genentech as head of development for 10 years in the early days and then was head of R&D at Arris), and Nikola Trbovic (ex–Pfizer Ventures partner) are the managing partners at SV Health Investors today and have recently been joined by Jamil Beg. Mike and Kate have worked together for almost 20 years now. Houman Ashrafian, an accomplished MD, head of experimental therapeutics at Oxford University, and company builder and entrepreneur, was a managing partner at SV until he was lured away by Sanofi recently to be Head of R&D, replacing John Reed who went to head up R&D at J&J.

The managing partners are joined by several experienced venture partners and a team of academic and industry advisors. The company is unusual in having both a strong U.K. and U.S. presence and is equally at home investing in companies in both locations. Schroder Ventures set up their first dedicated health-care fund in 1994. The SV group (as they are now known) manages several active venture capital funds, a publicly traded investment trust, and the specialist-focused Dementia Discovery Fund (DDF). The first DDF was a 15-year, $250 million fund dedicated to investing in start-ups and technology in the neurodegeneration space. It is truly unique: nothing quite like this exists elsewhere in biotech venture investing. SV have recently raised a second DDF with more conventional investment time lines.

The SV Group have invested in more than 50 biotech companies to date. Their sixth fund (SV6) closed in 2017, with capital commitments of more than $400 million. SV7 (called the IMF) closed in 2019 with $250 million committed by the LPs. Among their investments were Caraway Therapeutics, the lysosome company, and Catamaran Bio, the NK cell company, both of which I helped to start.

Other interesting SV investments include Convergence Pharmaceuticals, the ion channel company with a Nav1.7 blocker for pain sold to Biogen; AVROBIO, a gene therapy company in Cambridge (MA) focusing on Fabry's disease; and Bicycle Therapeutics, another Sir Greg Winter U.K.-based drug company. Also of note is Enara Bio, a T cell–focused immuno-oncology company in Oxford; Adimab, the New Hampshire-based mAb makers; Pionyr Immunotherapeutics, a myeloid cell immune-therapeutics company; and Artios, a successful oncology company making drugs that affect DNA repair mechanisms.

Pharma-Based VC Funds

Many of the large pharma companies have venture funds that invest in biotech companies. For example, AbbVie Ventures, Merck Ventures, Amgen, and Eisai were all investors in Caraway Therapeutics, and Takeda Ventures and Astellas were investors in Catamaran Bio. The pharma VCs usually have a slightly different brief to their noncorporate counterparts: their funds are used not only to bring in returns to the parent company but, perhaps more importantly, to keep an eye on technologies or drug discovery areas on behalf of the parent company. They can also help with due diligence should there be a potential acquisition opportunity. Sanofi Ventures, Pfizer Ventures, and Novartis Venture Fund, for example, are all part of the vibrant VC investing scene in Boston.

The Buy Side Firms

An equally important collection of investors is what is known as the "buy side." These are the very important firms that buy biotech stocks, either at the IPO or in the aftermarket. They are often specialist biotech funds or hedge funds consisting of generally very well-informed people (the principals often being trained scientists or MDs). They sometimes invest in private biotech companies as well.

Without these funds there to continue to support companies by buying their stock at the IPO or in subsequent public offerings, there would be no biotech companies or a biotech industry. The firms I briefly mention below are the ones that I am reasonably familiar with as they have people I know (or have met) in them. So the following is not an exhaustive list, nor is it meant to be.

I know the specialist fund Bain Capital largely because Adam Koppel, who I worked with at Biogen, is now a principal there.[cc] Bain Capital were spun out of the consulting practice that Bill Bain set up in 1984. Mitt Romney, U.S. Senator from Utah and former Presidential candidate, was the first CEO. Bain were responsible recently for setting up Cerevel with Pfizer. The company was recently acquired by AbbVie.

Mark Lampert founded the Biotechnology Value Fund (BVF) in 1993. Although I had met BVF people when I was in California, I have interacted with them more recently (they invested in Bionomics, the Australian company based in Adelaide, where I was an SAB member). BVF take a longer-term investing view than some other firms.

OrbiMed were founded in 1989 by Samuel Isaly as a public equity fund that invests in both private and public life sciences companies of all sizes: they have multibillions of dollars to invest. Kevin Koch, an SVP with me at Biogen and a founder of Array BioPharma in Boulder, Colorado (sold to Pfizer in 2019 for $11 billion), is a partner at OrbiMed. They also invested in Loxo Oncology, who developed an oral Ret kinase inhibitor and

---

[cc] In our Biogen days, we used to have interesting conversations about the efficacy of the Vertex cystic fibrosis (CF) drugs.

a Trk kinase inhibitor (larotrectinib), and were acquired by Lilly in 2019 for $8 billion. Like the Perceptive Advisors hedge fund covered just below, OrbiMed made a tonne of money from investing early in Acerta, the company that developed the BTK inhibitor acalabrutinib, a direct competitor of Imbruvica (ibrutinib) developed by Pharmacyclics.[21]

I know Otello Stampacchia at Omega Funds, who founded the Boston-based company in 2004, from many of the Greek meetings. He is ex–Goldman Sachs and had been at Lombard Odier in Geneva. Omega invest in both public and private biotech and health-care companies.

The hedge fund Perceptive Advisors is interesting because they are run by Joseph (Joe) Edelman, whose late father was chair of biochemistry at Columbia. The story goes that his father sold a company to Celltech.[dd] Joe started the hedge fund in 1999 with $6 million after a stint as a sell side analyst for Prudential Securities, to invest in large and small public biotech companies. Perceptive were an investor in True North Therapeutics, the company that Bioverativ acquired for their complement factor C1s mAb (sutimlimab) that is now approved for cold agglutinin disease. Edelman's fund also did very well from their investment in Acerta. Perceptive now have several billion dollars under management.

I have run into Peter Kolchinsky at RA Capital Management, a firm founded by Richard Aldrich in 2004, many times, including at the Greek meeting. Josh Resnick, who was a venture partner at SV Health Investors, is now at the firm. One of their claims to fame is that they have put together a comprehensive integrated map of the science behind the biotech industry, which was kept updated by various interns. The problem with such a map is that simply reviewing the abstracts of published papers does not tell you very much about the quality of the science, or whether the conclusions drawn are supported by the data.[ee]

As of December 2019, RA Capital have made more than 100 investments, leading 28 of them, and they have made more than 60 exits. Investments that you will recognise include bluebird bio, Moderna, Orchard Therapeutics, and Global Blood Therapeutics, who were acquired by Pfizer in 2022 for $5 billion for their sickle cell disease drug (Chapter 13).

The Baker Brothers are also well-known biotech investors. Julian Baker has a business background from Harvard and Felix Baker has a PhD in

---

[dd] Presumably after I left Celltech, because I do not remember it.

[ee] It looks sexy at first blush, though.

Immunology from Stanford. Their expertise has generated superior returns from their biotech investments by focusing on the "fundamentals of the companies they invest in"—code for understanding the science of what they are going to invest in. Assets under management grew from $250 million in 2003 to more than $23 billion, 10 years later. Their investments have included Incyte, Biomarin, and BeiGene, the Chinese "gene-based" biotech company. They also made a killing from investing in Pharmacyclics and the takeout of that company by AbbVie.[21]

Other buy side names to note are Fidelity, T. Rowe Price, and Janus Capital Group, who have all been investing in biotech for many years, and Deerfield, who have recently started investing in earlier stage (small) private biotech companies. You may also have heard of Lone Pine Capital, Viking Global, Pershing Square, Appaloosa Management, and Bridgewater Advisors, all active in biotech.

Relatively more recent entrants into the biotech hedge fund space include Rock Springs Capital in Baltimore founded by Kris Jenner (ex–T. Rowe Price) with $100 million in 2013. Alexander (Alex) Denner[ff] founded Sarissa Capital with $200 million at about the same time. Denner worked with Carl Icahn and was instrumental in getting ImClone Systems acquired for their epidermal growth factor (EGF) receptor mAb Erbitux by Eli Lilly for $6.5 billion. He was also an important player in the Genzyme acquisition by Sanofi and in stirring up Biogen-Idec, leading to the appointment of George Scangos as CEO in 2010.[gg]

Denner was an investor in and on the Board of Bioverativ before we were acquired by Sanofi.

SINGULAR EVENTS DEFINE BIOTECH COMPANIES

Stelios Papadopoulos once told me that for most biotech companies there are only one or two decisions made in a year that make a difference to the outcome of the company. Most Board decisions are housekeeping kinds of decisions and often—although by no means always—follow the guidance

---

[ff]Denner was brought up in Lynn, just across the causeway from Nahant where I now live. George Scangos lived for a while in Lynn too as a child, and both apparently spent some summer days on Nahant Beach.

[gg]Denner left the Board of Biogen and was replaced by his romantic partner Susan Langer, the daughter of serial entrepreneur and billionaire Bob Langer, a Professor at MIT. A somewhat curious choice that led to considerable commentary in the press.

of the management team. But there are times when a company will face a big issue, and the decisions that are made in the face of it can lead to critical outcomes in either the positive or negative directions.

I refer to these momentous decision-requiring time points as "singular events." They are not singular in the sense of being a one-off, but they are singular enough and company-defining enough to be referred to that way. We have seen many during biotech's history. And most of these events are not necessarily about money—that commodity for a biotech company is always an issue.

Singular events are not usually under the control of management either. However, when they happen the decisions and actions that the Board and the management team make in response can be company-defining.

One of the first biotech company-defining singular events was the outcome of the EPO patent litigation between Amgen and Genetics Institute. This led to the beginning of Amgen's rise to the top of the industry and to GI falling into the arms of an opportunistic Wyeth. This was not such a bad outcome for GI shareholders, but if the patent decision had gone the way that was expected and the two companies had both ended up with rights to sell EPO in the United States, I think GI would be a major biotech company now and not just a forgotten name. They had very good people and technology and were easily the technical equal of Genentech and better than most of the other cloning companies (Chapter 2).

The recombinant t-PA sales for Genentech in 1988/1989 were not what they were expecting. Partly as a result of this, Kirk Raab, who succeeded Swanson as the CEO, thought it necessary to see if there were suitors out there who would buy the company. Roche bought a 60% interest in 1990 and acquired the company outright in 2009 for $46.8 billion shortly after Pfizer bought Wyeth for $68 billion, and Merck acquired Schering-Plough for $41 billion. This was a singular event. If t-PA sales had been what they had forecast, then Genentech might still be an independent company. Another Genentech singular event—this time positive—was the coincidence of Creutzfeldt–Jakob disease caused by treatment with pituitary gland–derived growth hormone around the same time as the launch of Genentech's recombinant growth hormone, leading to the rapid uptake of the recombinant product (Chapter 1). There is probably more to the story of Roche's acquisition of Genentech, but it serves to illustrate the point.

In the monoclonal antibody theatre, the failure of Centoxin was the beginning of the end for Centocor as an independent company. It could be argued that Centocor were acquired for infliximab (Remicade). Nevertheless, both are singular events (Chapter 4).

The virus contamination in the Allston Ceredase manufacturing facility of Genzyme directly led to the acquisition of Genzyme by Sanofi, even though it took a little time to happen. Had the contamination not happened, or they had invested in a backup facility, or had they used a fed batch rather than the continuous fermentation process to make the product, Genzyme might still exist today. Instead, they were subjected to what in essence was a fire sale, and the company was committed to the annals of biotech history forever.

In 2001, Vertex acquired Aurora Biosciences for their screening platform. Aurora's nascent CF drugs affecting the mutant forms of the CFTR causing cystic fibrosis were not considered valuable at the time but came as part of the deal. More than 85% of the current revenue for Vertex Pharmaceuticals comes from their CF franchise, developed from the Aurora assets. Revenue in 2022 was $8.9 billion, up from $7.6 billion in 2021. If they had not acquired Aurora, the whole existence of Vertex would have been at risk because their other products were just not competitive.

Millennium came to the singular realisation that if they wanted to become a drug company, then the time and money that it would take to do this, using the targets that they had discovered using genetics and genomics, would be too much to be successful and provide adequate shareholder value.[20] This led to their new drug acquisition strategy, which in turn allowed them to become a "real" therapeutics company. It ultimately led them to be bought by Takeda in May 2009 for nearly $9 billion.

The nuclear winter that affected Alnylam and the rapid exit of Big Pharma from antisense and siRNA technology around 2010 was a singular event for Alnylam that led to their "five products in five years" strategy (Chapter 12). This drove the company to be the success it is today.

You could also argue that without the COVID-19 pandemic and the need for a SARS-CoV-2 vaccine, Moderna and BioNTech would still be relatively small biotech companies struggling to make cancer vaccines, and vaccines for relatively obscure pathogens. Instead, they are now both household names with multiple billions of dollars in revenue. This is not a criticism: it is a fact. The pandemic was a singular event that made these companies and others like Vir Biotechnology successful—much more quickly than they would otherwise have been.

My take on the Biogen aducanumab debacle, despite the renewed hope of lecanemab (Leqembi), is that it could easily be the singular event that determines the future of Biogen, one of the oldest and most venerated of the original cloning companies. I might be a tad biased but it would be a big loss if Biogen were to disappear into the arms of a company like J&J or, worse still, a Vertex. Biogen may have blazed a trail with their multiple sclerosis franchise and with their new spinal muscular atrophy (SMA) antisense oligonucleotide (ASO) drug, but there seems to have been a lack of innovation and imagination since Doug Williams and George Scangos left. Their future is inevitably tied to the success of their mAbs for treating Alzheimer's disease and their cash assets. For more on Biogen's future see the next chapter.

THE END OF THE JOURNEY

The biotech industry faces many challenges in the future.[22] It lost some ground in Washington when Congress passed the new drug pricing legislation in 2022, but this is by no means a terminal outcome. The rules that apply to new drug applications (NDAs) are now different to biologics licence applications (BLAs) and that may affect the categories of medicine that the industry focuses on for maximum opportunity in the market.

The focus on patients and unmet medical needs both in rare and in common diseases will continue. There are still many people in the industry passionate about making a difference by developing new technologies and applying them to drug discovery and taking those drugs through the development process. There are also both the VCs and the later investors around with lots of money to invest to make great returns, by continuing to support the process.

I also believe the current environment will inevitably get better. In 2020 and 2021, scores of small companies went public, and many of these are now running out of money with no means to raise any more. In 2022 and 2023, one can count the number of biotech IPOs on the fingers of both hands. It is a cyclical swing away from overheated valuations, exacerbated by the external environment. The industry has been here before and will emerge stronger for it. One hundred and seventy years ago, Jean-Baptiste Alphonse Karr remarked "*plus ça change, plus c'est la même chose*"—the more things change, the more they stay the same. This aphorism applies to biotech as well as any other human endeavour. It may seem like biotech has changed in 50 years, but its essence has not. I hope the development of new technologies and products that define this industry continues, and that "biotech" continues the trajectory it has been on since it started in the late 1970s.

It is a long time since I met the Nobel Prize in Chemistry winner Sir Robert Robinson walking up Grimms Hill. With that thought, it is good to reflect on all the water that has flowed beneath the biotech bridges and on the people who have created this business and the successful companies that comprise it. Some of these people are no longer here to see the fruits of their endeavours, but we salute them, nevertheless. As Herodotus (the Father of History) puts it[23]:

> The purpose (of history) is to prevent the traces of human events from being erased by time, and to preserve the fame of the important and remarkable achievements produced... .

Nowhere is that quote more apposite than for the biotech industry and the people who made it.

REFERENCES AND NOTES

1. Bayh–Dole Act 1980: United States Code, 2011 Edition: Title 35—PATENTS. PART II—PATENTABILITY OF INVENTIONS AND GRANT OF PATENTS CHAPTER 18—PATENT RIGHTS IN INVENTIONS MADE WITH FEDERAL ASSISTANCE. From the U.S. Government Publishing Office: CHAPTER 18— PATENT RIGHTS IN INVENTIONS MADE WITH FEDERAL ASSISTANCE.

2. Norrby E. 2010. *Nobel Prizes and life sciences*. World Scientific, Singapore.

3. https://www.biospace.com/article/biospace-launches-2022-hotbed-maps-to-high-light-thriving-life-sciences-clusters/

4. Conversation with Joel Marcus, Chairman and CEO of Alexandria Real Estate Equities, 2022.

5. Porter ME. 1998. Clusters and the new economics of competition. *Harvard Business Review*, November–December 1998.

6. 2021. Just what the doctor ordered. *The Economist*, August 14, 2021.

7. Conversation with Doug Williams, 2021, 2022.

8. Bio Britannia. 2022. The life sciences industry is a jewel in Britain's economy. It needs to sparkle. *The Economist*, July 23, 2022.

9. Drakeman D, Drakeman L, Oraiopoulos N. 2022. *From breakthrough to blockbuster: the business of biotechnology*. Oxford University Press, New York.

10. Walsh G. 2000. Biopharmaceutical benchmarks. *Nat Biotechnol* **18:** 831–833.

11. Sagonowsky E. 2021. The top 20 drugs by worldwide sales 2020. *Fierce Biotech*, May 3, 2021.

12. Buntz B. 2023. The fifty best-selling pharmaceuticals of 2022: COVID vaccines poised to take a step back. *Drug Discovery and Development*, April 18, 2023.

13. Dunleavy K. 2022. The top 20 drugs by worldwide sales in 2021. *Fierce Biotech*, May 31, 2022.

14. Duncan DE. 2021. *A philosopher on Wall Street: how creative financier Fred Frank forged the future.* Greenleaf, Austin, TX.

15. Conversation with Brook Byers, 2021.

16. Hawkins J. 2019. *Conscience and courage: how visionary CEO Henri Termeer built a biotech giant and pioneered the rare disease industry.* Cold Spring Harbor Laboratory Press, Cold Spring Harbor, NY.

17. An example of the Greek meeting agenda in 2001: "The Globalization of Biotechnology," held on Mykonos, July 11–14.

18. 2019. The Computer History Museum: Oral History with Anthony B Evnin. National Venture Capital Association (NVCA) Oral History Collection X8628.208.

19. Sheridan K. 2021. Flagship, Orbimed, ARCH, Alta top list of high-performing biotech venture capital firms. *STAT News*, August 23, 2021.

20. Conversation with Mark Levin and Third Rock Ventures website, 2022.

21. Vardi N. 2023. *For blood and money: billionaires, biotech and the quest for a blockbuster drug.* W.W. Norton, New York.

22. 2022. Biotechnology: more cash, stat. *The Economist*, August 13, 2022.

23. Herodotus. 2008. *The histories* (Oxford World's Classics) (tr. Waterfield R). Oxford University Press, Oxford.

CHAPTER 19

Dystopian Times

Dystopia is a strange place.[a] In the context of this chapter, it refers to where the biotech industry finds itself in late 2025 as the result of the convergence of several forces. Things are not entirely different nor are they all bad, but the context in which we work has changed. There are several reasons for these changes, and many consequences arise from them. These form the backdrop for this new chapter.

The attitude and actions of the present U.S. government administration towards academia and life science research in general, for political ideology reasons, is actively discouraging academics from investing time in new companies as they struggle to find their cancelled grants or get new grants or able students with the appropriate qualifications to fill graduate or postgraduate positions.[1] In addition, there is the prospect of changes being made to the Bayh–Dole Act that compromise the ability of universities to patent discoveries made from federally funded research. The Bayh–Dole legislation drove much of the previous innovation in the biotech industry (see p. 402).

You can call it political expediency or pragmatism if you like, but I and others believe these changes and the present climate will have an enduring effect on innovation in academic institutions in the United States and a knock-on effect on the number of United States–based start-up companies that can be invested in, if there is money available to invest in them.[2]

Another force for the industry to reckon with is the consequences of the continuing tariff debacle that has driven stock market volatility. This uncertainty has affected the tolerance of both venture capital and follow-on investors, effectively removing the IPO market for private biotech companies.

Although the reigniting of the drug prices debate has yet to play out, it is also bound to have consequences. As the IPO market is virtually nonexistent

---

[a]Dystopia: an imagined state or society in which there is great suffering or injustice, typically one that is totalitarian or postapocalyptic.

presently, biotech companies and their investors rely on Big Pharma to buy them as part of their exit scenario planning. The large companies might become more reluctant to acquire companies or their products as the prices they can charge in the United States for their drugs may have to match the prices they charge elsewhere. Advertised drug prices are rarely those that are actually paid anyway, especially when it comes to medicines that are technically challenging to make or use, such as cell and gene therapies. Furthermore, the reimbursement situation is very complicated in the United States, different to other countries and not likely to become any simpler. Any "most favoured nation legislation" will have a profound effect on the industry and expected return on investment for investors.

HEALTH AND HUMAN SERVICES

The third element of our current dystopia is the continuing turmoil at the U.S. Food and Drug Administration (FDA) and the Centers for Disease Control and Prevention (CDC) under the oversight of the "uniquely" qualified Health and Human Services Secretary Robert F. Kennedy, Jr. Several important and experienced people have decided to leave from both organisations or have been forced to do so. One of the most influential was Peter Marks, the head of the Center for Biologics Evaluation and Research (CBER), whose departure rattled the industry and led to biotech stock prices dropping after the news broke. His replacement Vinay Prasad only lasted a couple of months in the job before he left (or was removed) after issues with the gene therapy company Sarepta Therapeutics (chronicled below) only to be reinstated later.[b]

Things have not been quiet on the vaccine front either. Misplaced vaccine scepticism, especially promoting a false connection between measles vaccination and autism, led to the removal of all 17 members of the CDC's Advisory Committee on Immunization Practices and their replacement with much less qualified people. This antivax activism has culminated most recently in the defunding of mRNA-based vaccine research even though mRNA-based vaccines were fundamental in the fight against COVID-19 (see Chapter 16). The Biomedical Advanced Research and Development Authority (BARDA) terminated grants supporting mRNA vaccine development owing to "issues with the safety and effectiveness of the technology" and because mRNA vaccines were not "trusted by the public."[c]

---

[b] These moves do not provide much confidence in either the FDA or the CDC.

[c] This is, of course, a political statement and complete nonsense.

The worrying thing generally is that vaccination rates for several pathogens that can kill people are dropping, owing to misinformation propagated via so-called "social" media. Fortunately, the FDA did recently approve the emergency use of the Novavax protein-based vaccine for COVID-19 and the existing mRNA-based boosters from Pfizer and Moderna that target the K 2 strain of the SARS virus. But they are not available to every adult in every state in the United States right now, despite claims to the contrary. We are in danger of going backwards in vaccine research.[d]

Moderna (see p. 367) has not found it easy going since the end of the pandemic and have made several cuts to staff, pushed out the break-even date for the company to 2028 after a cut of $1 billion[e] in its revenue estimate for 2025. The company has also seen a 90% decline in stock price over the last three years.[f]

The lack of support for mRNA vaccines comes at a particularly inauspicious time given that the recent data on protection from the highly pathogenic avian flu virus H5N1 (so called bird flu) and HIV suggests that this (mRNA) is an excellent way to provide immunity to these infectious disease threats.[3,4,5] These actions also fail to appreciate that the mRNA-based vaccines for cancer being developed by, for example, BioNTech and Moderna (with Merck) are showing good efficacy in the clinic.

CHINA

The fourth difference in our current situation is the rise of Chinese life sciences R&D. China's biopharmaceutical sector has grown much faster since the pandemic compared to the rest of the world, including the United States and Europe.

I am told that the atmosphere in China now is as it used to be here in the 1980s at the start of the building of the whole biotech industry, with many very smart people doing groundbreaking innovative research and taking potential new drugs of many different types and modalities into the clinic rapidly.[g] The Chinese government provides many incentives in terms

---

[d] I made my views on vaccines quite clear in Chapter 16.

[e] This is a good deal of money for any company to miss its numbers by!

[f] How the once vaccine-mighty are fallen.

[g] There are anecdotal examples in which U.S. and Chinese companies were developing the same molecules at the same time. It took the United States–based company seven years to file an IND, yet the Chinese company started phase 3 trials within five years. This may be an extreme example, but you get the point.

of facilities and space for biotech. China was once viewed only as a maker of "me-too" drugs but has since emerged as a source of competitive and innovative new medicines.[6] Their ascendance has encouraged many licensing deals with pharmaceutical companies who are always on the lookout for assets with which to build up their pipelines. For example, Merck recently licensed a heart disease drug from Jiangsu Hengrui in a deal worth up to $2 billion, and Pfizer has announced that they will spend "big money" acquiring Chinese assets. Even though China is not immune from both their own and U.S. government (trade) interference, these deals have unsettled the U.S. biotech industry.[h]

European companies have also enjoyed positive relationships with Chinese companies. In early 2024, GSK acquired Aiolos Bio, a clinical-stage biopharmaceutical company, for its phase 2–ready anti-TSLP (thymic stromal lymphopoietin protein) antibody for the treatment of asthma for a $1 billion upfront payment and up to $400 million in success-based regulatory milestone payments. The antibody was licensed to Aiolos from Jiangsu Hengrui. It is unclear where the mAb is in the development process: finding a balance between dose and efficacy over time is probably the issue. They have competition as Amgen also have an anti-TLSP antibody (tezepelumab, brand name Tezspire) co-developed with AstraZeneca already on the market. There are also anti-IL-5 mAbs for the same asthma indication. Mepolizumab, reslizumab, and benralizumab are all designed to bind to IL-5 or its receptor IL-5-Rα and prevent activation of eosinophils that drive airway inflammation.

For TLSP, the difference between targeting the ligand or its receptor may be fundamental, as the ligand concentration far exceeds that of the receptor on cells. Upstream Bio, a company in Waltham run by CEO Rand Sutherland,[i] is developing a mAb against the TLSP receptor. They were with CAMP4 Therapeutics, one of the few companies to go public in 2024. The antibody was bought from Astellas but had its origins in the Regeneron VelocImmune mouse antibody programme (see pp. 75 and 118). The Upstream Bio antibody (Verekitug) is currently in phase 2 trials.

Chinese biotech companies themselves are doing well. BeiGene (see p. 434), formed in 2010 and now called BeOne Medicines, is worth a staggering $27 billion with more than 40 clinical and commercial stage assets. It

---

[h] It is quite possible—likely, even—that the present U.S. government will attempt to interfere with the positive China biotech dynamics that are emerging.

[i] I worked with Rand for a short time while at Sanofi.

should post its first operating profit this year. As further evidence of Chinese innovation, BioNTech recently licensed a VEGF-PDL1-bispecific antibody (BNT 327) that is now in phase 2 from the Chinese company Biotheus, which BioNTech subsequently acquired and now operates as its Chinese subsidiary.

There are many more bispecific antibodies and antibody-based drugs (Chapter 4) being developed in China. Their strength may not yet be new technologies (like CRISPR-Cas9), but they are engineering new products from existing technologies quickly and efficiently. Some of these products will inevitably end up in new Chinese-funded companies in the United States should the politics allow it.

CAR T UPDATE

As further evidence of clinical innovation, Chinese biotech Genocury Biotech has an experimental CD19 CAR T therapy for lymphoma (for CAR T therapy see Chapter 14, p. 300). This *in vivo* CAR T (i.e., the therapy is administered directly to the patient rather than by *ex vivo* manipulation of their T cells) resulted in complete remission in one patient after one month and the response was maintained, despite the absence of the preconditioning and lymphoablation needed for *ex vivo* CAR T therapies. This reinforces the view that the future of CAR T therapy is going to be by the direct application of a T-cell-modifying construct in a virus vector or in a liquid nanoparticle rather than by *ex vivo* manipulation of T cells.

It came as no surprise to see that the leading United States–based *in vivo* CAR T company, Capstan Therapeutics (p. 308, Box 4), was bought by AbbVie in a deal that closed in August 2025.[7] The deal was for up to $2.1 billion in cash, and it took place after Capstan started a phase 1 study in healthy volunteers. Their anti-CD19 CAR mRNA is designed to deplete B cells for the treatment of autoimmune diseases. For AbbVie, not only did they obtain the LNP-based CD19 drug but also a technology platform for the delivery of other RNA-based gene constructs. In a similar vein, Gilead Sciences recently acquired the lentivirus-based *in vivo* CAR T company Interius Biotherapeutics. Adding to this enthusiasm, BMS paid $1.5 billion to acquire ARCH-backed Orbital Therapeutics in October 2025 for their *in vivo* CAR-T also for B cell depletion.

The 2025 Nobel Prize for Physiology or Medicine was awarded to investigators (Fred Ramsdell, Mary Brunkov, and Shimon Sakaguchi), who characterised a class of T cells called T regulatory cells or Tregs for short.

The recipients have interesting historical connections to the biotech industry in the United States and the United Kingdom. Fred Ramsdell, early in his career, worked at Immunex (see p. 57) and then at Darwin Molecular with Mary Brunkov in Seattle. That company got acquired by Chiroscience in a stock swap deal (worth a~$120 million) in 1996, and Chiroscience later merged with Celltech (see p. 28) in 1999, five years before UCB bought the combined enterprise. Fred also worked at Zymogenetics (p. 60) before it was bought by BMS in 2010 for nearly $900 million.

Sonoma Biotherapeutics, a company developing CAR Tregs to treat autoimmune and inflammatory disease, was cofounded by Jeff Bluestone, Fred Ramsdell, and others in 2020. One of their competitors is Quell Therapeutics started in the Unted Kingdom by Syncona, Fidelity, SV Health Investors, and others in 2019. They raised $156 million in an oversubscribed series B financing in 2021 and have an alliance with AstraZeneca. Quell are developing CAR Treg–based products for rheumatoid arthritis and liver transplant rejection.

Not all pharma companies see the immunotherapy world the same way. Genentech recently closed their whole immunotherapy and cancer immunology team with its leader Ira Mellman leaving to join the Parker Institute for Cancer Immunotherapy. Dispatch Bio, CAR T pioneer Carl June and colleagues' latest CAR T company, which is focused on solid tumours, came out of stealth mode, recently raising $216 million, and is working with the Parker Institute and financed by ARCH Venture Partners.

In stark contrast, Lyell Immunopharma's (p. 391) recently acquired CAR T product that targets CD19/CD20 (in lymphoma) did not seem to impress anyone despite inducing reasonable responses in several patients. The therapy, previously known as IMPT-314, was acquired by Lyell in its $30 million buyout of ImmPACT Bio. The problem is that the "new" therapy does not seem to provide any benefits over the existing CAR T products.

ARTIFICIAL INTELLIGENCE (AI)

Okay, so I was wrong. I said that I did not see a new technology that would fundamentally change the biotech scene on the horizon in the way that other technologies such as antisense or CRISPR gene editing have. I was also thoroughly dismissive of the role of AI in drug discovery and in the formation of "AI biotechnology companies," but I did say that it was likely to command large sums of money in investment, even before the advent of generative AI and ChatGPT (see Chapter 10 and p. 133).

One of the problems is that it is difficult to define just what an AI-driven company is. Although some companies like to promote their use of AI, it is not just using computers in a conventional way to analyse large amounts of data. The definition is still, to some extent, like beauty being in the eye of the beholder.

In the last two years several companies have formed that are, at heart, "AI companies" that feed off both the general AI computational advances and in particular those in protein folding and protein–protein interactions. As I have alluded to before for other technology-driven biotech start-ups, the origin of the most well-funded examples of AI companies are those founded by Nobel Prize winners.

The 2024 Nobel Prize in Chemistry was awarded to John Jumper and Demis Hassabis at Google DeepMind in London, for developing a game-changing AI tool for predicting protein structures (AlphaFold) and David Baker (see p. 229, Box 3) at the University of Washington in Seattle, for his work on computational protein design, including the development of his Rosetta suite of protein folding programmes (see Box 1). DeepMind released AlphaFold 2 in 2020. It was an upgrade to the original programme that won the biennial protein-structure prediction contest CASP (Critical Assessment of Protein Structure Prediction), which began in the era of structural genomics (Chapter 10).

Many new companies have now been launched to exploit these and other predictive computational tools. It is useful to look at these in two ways. There are companies dedicated to developing computational tools for use in drug discovery such as Chai Discovery, Latent Labs, and EvolutionaryScale,

---

BOX 1. ALPHAFOLD AND ROSETTAFOLD

Hassabis, DeepMind's co-founder and chief executive, and Jumper, head of the AlphaFold team, led the development of AlphaFold2 (p. 214). To predict protein structures, the neural network incorporates data from libraries of protein 3D structures contained in the protein data bank (PDB), which contains more than 200,000 protein structures determined by X-ray crystallography or cryo-EM. AlphaFold3 incorporates molecular dynamics calculations to look at interactions between proteins and small molecules or nucleic acids.

　　David Baker's laboratory, who originally developed the protein folding programme Rosetta, created new versions of it (e.g., the RoseTTAFold network) that worked like AlphaFold2. Baker's team has combined RoseTTAFold with image-generating diffusion neural networks to create new proteins.[8]

and more fully integrated R&D organizations using AI that are hybrids of computational and wet-lab techniques (Box 2), with the intent to develop their own medicines. The two most notable companies, Isomorphic Labs with Hassabis and Jumper as founders and Xaira Therapeutics with Baker as a cofounder, have raised extraordinary amounts of money with which to pursue their dreams.

BOX 2. NOTABLE COMPANIES IN THE AI SPACE

"Fully Integrated" R&D Companies

Xaira Therapeutics ($1B Series A): founders are Foresite Capital and Arch Venture and included contributions from F-Prime, NEA, Sequoia Capital, Lux Capital, Lightspeed Venture Partners, Menlo Ventures, Two Sigma Ventures, The Parker Institute for Cancer Immunotherapy, Byers Capital, R Square Technology, and others. Protein–protein interaction mass-spec comes from Interline Therapeutics and computational capability from the Baker Lab. The company is based in the Gateway complex in San Francisco.

Isomorphic Labs ($600M Series A): founded by Demis Hassabis and John Jumper. Investors are Thrive Capital and Alphabet. Their drug development talent includes experienced drug hunters and suggests perhaps an initial focus on small-molecule kinase inhibitors.

Divergence Labs is a next-generation biopharmaceutical company focused on transforming "bold biologic insights into breakthrough medicines." It was seeded by venture firm Dimension Capital, which has also invested in Chai Discovery (below). It is led by an experienced life science investing team. They have people in San Francisco, Boston, and Cambridge, UK.

Polaris Partners led a $40 million Series A round for *Noetik*, a Bay Area start-up developing artificial intelligence models of cancer. The company was founded in 2022 by ex-Recursion executives (p. 229). The company is studying proteins and nucleic acids from thousands of tumours from non-small cell lung cancer patients. These data are fed into a self-supervised transformer model called "OCTO" to look for points of potential therapy. Other investors in the Series A include Khosla Ventures, Wittington Ventures, and Breakout Ventures.

Archon Biosciences is another David Baker biotech start-up in Seattle. They have raised $20 million in a seed round from Madrona Ventures to develop new classes of antibody-based drugs called "antibody cages,"[9] where they attach AI-designed proteins to standard antibodies to give them new properties.

CHARM Therapeutics is yet another David Baker company set up in 2021 in the United Kingdom (Box 3, p. 229) using AI to design new cancer drugs. They closed a series B recently to develop a drug targeting menin in AML. They have an AI model, DragonFold, to predict how proteins and small molecules interact and bind.

VantAI was spun out of Roivant Sciences in 2019 and is focusing on protein degraders and proximity-enabling small molecules.

Computational Tool Development

Chai Discovery has raised $70 million in their Series A in 2025 and more than $100 million since inception where DCVC led the charge. New investors include Menlo Ventures, DST, Thrive Capital, Dimension, OpenAI, and Anthropic. Mikael Dolsten (ex-Pfizer R&D head) is on the Board. The founders are Joshua Meier (formerly of Absci, OpenAI) and Jack Dent (formerly of Stripe, Facebook) and the company is based in San Francisco. Chai is using AI to predict and reprogram protein–protein, protein–nucleic acid, and nucleic acid–nucleic acid interactions.

Latent Labs ($50 million Series A), whose founder is Simon Kohl (former co-lead of AlphaFold at DeepMind), is the European/London version of Chai and investors are Sofinnova (Paris), Isomer Capital, 8VC, and Pillar VC. The company intends to build AI models for designing and optimizing proteins.

EvolutionaryScale raised a $142 million seed round. The founder is Alex Rives (former project lead at Facebook's evolutionary scale modelling lab) and investors are Nat Friedman/Daniel Gross, Lux Capital, Amazon, and Nvidia. According to their website, the company understands the biology derived from the genetic code (i.e., proteins) and uses it as a generative tool.[j]

Isomorphic was launched in November 2021 and signed research deals with both Novartis and Eli Lilly very early in its existence. Xaira—an "integrated biotechnology company"—was formed in 2023. According to their website, Xaira will "bring AI tools together with expansive data generation and analysis to find new medicines." It is led by the notorious CEO Marc Tessier-Lavigne, ex-President of Stanford University.

The excessive amounts of money these companies have managed to raise is unusual, especially when it is entirely unclear how or even if the computational tools will speed up the drug discovery and development process and, more importantly, what their products will look like and how they will be developed. But the companies have plenty of money and the time to develop their drug discovery and development strategy. I would rather be in the position of having hundreds of millions of dollars at my disposal and the time to find my drugs than to have a target and a drug but no money! However, it is wise to remember that this money will come with milestone strings attached and they will certainly not get it all at once. It is reminiscent of the large Sana Biotechnology and Lyell Immunopharma and other financings I referred to earlier (p. 390). I hope these new companies do not end up in the same place.

---

[j]Despite being on the Board of Divergence Labs I have no idea what that statement actually means.

Xaira Therapeutics and Isomorphic Labs, although the most prestigious AI newcomers owing to their founders, are not the only new kids on the block, and this time Big Pharma is not standing on the sidelines watching: they are active players in this new game.

Big Pharma and AI

There are other AI-orientated companies that are worth a mention in the context of the moves being made by Big Pharma to avoid being left behind in the AI race. Lilly has formed a relationship with Superluminal Medicines, a machine learning start-up with a focus on G protein–coupled receptors involved in obesity and metabolic disease—after Superluminal's $120 million series A in 2024.

Septerna (p. 226) recently partnered with Novo Nordisk on the same subject in a lucrative deal (worth up to $2.2 billion). In the same therapeutic area, Fauna Bio is using AI to identify targets in their biobank that are highly conserved across species including humans. They are focussing on metabolism in ground squirrels. (Bears, who hibernate more deeply and for much longer times would probably be a better model of insulin resistance and sensitivity.[k])

Sanofi has invested in Earendil Labs, an affiliate of the Chinese company Helixon Therapeutics founded by pharma-experienced Chinese nationals working in the United States. The companies entered into a license agreement in the spring of 2025 that gives Sanofi rights to two bispecific antibodies, HXN-1002 and HXN-1003, as well as access to Earendil's proprietary AI drug discovery platform. They say this platform "enables precise optimization of functionality, manufacturability, and developability of protein-based biologics." HXN-1002 is a bispecific antibody that targets both integrin $\alpha 4\beta 7$ and TL1A (TNF-like ligand 1A), and HXN-1003 targets TL1A and IL-23: all drivers of inflammation in various human autoimmune diseases. Earendil's own subcutaneous, long half-life anti-TLA 1 antibody has just entered phase 1 studies in human volunteers.

Under the license agreement, Earendil obtained a $125 million upfront payment in the Sanofi deal and is eligible to receive (up to) a total of $1.72 billion in development and commercial milestone payments and interim payments. Earendil's initial investors were 5Y Capital and Hillhouse Investment

---

[k]Maybe not as laboratory-friendly, though.

Management Ltd. both coming out of China—yet another example of the influence that China is having on the biotech scene in the United States.

The acquisition of United Kingdom–based Exscientia by Recursion (p. 229), one of the first AI companies, for nearly $700 million last autumn only provides support to the view that "more AI is better." Having some Big Pharma partners in the mix to underpin the drug discovery programmes being entertained is better still.

You could argue that the new companies in biotech based on AI approaches such as Isomorphic Labs are following that same template as former biotech companies—but they are not. They are betting (and it is a bet) that their use of AI and raw compute power will eliminate most of the time it currently takes to do both drug discovery and drug development. It may do that eventually, but it is going to take a while and a lot of electricity to get to that point.[l]

Despite the constant hype around AI and drug discovery, I admit that I am slightly more convinced now than I was before that it is a new direction for biotech and a new drug discovery technology wave to surf. Investors clearly think so, demonstrated by their big bets in several very well-funded companies. But like all biotechnology companies, they will be judged eventually on the compounds or antibodies or other modalities they find that become drugs in humans. As for gene editing (p. 334), technology is not enough.[m]

I think I am where George Yancopoulos apparently is—a sceptic who suspects it is more likely that the biggest successes will come from sequencing very large numbers of human genomes to find genetically driven solutions and new drugs. Regeneron continues to do this, sequencing the genomes of millions of people and with access to public databases such as the UK Biobank, which now contains more than 450,000 fully sequenced and phenotyped individuals.[11] To add to their data, Regeneron tried to buy 23andMe (see below) out of bankruptcy.

As Craig Venter told me a few months ago, most of the sequences that are produced and analysed are not "phased," being mixtures of shorter

---

[l] I would not be surprised to see an antibiotic-focussed AI-driven company in the cards either after the announcement by a laboratory at MIT in August 2025 that they have used AI to design two new antibiotics that can kill MRSA from looking at thousands of compounds in silico for antibiotic properties.[10]

[m] The hiring by Genentech of a computational biologist in 2020 to run their R&D (instead of someone with a more conventional background) does not seem to have increased their productivity much, and many companies that started life as AI companies (e.g., Recursion Pharmaceuticals) ended up licensing compounds so that they had something in the clinic.

sequences (from a combination of the two strands of DNA in a chromosome), and thus do not really exist in the wild. To determine properly the sequence of a genome from telomere to telomere one needs to use long-read sequencing (like PacBio) and DNA from sperm, so one is only sequencing a single (i.e., proper) copy of a contiguous genome (see Chapter 7).

23ANDME

It is worth reflecting on what has been going on at Illumina and 23andMe, two companies I covered in Chapters 7 and 8. You may recall that as part of what was always a rather unconvincing combined strategy of consumer genetics and drug discovery, 23andMe was trying to use the personal genetic information it collected to discover new therapeutic targets and drugs to affect them, as well as following an Ancestry-like consumer genetics business. The reliability of some of the self-reported phenotype data was always going to be a problem when attempting to link it to target identification.

The company once had a market cap of $3.5 billion but it fell to $100 million and finally to only $20 million. It had formed partnerships with GSK and Genentech that brought in more than $300 million to pursue its drug discovery ambitions. 23andMe reported pronounced revenue declines over the last two years and, in November 2024, shut down its therapeutics "division" and eliminated more than 200 jobs. Coupled with some earlier unfortunate data leaks, the company ended up filing for Chapter 11 bankruptcy. Following the resignation of most of the Board in September 2024, the CEO, Anne Wojcicki, resigned after failing multiple times to take the public company private.

Regeneron, with its commitment to genetics (p. 195), stepped up in May 2025, entering into an agreement to buy 23andMe for $256 million. Their bid was deliberately more than the former CEO had offered to buy the company out of bankruptcy. Wojcicki made a counteroffer for the company through a California-based entity that she had founded called TTAM Research Institute. She eventually prevailed and bought the company she had founded back in 2006 for $305 million in June 2025. It will be interesting to see what happens next, as the basic strategic dilemma remains.[12]

ILLUMINA

SomaLogic (p. 57) was acquired by Standard BioTools in late 2023. It was not a very successful or happy arrangement and in June 2025, Illumina announced that it had decided to buy SomaLogic (or rather, its proteomics technology) for $350 million plus performance milestones, amounting to as much as $75 million, ostensibly to increase Illumina's presence in the protein biomarker discovery market. This move came after Illumina finally off-loaded GRAIL (p. 392), their DNA-based diagnostics effort, which was always a promising but troubled adolescent company that they tried to buy (and not buy) after many regulatory hassles. This divestment of their interest was described as being the "final chapter in one of the most disastrous attempted mergers in biotech history."[n] It has resulted in considerable restructuring at GRAIL and is another good lesson for new biotech entrepreneurs to remember.

MICRORNA-BASED COMPANIES

The same year the Nobel Prize in Chemistry was awarded for protein AI, the 2024 Physiology or Medicine Prize was awarded to Victor Ambros and Gary Ruvkun for finding and characterizing microRNAs (MiRs) in *C. elegans* (Chapter 9). By binding to mRNA, microRNAs were shown to control translation and add another level of gene regulation after transcription and splicing.

MicroRNAs are small single-stranded RNAs of about 22 nucleotides. The first microRNA, lin-4, was discovered by Ambros in 1993 followed by the discovery of let-7 by the Ruvkun Lab in 2000. There are now hundreds of microRNAs identified in the human genome, the biological functions of which are mostly unknown. There have been some attempts to use microRNAs directly as drugs. *MiR-34,* a human MiR that is not dissimilar to let-7, was tested in mouse cancer models, leading to high doses administered to humans. Some of the trial participants died for immunological reasons so the trial was stopped. Santaris Pharma in San Diego tried to affect hepatitis C infection by controlling the expression of a microRNA and Novo Nordisk bought a company called Cardior Pharmaceuticals in Germany to access their microRNAs controlling heart failure. Neither of these projects was successful. Finding selective microRNAs is not as straightforward as was initially hoped.

Regulus Therapeutics and uniQure

Regulus (p. 266), a gene regulation company based in La Jolla, California, developing a drug called farabursen, has had a more positive outcome: Novartis

---

[n] A 2023 quote from Matthew Herper at STAT news.

has acquired them for more than $800 million. It has been a long road to an exit for Regulus. Alnylam and Ionis created Regulus in 2007, and it went public in 2012. Despite deals with Sanofi, GSK, and AstraZeneca, nothing much emerged from their R&D until they found an oligonucleotide that targets *MiR-17*. It is now being tested in autosomal polycystic kidney disease, a condition with considerable unmet medical need for the patients and a potential for accelerated approval by regulators.

Rather unexpectedly—and to the benefit of both Huntington's patients and uniQure (pp. 282–283) stockholders—it was reported in September 2025 that an AAV5 gene therapy construct containing an MiR designed to switch off the translation of the mutant huntingtin mRNA markedly slowed the progression of the disease compared to historical controls. The DNA was delivered by MRI-guided microcatheter in a lengthy surgical procedure (>12 h) to the caudate nucleus and the putamen of the brain of several Huntington's disease patients. This is most encouraging for the patients with this devastating disease, but the results will need to be peer-reviewed and published somewhere other than in a press release to understand the real value. uniQure anticipates a BLA filing in early 2026. It will be interesting to see how much the FDA appreciates the natural history–based "controls" and what the price of the one-time therapy will be.

CAMBRIDGE BIOTECH

Alnylam Pharmaceuticals

Alnylam (Chapter 12), which also makes drugs that target RNA, continues to go from strength to strength. In the "old days," Biogen and Cambridge-based biotechnology used to be mentioned in the same breath. Biogen would be lucky now to get a reference at all, or at least a positive one. Today the Cambridge buzz is all about Alnylam Pharmaceuticals and Vertex Pharmaceuticals (see next section), and their valuations reflect that reality.[°]

I did not cover what Vertex was up to in the book before because two books that are worth a read have been written about the company already (p. 235). However, these books are now a little dated so it's worth bringing

---

[°]Vertex now has a market cap of >$100 billion, Alnylam of ~$50 billion, and Biogen only $20 billion. In December 2012 (when I was first at Biogen) the Vertex market cap was $9 billion compared to Biogen at $32 billion.

readers up to speed on Vertex and contrast their success with Biogen's lack-lustre performance. I will do this after covering Alnylam.

First, it is worth saying that Alnylam is not yet profitable! Nevertheless, the products they have created are successful with some being likely to be quite important for patients. Amvuttra, an RNAi made against trans-thyretin mRNA—already approved for hereditary transthyretin amy-loidosis (hATTR)—has now been approved for treating transthyretin amyloid cardiomyopathy (ATTR-CM), potentially a much larger mar-ket. It comes with a broad label which should be an advantage over its competitors from Pfizer and BridgeBio with their drugs, despite the fact that these drugs are taken orally rather than by injection. As for hAT-TR, the drug is going to continue to cost patients more than $450,000 per year. Amvuttra will be the revenue driver for Alnylam in the future. With a label expansion the drug brought in nearly $500 million in the second quarter of 2025 and provided some three-quarters of overall sales for Alnylam (amounting to more than $900 million) in the first half of 2025. They will be profitable soon.

Many people including John Maraganore, until recently the CEO of Alnylam, think that the company is the undisputed leader in RNA thera-peutics and will become the next biotech company worth more than $100 billion—up there with other independent companies with biotech heritage like Amgen, Gilead Sciences, Regeneron, and Vertex.

So much so that Maraganore has started a new RNA interference ther-apeutics company called City Therapeutics with ARCH Venture Partners as lead investor providing $135 million. City features a stellar class of other investors including Fidelity Investments, Invus, Rock Springs Capital, and Regeneron Ventures. The differentiator from others in the field is that they will deliver the drugs using proprietary and sophisticated delivery vehicles.

John has also been busy setting up his next play: a cardiovascular com-pany called Corsera Health with his old collaborator Clive Meanwell who was CEO of The Medicines Company. With $50 million as a start, they are targeting RNAi to angiotensinogen mRNA and to PCSK9 mRNA to treat hypercholesterolaemia and lower LDL cholesterol with the help of an AI-driven heart health application.

In the same space but with a different therapeutic modality, Ionis Pharmaceuticals (p. 250) announced that Tryngolza, their ASO targeting se-vere hypertriglyceridemia (sHTG) and the accompanying acute pancreatitis succeeded expectations in two phase 3 trials for both endpoints. Tryngolza is already approved for familial chylomicronemia syndrome (FCS). Ionis have competition in the form of Arrowhead Pharmaceuticals, an RNAi company that started its RNA work by acquiring assets from Novartis in 2015. Their major focus is on cardiovascular disease and metabolism, but they have just announced a multibillion-dollar deal with Novartis targeting α-synuclein for PD.

Alongside these start-ups, Atlas Venture, The Column Group (TCG), and Droia Ventures announced they had co-led a $100 million Series A investment in a new gene-silencing company called Judo Bio to focus on delivering interfering RNA to the kidney. Maraganore, being the doyen of RNAi, is also an advisor to Judo.

Vertex

If you take the ferry from Lynn to Boston in the summer as I sometimes do, you pass the Vertex building in the Seaport district. If you look carefully, you will appreciate that their once-unobstructed view of the Boston waterfront and Logan airport has been spoiled by a tall building constructed between Vertex and the sea. I muse whether obstruction is a view of things to come for Vertex commercially. Their cystic fibrosis (CF) franchise continues to provide enormous revenues for the company. In 2024, it brought in $11 billion in sales, a 12% increase over 2023. One wonders how long this can go on, and what other revenue drivers might be coming. The company is getting label expansions for Trikafta and has developed a new triple combination for adults with other mutations in the CFTR, with a drug called Alyftrek, also known as "vanza triple" for paediatric use. However, this may not be enough to keep sales moving at the rate that they have been to justify their stock price and P:E ratio.

It is possible that Vertex's long-term future will be in the pain business. The company is building a pain franchise on the back of its new drug VX-548 (suzetrigine, trade name Journavx), but the revenue projections for this potential therapeutic are nowhere close to that of the CF business. Suzetrigine works by selectively blocking the Nav1.8 sodium channel on pain-mediating neurons and has now been approved for short-term acute pain indications—the first newly approved painkiller for some 20 years.[13,14] Contrast this with the failure of the Biogen–Convergence collaboration on drugs affecting Nav1.7, a sister sodium channel to 1.8 (see p. 240). The effect of blocking Nav1.8 with suzetrigine for chronic pain indications has not been so clear-cut, largely owing to the pronounced placebo effect that plagues chronic pain studies.[P]

There is increasing competition from the likes of Latigo Biotherapeutics in California for Nav1.8 inhibitors with faster onset of action. Both companies will have to see what happens in much larger phase 3 trials in many more patients to find out whether they have a drug. There is also competition for Vertex in the CF space. Sionna Therapeutics, another one of the very few companies to go public in the last two years, obtained a set of CFTR modulators from Sanofi where there was a CF research group and in licensed others from AbbVie, focussing on the NBD1 domain (Chapter 10). Their series C round raised close to $200 million. RA Capital owned ~30% of the company with TPG, Atlas Venture, and OrbiMed being some of the other investors. AbbVie got $5 million upfront and $8.6 million worth of common stock for their compounds as part of the deal.

Biogen

I would not like to be the CEO of Biogen, the once widely venerated biotech company. It is interesting to contrast the successes of Vertex and Alnylam to the present state of Biogen. This company was one of the first biotech companies, and one that remains independent helped by the acquisition of Idec Pharmaceuticals in 2003 (p. 85). The drugs Rituxan and Ocrevus that they got as part of the Idec merger still bring in nearly 20% of Biogen's less-than-stellar revenue, now that their MS franchise has deteriorated so significantly.

Biogen's value has declined markedly since CEO George Scangos' era. Where has all that value gone? The Bioverativ spinout (p. 243) was a success in that the sales of the two products Eloctate and Alprolix were much higher when in Bioverativ than they were as part of Biogen. Of course, Biogen participated in the upside of the Sanofi acquisition. But there has been and still is a lack of strategic clarity, especially with respect to what therapeutic areas to focus on.

Biogen made a mistake in reducing their interest in immunology when inflammation mechanisms are so relevant to all neurodegenerative disease, especially when they had a very good immunology research and development group. The anti-β-amyloid drug Aducanumab was not a success, and neither was their antibody designed to remove α-synuclein in Parkinson's

---

[P] In their phase 2 study those patients on a sham pill also felt that their pain was reduced.

disease.[15] Their licensing deal with Eisai for lecanemab (Leqembi) has been a small salvation, but sales are modest (expected to be ~$200 million in 2025), and there are competitors such as donanemab from Lilly and, more recently, trontinemab from Roche, an antibody designed to be transported across the blood–brain barrier using the transferrin receptor.[16]

Shuttling Companies

Shuttling antibodies and oligonucleotides using the transferrin and other receptors has become a hot pursuit. It is not entirely new: Dyne Therapeutics was already using the transferrin receptor for transporting their oligonucleotides for DMD and DM1 (see below). But what kicked it into high gear was the purchase of Aliada Therapeutics by AbbVie for $1.4 billion in December 2024. AbbVie's main interest was in ALIA-1758, an antipyroglutamate amyloid β antibody currently in phase 1. Aliada had developed vectors targeting the transferrin receptor and CD98 for the purpose of shuttling molecules across the blood–brain barrier.

Now there are several "shuttling" companies. A prominent one is Manifold Bio, founded in 2021 by bioengineers Gleb Kuznetsov, Pierce Ogden, and Shane Lofgren, plus George Church (of woolly mammoth fame). They are a "protein therapeutics company with an AI-driven *in vivo* biologics design platform" primarily being used to find novel receptors that can be used as shuttle targets for antibodies carrying different payloads to different tissues, especially the brain. These *in vivo* studies are done in both mice and in nonhuman primates. They are unusual for a multiplexed NHP approach, as well as the fact that they have been financed so far exclusively by investors from the tech space.

Their seed round of $5.4 million was led by Playground Global and their Series A financing round by Triatomic Capital joined by Section 32, FPV Ventures, Horizons Ventures, and Tencent. The round included the existing investors Playground Global, Fifty Years, and FAST by GETTYLAB. There are some notable people on the Board including Jeff Huber, once the CEO of GRAIL, and Steve Holtzman previously at Millenium Pharmaceuticals (p. 148) and Biogen and the CEO of Infinity Pharmaceuticals and Decibel Therapeutics. Doug Williams is a strategic advisor to the company.[q] Manifold announced in November 2025 a large strategic alliance with Roche to develop other shuttle vectors. The deal is very reminiscent of the megadeals that Steve Holtzman did while he was at Millennium!

---

[q]In the interest of full disclosure, I am also a consultant to Manifold Bio.

Novartis, BMS, and Eisai are also on the shuttling vectors bandwagon and have formed alliances with BioArctic to work on getting antibodies across the blood–brain barrier. It is rumoured that Atlas Venture is also starting a new AI-driven shuttling company, suggesting that this technology could almost be classed as the next new thing in biotech. There are certainly many different tissue types that need to be targeted specifically for different diseases. Gene therapy company Voyager Therapeutics (p. 289), run by ex-Biogen veteran Al Sandrock, have also developed brain shuttles based on modifications to the capsid of their AAV9- and AAV5-derived gene therapy vectors.

Biogen has its own transporter antibody programme in a deal with Denali Therapeutics..[17] If history tells us anything, a future where most research and development is done outside of the company—as Biogen apparently aims to do—is not going to be very successful. Collaborations are important, of course, but you need to have your own R&D to be able to find and judge good collaborations in the first place. This is usually not done very well by business development–driven initiatives. It's a shame: Biogen was and still is the source of many very influential and successful industry veterans.

GENE EDITING AND GENE THERAPY

The other current product of interest from Vertex is Casgevy (p. 331), the sickle cell disease (SCD) gene editing therapy that they licensed from CRISPR Therapeutics. Casgevy (exagamglogene autotemcel, or exa-cel) is a genome-edited cellular therapy consisting of autologous $CD34^+$ hematopoietic stem cells edited by CRISPR-Cas9 at the erythroid-specific enhancer region of the *BCL11A* gene (Chapter 15). CRISPR Therapeutics received more than $1 billion from Vertex for the asset. Casgevy was approved in December 2023, but it took a while before more than 20 SCD patients began the cell collection process. Vertex has spent more than $500 million in rolling out Casgevy, and it is going to take several hundred more patients for therapy to start turning a profit.

bluebird bio (p. 287) is providing patients with a gene therapy alternative to Casgevy called Lyfgenia (lovotibeglogene autotemcel, aka lovo-cel). This gene therapy works by adding an altered functional β-globin gene (T87Q) to erythrocytes via CD34 cells. Unfortunately, sales of this drug have not gone well, and the company has restructured and significantly reduced head count. Lyfgenia treatment costs $3.1 million compared to the $2.2 million for Casgevy.

The other two bluebird bio gene therapy products—Zynteglo (for β-thalassemia) and Skysona (for cerebral adrenoleukodystrophy)—are in the same predicament. Out of almost 70 children who received Skysona, several have developed myeloblastic syndrome, probably a result of the genomic integration of the vector, and, sadly, one has died.[18] The long-term future of bluebird bio is unclear and there may be a completely empty nest quite soon. Rather tellingly perhaps, the company has changed its name back to the original one, Genetix Biotherapeutics.

2seventy bio, a bluebird bio spinout (p. 307) was recently acquired by BMS for $286 million to obtain Abecma, their CAR T therapy. 2seventy had fallen on hard times of late as well, having sold off most of its R&D work to Regeneron and laying off staff. But it still appears to have fared rather better than bluebird bio.[r] Adding to general gene therapy concerns, Intellia have reported incidences of high liver toxicity and the death of one patient with Nexiguran Ziclumeran (nex-z), their CRISPR-Cas9 gene editor targeting the transthyretin gene to treat ATTR amyloidosis.

Beam and Prime

Other gene editing companies (Chapter 15) have faced cutbacks as well. Prime Medicine (p. 338), who signed a partnership with BMS and received $110 million in upfront and equity payments and up to $3.5 billion in promised milestones, has also undergone restructuring and layoffs. They have also finally reduced the unsustainable numbers of projects they were pursuing. In May 2025 they replaced their CEO, Keith Gottesdiener, laid off some more staff (including CSO Jeremy Duffield), and terminated their only clinical programme for chronic granulomatous disease (CGD). Their CBO Richard Brudnick retired. Prime's market cap value is now roughly 10% of what it was after the IPO. The company is still pursuing programmes for two liver diseases, Wilson's disease and α-1 antitrypsin deficiency.

Beam Therapeutics, the first gene editing company, continues clinical development of its product targeting SCD with patients' cells being edited to express the foetal gamma globin chains (p. 342). Their gene editing approach is much more precise than CRISPR-Cas 9-based gene editing but still requires heavyweight conditioning before the CD34 cell transplants.[19] Future SCD treatments will have to be with technology that does not require

---

[r] Maybe they should have continued to fly together.

preconditioning regimens. To that end, Vertex is working with San Diego company Enlaza Therapeutics to develop gentler preconditioning approaches.

It looks like new treatments for SCD are not going to come very soon from other approaches, including small molecules. Pfizer pulled Oxbryta (p. 287) from the market after complications in patients following treatment. Pfizer obtained this drug in its $5.4 billion acquisition of Global Blood Therapeutics. As I noted before, I was never a big fan of GBT, the drug, or the approach (p. 283).

Another drug from GBT may not make it either. Pfizer obtained inclacumab, a P selectin–directed inhibitory antibody from them, but the phase 3 trial of it has been terminated owing to "low recruitment of patients"—whatever that means.[s] Pfizer also has a HbS polymerisation inhibitor in phase 3, but I am sceptical about that too, on mechanistic grounds.[t] Incidentally, Pfizer also ditched its haemophilia A gene therapy partnership with Sangamo BioSciences (pp. 283 and 319).

Even in the face of the failure of several high-profile gene editing companies, David Liu—Prime Medicine founder, gene editing doyen, and KOL in the field of base editing—continues to wax lyrical about the prospects for the technology, specifically for rare disease and even generally for more common diseases.[20] He may be right, but it is going to take a long time to get there and investors—as they are wont to do—are running out of patience.

Sarepta

Another disease for which gene therapy approaches are important is Duchenne muscular dystrophy (DMD) (Chapter 7). One of the big players in DMD gene therapy is—or was—Sarepta (p. 290).[21] They have a gene therapy product based on an AAV vector called Elevidys (delandistrogene moxeparvovec-rokl), which was approved in 2023. It is delivered via intravenous infusion to produce a functional mini dystrophin protein in the muscles of boys with DMD. More than 800 patients have received the gene therapy in trials and post-approval. What has happened most recently is that the liver toxicity concerns that were always there (as the treatment elevates liver enzymes) have intensified, and several boys have died from liver failure

---

[s] In my experience, this usually means physicians don't want to enter their patients into the clinical trials.

[t] It would be good to be proved wrong!

and immune-mediated myositis. This led to the suspension of distribution and of clinical trials testing of the gene therapy in July 2025.

The story is not quite so simple as the relationships of the company with the FDA have been rather difficult and unproductive.[u] The FDA asked Sarepta to stop selling Elevidys. Sarepta said it would continue to sell the therapy to ambulatory patients, as the deaths were in nonambulatory individuals. The FDA, with the stroke of Vinay Prasad's pen (before he got removed and then re-instated), narrowed the approval to patients aged 6 months through 11 years old at risk of more severe disease. Sarepta stopped selling the product entirely, leading to a great deal of consternation with the patients and their parents. Bowing to pressure applied by advocacy groups, the drug is still available. There is also the prospect of yet further clinical trials to get a re-approval including studies with a new immunosuppression regimen.

To save money, Sarepta has laid off 500 employees (nearly 40% of the work force), and it will sell all the stock it owns in Arrowhead Pharmaceuticals—its partner on several of its rare disease programmes—to raise the money needed to pay an Arrowhead milestone that the company still owes. Sarepta's future is far from assured.

Dyne(amo)

The woes of Sarepta have given others in the DMD gene therapy field some momentum. One such company is Dyne Therapeutics in Waltham, MA, who have presented data on their experimental DMD drug. The company, with new CEO John Cox at the helm (ex-Biogen, -Bioverativ, and -Repertoire Immune Medicines), has hired a new management team that includes individuals from John's former teams, including Doug Kerr as CMO (p. 292).[v] Dyne's DMD treatment (DYNE-25) is an exon-skipping oligonucleotide-based therapy. At a dose of 20 mg/kg, this drug has shown efficacy in DMD patients that exceeds the Sarepta drug but without the liver toxicity. Dyne is also developing a similar therapy for myotonic dystrophy type 1 (DM1). Novartis bought San Diego–based Avidity Biosciences, a company with a portfolio of antibody-oligonucleotide conjugates quite similar to those of Dyne Therapeutics, in October 2025 for $12 billion, one of the biggest acquisitions in 2025 after J&J

---

[u] An example of how *not* to do things and yet another classic biotech history lesson similar to the ones you have read about before.

[v] A move straight out of the biotech playbook—better the devils you know than those that you do not.

bought Intra-cellular Therapies earlier in the year. This is designed to strengthen their gene therapy efforts in neuromuscular disorders following the acquisition of AAV-based gene therapy company Kate Therapeutics, which also has preclinical gene therapy programs aimed at DMD, DM1, and facioscapulohumeral muscular dystrophy (FSHD), in 2024 for $1.1 billion.

Rocket Pharmaceuticals

I was once a fan of Rocket Pharmaceuticals (p. 291), but this company, too, has been going through some difficult times. The phase 2 pivotal study of RP-A501, their AAV gene therapy for Danon disease, was put on clinical hold after the death of a patient caused by an infection following capillary leak syndrome. Danon disease results from inherited mutations in the *LAMP2* gene, leading to muscle weakening and intellectual disability. In August 2025, the FDA lifted the clinical hold for the therapy to be continued, at a lower dose of vector and with changes to the immunosuppression regimen away from the C3 inhibitor used in the original phase 2 trial. The new study will recruit 12 patients, six of whom have been treated already. The stock price, which had taken a hit owing to the clinical hold, has begun to rise again as a result.[w]

One of Orchard Therapeutics' (the beneficiary of the GSK gene therapy asset auction; p. 280) claims to fame is to have developed the most expensive medicine in history. The list price of $4.25 million for Lenmeldy, their gene therapy for metachromatic leukodystrophy, a Mendelian inherited neurodegenerative disease, easily tops the list prices of Lyfgenia and Casgevy. The company justifies the eye-popping price on the basis that it reflects "its clinical, economic, and societal value."[x]

Other gene therapy approaches for Rett syndrome and autosomal deafness (type 9), replacing a faulty *OTOF* gene using an AAV vector, have moved forward and some success has been reported.[22] Roctavian, the Factor VIII gene therapy product (valoctocogene roxaparvovec-rvox) from BioMarin, has demonstrated long-term efficacy and safety, but slow uptake and reimbursement issues have really affected sales of one of the industry's first gene therapy products (see p. 283). Sales of Carvykti—a CAR T directed against BCMA for myeloma (see p. 308)—were close to $700 million in 2024 for

---

[w] But I would not say it is "taking off like a rocket."
[x] Well, it would of course.

Johnson & Johnson and Legend Biotech.[23] With long-term "cures" being reported for this molecule, gene therapy is an important new sector for biotech that needs to be supported in the future.

DEALS

In Chapter 6, I mentioned the building of the company Cerevel Therapeutics. This 2018 vintage company was not really a typical biotech because it was not based on any technological advance. It was a way that Pfizer and Bain Capital decided to exploit some Pfizer neuropsychiatric drug assets that were not a strategic focus for Pfizer. It turned out to be rather successful. The company raised a boatload[y] of money and were able to advance the compounds into some new clinical trials. Better still for Pfizer and the investors, Cerevel was acquired in late 2023 for nearly $9 billion by AbbVie. The most advanced drug in the pot was emraclidine, a positive allosteric modulator (PAM) of the muscarinic M4 receptor and potentially a next-generation antipsychotic (see Box 3). Unfortunately, the drug has failed in two pivotal phase 2 trials, significantly reducing the overall value of the acquisition.

The second drug, Tavapadon, a selective dopamine D1/D5 partial agonist for Parkinson's disease, has fared better and is showing efficacy in a phase 3 clinical trial. Darigabat, a third acquired asset, is an α-2/3/5-selective $GABA_A$ receptor PAM for treating epilepsy and panic disorder.

Emraclidine had been expected to compete with Cobenfy, a clever combination drug that BMS obtained by acquiring Karuna (p. 129). This was a smarter move (in my opinion) than acquiring Cerevel, and Cobenfy was duly approved in September 2024. It is going to be an important new medicine to treat schizophrenia: it will surely rack up billions of dollars in future sales.

BOX 3. ALLOSTERIC MODULATORS

Allosteric modulator drugs bind to an alternative site on a receptor rather than to the orthosteric site where the natural ligand binds. In doing so, they alter the activity of the receptor. There are positive allosteric modulators (PAMs), which increase the response of a receptor, and negative allosteric modulators (NAMs), which do the opposite. Greater selectivity can potentially be obtained by targeting allosteric sites rather than an active site.

---

[y] It is unclear what a "boatload" of money is exactly. For a discussion about what a tonne of money is, see p. 398.

On the back of the Karuna deal and mindful no doubt of the Cerevel paradigm (which made a lot of money for Pfizer on some potential drugs that might never have amounted to anything), BMS is now spinning out five of their immunological assets into a new Bain Capital–sponsored company.[24] Three of the potential drugs are already in clinical trials. The most advanced is a Toll-like receptor (TLR) 7/8 antagonist (afimetoran) in phase 2 for SLE. Next up is an oral TYK2 inhibitor that completed a trial in psoriasis, and the third is an IL2 fusion protein in phase 1 for atopic dermatitis. The other compounds are two biologics targeting the IL-18 and IL-10 pathways. BMS has 19% of the equity of the new company, and Robert Plenge, head of R&D for BMS, will have a Board seat.

Welsh Dragons

Another thing that the Cerevel transaction did was to stimulate the formation of a new company in the United Kingdom financed by SV Health Investors and Sanofi Ventures. Draig Therapeutics' founders are ex–industry scientists from Cardiff University's Medicines Discovery Institute. Draig is also developing new neuropsychiatric drugs. They raised a large (for the United Kingdom) $140 million Series A in mid-2025 to support a phase 2 trial of DT-101 for major depressive disorder. DT-101 is a drug designed to modulate the AMPA receptor, a difficult one to target selectively. In a phase 1a study in 60 healthy volunteers, DT-101 showed clear target engagement and was well-tolerated.

Draig is not alone in targeting AMPA. Neurocrine (see p. 126), a long-standing biotech company in San Diego with a richly textured past, is also developing a positive allosteric modulator of AMPA (see Box 3) for the treatment of major depressive disorder (MDD). Their drug candidate (osavampator) is currently in phase 3 trials.

Neurological disease is a very tough therapeutic area in which to develop drugs, as Sage Therapeutics can testify. Sage's positive allosteric modulator targeting the NMDA receptor (dalzanemdor) failed in phase 2 trials for cognitive impairment in AD, having already failed in PD. They have had hiccups in their commercial plans for zuranolone, a positive allosteric modulator of the GABA receptor and the first oral drug for postpartum depression (PPD). Coupled with a lawsuit with Biogen over the failure of SAGE-324 for essential tremor, Sage has had to take some drastic measures, including the inevitable staff layoffs. This once "high-flying" biotech company, formed in 2010, was put out of its misery by being acquired by Supernus Pharmaceuticals for $561 million.

Getting exits for small biotech companies over the last two years has been far from easy. There have been very few IPOs, so investors are not getting their returns by that means. There have been many company failures owing to not being able to find money with which to continue to do business. Many companies are valued below their capital on hand, and seed rounds for start-ups are very difficult to do. There are too many people chasing too few corporate deals with Big Pharma. This is a time-consuming pastime, and usually unsuccessful: if it is successful, the company often has to give away its crown jewels—but needs must.

ENTRANCES AND EXITS

It used to be an unusual situation, like seeing a black-footed ferret, for there to be empty lab space available in and around Kendall Square in Cambridge, MA, but that is now the new reality. The once majestic Genzyme manufacturing building (Plate 22) on the Charles River just looks sad today. National Resilience, who acquired the building in 2021, has announced several facility closures recently including the Allston site (which had already been earmarked for shutdown; p. 292).

Fortunately, there is continuing M&A activity in the sector that provides some return on investment. Examples involving some of the companies mentioned before are described below.

Following the BMS-Karuna deal, as mentioned above one of the biggest acquisitions was J&J's purchase of Intra-Cellular Therapies in early 2025 for $14.6 billion, mainly for its depression and schizophrenia drug Caplyta. Meanwhile Lilly is on a buying spree,[z] buying Scorpion Therapeutics for $2.5 billion—a company that shares the same founder as Loxo (p. 432). They also acquired Verve Therapeutics, a base editing company (p. 339). VERVE-102, their leading opportunity, targets the *PCSK9* gene. Given the existing antibody-based drugs that target *PCSK9*, I wonder what the commercial reality of this product will be. Lilly also bought Morphic Therapeutic for their oral integrin therapies, in particular MORF-057, which inhibits the function of $\alpha 4\beta 7$ integrin for the treatment of inflammatory bowel disease (IBD). This was a company that I was keen to invest in when SV Health Investors got to see it during its capital raising.[aa]

---

[z] Given their sales, they can afford it.
[aa] In VC speak, we "passed on the opportunity." You don't get them all right.

In another large deal, Sanofi acquired Blueprint Medicines for $9.1 billion to obtain the potential mastocytosis drug Ayvakit. Sanofi also bought Vigil Neuroscience (p. 109) for $470 million for its Alzheimer's disease–treating TREM2 agonist. At the same time, Enjaymo, the antibody for anaemia that Sanofi obtained as part of the Bioverativ acquisition, was off-loaded to Recordati for $825 million.[bb]

Arvinas, a company that was grown in the classical biotech way with several venture rounds at increasing valuations before going public, is run by John Houston (who I worked with at Glaxo). I have contrasted it earlier with companies that raised hundreds of millions of dollars only to fail (pp. 395 and 398). Pfizer and Arvinas are partnered on their protein degrader technology (PROTACs). Pfizer paid Arvinas $1 billion upfront in 2021 for the programme, but they have not yet acquired Arvinas, as one might have predicted. On the contrary, it could be that their relationship will not continue. The reason is that vepdegestrant, their PROTAC for the oestrogen receptor, did not extend progression-free survival in a phase 3 oestrogen receptor–positive group of breast cancer patients in the "intent-to-treat population."[cc]

Vepdegestrant did cut the risk of progression or death by at least 40% compared to hormone therapy so it may yet go forward: the drug has been filed as a second-line treatment for ER-positive, HER2-negative breast cancer. In the latest vepdegestrant storyline twist, Arvinas and Pfizer have decided to out-license the potential drug to a third party, apparently owing to the complex clinical and commercial environment that would await such a drug in the breast cancer space. I suspect the CEO will be retired before the drug hits the market. In other PROTAC company news, Kymera Therapeutics (p. 397) licensed their CDK 2 degrader to Gilead Sciences.

BACK TO VENTURE

Some of the more bullish venture capital groups (Chapter 18) such as RA Capital and ARCH Venture Partners continue to invest at an impressive rate, especially in big financings (p. 426). ARCH raised $3 billion for their

---

[bb] I guess it was too small a product for Sanofi to be bothered with.

[cc] The intent-to-treat (ITT) population in a clinical trial includes all participants who were randomized to a treatment group, regardless of whether they received the assigned treatment, adhered to the protocol, or completed the study.

XIII fund, a similar amount to their XII fund that closed in 2022. Bain Capital and Flagship Pioneering have done likewise, raising billions of dollars for their latest funds. Some of the smaller VCs are not finding it so easy. Apple Tree Ventures, run by Seth Harrison (see p. 210), is currently in a dispute with Rigmora Holdings (one of its principal limited partners), putting several Apple Tree portfolio companies at risk.

The focus of attention for the VCs are almost always companies in the dominant biotech hubs of Boston, South San Francisco, San Diego, and Seattle, and despite efforts in other locations, these remain *the* places to be. The United Kingdom is making concerted efforts to be a desirable place for small biotech companies, with government backing for the sector and as the home of two of the biggest pharmaceutical companies AstraZeneca and GSK. The science being done in the United Kingdom is, as always, first rate and competitive, and may be helped by U.S. academia's present difficulties. But it could all be derailed. The recent vote of no confidence in the U.K. research base by Merck, Eli Lilly, and AstraZeneca by cancellation of large building projects has not helped, and the spectre of "most favoured nation" (MFN) drug-pricing policy when the United Kingdom's own drug pricing strategy is in complete disarray, makes things even more difficult.

Access to enough (venture) capital solely in the United Kingdom to build a big biotech company is virtually impossible. Among the sector's largest investors in the United Kingdom is Novo Holdings, the owner of Novo Nordisk, the Danish company that has the anti-obesity drug Wegovy, Lilly's Zepbound competitor. Lilly recently showed that Zepbound (tirzepatide, the same drug substance as in Mounjaro) was better than Novo Nordisk's rival Wegovy, in a phase 3 weight loss comparison trial, although both drugs showed good efficacy.

Obesity Drugs

Newer drugs for obesity include Novo's CagriSema, a medication that combines the glucagon-like peptide (GLP-1) receptor agonist semaglutide with the amylin analog cagrilintide. Lilly is also developing orforglipron, an oral GLP-1 receptor agonist. It is likely to receive approval soon but will have competition from Structure Therapeutics, the latest Ray Stevens (see Syrrx on p. 228) creation which has the oral compound aleniglipron in phase 2 for weight loss and diabetes.

Given the huge market, there are multiple new entrants into the obesity field. Viking's clinical programmes include VK2735, a novel dual agonist of

GLP-1 and glucose-dependent insulinotropic polypeptide (GIP) receptors: it has some headwinds. Verdiva Bio in turn has raised more than $400 million to advance a portfolio of assets obtained from Sciwind Biosciences including ecnoglutide, a once-weekly oral GLP-1 drug. The company is backed by Forbion, General Atlantic, RA Capital, OrbiMed, and others.[dd] Ambrosia Biosciences is a next-generation obesity biotech in Boulder, Colorado chaired by Kyle Lefkoff, the cofounder of Array BioPharma (p. 56), which Pfizer bought for $11.4 billion in 2019.

Deep Apple Therapeutics is working on various drugs for obesity, including an MC4R agonist: at least one of its obesity assets may enter the clinic next year. Metsera, a 2022 vintage company, licensed a long-acting injectable GLP-1 agonist from Korean company D&D Pharmatech in 2024 and went public, raising more than $300 million—one of the very few companies to go public in 2025. Metsera is advancing MET-097i, the phase 2b injectable GLP-1 receptor agonist designed for monthly dosing. Their clinical pipeline also features a monthly injectable amylin analog and an oral GLP-1 peptide derivative. Pfizer was in the process of acquiring the company but Novo Nordisk (with a new CEO) has now thrown a spanner in the works by persuading Metsera's board that they are a better suitor for the assets than Pfizer and are attempting to outbid them with a better offer. Metsera had already approved the sale of the company to Pfizer for $4.9 billion, plus milestones.

As a result, Pfizer is suing Metsera for "breach of contract, breach of fiduciary duty, and tortious interference in contract arising." The outcome will be interesting and the price that the company is bought for will inevitably be higher. The moral of the story—don't mess with the big boys.

BEST-SELLING DRUGS

As most of you are aware from the TV advertisements,[ee] the battle between Novo and Lilly for their weight loss drugs has dominated thinking about best-selling drugs. Box 4 provides a list of the actual top 15 best-selling drugs for 2023 and 2024 courtesy of *Genetic Engineering and Biotechnology News*. This follows on from those in 2021 and 2022 shown earlier. As you will see Novo Nordisk's Ozempic (semaglutide) at #3 is presently ahead of Lilly's Mounjaro at #9 which has been used mainly to

---

[dd] In the usual spirit of biotechnology, many of Verdiva's team come from Aiolos Bio.

[ee] TV advertisements for drugs still give me heartburn!

BOX 4. BEST-SELLING DRUGS IN 2023 AND 2024[ff]

Merck's Keytruda continues to dominate the oncology drugs space. Pembrolizumab had sales of $25 billion in 2023 and $29.5 billion in 2024. Representing the other big market (i.e., cardiovascular disease), Eliquis from BMS and Pfizer for atrial fibrillation and prevention of stroke made $18.9 billion in 2023 and $20.7 billion in 2024. Ozempic (semaglutide) for weight loss comes in at #3, making nearly $15 billion in 2023 and $18.6 billion in 2024 for Novo Nordisk. More than justifying the Sanofi–Regeneron deal, #4 seller Dupixent (dupilumab) had sales of $12.4 billion in 2023 and $15 billion in 2024. The anti-HIV drug Biktarvy (a cocktail of bictegravir, emtricitabine, and tenofovir alafenamide) from Gilead Sciences[gg] is in fifth place with 2023 sales of $11.8 billion and 2024 sales of $13.4 billion.

The Jardiance[hh] family of drugs for inflammation is #6 on the list, with BI and Lilly sharing the spoils ($11.3 billion in 2023 and $13 billion in 2024). Used for heart failure and diabetes, its main ingredient is empagliflozin, a sodium-glucose cotransporter-2 (SGLT2) inhibitor. In seventh place—advertised to death on the television for inflammatory bowel disease—is Skyrizi (risankizumab), which garnished AbbVie sales of $7.7 billion in 2023 and $11.7 billion in 2024. Darzalex (daratumumab; #8) from J&J and Genmab and used to treat myeloma, had sales of $9.7 billion in 2023 and sales of $11.7 billion in 2024. As previously mentioned, Eli Lilly's drug Mounjaro (tirzepatide) for type 2 diabetes was #9, having 2023 sales of $5.2 billion, rising to $11.5 billion in 2024, a steep increase. Stelara (ustekinumab; #10), a Skyrizi competitor from Janssen Biotech (J&J), made $10.9 billion in 2023 and $10.4 billion in 2024.[ii] Treatments ranked #11–#15 in 2024 generated between ~$8.6 billion and $10.2 billion in revenues. These include Trikafta/Kaftrio (elexacaftor/tezacaftor/ivacaftor and ivacaftor) from Vertex Pharmaceuticals for CF, check point inhibitor Opdivo (nivolumab) from Bristol Myers Squibb, Eylea/Eylea HD (aflibercept) for age-related macular degeneration of the eye from Regeneron Pharmaceuticals, and

---

[ff] Top-selling drugs are ranked based on sales or revenue reported for 2024 by biopharma companies in press announcements, annual reports, and in investor materials. The total 2024 aggregate value of the top 10 best-selling drugs was $154.8 billion, up 6.5% from $145.9 billion in 2023—and up 53% over five years from the $101.15 billion generated in 2019.

[gg] The Gilead Sciences drug lenacapavir (Yeztugo)—which interferes with virus assembly, was approved in 2022, and is now available for pre-exposure prophylaxis (PrEP)—is the best anti-HIV drug available today.

[hh] The family consists of Jardiance (empagliflozin), Synjardy (empagliflozin and metformin), Synjardy XR (empagliflozin and metformin extended-release), Glyxambi (empagliflozin and linagliptin), and Trijardy XR (empagliflozin, linagliptin, and metformin extended-release).

[ii] Maybe TV adverts do make a difference after all.

Humira (adalimumab) from AbbVie (which until recently was the top-selling drug). The vaccine Gardasil/Gardasil 9, a quadrivalent human papillomavirus (Types 6, 11, 16, and 18) vaccine and its nine-virus strain successor (Chapter 16), is the 15th best-selling drug.

treat type 2 diabetes patients. Lilly's overall sales of Tirzepatide reached nearly $3.6 billion in the third quarter of 2025, growing nearly 200% from last year. It will not be long before this obesity drug beats Keytruda as the most valuable drug. Tirzepatide in its various forms is predicted to have sales of more than $40 billion/year in 2026, indicative of the number of people using it and paying for it themselves.

EPILOGUE

Despite the present environment, the industry continues to develop in a generally positive direction. There is a shortage of investment capital for new companies, and good exits have been hard to come by. There have been some gene therapy success and some setbacks, but by-and-large the biotech business continues to progress. The industry has been here before, although the geopolitical context is markedly different, and the industry will have to adapt to that. An excellent book has been written recently describing this shift.[25] Biotech and Pharma executives will need to appreciate that the game has changed.

The AI examples I described indicate that there is no lack of investment if the "perceived" value is there and can be increased. These investments have been a major value driver over the last two years for the industry. In September 2025, the XBI—the index for biotech stocks —moved into positive territory after a dismal first 8 months of the year and it continues to rise, which is encouraging for investors and might trigger some IPO activity in early 2026 (and a more enthusiastic J.P. Morgan conference in January).

If we are not very careful though, in 20 years we may be buying some (or maybe most) of our new innovative drugs from China—not discovering and making them here.[6] No new "biotech initiatives" will fix this in the short term because it is the political situation that is the problem. We are in danger of throwing out the biotech baby with the academic bathwater.

I ask you to please reflect on two things as you read this new chapter of the history of the biotech industry. First, most of the biotech industry was developed in the United States and included many people with very diverse backgrounds. Second, government hubris is fundamentally very dangerous.[jj]

REFERENCES AND NOTES

1. Thorp H. 2025. The Columbia deal is a tragic wake-up call. *Science* **389**: 551. doi:10.1126/science.aeb0424

2. Beier D, Osbourn J. 2025. The university–industry engine that drives U.S. innovation is under attack. *Biocentury* August 22, 2025.

3. Hatta M, Hatta Y, Choi A, Hossain J, Feng C, Keller MW, Ritter JM, Huang Y, Fang E, Pusch EA, et al. 2024. An influenza mRNA vaccine protects ferrets from lethal infection with highly pathogenic avian influenza A(H5N1) virus. *Science Translat Med* **16**: 778. doi:10.1126/scitranslmed.ads1273

4. Mallapaty S. 2025. mRNA vaccines for HIV trigger strong immune response in people. *Nature* **644**: 311–312. doi:10.1038/d41586-025-02439-4

5. Shooting the messenger: RFK Jr's attack on mRNA technology endangers the world. *The Economist*, Aug 23, 2025.

6. Vokinger KN, Li G, Wouters OJ. 2025. The rise of drug innovation in China—implications for patient access in the United States and globally. *N Engl J Med* **393**: 839–841. doi:10.1056/NEJMp2505821

7. Hunter TL, Bao Y, Zhang Y, Matsuda D, Riener R, Wang A, Li JJ, Soldevila F, Chu DSH, Nguyen DP, et al. 2025. *In vivo* CAR T cell generation to treat cancer and autoimmune disease. *Science* **388**: 1311–1317. doi:10.1126/science.ads8473

8. Watson JL, Juergens D, Bennett NR, Trippe BL, Yim J, Eisenach HE, Ahern W, Borst AJ, Ragotte RJ, Milles LF, et al. 2023. De novo design of protein structure and function with RFdiffusion. *Nature* **620**: 1089–1100. doi:10.1038/s41586-023-06415-8

9. Divine R, Dang HV, Ueda G, Fallas JA, Vulovic I, Sheffler W, Saini S, Zhao YT, Raj IX, Morawski PA, et al. 2021. Designed proteins assemble antibodies into modular nanocages. *Science* **372**: eabd9994. doi:10.1126/science.abd 9994

10. Incorvaia D. 2025. Generative AI models build new antibiotics starting from a single atom. *Fierce Biotech*, August 14, 2025.

11. The UK Biobank Whole-Genome Sequencing Consortium. 2025. Whole-genome sequencing of 490,640 UK Biobank participants. *Nature* **645**: 692–701. doi:10.1038/s41586-025-09272-9

12. Kwon D. 2025. What went wrong at 23andMe? Why the genetic-data giant risks collapse. *Nature* **638**: 14–15. doi:10.1038/d41586-025-00118-y

---

[jj]Definition of hubris: excessive pride or self-confidence. The definition of hubris in Greek tragedy: excessive pride toward or defiance of the gods, leading to nemesis.

13. Jones J, Correll DJ, Lechner SM, Jazic I, Miao X, Shaw D, Simard C, Osteen JD, Hare B, Beaton A, et al. 2023. Selective inhibition of $Na_V1.8$ with VX-548 for acute pain. *N Engl J Med* **389**: 393–405. doi:10.1056/NEJMoa2209870

14. Wosen J. 2023. Inside Vertex's decades-long quest to develop a new class of non-opioid painkillers. STAT+, December 11, 2023.

15. Lang AE, Siderowf AD, Macklin EA, Poewe W, Brooks DJ, Fernandez HH, Rascol O, Giladi N, Stocchi F, Tanner CM, et al. 2022. Trial of Cinpanemab in early Parkinson's disease. *N Engl J Med* **387**: 408–420. doi:10.1056/NEJMoa2203395

16. Grimm HP, Schumacher V, Schäfer M, Imhof-Jung S, Freskgård P-O, Brady K, Hofmann C, Rüger P, Schlothauer T, Göpfert U, et al. 2023. Delivery of the Brainshuttle™ amyloid-β antibody fusion protein trontinemab to non-human primate brain and projected efficacious dose regimens in humans. *Mabs* **15**: 2261509. doi:10.1 080/19420862.2023.2261509

17. Pizzo ME, Plowey ED, Khoury N, Kwan W, Abettan J, DeVos SL, Discenza CB, Earr T, Joy D, Lye-Barthel M, et al. 2025. Transferrin receptor–targeted anti-amyloid antibody enhances brain delivery and mitigates ARIA. *Science* **389**: eads3204. doi:10.1126/science.ads3204

18. Duncan CN, Bledsoe JR, Grzywacz B, Beckman A, Bonner M, Eichler FS, Kühl J-S, Harris MH, Slauson S, Colvin RA, et al. 2024. Hematologic cancer after gene therapy for cerebral adrenoleukodystrophy. *N Engl J Med* **391**: 1287–1301. doi:10.1056/NEJMoa2405541

19. Orkin SH. 2025. The fetal-to-adult hemoglobin switch—mechanism and therapy. *N Engl J Med* **392**: 2135–2149. doi:10.1056/NEJMra2405260

20. Ledford H. 2025. Brain editing now 'closer to reality': the gene-altering tools tackling deadly disorders. *Nature* **644**: 847–848. doi:10.1038/d41586-025-02578-8

21. Intelligent Medical Objects. 2023. *Advances in rare disease research & therapeutics.* STAT E-BOOK, Boston. www.statnews.com/advances-in-rare-diseases

22. Lv J, Wang H, Cheng X, Chen Y, Wang D, Zhang L, Cao Q, Tang H, Hu S, Gao K, et al. 2024. AAV1-hOTOF gene therapy for autosomal recessive deafness 9: a single-arm trial. *Lancet* **403**: 2317–2325. doi:10.116/S0140-6736(23)02874-X

23. Bird S, Pawlyn C. 2025. Immunotherapy using CAR T cells shows promising long-term outcomes for people with the blood cancer myeloma. *Nature* **645**: 44–45. doi:10.1038/d41586-025-02592-w

24. Ramírez-Valle F, Maranville JC, Roy S, Plenge RM. 2024. Sequential immunotherapy: towards cures for autoimmunity. *Nat Rev Drug Discov* **23**: 501–524. doi:10.1038/s41573-024-00959-8

25. Gerstle G. 2022. *The rise and fall of the neoliberal order: America and the world in the free market era.* Oxford University Press, New York.

Biotechnology Product Time Line

Year	Technology Advances, Nobel Prizes, and IP	Companies Set Up	New Drugs Approved
1973	Recombinant DNA: Cohen & Boyer paper	Cetus (1971)	
1975	Monoclonal antibodies Asilomar Conference		
1976		Genentech	
1977	DNA sequencing methods developed Moratorium on recombinant DNA lifted in Cambridge (MA)	Genex Corporation	
1978	Nobel Prize for restriction enzymes	Biogen Hybritech	
1979	Wistar mAb patents	Centocor Molecular Genetics Inc. (MGI)	
1980	Positional cloning Nobel Prize for DNA sequencing Stanford Rec DNA patent	Celltech DNAX Genetic Systems	
1981	Cryo-EM	Amgen Applied Biosystems Chiron Genetics Institute Genzyme Immunex Synergen ZymoGenetics	

(ADC) Antibody drug conjugate; (ASO) antisense oligonucleotide; (c) small-molecule drug; (CT) cell therapy; (GT) gene therapy; (mAb) monoclonal antibody; (rec) recombinant protein; (siRNA) small interfering RNA.

Year	Technology Advances, Nobel Prizes, and IP	Companies Set Up	New Drugs Approved
1982	ABI Model 470A Protein Sequencer available		Humulin (rec insulin)
1983	Basic AAV vectors Boss & Cabilly patents ABI DNA synthesizer: Model 380A	Oncogen	
1984	TCRs cloned Nobel Prize for mAb technology	Agouron ImClone Systems	
1985	Humanisation of mAbs by phage display and CDR grafting		Protropin (rec human growth hormone)
1986	EST sequencing ABI 370A DNA sequencer	Athena Therapeutics British Biotech Idec Protein Design Labs (PDL)	Intron A (rec interferon) Muronomab (mAb-OKT 3) Recombivax (Hep B SAg vaccine) Roferon (rec interferon)
1987	YACS developed t-PA patent litigation (UK) GI/Amgen EPO patent dispute	Amylin Cephalon Gilead Sciences Invitrogen Ligand Pharmaceuticals Medarex Mercator Novavax	Humatrope (rec insulin)
1988	First gene therapy clinical trial	Affymax Genetic Therapy Inc. MedImmune Regeneron Systemix Vical	Eprex (rec erythropoietin-EPO)
1989	HCV cloned mAbs made in humanised mice Nobel Prize for oncogenes	Cambridge Antibody Technology (CAT) Cell Genesys Genpharm Icos Ionis Perseptive Biosystems Vertex Pharmaceuticals	Epogen (rec EPO) Engenix B (Hep B SAg vaccine) Procrit (rec EPO)

Year	Technology Advances, Nobel Prizes, and IP	Companies Set Up	New Drugs Approved
1990	HGP started Aptamers developed *Lancet* TNF mAb paper		Actimmune (rec interferon γ)
1991	Lentivectors made Venter EST paper	Incyte Myriad Genetics Nexagen Pharmacyclics Somatix Sugen Tularik	Ceredase (GCase) Leukine (rec GM-CSF) Neupogen (rec G-CSF)
1992	BAC technology developed	Avigen Aviron Dendreon (ACT) HGS Lynx Mercator Neurocrine Biosciences Nemapharm Onyx Solexa TIGR	Recombinate (rec Factor VIII) Proleukin (rec IL2)
1993	CAR T cells developed Nobel Prize for PCR	Affymetrix Millennium Sequana Therapeutics Signal Pharmaceuticals	Kogenate (rec Factor VIII) Betaseron (rec β-interferon)
1994	*BRCA1* gene cloned	Alexandria Real Estate Equities (ARE) Exelixis	ReoPro - GP IIb/IIIa mAb Cerezyme (rec GCase)
1995	Shotgun sequencing of *H. influenzae* EPO patent ruled in Amgen's favour	Aurora Biosciences Infinity Pharma Sangamo BioSciences	
1996	First zinc finger nucleases (ZNFs) Bermuda Accords agreed in HGP	Bionomics CuraGen deCode Genetics	Avonex (rec β-interferon) Vistide (ASO for CMV)
1997	MegaBACE sequencers	BioMarin Genos Kudos Maxygen Syntonix	Rituxan (Rituximab) CD20 mAb

Year	Technology Advances, Nobel Prizes, and IP	Companies Set Up	New Drugs Approved
1998	ABI 3700s available RNAi paper published	Array BioPharma Celera PTC Therapeutics Seagen uniQure	Enbrel (etanercept)- rec TNFR/Fc Fomivirsen (vitravene) ASO for CMV Herceptin - Her2 mAb Remicade (Infliximab) anti TNF mAb Provigil (modafinil)-(c) Synagis (pavlivizumab) RSV mAb
1999	*Drosophila* genome-sequence jamboree Deaths cause gene therapy virtual moratorium	Astex Cellectis Galapagos Genmab Myriad Pharmaceuticals Rocket Pharmaceuticals Syrrx Structural GenomiX (SGX)	Tamiflu (oseltamivir) flu virus inhibitor (c)
2000	Draft human genome sequence completed	Affinium 454 Life Sciences CureVac SomaLogic	ReFacto (rec Factor VIII) Mylotarg (ADC)
2001	Cabilly II patent issues		Alemtuzumab (CAMPATH 1-CD3 mAb) Fabrazyme (rec agalsidase β) Gleevec (BCR-Abl kinase inhibitor) (c) Kineret (anakinra) rec IL1 antagonist Tenofovir (HIV antiviral) (c)
2002		Ensemble Therapeutics Incyte Corporation	Humira (adalimumab)- TNF-α mAb
2003		Proteolix	Aldurazyme (laronidase) for MPS 1 Hepsera (adefovir) (c) for Hep C Velcade (bortezomib) proteasome inhibitor (c)

Year	Technology Advances, Nobel Prizes, and IP	Companies Set Up	New Drugs Approved
2004	PAX oligo: Alnylam–Ionis cross-licencing deal	Arrowhead Pharmaceuticals PacBio Sirtris Pharmaceuticals	Avastin (bevacizumab) Her 2 mAb Erbitux (EGFR mAb) Tarceva (erlotinib) EGFR kinase inhibitor (c) Tysabri (natalizumab)— α-4 integrin receptor mAb
2005		Oxford Nanopore	Symlin (amylin peptide) Byetta (Exenatide)-GLP1 agonist peptide
2006	Illumina Genome Analyser launched IPSCs Nobel Prize for RNAi	Amunix Complete Genomics Praxis Pharma 23andMe	Atripla (HIV RTase inhibitor) (c) Gardasil (HPV vaccine) Myozyme (rec aglucosidase α) Vectibix (EGFR mAb)
2007	Stem cell transplants Nobel Prize for stem cell modifications	Adimab Dicerna Pharmaceuticals Fate Therapeutics Foundation Medicine Heptares Therapeutics Ion Torrent Regulus Therapeutics Receptos	Soliris (Eculizumab)- complement C5 mAb
2008	Bispecific mAbs (BiTEs) GVAX cancer vaccine Nobel Prize for fluorescent proteins	Adaptimmune BioNTech Immunocore	Arcalyst (rilonacept) IL-1 trap rec protein Cimzia (rec TNF blocker protein)
2009	CAR T manufacturing process developed	Bicycle Therapeutics ConfometRX Karuna Kite Pharma Kymab	Cervarix (HPV vaccine)
2010	TALENs	bluebird bio Convergence Moderna	Provenge: cancer vaccine (CT)
2011	CRISPR-Cas9	Caribou Biosciences Global Blood Therapeutics (GBT)	Adcetris (ADC) Aflibercept (Eylea) rec VEGF antagonist

Year	Technology Advances, Nobel Prizes, and IP	Companies Set Up	New Drugs Approved
		Blueprint Medicines Translate Bio	Orencia (abatacept) rec CTLA4-Fc fusion Yervoy (ipilimumab) CTLA 4 mAb
2012	AbbVie split from Abbott Nobel Prize for GPCR structure CRISPR-Cas 9 IP filed	Avidity Biosciences CRISPR TX Sarepta Therapeutics (for- merly AVI BioPharma) Viking Therapeutics	Cometriq-MET/VEGF kinase inhibitor (c) Kalydeco-CFTR regula- tor (c) Truvada-HIV antiviral (c) Glybera (ex US) LPL deficiency (GT)
2013	Direct genome editing in humans Nobel Prize for vesicle trafficking	Acerta Pharma Alector Therapeutics Amylyx Pharmaceuticals Arvinas Denali Therapeutics Editas Medicine Juno Therapeutics Loxo Oncology Recursion Rubius Therapeutics Spark Therapeutics Valneeva Voyager Therapeutics	Kadcyla - transtuzumab emtansine (ADC) Simponi (golimumab) TNF mAb Sovaldi (sofosbuvir) Hep C antiviral (c)
2014		Genomics PLC Freeline Therapeutics Intellia Therapeutics Legend Biotech Morphic Therapeutic	Alprolix (rec Factor IX) Blincyto (blinatumomab) -bispecific mAb Eloctate (rec Factor VIII) Keytruda (anti-PD1 mAb) Kynamro (mipomersen) -ApoB (ASO) Lemtrada (alemtuzumab) CD52 mAb Olaparib-PARP inhibitor (c) Opdivo (nivolumab) anti PD1 mAb
2015	ALS exome sequencing	AVROBIO Arrakis Therapeutics C4 Therapeutics Codiak Biosciences Gossamer	Cotellic (cobimetinib)- B-raf inhibitor (c) Dengvaxia (dengue fever vaccine)

Year	Technology Advances, Nobel Prizes, and IP	Companies Set Up	New Drugs Approved
		GRAIL Nkarta Kymera Therapeutics Ribometrix Orchard Therapeutics Pionyr Immunotherapeutics Verge Genomics	Praluent (alirocumab) PCSK 9 mAb Repatha (evelocumab) PCSK 9 mAb
2016		Artios Pharma Bioverativ CAMP 4 Bio Enara Bio Forty Seven Inc. Skyhawk Therapeutics Structure Therapeutics (originally ShouTi, Inc.) Vaccitech Vir Biotechnology	Cabometyx-Met/VEGFR inhibitor (c) Spinraza (nusinersen)-SMA (ASO) Strimvelis-ADA deficiency (GT)
2017	Base editing Nobel Prize for cryo-EM	Beam Therapeutics Dyne Therapeutics Laronde 2Seventy Bio	Besponsa (CD22 ADC) Dupixent (dupilumab) IL4 receptor mAb Kevzara (sarilumab) IL6 mAb Luxturna (AAV GT for RPE65) Ingrezza (c) Kymriah (CD19 CAR T) (CT) Yescarta (CD19 CAR T) (CT)
2018	Nobel Prizes for T cells (checkpoints) and phage display	Allogene Caraway Therapeutics Cerevel Therapeutics Casma Therapeutics Latigo Therapeutics Lyell Immunopharma Metagenomi Rejuvenate Bio Sana Biotechnology Tessera Therapeutics	Hemlibra (bispecific mAb for haemophilia) Libtayo (anti-PD1 mAb) Onpattro (patisiran) siRNA for hATTR amyloidosis Orilissa (elagolix) (c) Palynziq (rec enzyme for phenylketonuria)

Year	Technology Advances, Nobel Prizes, and IP	Companies Set Up	New Drugs Approved
2019	Base editing developed Direct in vivo human gene editing	Catamaran Bio Interius Bio Manifold Bio Prime Medicine Sionna Therapeutics Verve Therapeutics	Adakveo (crizanlizumab) P selectin mAb Givlaari (givosiran) (siRNA) Oxbryta (voxelotor) for SCD (c) Trikafta CFTR modulator (c) Evrysdi (Zolgensma)-SMA (AAV GT)
2020	Nobel Prize for CRISPR-Cas9 Illumina bids for GRAIL	Kate Therapeutics Resilience Biotherapeutics Scorpion Therapeutics Shoreline Biosciences Repertoire Immune Medicines Vigil Neuroscience	COMIRNATY (BNT 162b2)-SARS CoV2 mRNA vaccine Libmeldy (GT) Oxlumo (lumasiran) (siRNA) Ozanimod ($S1P_1$ agonist) (c) Spikevax (mRNA-1273)-SARS CoV 2 mRNA vaccine Tecartus CD19 CAR T (CT)
2021	Structure prediction using DeepMind/AlphaFold	Actio Biosciences Aliada Therapeutics Capstan Therapeutics CHARM Therapeutics Deep Apple Therapeutics Isomorphic Labs Upstream Bio	Abecma (BCMA CAR T) (CT) Aduhelm (aducanumab) β amyloid mAb Breyanzi (CD19 CAR T) (CT) Leqvio (inclisiran) PCSK9 (siRNA) Rybrevant (amivantamab) bispecific mAb (EGFR/MET)

Year	Technology Advances, Nobel Prizes, and IP	Companies Set Up	New Drugs Approved
2022	Human genome sequenced CRISPR-Cas 9 IP litigation	Aera Therapeutics Altos Labs Enlaza Therapeutics Metsera Therapeutics Noetik Nvelop Therapeutics Orbital Therapeutics Septerna Superluminal Medicines	Carvykti (BCMA CAR T) (CT) Enjaymo (sutimlimab) C1s mAb Hemgenix (Factor IX GT) Lunsumio (mosunetuzumab) bispecific mAb Nuvaxovid protein SARS-CoV2 vaccine Pyrukynd (PK deficiency) (c) Relyvrio (c) for ALS Roctavian Factor VIII AAV GT Tebentafusp (T-cell engager–specific mAb) Upstaza (GT) Vabysmo-bispecific mAb for AMD Vutrisiran (siRNA-transthyretin) Zynteglo (GT) (β-thalassaemia)
2023	Nobel Prize for mRNA vaccine	Aiolos Bio Archon Bio City Therapeutics Latent Labs	ABRYSVO (RSV vaccine) Altuviiio (rec Factor VIII derivative) AREXVY (RSV vaccine) Elevidys (GT for DMD) Leqembi (β amyloid mAb) Qalsody (tofersen) (ASO for ALS) Roctavian (AAV GT Factor VIII) in USA

Year	Technology Advances, Nobel Prizes, and IP	Companies Set Up	New Drugs Approved
2024	Nobel Prizes for protein structure prediction and microRNA	Ambrosia Biosciences Candid Therapeutics Chai Discovery Draig Therapeutics Earendil Therapeutics Topo Therapeutics Xaira Therapeutics (AI)	Lenmeldy (atidarsagene autotemcel) for MLD Beqvez (fidanacogene elaparvovec-dzkt) (AAV for Heme B) (discontinued) mRESVIA (mRNA vaccine for RSV) Hympavzi (marstacimab-hncq): TFPI antagonist mAb Kisunla (donanemab-azbt) amyloid β-directed mAb for AD Tecelra (afamitresgene autoleucel) (MAGE-A4)-directed autologous T-cell immunotherapy Cobenfy (xanomeline and trospium chloride) for schizophrenia Alyftrek (deutivacaftor, tezacaftor, and vanzacaftor) triple combination CFTR modulator
2025	Nobel Prize in Physiology or Medicine for Tregs	Corsera Health Dispatch Bio Divergence Labs Verdiva Bio	Alhemo (concizumab-mtci) for heme A & B Journavx (suzetrigine) NaV1.8 blocker Qfitlia (fitusiran) anti-thrombin III-directed siRNA (heme A & B) Dawnzera (donidalorsen) prekallikrein-directed ASO for hereditary angioedema Lynozyfic (linvoseltamab-gcpt) bispecific B-cell maturation antigen (BCMA)-directed CD3 T-cell engager

This list focuses on technology, IP, companies, and products that are mentioned in the book. It is not comprehensive and is not intended to be. Apologies to those companies not mentioned.

Relevant Nobel Prizes

PRIZES IN PHYSIOLOGY OR MEDICINE

1923	Frederick G. Banting and John Macleod "for the discovery of insulin"
1933	Thomas H. Morgan "for his discoveries concerning the role played by the chromosome in heredity"
1951	Max Theiler "for his discoveries concerning yellow fever and how to combat it"
1954	John F. Enders, Thomas H. Weller, and Frederick C. Robbins "for their discovery of the ability of poliomyelitis viruses to grow in cultures of various types of tissue"
1959	Severo Ochoa and Arthur Kornberg "for their discovery of the mechanisms in the biological synthesis of ribonucleic acid and deoxyribonucleic acid"
1962	Francis Crick, James Watson, and Maurice Wilkins "for their discoveries concerning the molecular structure of nucleic acids and its significance for information transfer in living material"
1968	Robert W. Holley, Har Gobind Khorana, and Marshall W. Nirenberg "for their interpretation of the genetic code and its function in protein synthesis"
1972	Gerald M. Edelman and Rodney R. Porter "for their discoveries concerning the chemical structure of antibodies"
1975	David Baltimore, Renato Dulbecco, and Howard Temin "for their discoveries concerning the interaction between tumour viruses and the genetic material of the cell"
1976	Baruch S. Blumberg and D. Carleton Gajdusek "for their discoveries concerning new mechanisms for the origin and dissemination of infectious diseases"
1978	Werner Arber, Daniel Nathans, and Hamilton O. Smith "for the discovery of restriction enzymes and their application to problems of molecular genetics"
1983	Barbara McClintock "for her discovery of mobile genetic elements"
1984	Niels K. Jerne, Georges J.F. Köhler, and César Milstein "for theories concerning the specificity in development and control of the immune system and the discovery of the principle for production of monoclonal antibodies"

1985	Michael S. Brown and Joseph L. Goldstein "for their discoveries concerning the regulation of cholesterol metabolism"
1989	J. Michael Bishop and Harold E. Varmus "for their discovery of the cellular origin of retroviral oncogenes"
1990	Joseph E. Murray and E. Donnall Thomas "for their discoveries concerning organ and cell transplantation in the treatment of human disease"
1993	Richard J. Roberts and Phillip A. Sharp "for their discoveries of split genes"
1994	Alfred G. Gilman and Martin Rodbell "for their discovery of G-proteins and the role of these proteins in signal transduction in cells"
1996	Peter C. Doherty and Rolf M. Zinkernagel "for their discoveries concerning the specificity of the cell mediated immune defence"
2001	Leland H. Hartwell, Tim Hunt, and Sir Paul M. Nurse "for their discoveries of key regulators of the cell cycle"
2002	Sydney Brenner, H. Robert Horvitz, and John E. Sulston "for their discoveries concerning genetic regulation of organ development and programmed cell death"
2004	Richard Axel and Linda B. Buck "for their discoveries of adorant receptors and the organization of the olfactory system"
2006	Andrew Z. Fire and Craig C. Mello "for their discovery of RNA interference—gene silencing by double-stranded RNA"
2007	Mario R. Capecchi, Sir Martin J. Evans, and Oliver Smithies "for their discoveries of principles for introducing specific gene modifications in mice by the use of embryonic stem cells"
2008	One half awarded to Harald zur Hausen "for his discovery of human papilloma viruses causing cervical cancer" and the other half to Françoise Barré-Sinoussi and Luc Montagnier "for their discovery of human immunodeficiency virus"
2009	Elizabeth H. Blackburn, Carol W. Greider, and Jack W. Szostak "for the discovery of how chromosomes are protected by telomeres and the enzyme telomerase"
2011	One half to Bruce A. Beutler and Jules A. Hoffmann "for their discoveries concerning the activation of innate immunity" and the other half to Ralph M. Steinman "for his discovery of the dendritic cell and its role in adaptive immunity"
2012	Sir John B. Gurdon and Shinya Yamanaka "for the discovery that mature cells can be reprogrammed to become pluripotent"
2013	James E. Rothman, Randy W. Schekman, and Thomas C. Südhof "for their discoveries of machinery regulating vesicle traffic, a major transport system in our cells"

2018	James P. Allison and Tasuku Honjo "for their discovery of cancer therapy by inhibition of negative immune regulation"
2020	Harvey J. Alter, Michael Houghton, and Charles M. Rice "for the discovery of Hepatitis C virus"
2023	Katalin Karikó and Drew Weissman "for their discoveries concerning nucleoside base modifications that enabled the development of effective mRNA vaccines against COVID-19"
2024	Victor Ambros and Gary Ruvkun for the "discovery of microRNA and its role in post-transcriptional gene regulation"
2025	Mary Brunkow, Fred Ramsdell, and Shimon Sakaguchi for their "discoveries concerning peripheral immune tolerance"

PRIZES IN CHEMISTRY

1947	Sir Robert Robinson "for his investigations on plant products of biological importance, especially the alkaloids"
1951	Edwin M. McMillan and Glenn T. Seaborg "for their discoveries in the chemistry of the transuranium elements"
1958	Frederick Sanger "for his work on the structure of proteins, especially that of insulin"
1962	Max F. Perutz and John C. Kendrew "for their studies of the structures of globular proteins"
1964	Dorothy Hodgkin "for her determinations by X-ray techniques of the structures of important biochemical substances"
1980	One half to Paul Berg "for his fundamental studies of the biochemistry of nucleic acids, with particular regard to recombinant-DNA" and the other half to Walter Gilbert and Frederick Sanger "for their contributions concerning the determination of base sequences in nucleic acids"
1982	Aaron Klug "for his development of crystallographic electron microscopy and his structural elucidation of biologically important nucleic acid-protein complexes"
1984	Robert Bruce Merrifield "for his development of methodology for chemical synthesis on a solid matrix"
1993	"For contributions to the developments of methods within DNA-based chemistry" with one half to Kary B. Mullis "for his invention of the polymerase chain reaction (PCR) method" and with one half to Michael Smith "for his fundamental contributions to the establishment of oligonucleotide-based, site-directed mutagenesis and its development for protein studies"

2008	Osamu Shimomura, Martin Chalfie, and Roger Y. Tsien "for the discovery and development of the green fluorescent protein, GFP"
2012	Robert J. Lefkowitz and Brian K. Kobilka "for studies of G-protein-coupled receptors"
2017	Jacques Dubochet, Joachim Frank, and Richard Henderson "for developing cryo-electron microscopy for the high-resolution structure determination of biomolecules in solution"
2018	One half to Frances H. Arnold "for the directed evolution of enzymes" and the other half to George P. Smith and Sir Gregory P. Winter "for the phage display of peptides and antibodies"
2020	Emmanuelle Charpentier and Jennifer A. Doudna "for the development of a method for genome editing"
2022	Carolyn R. Bertozzi, Morten Meldal, and K. Barry Sharpless "for the development of click chemistry and bioorthogonal chemistry"
2024	David Baker "for computational protein design": the other half jointly to Demis Hassabis and John Jumper "for protein structure prediction"

Recommended Reading

BOOKS REFERENCED IN THE TEXT

Bazell R. 1998. *Her-2: the making of Herceptin, a revolutionary treatment for breast cancer.* Random House, New York.

Bingham K, Hames T. 2022. *The long shot: the inside story of the race to vaccinate Britain.* Oneworld, London.

Canavan N. 2018. *A cure within: scientists unleashing the immune system to kill cancer.* Cold Spring Harbor Laboratory Press, Cold Spring Harbor, NY.

Cobb M. 2022. *As gods: a moral history of the genetic age.* Basic Books, New York.

Davies K. 2001. *Cracking the genome: inside the race to unlock human DNA.* The Free Press, New York. In the United Kingdom: Davies K. 2000. The sequence: inside the race for the human genome. Weidenfeld & Nicolson, London.

Davies K. 2010. *The $1000 genome: the revolution in DNA sequencing and the new era of personalized medicine.* Free Press, New York.

Davies K. 2021. *Editing humanity: the CRISPR revolution and the new era of genome editing.* Pegasus Books, New York.

Davis DM. 2014. *The compatibility gene: how our bodies fight disease, attract others, and define ourselves.* Penguin, New York.

Drakeman DL, Drakeman LN, Oraiopoulos N. 2022. *From breakthrough to blockbuster: the business of biotechnology.* Oxford University Press, New York.

Duncan DE. 2021. *A philosopher on Wall Street: how creative financier Fred Frank forged the future.* Greenleaf Book Group Press, Austin, TX.

Farrar J, Ahuja A. 2021. *Spike: the virus vs the people.* Profile Books, London.

Field D, Davies N. 2015. *Biocode: the new age of genomics.* Oxford University Press, Oxford.

Friedberg EC. 2010. *Sydney Brenner—a biography.* Cold Spring Harbor Laboratory Press, Cold Spring Harbor, NY.

Geraghty JA. 2022. *Inside the orphan drug revolution: the promise of patient-centered biotechnology.* Cold Spring Harbor Laboratory Press, Cold Spring Harbor, NY.

Gerstle G. 2022. *The rise and fall of the Neoliberal Order. America and the world in the free market era.* Oxford University Press, New York.

Giroir B. 2023. *Memoir of a pandemic—fighting COVID from the front lines to the White House.* Texas A&M University Press, College Station, TX.

Goldstein D. 2022. *The end of genetics: designing humanity's DNA.* Yale University Press, New Haven, CT.

Goozner M. 2004. *The $800 million pill: the truth behind the cost of new drugs*. University of California Press, Berkeley.

Hall S. 1987. *Invisible frontiers: the race to synthesize a human gene*. AbeBooks, Victoria, BC.

Hammond PM, Carter GB. 2002. *From biological warfare to healthcare: Porton Down 1940–2000*. Palgrave Macmillan, London.

Hawkins J. 2019. *Conscience and courage: how visionary CEO Henri Termeer built a biotech giant and pioneered the rare disease industry*. Cold Spring Harbor Laboratory Press, Cold Spring Harbor, NY.

Hughes SS. 2011. *Genentech: the beginnings of biotech*. University of Chicago Press, Chicago.

Isaacson W. 2021. *The code breaker: Jennifer Doudna, gene editing, and the future of the human race*. Simon & Schuster, New York.

Jóhannesson GTh. 1999. *Kari I jotunmod. Saga Kara stefannsonar og isleskrar erfoagreiningar*. Bókafélagið, Reykjavik, Iceland.

Jones S. 1992. *The biotechnologists: and the evolution of biotech enterprises in the USA and Europe*. Macmillan Press, London.

Judson HF. 1996. *The eighth day of creation*, expanded edition. Cold Spring Harbor Laboratory Press, Cold Spring Harbor, NY.

Kakkis E. 2022. *Saving Ryan: the 30-yr journey into saving the life of a child*. Impositivity Media LLC, Burbank, CA.

Kinch MS. 2016. *A prescription for change: the looming crisis in drug development*. UNC Press, Chapel Hill, NC.

Loftus P. 2022. *The messenger: Moderna, the vaccine, and the business gamble that changed the world*. Harvard Business Review, Brighton, MA.

Norrby E. 2010. *Nobel Prizes and life sciences*. World Scientific, Singapore.

Quammen D. 2022. *Breathless: the scientific race to defeat a deadly virus*. Simon & Schuster, New York.

Rasmussen N. 2014. *Gene jockeys: life science and the rise of biotech enterprise*. Johns Hopkins University Press, Baltimore.

Ridley M, Chan A. 2021. *Viral: the search for the origin of COVID-19*. Harper, New York.

Robbins-Roth C. 2000. *From alchemy to IPO: the business of biotechnology*. Perseus, New York.

Rose S. 1966. *The chemistry of life*, 1st ed. Penguin Books, New York.

Shreeve J. 2004. *The genome war: how Craig Venter tried to capture the code of life and save the world*. Alfred A. Knopf, New York.

Silver LM. 2007. *Remaking Eden: how genetic engineering and cloning will transform the American family*. Ecco Books, New York.

Teitelman R. 1989. *Gene dreams: Wall Street, academia and the rise of biotechnology*. Basic Books, New York.

Vardi N. 2023. *For blood and money: billionaires, biotech and the quest for a blockbuster drug*. W.W. Norton, New York.

Vasella D, Slater R. 2003. *Magic cancer bullet: how a tiny orange pill is rewriting medical history*. Harper Business Books, New York.

Venter JC. 2007. *A life decoded: my genome: my life*. Penguin, New York.

Werth B. 1995. *The billion-dollar molecule: one company's quest for the perfect drug*. Simon & Schuster, New York.

Werth B. 2014. *The antidote: inside the world of new pharma*. Simon & Schuster, New York.

OTHER BOOKS TO NOTE

Armstrong S. 2014. *p53: the gene that cracked the cancer code*. Bloomsbury Sigma, London.

Ashburner M. 2006. *Won for all: how the* Drosophila *genome was sequenced*. Cold Spring Harbor Laboratory Press, Cold Spring Harbor, NY.

Brownlee GG, Southern E. 2014. *Fred Sanger—double Nobel laureate: a biography*. Cambridge University Press, Cambridge.

Holmes FL. 2001. *Meselson, Stahl and the replication of DNA: a history of 'the most beautiful experiment in biology'*. Yale University Press, New Haven, CT.

Lefkowitz R, Hall R. 2021. *A funny thing happened on the way to Stockholm: the adrenaline-fueled adventures of an accidental scientist*. Pegasus Books, New York.

Marks LV. 2015. *The lock and key of medicine: monoclonal antibodies and the transformation of healthcare*. Yale University Press, New Haven, CT.

Mukherjee S. 2016. *The gene: an intimate history*. Scribner Books, New York.

Nurse P. 2020. *What is life?: understand biology in five steps*. David Fickling Books, Oxford.

Portugal FH. 2015. *The least likely man: Marshall Nirenberg and the discovery of the genetic code*. MIT Press, Cambridge, MA.

Ptashne M, Gann A. 2002. *Genes and signals*. Cold Spring Harbor Laboratory Press, Cold Spring Harbor, NY.

Ridley M. 1999. *Genome: the autobiography of a species in 23 chapters*. Harper, New York.

The Rockefeller University Press. 1971. *A notable career in finding out: Peyton Rous 1879–1970*. The Rockefeller University Press, New York.

Watson JD. 1968. *The double helix: a personal account of the discovery of the structure of DNA*. Touchstone Books, New York.

Zerhouni E. 2025. *Disease knows no politics*. Prometheus Press, Essex, CT.

Acknowledgements

Many people have helped me in the construction of this book. I would like to start off by thanking Fintan Steele, my editor and co-conspirator, for his editorial nous and pragmatic advice, whilst maintaining a great sense of humour. A more appropriately erudite and qualified person would be impossible to find. I also thank Kevin Davies for his inspired suggestion of Fintan as an editor and to Larry Gold for helping to persuade Fintan to read a draft of the book in the first place.

I thank John Inglis at Cold Spring Harbor Laboratory (CSHL) Press for taking a punt on whether the book would be worth publishing and for believing in me and that the book might actually be a useful history of the industry—and a little fun at the same time, based on the personal anecdotes. I also thank the staff and everyone at CSHL Press for helping to turn the manuscript into a real book. In particular, I would like to thank Barbara Acosta, Carol Brown, Kathleen Bubbeo, Danett Gill, and Denise Weiss for their editorial and production prowess and for being so easy to work with in making the book a reality.

It is so appropriate for it to be CSHL Press publishing the book because this laboratory and the scientists who worked there have had their own unique and very important parts to play in the history of the biotech industry. In addition, the CSHL library includes many archives of some of the key biotech players like Charlie Weissmann and Sydney Brenner.

I would like to acknowledge three individuals in particular: Stelios Papadopoulos, John Maraganore, and Jeremy Levin, who undertook to read a near-final version of the manuscript for factual correctness (as well as giving me advice earlier on, when asked questions about various times or companies). I can think of few other individuals as qualified as them to help in that regard—biotech industry–wide. But any remaining inconsistencies and factual errors are mine and mine alone, and I take full responsibility for them.

In the course of writing the book, Alan Buckler, Chris Henney, George Poste, and Doug Williams all read early versions of a "semicomplete" manuscript and provided feedback on what they liked (or did not like). As they will see, I took notice of most of the comments that they made. They also helped to clarify some sections of the book where their first-hand knowledge of different companies was of enormous value and provided a narrative from which I could work.

I am indebted to many other people who provided information that helped me to write the story of many of the other companies that I cover. They know who they are, but I would like to thank in particular the following individuals (in alphabetical order) who all contributed and helped to make the book what it is: Spyros Artavanis-Tsakonas, Elizabeth Ashforth, Frank Bennett, Karen Bernstein, Kate Bingham,

Jens Boch, Mark Bodmer, Brook Byers, John Cox, Tony Evnin, Jean-François Formela (JFF), Simba Gill, David Goldstein, Linda Grais, Seth Harrison, Steve Holtzman, Robin Jacob, Martin Leach, Glenn Larsen, Mark Levin, Ron Lindsay, Peter Linsley, Stefan Lohmer, Michael Kinch, Tom Maniatis, Joel Marcus, Robert McBurney, Frank McCormick, Rick Myers, Tina Nova, Bette Phimister, John Reed, Peter Rigby, Mike Riordan, Mike Ross, Vadim Sapiro, Gabe Schmergel, Steve Sherwin, Bob Stein, and Minka vanBuezekom. I would also like to call out Deborah Ann and Theta Tsu specifically for "inviting" me to the Oncogen reunion on Queen Anne Hill in Seattle in June 2022 so I could talk with some of the scientists and their colleagues who worked there in the early days. If I have left anyone else out that I also contacted and spoke to on the phone or communicated with by e-mail, then please forgive me.

Finally, I should like to thank my wife and family for their forbearance and for accepting my frequent disappearances into the back room during the latter stages of the SARS-CoV-2 pandemic (and afterwards) to get some peace and quiet (P&Q)[a] so that I could concentrate and really write in earnest—something I had not done for quite a while.

---

[a] My grandson likes to listen to the "P&Q Band," which is what I referred to was playing when the car radio was turned off, as in "let's listen to the P&Q Band."

Index

Page numbers followed by *f* or *b* denote a figure or box respectively on the corresponding page. Page references for notes are shown as the page number followed by *n* and the individual note letter designation. References for Plates in the photo insert are listed by Pl#.

About the Author

Dr. Tim Harris is a molecular biologist and biochemist. He started work in the biotech industry almost at its inception. He began his career in 1974, working on animal viruses, and was one of the first scientists to join the U.K. biotech company Celltech in 1981. Subsequently he spent nearly five years at Glaxo Group Research (now GSK) in the United Kingdom as Director of Biotechnology.

Tim has founded several biotech companies since moving to the United States in 1993 to be head of R&D at Sequana Therapeutics, a genomics company in San Diego. He founded and ran SGX Pharmaceuticals (acquired by Eli Lilly) and served briefly as the Chief Technology Officer (CTO) and Director of the Advanced Technology Program (ATP) at SAIC-Frederick, Inc. in Maryland. Latterly, Tim moved to Biogen as SVP Translational Medicine in 2011, before joining the haematology spinout Bioverativ, as head of R&D in March 2017 before the company was acquired by Sanofi.

Tim is a presently an operating partner at SV Health Investors and actively works on behalf of several small biotechnology companies, some of which he founded. He has published more than 100 peer-reviewed research papers and reviews and has written many commentaries for *BioCentury* and other industry magazines.

www.ingramcontent.com/pod-product-compliance
Lightning Source LLC
Chambersburg PA
CBHW061229220326
41599CB00028B/5377